SYNTHESE LIBRARY

STUDIES IN EPISTEMOLOGY,

LOGIC, METHODOLOGY, AND PHILOSOPHY OF SCIENCE

VOLUME 259

LOGIC AND SCIENTIFIC METHODS

LOGIC AND SCIENTIFIC METHODS

Volume One of the Tenth International Congress of Logic, Methodology and Philosophy of Science, Florence, August 1995

Edited by

MARIA LUISA DALLA CHIARA
University of Florence

KEES DOETS
University of Amsterdam

DANIELE MUNDICI
University of Milan

and

JOHAN VAN BENTHEM
University of Amsterdam and *Stanford University*

KLUWER ACADEMIC PUBLISHERS
DORDRECHT / BOSTON / LONDON

A C.I.P. Catalogue record for this book is available from the Library of Congress

ISBN 0-7923-4383-2 (Vol. 1)
ISBN 0-7923-4384-0 (Vol. 2)
ISBN 0-7923-4385-9 (Set)

Published by Kluwer Academic Publishers,
P.O. Box 17, 3300 AA Dordrecht, The Netherlands.

Kluwer Academic Publishers incorporates
the publishing programmes of
D. Reidel, Martinus Nijhoff, Dr W. Junk and MTP Press.

Sold and distributed in the U.S.A. and Canada
by Kluwer Academic Publishers,
101 Philip Drive, Norwell, MA 02061, U.S.A.

In all other countries, sold and distributed
by Kluwer Academic Publishers Group,
P.O. Box 322, 3300 AH Dordrecht, The Netherlands.

Printed on acid-free paper

Printed in the Netherlands

TABLE OF CONTENTS

EDITORIAL

These volumes continue a long tradition. They document part of the scientific program of the 10th International Congress of Logic, Methodology and Philosophy of Science, held at Florence, August 1995. The founding fathers of the Congress, including Alfred Tarski, Evert Beth, and Abraham Robinson, left us a legacy of cooperation between logicians and philosophers of science, on an international scale which is still unique. At the time, this alliance seemed natural. Philosophers and mathematicians in the fifties still remembered the heighdays of logical positivism, and their interests were broad. Not accidentally, Tarski's best-known textbook is a joint introduction to logic and methodology of the exact sciences. In the meantime, many things have changed. Logicians have developed new contacts with computer science and linguistics, philosophers of science also look toward history and sociology. Nevertheless, our Congress is still there, like a Grand Hotel, with perhaps a somewhat faded splendour, but with a distinctive style, elegance and comfort. And it serves an important purpose.

The invited lectures published here demonstrate much of what goes on in the fields of the Congress. Through an extensive selection process, General and Sectional Program Committees have chosen speakers whose excellence in research deserves recognition. How much this honour is appreciated is shown by a remarkably high acceptance rate. The sectional organisation covers the traditional subdisciplines of mathematical logic, philosophical logic, as well as their interfaces with computer science, linguistics, philosophy. Philosophy of science is broadly represented, too, including both general issues of methodology and ethics, and specific foundational issues of natural sciences, social sciences and humanities. In addition to this planned part, the Congress also serves as the meeting place for an international community from all over the world. There have been a host of interesting contributed talks, often leading to osmosis through the sectional barriers of the program. Selections of these papers will appear in a number of satellite volumes, edited by the Florence local organizers.

The program of the Congress has further special features. Traditionally, the opening and closing lectures are given by eminent scholars and scientists, who give us a broad perspective on the intellectual environment that we are working in. We are pleased to include the lectures by professors Beltrametti and van Fraassen, who did just that. Also, the general Program Committee has organized a number of special Symposia, to high-light some topics of current interest, and to enhance the connectivity of our field. In Florence, these were "Semantics and Semiotics", "Logic and Philosophy of Science", "Logic in Eastern Europe" and "Logic in the Far East". These symposia are represented here, too. Not included are the so-called Affiliated Meetings on "Model Theory" (organized jointly by the Kurt Gödel Society and the Italian Association of Logic and its Applications, AILA) and "Proof Theory for Natural Language" (European Association for Logic, Language and Information), whose materials will be published elsewhere by the responsible organizers.

There remains the pleasant task of expressing thanks where these are due. We have been assisted very generously by Mrs. Annie Kuipers (Kluwer Academic Publishers). Most of the editorial logistics have been taken care of by Kees Doets at the University of Amsterdam, assisted by Jeroen Haak and Joeri van Ruth. The four editors themselves see this Volume as the visible finished product of several years of effective and pleasant collaboration in a joint Dutch-Italian team. Readers with even the slightest knowledge of contemporary Europe will know that this is indeed a winning combination ...

Maria Luisa Dalla Chiara (Firenze)
Kees Doets (Amsterdam)
Daniele Mundici (Milano)
Johan van Benthem (Amsterdam and Stanford)

SECTIONS, PLENARY LECTURES AND SPECIAL SYMPOSIA

LOGIC

Section 1. Proof Theory and Categorical Logic

Section Program Committee:

G.E. Mints, Stanford (chair)
V. Orevkov, St. Petersburg
J. Lambek, Montréal

Invited speakers:

A. Cantini, Proof-theoretic aspects of self-referential truth
Università degli Studi di Firenze

A. Joyal, Free lattices, communication and money games
Université du Quebec à Montréal

J. Krajíček, On methods for proving lower bounds in propositional logic
Mathematical Institute of Science, Prague

V.Yu. Sazonov, On bounded set theory
Russian Academy of Sciences, Pereslavl-Zalessky

Section 2. Model Theory, Set Theory and Formal Systems

Section Program Committee:

W.A. Hodges, London (chair)
X. Caicedo, Bogota
C.A. Di Prisco, Caracas

Invited speakers:

Ph.G. Kolaitis, Infinitary logic in finite model theory
University of California, Santa Cruz

A. Scedrov, Decision problems for second-order linear logic
(with P.D. Lincoln and N. Shankar)
University of Pennsylvania, Philadelphia
and
SRI International

S. TODORČEVIĆ,
Comparing the continuum with the first two uncountable cardinals
University of Toronto
and
Matematički Institut, Belgrade

Section 3. Recursion Theory and Constructivism

Section Program Committee:
W. SIEG, Pittsburgh (chair)
P. ODIFREDDI, Torino
K. AMBOSS-SPIES, Heidelberg

Invited speakers:
K. KELLY, Church's thesis and Hume's problem
(with O. SCHULTE)
Carnegie Mellon University, Pittsburgh

Y.N. MOSCHOVAKIS, The logic of functional recursion
Department of Mathematics, University of Los Angeles

H. SCHWICHTENBERG, Computability for partial continuous functionals
(with K. STROETMAN)
Universität München

R. SOARE, Computability, enumerability, macros, and meaning
University of Chicago

G. TAMBURRINI,
Mechanistic theories in cognitive science: the import of Turing's thesis
Instituto di Cibernetica del C.N.R.

Section 4. Logic and Computer Science

Section Program Committee:
C. BÖHM, Roma (chair)
J.A. BERGSTRA, Amsterdam and Utrecht
P.G. CLOTE, Boston

Invited speakers:
P. MARTIN-LÖF, The semantic significance of computation
Academy of Sciences, Stockholm, Sweden

G.D. PLOTKIN, Type theory and recursion
University of Edinburgh

D.S. SCOTT, 25 years of domain theory

Carnegie Mellon University, Pittsburgh

Section 5. Philosophical Logic

Section Program Committee:

P. GÄRDENFORS, Lund (chair)
G. SAMBIN, Padova
M. DETLEFSEN, Notre Dame

Invited speakers:

G. BOOLOS (†), Frege's theorem and the Peano postulates

MIT, Cambridge MA

E. CASARI, Conjoining and disjoining on different levels

Centro Fiorentino di Storia e Filosofia della Scienza

K. DOŠEN, Logical consequence, a turn in style

Matematički Institut, Belgrade
and
IRIT, University of Toulouse III

D. MAKINSON, Applying normative rules with restraint
(with S.O. HANSON)

Ville d'Avray, France
and
Uppsala University

GENERAL PHILOSOPHY OF SCIENCE

Section 6. Methodology

Section Program Committee:

A.E. MUSGRAVE, Otago (chair)
C.A. HOOKER, Newcastle
N.C.A. DA COSTA, Sao Paulo

Invited speakers:

P. ACHINSTEIN, The empirical character of evidence

Johns Hopkins University, Baltimore

S. FRENCH, Partiality, pursuit and practice

University of Leeds

Section 7. Probability, Induction and Decision Theory

Section Program Committee:

R.C. JEFFREY, Princeton (chair)
T. SEIDENFELD, Pittsburgh
H. GAIFMAN, New York

Invited speakers:
M. GOLDSTEIN, Prior inferences for posterior judgements
Science Labs, Durham

B. SKYRMS, Game theory, rationality and evolution
University of California, Irvine

Section 8. History of Logic, Methodology and Philosophy of Science

Section Program Committee:
N. CARTWRIGHT, London (chair)
P. MADDY, Irvine
L. LAUDAN, Hawaii

Invited speakers:
U. GÄHDE, Anomalies and the revision of theory-elements: notes on the advance of Mercury's perihelion
Universität Bayreuth

G. MOORE, The prehistory of infinitary logic: 1885–1955
McMaster University, Hamilton, Ontario

Section 9. Ethics of Science and Technology

Section Program Committee:
K. SHRADER-FRECHETTE, Tampa (chair)
B. HANSSON, Lund
R. QIU, Beijing
W. SHEA, Montréal

Invited speakers:
S. BEDER, Engineers, ethics and sustainable development
University of Wollongong

L. JONATHAN COHEN,
Are there ethical reasons for being, or not being, a scientific realist?
Oxford

D. FAN, Rights and autonomy of scientists: cases in P.R. China
Beijing

I. HRONSZKY, On ethics and changing expertise in technology assesment
University of Budapest

M. KAISER,
Unpopular and precautionary science: towards a new ethics of science
The Research Council of Norway, Oslo

N. ORESKES, Testing models of natural systems: can it be done?
Darthmouth College, Hanover

PHILOSOPHICAL AND FOUNDATIONAL PROBLEMS ABOUT THE SCIENCES

Section 10. Philosophy of Logic, Mathematics and Computer Science

Section Program Committee:
H. FIELD, New York (chair)
G. LONGO, Paris
J. ETCHEMENDY, Stanford

Invited speakers:
CH.D. PARSONS, What can we do "in principle"?
Harvard University, Cambridge MA

J. PEARL, Causation, action, and counterfactuals
University of California, Los Angeles

Section 11. Philosophy of the Physical Sciences

Section Program Committee:
J.S. EARMAN,
and
K.L. MANDERS, Pittsburgh (chairs)
G. TORALDO DI FRANCIA, Firenze

Invited speakers:
D. ALBERT, On the character of statistical-mechanical probabilities
Columbia University, New York City

P. GALISON, unable to attend
Harvard University, Cambridge MA

G. GHIRARDI, Macroscopic reality and the dynamical reduction program
Universita degli Studi di Trieste

Section 12. Philosophy of Biological Sciences

Section Program Committee:

M. Ruse, Guelph and Montpellier (chair)
P. Kitcher, Santa Dominga
W.J. van der Steen, Amsterdam

Invited speakers:

R. Burian, On conflicts between genetic and developmental viewpoints and their attempted resolution in molecular biology
Virginia Polytechnic Institute and State University, Blacksburg VA

J. Gayon, The "paramount power of selection": from Darwin to Kauffman
Université de Bourgogne, Dijon
and
Institut d'Histoire et de Philosophie des Sciences et des Techniques, Paris

C.N. van der Weele,
Environmental causes in ontogeny and an ethics of attention
University of Utrecht

Section 13. Philosophy of Cognitive Science and AI

Section Program Committee:

N. Block, Cambridge MA (chair)
M. Davies, Oxford
A. Clark, St. Louis

Invited speakers:

D. Chalmers, The extended mind
University of California, Santa Cruz

F. Jackson, Mental causation without the language of "thought"
The Australian National University

J. Kim, What is the problem of mental causation?
Brown University, Providence RI

B.P. McLaughlin, Classical constituents in Smolensky's ICS architecture
Rutgers University, New Brunswick NJ

Section 14. Philosophy of Linguistics

Section Program Committee:

S. Soames, Princeton (chair)
H. Schnelle, Bochum
G. Chierchia, Milano

Invited speakers:

D.R. DOWTY, Non-constituent coordination, wrapping, and multimodal categorical grammars
Ohio State University, Columbus OH

H. KAMP,
Cognitive structure, shared cognitive structure and lexical semantics
Universität Stuttgart

T.D. PARSONS, Meaning sensitivity and grammatical structure
University of California at Irvine

Section 15. Social Sciences

Section Program Committee:

PH. PETTIT, Canberra (chair)
J. ELSTER, Oslo and Chicago
A. ROSENBERG, Riverside

Invited speakers:

R.M. DAWES, Qualitative consistency masquerading as quantitative fit
Carnegie Mellon University, Pittsburgh

D.M. HAUSMAN, Why does evidence matter so little to economic theory?
University of Wisconsin

D. SPERBER, Selection and attraction in cultural evolution
CNRS and École Polytechnique, Paris

PLENARY SPEAKERS

Opening

E.G. BELTRAMETTI, Trends in the foundations of physics
Department of Physics
University of Genoa

Closure

B.C. VAN FRAASSEN,
Structure and perspective: philosophical perplexity and paradox
Princeton University

SPECIAL SYMPOSIA

Symposium 1. Logic and Philosophy of Science: current interfaces

Section Program Committee:
T.A.F. KUIPERS, Groningen (chair)
J. BUTTERFIELD, Cambridge UK
G. HELLMAN, Minneapolis

Invited speakers:
K. KELLY, Reliable belief revision
(with O. SCHULTE and V. HENDRICKS)
Carnegie Mellon University, Pittsburgh
and
University of Copenhagen

V. RANTALA,
Explanatory translation: beyond functionalism and reductionism
Tampere University

P. THAGARD, Abductive reasoning: logic, visual thinking, and coherence
(with C. SHELLEY)
University of Waterloo

P. WEINGARTNER, Can the laws of nature (physics) be complete?
Salzburg University

Symposium 2. Semantics and Semiotics

Section Program Committee:
J. PELC, Warsaw (chair)
G. CHIERCHIA, Milano
P. LEONARDI, Venezia

Invited speakers:
U. ECO, On meaning, logic and verbal language
Milano

D. FØLLESDAL, Semantics and semiotics
Stanford University
and
University of Oslo

Symposium 3. Logic and Philosophy of Science in the Far East

Section Program Committee:
R. QIU, Beijing (chair)

T. FUJIMOTO, Tokyo
H. HSIU HWANG, Hong Kong
S. XIAOLI, Beijing

Invited speakers:

W. LU, Logic in China

Beijing

H. SAKAMOTO, Recent develepments of logic, computer science and philosophy of science in Japan

Nihon University, Tokyo

S-Y. SONG, Philosophy of science in Korea, 1950–1990

Hallym University, South Korea

C-H. LIN, Philosophy of science in Taiwan
(with D. FU)

National Taiwan University, Taipei
and
National Tsing-Hua University, Hsinchu

Symposium 4. Logic in Central and Eastern Europe

Section Program Committee:

P. HÁJEK, Prague,
and
S. ARTEMOV, Moscow (chairs)
L. TONDL, Prague

Invited speakers:

L. BUKOVSKÝ, Logic in Czechoslovakia and Hungary

Katedra matematickej informatiky, Kosice

V.A. USPENSKY, Mathematical logic in the former Soviet Union: brief history and current trends

Moscow State University

D. VAKARELOV, Logic in Central and Eastern Europe: Balkan region

Sofia University

R. WÓJCICKI, The postwar panorama of logic in Poland

PAN, Warszawa

Executive committee of the Division of Logic, Methodology and Philosophy of Science, International Union of History and Philosophy of Science, 1991–1995:

President J.E. FENSTAD (Norway), 1st Vice-President D. PRAWITZ (Sweden), 2nd Vice-President R. WÓJCICKI (Poland), Secretary General E. SOBER (USA), Treasurer A. PRELLER (France), Past President L.J. COHEN (UK), Assessors M. BOFFA (Belgium), R. CHUAQUI (Chile, †), R.P.H. FORREST (Australia), M. HALLETT (Canada), H. ISHIGURO (Japan), V. LEKTORSKY (Russia), G.E. MINTS (USA), G. SAMBIN (Italy).

General Program Committee:

J. VAN BENTHEM (Chair, the Netherlands and USA), R. CHUAQUI, (Chile, †), P. HÁJEK (Czech Rep.), D. MUNDICI (Italy), R. QIU (China), E. SOBER (USA).

Local Organizing committee:

M.L. DALLA CHIARA (chair), L. AMADUCCI, T. ARECCHI, A. BENEDETTI, S. BERNINI, M. BRESCIANI, S. CALIFANO, A. CANTINI, E. CASARI, E. CASTELLANI, G. CORSI, P. CORSI, F. DESSI-FULGHERI, A. FONNESU, P. GALLUZZI, R. GIUNTINI, E. GIUSTI, F. LAUDISA, P. MANGANI, A. MARCJA, P. MINARI, M. MUGNAI, A. PAGNINI, E. PANCONESI, P. PARRINI, A. PERUZZI, G. PERUZZI, S. POGGI, M. POLSINELLI, P. ROSSI, D. SEARS, G. TORALDO DI FRANCIA.

Financial sponsors of the congress:

The 10th International Congress of Logic, Methodology and Philosophy of Science was arranged by the International Union of History and Philosophy of Science (IUHPS), Division of Logic, Methodology and Philosophy of Science (DLMPS), in collaboration with the University of Florence and the Region of Tuscany. Other supporting institutions are: the European Union (European Commission, Directorate General XII for Science, Research and Development), the National Research Council (the Presidence and the Committees for Philosophy and Mathematics), the University of Siena, the Italian Association for Logic and its Applications (AILA), the Italian Society for Logic and Philosophy of Sciences (SILFS), the Kurt Gödel Society (KGS), the Carlo Marchi Foundation (Florence), the Florence Center for the History and Philosophy of Science, the Italian Institute for Philosophical Studies (Naples), the Center for Epistemology and History of Science F. Enriques (Bologna, Cesena), the Norwegian Research Council, the City of Florence and the Province of Florence. The Congress was held under the auspices of the President of the Italian Republic.

PRESIDENT'S ADDRESS

Dear Members of the Congress and Distinguished Guests! It is a great pleasure and privilege to join in a Welcome to all of you to the 10th International Congress of Logic, Methodology and Philosophy of Science, here in the city of Florence.

It is my special privilege to acknowledge the great and very successful work of the Local Organizing Committee under the engaged and effective leadership of Professor Maria Luisa Dalla Chiara and of the various programme committees under the able leadership of Professor Johan van Benthem. Their dedicated work has been a necessary condition for the success of the Congress. It is now up to you to extend this necessary condition to an also sufficient one through your participation and engagement during the week in front of us.

Let me, as an introduction to the work of this Congress, make a few remarks on the history of our organization and congresses. The International Union of the History and Philosophy of Science was created in 1956 by the federation of the International Union of History of Science, which was founded in 1947, and the International Union of Philosophy of Sciences, founded in 1949. The two parent bodies became, respectively, the Division of History of Science and the Division of Logic, Methodology and Philosophy of Science, which is our organization. You may be surprised by the somewhat late founding date, but there is a prehistory. During the 1930s we saw the emergence of a unity-of-science movement, chiefly inspired by the school of logical empiricism. There was a series of congresses devoted to this theme; the first one in Praha in 1934. If you look at the International Committee which had the scientific responsibility for these congresses, you will find almost all the great names of the 1930s in logic, philosophy of science, physics and mathematics, and even a few in the social sciences. There was in a certain sense — and despite the threatening political catastrophy — a widely shared feeling of progress and victory; the third congress in Paris in 1937 took the form of a conference devoted to a project for an International Encyclopedia of Unified Science, where the foundation for science and rationality was once and for all to be inscribed. We know what happened instead.

After the war there was a need to rebuild. Part of the unity-of-science movement survived in the United States where many of the main participants had found a new home. But the unity-of-science movement had a universal goal and there was an urgent need to reestablish international links and co-operation. This was one of the motives behind the creation of our union; one person who played a prominent role in this task was the Polish-American lo-

M.L. Dalla Chiara et al. (eds.), Logic and Scientific Methods, 1–4.

gician Alfred Tarski. This is, however, not the place to tell a full story. The first President of our Divison was the great logician S.C. Kleene and the first congress was held at Stanford in 1960. Now we are at the start of the 10th, and if we look at the collected proceedings from these congresses, we may claim some success in achieving our aims in strenghtening the science and promoting international collaboration. But there is another part of the story which we need to recall.

Our Union is part of a larger family, the International Council of Scientific Unions (ICSU). ICSU was created in 1931. It also has a prehistory, and its new start in 1931 came as the end of a long effort to reestablish international cooperation within the broader scientific community, which had been disrupted by the first world war and the immediate turbulent afterperiod. After the second world war, in the period when our Union was created, we saw an increased importance of ICSU. Since its start ICSU has with great determination pursued a policy of non-discrimination, affirming the rights and freedom of scientists throughout the world to engage in international scientific activities. Thus scientists have been able to meet and discuss without political constraints. This was of outmost importance during the cold war period, and scientists have not always been aware that without ICSU and the unions freedom in the pursuit of science would have suffered.

Focusing on "freedom of pursuit" stresses a certain internal or self-centered perspective on the scientific enterprise. There is perhaps an explanation. After the war there has been an implicit social contract between science and society. Science emerged from the war with great status and trust — it had delivered on its promises. This trust was in the post-war years transformed into a belief in the ability of science to solve the urgent problem nations were faced with. And the belief seemed justified. The wealth of nations increased thanks to the strength of science-based technology, health improved thanks to progress in medical science, and scientists made their contribution to security in the cold war period. So science prospered and enjoyed a great deal of internal freedom in setting its agenda.

The end of the cold war changed all this. In fact, signs were already visible, but the dramatic events of 1989 made it urgent. The social contract seems to fall apart. Health and welfare, perhaps extended to include the right to work, are still words of honor, but security today has acquired a new meaning reflecting our concerns with environment and sustainable development. And there is no longer an unquestioned trust in science; accountability and responsibility are the new words. This means changed conditions for science in almost all nations, in some places rather drastic. As a consequence there are new challenges for international collaboration. It is once more of urgent importance to keep open and strenghten our links to all parts of the world. This is one of the reasons that we have organized two special symposia at this Congress, one on Logic and Philosophy of Science in the Far East, and one on Logic in Central and Eastern Europe. We want all parts to be strong and equal partners in our endeavors. We have also seen that freedom-of-pursuit at ICSU level has been extended to freedom-and-responsibility-in-pursuit, and our Division has played a role in this change. At our Congress it is reflected through an expanded section on the Ethics of Science and Technology.

Earlier this year the Swedish Natural Science Research Council held a meeting on the survival of sciences in the post cold war period. Such questions are obviously rhetorical and not meant to be literally read. But as I have tried to indicate above there is a real substance to this rhetoric. Translated to our more narrow context, the question is whether the philosophy of science will survive. If we read our history, we will see how closely the unity-of-science movement of the 1930s was associated with formal logic and the methodologies of the exact natural sciences.

Let me take the study of language as an example. Language is a complex phenomenon and no science and technology have an exclusive claim on it. I mentioned above that the International Encyclopedia of Unified Science was meant to be the platform of any future movement, perhaps, also intended as the confessional tract for our Congress?

In this tract we read in an article by C.W. Morris that "... it has become clear to many persons today that man — including scientific man — must free himself from the web of words that he has spun and that language — including scientific language — is greatly in need of purification, simplification and systematization. The theory of signs is a useful instrument for such debabelization". Let me add that logical syntax and semantics in the style of Carnap were to be the main tools. In another article of L. Bloomfield we see that the study of natural languages is basically reduced to language description and that meaning at most is a stimulus-response addendum.

At the 1960 Congress at Stanford N. Chomsky is a main speaker, and a new science of language is the center piece. Some congresses later in 1975 we see the rebirth of the old alliance of logic and natural language through the work of R. Montague and important predecessors such as Curry and Reichenbach. In recent congresses, perhaps also at this one, we see increased links between logic, language and cognitive sciences. The statistical procedures of structuralism, the recursion theory of generative grammar, the intensional higher order logic of the logicians have been extended with non-linear dynamical system theory and cognitive modeling.

There are a number of lessons to be learnt. One which is close to my heart, and which I really have to resist expanding on at this occasion, is the interplay between the mathematics you master and the larger conceptual frameworks you build. I have touched upon language. I could have made a reference to a witty but serious lecture at the 1960 Congress by J. Schwartz on "the pernicious influence of mathematic on the sciences", where, in particular, the physical sciences are in focus. Or I could have chosen economics and mentioned the first Nobel Price winner J. Tinbergen, who was a student of Ehrenfest, who himself was a student of Boltzman; the thesis of Tinbergen was on “minimum-problemen in de natuurkunde en de ekonomie”, using a type of mathematics completely at odds with the currently fashionable trend of seeing the economy as an evolving complex system, where you could look at a steady state economy as a steady state fluid weakly heated from below.

I shall not dwell further on these examples, but return to another lesson to be learned. Be careful with confessional tracts. The International Encyclopedia of Unified Science is interesting, admirable — but dated and largely dead. Science, in the broad sense of Wissenschaft, is an ever unfolding story. We need to reflect on the history, the logic and structure, and the social context of our enterprise. We need to fight to protect the freedom-of-pursuit, we need also to be concerned about the responsibilities. We need to stand united.

I would like to add one remark. — Homo ludens, the playful attitude! — What we do — the subject matter — is important, but how we do — the style — should be infused with enthusiasm and engagement — even joy. If you at all have looked at the programme for the next week, you should be convinced that it will be a week in which to participate, to be engaged and to enjoy.

Jens Erik Fenstad
University of Oslo, Norway

1
Proof Theory and Categorical Logic

ANDREA CANTINI

PROOF-THEORETICAL ASPECTS OF SELF-REFERENTIAL TRUTH

1. INTRODUCTION

Since Tarski's fundamental work in 1932, a traditional problem in formal semantics is to find non-trivial models of formal languages "containing their own truth predicate". Typically, the motivation is philosophical; one is concerned with potential applications to the analysis of circularity and self-reference in natural languages. In this direction, a paradigmatic work is certainly Kripke's seminal paper *Outline of a theory of truth*, which marks the starting point for a considerable amount of new research (cf. the recent survey of Sheard [SH]).

However, we can legitimately conceive different motivations and alternative routes for studying self-referential truth, which drive us naturally into the realm of (reductive) proof theory. First of all, we may wish to build *consistent theories*, instead of *models*: the truth predicate T is then assumed as *primitive* and characterized by means of suitable axioms establishing its closure conditions. So we are also left with the question: what can we do in such theories? What can we use them for? Important results in axiomatic vein are contained in the papers by Friedman and Sheard [FS] and Feferman [F4].

Another stimulus for a proof-theoretic analysis is to observe that formal semantics produces notions with increasing logical complexity. This is clearly exemplified by the core concepts, which emerge from Tarskian semantics, Kripke's approach and the Herzberger-Gupta theory of truth revision. In fact, it is well-known that, while Tarskian truth (for the standard arithmetical language) leads to a non-arithmetical set, Kripke's truth (in the least fixed point model over the structure $\mathcal{N}$ of natural numbers) already reaches Π^1_1-completeness and stable truth with respect to all revision sequences over $\mathcal{N}$ is even Π^1_2-complete (Burgess [Bu]; Mc Gee). Since the mathematical objects arising from semantical investigation are quite complex (of a complexity which is somewhat uncommon even in mathematical practice), it may be rewarding to axiomatize their general features and to see how the corresponding systems reflect the increasing logical strength of the intended truth predicate.

In this paper we are primarily concerned with theories of partial truth over (suitably expanded) combinatory algebras. The formal frameworks we will introduce can be regarded as abstract semantical theories, which generalize Kripke's interpretation for self-referential truth; at the same time, they also yield an axiomatization of classical Frege structures (as described by Scott, Aczel, Flagg and Myhill; see [S], [A1], [FM]) and hence they contain a rich

M.L. Dalla Chiara et al. (eds.), Logic and Scientific Methods, 7–27.

theory of type-free abstraction and of partial and total classes in the sense of Feferman (see [F2],[F3]).

There are, however, a few differences with the cited sources:

1) we enrich our language with predicates $\prec$, $\preceq$, which allow to compare the "groundedness" levels, at which distinct truths are established;
2) we adopt suitable generalized induction principles (for truth);
3) we seriously consider the possibility that the predicate N of being a natural number (relative to a given combinatory algebra) is partial (with respect to truth; we cannot infer "it is true that $x \notin N$" from "$x \notin N$").

1)–2) convey the idea that the intended model for self-referential truth is inductive (i.e the minimal fixed point model over a given combinatory algebra). 3) implies that N is only potentially given and we show that the assumption produces a drastic weakening of proof-theoretic strength.

There are conceptual and technical reasons behind the proposed approach to self-referential truth. First of all, the Kripke-Feferman schema (KF in short, after [R]) embodies a natural conception of truth, based on a predicative intuition (the step from firmly grounded statements not involving semantical facts to less grounded statements). A second reason is that the formal systems based on KF, once strengthened with 1)–3), unify several approaches based on Π^1_1-monotone semantical schemata (see [C3]). A final point we make is that the basic theory we study can produce inner models for rich fragments of admissible set theory and second order arithmetic. Nevertheless, this can be done without going beyond the strength of Peano arithmetic PA and, when partiality of N is assumed, we get a system equivalent to primitive recursive arithmetic PRA.

As to the specific content, after introducing the basic language and its inductive semantics (§§2–3), we consider a sequent formulation of a ground system OTN^- for truth (§4). In §5 we give a proof-theoretic analysis of the system which extends OTN^- with full generalized induction and number-theoretic induction for total predicates; the main theorem is proved by embedding the theory in a non-standard infinitary (consistent but ω-inconsistent) system and by use of the asymmetric interpretation technique (see [G]). In §6 we reconsider the applicative axioms by adding a reflection principle for numbers N-ref, which implies that every operation with only number theoretic values is constant. In the subsequent section we show that N-ref is consistent with the truth axioms insofar as N is considered a partial predicate. Indeed, we prove that the resulting theory is not stronger than primitive recursive arithmetic. In the final section §8 we state a general theorem about the computational content of several truth systems.

2. PRELIMINARIES

The theories we have in mind describe (suitably expanded) combinatory algebras, having the form:

$$\langle \mathcal{M}, T(\mathcal{M}), \prec_{\mathcal{M}}, \preceq_{\mathcal{M}}, N(\mathcal{M})\rangle.$$

Here $\mathcal{M}$ is a combinatory algebra, possibly expanded with pairing, projections and certain operations which depend on the additional structure $N(\mathcal{M})$. The idea is that $\mathcal{M}$ can be regarded as a pure computational universal machine; the elements of $\mathcal{M}$ are operations which combine together according to the laws of combinatory logic, without explicit logical or mathematical structure. For instance, numbers can be certainly identified in $\mathcal{M}$, but we do not have means to talk about the set $N(\mathcal{M})$ of (those objects of $\mathcal{M}$ representing) natural numbers, nor we can classify objects in $\mathcal{M}$ according to their functional behaviour. These intrinsic limitations of $\mathcal{M}$ are met in two ways. First of all, $\mathcal{M}$ is endowed with a mathematical structure $N(\mathcal{M})$; for simplicity, we choose for $N(\mathcal{M})$ the structure of natural numbers (but any other well defined type, depending on the kind of applications we have in mind, will do). Second, $\mathcal{M}$ is assigned a logical structure which consists of an abstract truth predicate $T(\mathcal{M})$ and a pair of witness comparison relations $\prec_{\mathcal{M}}$, $\preceq_{\mathcal{M}}$. Roughly speaking, $T(\mathcal{M})$ allows to introduce the notions of proposition , predicate application, predicate abstraction and class (in the intensional sense), while $\prec_{\mathcal{M}}$, $\preceq_{\mathcal{M}}$ make explicit the idea that $T(\mathcal{M})$ is inductively generated. So $a \preceq_{\mathcal{M}} b$ means that the truth of a is verified not later than the truth of b (whenever b happens to be true), while $\prec_{\mathcal{M}}$ is the strict ordering associated to $\preceq_{\mathcal{M}}$.

Conceptually, there is a pleasant symmetry between operations and classifications: the step from $\mathcal{M}$ to $\langle \mathcal{M}, T(\mathcal{M}), \prec_{\mathcal{M}}, \preceq_{\mathcal{M}}, N(\mathcal{M})\rangle$ provides a framework where we can *classify operations*, which are the intended objects of $\mathcal{M}$; but on the other hand, using the combinatory structure, we can *operate on classifications* (if $A(x,v)$ is any condition, our model will contain not only $\{x : A(x,v)\}$, i.e., the possibly partial classification defined by A uniformly in v, but also $\lambda v.\{x : A(x,v)\}$, the operation expressing such uniformity).

The basic language $\mathcal{L}_T$ contains:

- countably many individual variables $x_1, x_2, x_3, \ldots$;
- the logical constants $\neg$, $\wedge$, $\vee$, $\exists$, $\forall$;
- the individual constants K (constant function combinator), S (composition combinator), *SUC* (successor), *PRED* (predecessor), *PAIR* (ordered pair operation), *LEFT*, (left projection), *RIGHT* (right projection), $\overline{0}$ (zero), D (definition by cases on numbers), *LEQ* (test for the $\leq$-relation on natural numbers);
- the binary function symbol *Ap* (application operation) and the predicate symbols N (natural numbers), T (truth),$\prec$,$\preceq$ (witness comparison), $=$(equality).

Terms are inductively defined from variables and constants via application *Ap*. We use x, y, z, u, v, w, f, g as metavariables for variables; t, t', s, s', r, r', etc., range over terms. We write (ts) instead of $Ap(t, s)$; outer brackets are usually omitted, while missing ones are restored by associating to the left. We adopt a few shorthands for special terms: $t + 1 := SUCt$ (=the successor of t); $\langle t, s\rangle := PAIRts$ (=the ordered pair composed by t and s); $(t)_1 := LEFTt$ (=the left projection of t) and $(t)_2 := RIGHTt$ (= the right projection of t). As usual, a numeral is any term obtained from the constant zero by means of a finite number of successor applications; if $n \in \omega$ (ω =the set of natural numbers), $\overline{n}$ stands for the n-th numeral, i.e. the term built up from $\overline{0}$ with n applications of *SUC*. If t is an arbitrary term of $\mathcal{L}_T$, $\lambda x.t$ is introduced by induction on the notion of $\mathcal{L}$-term: $\lambda x.x := SKK$; $\lambda x.t := Kt$, whenever x is not free in t; $\lambda x.(ts) := S(\lambda x.t)(\lambda x.s)$, if x is free in ts.

Atomic formulas (atoms in short) have the form $t = s$, $\neg t = s$, Nt, $\neg Nt$, Tt, $\neg Tt$, $t \prec s$, $\neg t \prec s$, $t \preceq s$, $\neg t \preceq s$. An atom which does not contain negation is called positive; otherwise, it is called negative. Formulas are inductively generated from atoms by means of $\vee, \wedge, \exists, \forall$; so they are always in negation normal form. Negation $\neg$ is inductively extended to arbitrary formulas by means of the classical De Morgan laws for connectives and quantifiers, and by stipulating that $\neg\neg A = A$ on atoms. The remaining missing connectives $\rightarrow$ and $\leftrightarrow$ are defined classically.

Basic fragments of $\mathcal{L}_T$:
$\mathcal{L}_{OP} := \mathcal{L}_T - \{T, \prec, \preceq, N\}$;
$\mathcal{L}_{ON} := \mathcal{L}_T - \{T, \prec, \preceq\}$;
$\mathcal{L}_{OT} := \mathcal{L}_T - \{N\}$.

Further Notations:
$x \leq y := LEQxy = \overline{0}$; $x \in N := Nx$;
$\forall x \in N.A := \forall x(x \in N \rightarrow A)$; and $\exists x \in N.A := \exists x(x \in N \wedge A)$
$f : N \rightarrow N := \forall x \in N.\ fx \in N$.

Following [S] , we can assume that there are provably disjoint closed terms *ID*, *TR*, *NAT*, *NEG*, *AND*, *ALL*, *PORD*, *SORD*, which represent the basic formula constructors of $\mathcal{L}_T$. Then we can define the map $A \mapsto [A]$ by induction on the notion of $\mathcal{L}_T$-formula:
$[t = s] = (IDt)s;[Ns] = NATs;\ [Tt] = TRt$;
$[t \prec s] = SORDts;\ [t \preceq s] = PORDts$;
$[\neg A] = NEG[A];\ [A \wedge B] = AND[A][B];\ [\forall xA] = ALL(\lambda x[A])$.

We stress that $[A]$ and A have the same free variables in common.

$TA := T[A]$

To increase readability, we keep using $\neg$, $\wedge$, $\forall$, etc., and infix notation, instead of the terms *NEG*, *AND*, *ALL*, etc. Thus $t \wedge s$, $\forall f, \forall x.t$, $\neg t$ stand for the terms $(ANDt)s$, $ALLf$, $ALL(\lambda x.t)$, $NEGt$ (in the given order).

3. INDUCTIVE MODELS

3.1. *DEFINITION*

A N-expanded combinatory algebra (n.c.a) is any first order structure for the language $\mathcal{L}_{ON}$ which satisfies the axioms:

COMB $\forall x \forall y \forall z(Kxy = x \wedge Sxyz = xz(yz))$;
PAIR $\forall x \forall y((\langle x, y\rangle)_1 = x \wedge (\langle x, y\rangle)_2 = y)$;
NAT.1 $N\overline{0} \wedge \forall x(Nx \rightarrow (N(x+1) \wedge \neg(x+1) = \overline{0} \wedge PRED(x+1) = x))$;
NAT.2 $\forall x \forall y \forall u \forall v(Nx \wedge Ny \wedge \neg x = y \rightarrow Dxxuv = u \wedge Dxyuv = v)$.

As to the existence of n.c.a's, it is well-known that every combinatory algebra (i.e., any model of COMB) can be enlarged to a n.c.a with the same universe and application operation, where N is an adequate set of numerals (see [B]) and is isomorphic to the set of natural numbers.

We now quickly recall how to extend any given n.c.a to a classical Frege structure satisfying the additional witness comparison axioms (and even more). The construction is standard and can be traced back to several authors (the history is discussed in [F3]). The only difference here is that we also have to cope with $\preceq, \prec$. The intended interpretation of T is the least fixed point over a given n.c.a of a positive elementary operator (defined by) $\mathcal{T}(x, P)$, in the sense of Moschovakis. $\mathcal{T}(x, P)$ assembles the obvious clauses for T-introduction, according to Kleene's strong three-valued logic, under the assumption that N and $=$ are total predicates with respect to T. The specific formula is written in the language $\mathcal{L}_T \cup \{P\}$, P being a fresh predicate symbol; note that $\mathcal{T}(x, P)$ positively depends on P, $\preceq, \prec$:

$$\begin{aligned}
&\mathcal{T}(x,P) := \exists u \exists v((x = (\neg\neg u) \wedge Pu) \vee (x = [u = v] \wedge u = v) \vee \\
&\vee(x = [\neg(u = v)] \wedge \neg(u = v)) \vee (x = [Nu] \wedge Nu) \vee (x = [\neg Nu] \wedge \neg Nu) \vee \\
&\vee(x = [u \prec v] \wedge (u \prec v)) \vee (x = [u \preceq v] \wedge (u \preceq v)) \vee \\
&\vee(x = [\neg u \prec v] \wedge (v \preceq u)) \vee (x = [\neg u \preceq v] \wedge (v \prec u)) \vee \\
&\vee(x = [\neg Tu] \wedge P(\neg u)) \vee (x = [Tu] \wedge Pu) \vee \\
&\vee(x = (u \wedge v) \wedge Pu \wedge Pv) \vee (x = [\neg(u \wedge v)] \wedge (P(\neg u) \vee P(\neg v))) \vee \\
&\vee(x = (\forall u) \wedge \forall u P(uu)) \vee (x = \neg(\forall u) \wedge \exists u P(\neg(uu)))).
\end{aligned}$$

Let $\mathcal{M}$ be an n.c.a. We define uniformly in $\mathcal{M}$, for every ordinal α:

1. a structure $\mathcal{M}^\alpha$ for $\mathcal{L}_T$;
2. a subset $T(\alpha, \mathcal{M})$ of $|\mathcal{M}|$ (the universe of $\mathcal{M}$) and two binary relations $\preceq^\alpha_\mathcal{M}$ and $\prec^\alpha_\mathcal{M}$ on $|\mathcal{M}|^2$.

Since $\mathcal{M}$ is fixed, we write T^α, $\preceq^\alpha$, $\prec^\alpha$ for $T(\alpha, \mathcal{M})$, $\preceq^\alpha_\mathcal{M}$ and $\prec^\alpha_\mathcal{M}$ (respectively). $\mathcal{T}(a, T^\alpha, \preceq^\alpha, \prec^\alpha)$ is an abbreviation for " $\mathcal{T}(x, P)$ is satisfied in $\mathcal{M}_\alpha$, whenever x, P, $\preceq, \prec$ are interpreted by a, $T^\alpha, \preceq^\alpha, \prec^\alpha$ (in the given order)."

3.2. *DEFINITION*

(by simultaneous transfinite recursion on α):

1. $T^0 = \emptyset = \preceq^0 = \prec^0$;
2. if λ is a limit, $T^\lambda = \cup\{T^\beta : \beta < \lambda\}$;
   $\preceq^\lambda = \cup\{\preceq^\beta : \beta < \lambda\}$ and $\prec^\lambda = \cup\{\prec^\beta : \beta < \lambda\}$;

3.1. $a \in T^{\alpha+1} \Leftrightarrow \mathcal{T}(a, T^\alpha, \preceq^\alpha, \prec^\alpha)$;

3.2. $a \preceq^{\alpha+1} b \Leftrightarrow (a \preceq^\alpha b)$ or $(\mathcal{T}(a, T^\alpha, \preceq^\alpha, \prec^\alpha)$ and $b \notin T^\alpha)$;
   $a \prec^{\alpha+1} b \Leftrightarrow (a \prec^\alpha b)$ or $(\mathcal{T}(a, T^\alpha, \preceq^\alpha, \prec^\alpha)$ and not $\mathcal{T}(b, T^\alpha, \preceq^\alpha, \prec^\alpha))$.

4. $\mathcal{M}^\alpha = \langle \mathcal{M}, T^\alpha, \preceq^\alpha, \prec^\alpha \rangle$.
5. $|a|$=least α such that $a \in T^\alpha$, if it exists; else $|a|$=the least cardinal > the cardinal of $|\mathcal{M}|$.
6. $T(\mathcal{M}) = \cup\{T^\alpha : \alpha \in \Omega\}$;
   $\prec_\mathcal{M} = \cup\{\prec^\alpha : \alpha \in \Omega\}$, $\preceq_\mathcal{M} = \cup\{\preceq^\alpha : \alpha \in \Omega\}$ (Ω = class of ordinals).

3.3. *LEMMA*

Let S be any set among $T^\alpha, \preceq^\alpha, \prec^\alpha$.

1. Monotonicity: if $\alpha < \beta$, $S^\alpha \subseteq S^\beta$
2. Consistency: either $a \notin T^\alpha$ or $(\neg_\mathcal{M} a) \notin T^\alpha$ ($\neg_\mathcal{M}$= value of *NEG* in $\mathcal{M}$); not $(a \preceq^\alpha b$ and $b \prec^\alpha a)$.
3. $a \in T(\mathcal{M})$ and $|a| \leq |b| \Leftrightarrow a \preceq_\mathcal{M} b$.
4. $a \in T(\mathcal{M})$ and $|a| < |b| \Leftrightarrow a \prec_\mathcal{M} b$.

If we apply Lemma 3.3.1, we get a fixed point of the operator defined by $\mathcal{T}$ and hence we have:

3.4. *PROPOSITION (Truth axioms)*

Let $\mathcal{M}^+$ be the structure $\langle \mathcal{M}, T(\mathcal{M}), \preceq_\mathcal{M}, \prec_\mathcal{M} \rangle$. Then $\mathcal{M}^+$ verifies:

1. $TA \leftrightarrow A$, if $A = (x = y)$, Nx, $(\neg x = y)$, $x \preceq y$, $x \prec y$;
2. $T\neg(x \preceq y) \leftrightarrow (y \prec x); T\neg(x \prec y) \leftrightarrow (y \preceq x)$
3. $TTx \leftrightarrow Tx$; $T\neg Tx \leftrightarrow T\neg x$;
4. $T\neg\neg x \leftrightarrow Tx$;
5. $T(x \wedge y) \leftrightarrow Tx \wedge Ty$; $T\neg(x \wedge y) \leftrightarrow T\neg x \vee T\neg y$;
6. $T(\forall f) \leftrightarrow \forall x T(fx)$; $T\neg(\forall f) \leftrightarrow \exists x T\neg(fx)$;

CONS $\neg(Tx \wedge T\neg x)$;

N-Tot $\neg Nx \rightarrow T\neg Nx$.

On the other hand, lemma 3.3.3–3.3.4 easily implies:

3.5. *PROPOSITION (Witness comparison axioms WA)*

$\mathcal{M}^+$ verifies

1. $Tx \rightarrow x \preceq x$;
2. $x \preceq y \wedge y \preceq z \rightarrow x \preceq z$;
3. $x \preceq y \rightarrow Tx$;
4. $x \prec y \rightarrow x \preceq y$;
5. $Tx \wedge \neg Ty \rightarrow x \prec y$;

6. $Tx \wedge Ty \rightarrow x \preceq y \vee y \prec x$.

We conclude with the problem of representing predicate abstraction and application, and the fixed point operator for predicates. We set:
$\{x : A\} := \lambda x.[A]$; $x \in y := T(yx)$; $x \bar{\in} y := T\neg(yx)$;
$Cl(y) := \forall x(x \in y \vee x \bar{\in} y)$ (y is a class or a total predicate).
$Ixv.A(x,v) := Y(\lambda v.\{x : A(x,v)\}$ (where Y is the paradoxical combinator).

For simplicity, we write $I(A) := Ixv.A(x,v)$. Clearly $I(A) = \{x : A(x, I(A))\}$, modulo combinatory logic.

$A(x,v)$ (with the free variables shown) is called an operator form (in the variable v) iff $A(x,v)$ is inductively generated via $\vee, \wedge, \forall y, \exists z$ (with $y, z \notin \{x,v\}$) from atoms of the form $t = s$, $\neg t = s$, Nt, Tt, $t \in v$, provided v is not free in t, s. (We could have been more liberal, also allowing atoms $t \prec s$, $t \preceq s$; but this would have not result in a strengthening, as all the theories considered prove the equivalence $TA \leftrightarrow A$, for $A = t \prec s$, $t \preceq s$).

The choice of $I(A)$ is justified by the following theorem (proof by transfinite induction):

3.6. *THEOREM (Generalized induction)*
If $A(x,v)$ is an operator form in v and $B(x)$ is an arbitrary formula of $\mathcal{L}_T$, then $\mathcal{M}^+ \models \forall x(A(x,B) \rightarrow B(x)) \rightarrow \forall x(x \in I(A) \rightarrow B(x))$.

4. THE BASIC CALCULUS

Since we are interested in proof theoretic analysis, we outright present a basic formal system and its extensions as (classical) sequent calculi in the style of Tait. Sequents are finite sets of formulas; they are denoted by capital Greek letters $\Gamma, \Delta, \ldots$ The intended meaning of a (non-empty) sequent Γ is the finite disjunction over the formulas of Γ. The expression "Γ, Δ" stands for the set theoretic union of Γ with Δ.
In order to achieve a greater generality, we shall first consider a system in which N is conceived as a partial predicate. OTN$^-$ is an acronym for operations, truth and numbers; the$^-$-exponent reminds that induction principles are omitted and that N is considered only as partial predicate with respect to the truth predicate T.

4.1. *Axioms and rules of OTN^-*

Logical axioms:
$\Gamma, p = p$;
$\Gamma, \neg p = q$, $\neg A(p)$, $A(q)$, where A is atomic;
$\Gamma, \neg A$, A , where A is atomic.

Axioms for operations, witness comparison, numbers and consistency: they include sequent formulations of the axioms COMB, PAIR, NAT, CONS, WA (we leave them to the reader; for instance we shall have $\Gamma, \neg t \in N, (t+1) \in N$; $\Gamma, \neg Tt, \neg T\neg t$).
They always have the form Γ, Δ, where Δ is a suitable finite set of atomic formulas

Logical Rules: standard rules for introducing $\wedge, \vee, \forall$, $\exists$ and Cut.

Truth Rules:

$$\frac{\Gamma, \mathcal{T}^+(t, T)}{\Gamma, Tt;}$$

$$\frac{\Gamma, \neg\mathcal{T}^+(t, T)}{\Gamma, \neg Tt.}$$

Proviso: $\mathcal{T}^+(x, P)$ is the operator which is obtained from $\mathcal{T}(x, P)$ by omitting the clause "$(x = [\neg Nu] \wedge \neg Nu)$".

The basic system OTN$^-$ can be strengthened with induction rules for N and for T. Here is a list of the main inferences we are going to consider.

Full N-Induction Rule IND$_s$:

$$\frac{\Gamma, A(0) \qquad\qquad \Gamma, \forall x(A(x) \rightarrow A(x+1))}{\Gamma, \neg t \in N, A(t)}$$

Positive N–Induction Rule IND$_p$: as IND$_s$, except that $A(t)$ has the form $t \in b$.

Class N–Induction Rule IND$_c$: as IND$_p$, with the additional premise $Cl(b)$ (b is a class).

Generalized Induction GID$_s$:

$$\frac{\Gamma, \forall x(A(x, B) \rightarrow B(x))}{\Gamma, \neg t \in I(A), B(t),}$$

provided $A(x, a)$ is an operator form and $B(x)$ is arbitrary.

GID$_p$ (positive generalized induction): as GID$_s$, with the proviso that $B(t)$ has the form $t \in b$.

4.2. *Other Systems*

OTN=OTN$^-$+N-Tot; OT:= the fragment of OTN$^-$, formalized in the N-free fragment $\mathcal{L}_{OT}$.

ON$^-$ is the subsystem of OTN$^-$, which is formalized in the language $\mathcal{L}_{ON}$ (so $T, \prec, \preceq$ are omitted). If $\mathcal{S}$ is any given system in $\mathcal{L}_T$, $\mathcal{S}_c = \mathcal{S}$+IND$_c$, $\mathcal{S}_p = \mathcal{S}$+IND$_p$ and $\mathcal{S}_s = \mathcal{S}$+IND$_s$.

4.3. *Derivability relations*

If $\mathcal{S}$ is any system, $\mathcal{S} \vdash_n^m \Gamma$ ($m, n \in \omega$) is inductively defined by the following clauses:

1 if Γ is an axiom, $\mathcal{S} \vdash_n^m \Gamma$ for every m, n;
2 assume that Γ is the conclusion from the premises Γ_i of a logical or mathematical rule, or of a cut rule of rank$< n$; $\mathcal{S} \vdash_n^{m_i} \Gamma_i$ $(i \leq 3)$ and $m_i < m$. Then $\mathcal{S} \vdash_n^m \Gamma$.

Due to lack of space, we cannot develop detailed consequences of the axioms. However, before proceeding to the proof-theoretic investigation, we underline that most interesting consequences of the theory do not depend on N nor on the totality of N. So they go through in OT. It is possible to relate OTN_c with Feferman's explicit mathematics ([F2]). So, if $CL := \{x : Cl(x)\}$, CL is a model of Feferman's EM$\lceil$, provably in OTN_c. As to the witness comparison axioms, they imply generalizations of classical results from Recursion Theory: reduction, separation, the Myhill-Shepherdson theorem for extensional operations, a uniform version of Knaster-Tarski theorem. In addition, OT can interpret admissible set theory. For instance, pure sets are identified with elements of AD, satisfying $x \in AD \leftrightarrow Cl(x) \wedge \forall y(y \in x \rightarrow y \in AD)$.

It follows that OT+GID_s contains a good theory of ordinals with full transfinite induction. That ordinals are available, is also essential for the interpretation in OTN_c+GID_s of alternative systems, dealing with forms of type-free abstraction and self-reference. In particular, in OTN_c+GID_s , we can interpret Myhill's extension of classical Frege structures with levels of implication (see [FM]) and the system of [AF], based on a form of definitional equivalence. As to formal semantics, OTN_c+GID_s can also model theories of truth, which are closed under logical consequence (so that $T(A \rightarrow A)$ always holds); we can actually refine the results by Friedman and Sheard [FS] and [C2] (for details see [C3]).

5. A STRONG THEORY OF SELF-REFERENTIAL TRUTH WHICH IS PROOF-THEORETICALLY EQUIVALENT TO PA

We deal with the proof theoretic analysis of the theory of classical Frege structures, extended with full generalized induction, number-theoretic induction for classes and witness comparison axioms.

Indeed, let OTN_c+GID_s = OTN^-+N-Tot+IND_c+GID_s, where N-Tot is the axiom $\forall x(\neg(x \in N) \rightarrow T\neg(x \in N))$. Then we will prove that, if $A \in \mathcal{L}_{ON}$ and OTN_c+$\text{GID}_s \vdash A$, then $\text{ON}_s^- \vdash A$. The proof is constructive and it can be carried out in PRA with a suitable instance of transfinite induction on (the standard well-ordering of type) ϵ_0.

Step 1: Simulating a recursively saturated model.

Due to the rule GID_s, we have no hope to reduce cuts in the given system to atomic cuts. Thus, we embed $\mathrm{OTN}_c+\mathrm{GID}_s$ into a system OTN^∞, which contains an infinitary rule, implying that T is inductively defined by iterating the operator $\mathcal{T}$ along ω. Of course, this is generally false, unless we "live" in a non-standard model (actually a recursively saturated one). In fact, OTN^∞ is provably ω-inconsistent; nevertheless, the system enjoys the required cut elimination property and can interpret the full system $\mathrm{OTN}_c+\mathrm{GID}_s$.

5.1. *Simultaneous Inductive definition of* $T^n, \prec^n, \preceq^n$ $(n \in \omega)$.

1. $T^0 t := (t \neq t)$; $t \prec^0 s := t \preceq^0 s := (ts \neq ts)$;
2. $T^{n+1} t := \mathcal{T}(t, T^n)$;
   $t \prec^{n+1} s := (t \prec^n s) \vee (\mathcal{T}(t, T^n) \wedge \neg \mathcal{T}(s, T^n))$;
   $t \preceq^{n+1} s := (t \preceq^n s) \vee (\mathcal{T}(t, T^n) \wedge \neg T^n s)$.

5.2. *The n-relativization of A.*

A^n is the formula which is obtained from A by replacing each occurence of $T, \prec, \preceq$ in the given order by $T^n, \prec^n, \preceq^n$.

5.3. *REMARK.*

For each $n \in \omega$, $T^n t$, $t \prec^n s$, $t \preceq^n s$, A^n are formulas of $\mathcal{L}_{ON}$. Also $\mathcal{T}(t, T^n)$ is actually $\mathcal{T}(t, T^n, \prec^n, \preceq^n)$; but we left the dependence on $\prec^n$, $\preceq^n$ implicit for notational simplicity.

5.4. *Definition of* OTN^∞.

The system OTN^∞ is obtained from $\mathrm{OTN}_c+\mathrm{GID}_s$ by substituting GID_s with the following infinitary rule (T^∞) :

$$\frac{\ldots \Gamma, \neg T^n t \ldots \quad (n \in \omega)}{\Gamma, \neg T t.}$$

5.5. *REMARK.*

a) For each $n \in \omega$, $\neg T^n t$ is a formula of $\mathcal{L}_{ON}$, whose logical complexity increases with $n \in \omega$.
b) Active formulas in all axioms and mathematical inferences —T^∞ included— are *atomic*.

5.6. *Inductive definition of the derivability relation* $OTN^\infty \vdash_k^\alpha \Gamma$ $(k \in \omega$, $\alpha < \epsilon_0)$.

1.1 If Γ contains (as a set) an axiom of OTN^∞, $\mathrm{OTN}^\infty \vdash_k^\alpha \Gamma$ for every α and every $k \in \omega$;

1.2 Assume that Γ is the conclusion from the premises Γ_i of a finitary rule, or of (T^∞), or else of a cut rule applied to formulas of rank $< k \in \omega$; assume also that $\mathrm{OTN}^\infty \vdash_k^{\alpha_i} \Gamma_i$ $(i \leq \omega)$ and $\alpha_i < \alpha$. Then $\mathrm{OTN}^\infty \vdash_k^\alpha \Gamma$.

$\mathrm{OTN}^\infty \vdash^{<\omega} \Gamma := \mathrm{OTN}^\infty \vdash_k^m \Gamma$, for some finite m, k.
$\mathrm{OTN}^\infty \vdash \Gamma := \mathrm{OTN}^\infty \vdash_k^\alpha \Gamma$, for some α and $k \in \omega$.
$\mathrm{OTN}_c \vdash_k^m \Gamma$, where k, $m \in \omega$, is the corresponding derivability notion for the finitary subsystem OTN_c.

5.7. *THEOREM (Weak Cut Elimination).*

1. If $\mathrm{OTN}^\infty \vdash \Gamma$, then $\mathrm{OTN}^\infty \vdash_1^\alpha \Gamma$, for some $\alpha < \epsilon_0$.
2. 1) extends to OTN_c (of course with α and ϵ_0 replaced in the given order by a finite number and by ω).

(Proof: adapt the usual arguments for ω -logic and for finitary logic).

5.8. *LEMMA (Embedding).*

If $\mathrm{OTN}_c+\mathrm{GID}_s \vdash A$, then $\mathrm{OTN}^\infty \vdash_k^\alpha A$, for some $k \in \omega$, $\alpha < \omega 2$.

PROOF (Hint). Let A be an operator form, $B(x)$ arbitrary. Then we can easily check that for each $n \in \omega$, there exists a $k(n)$ such that, for every t :

$$\mathrm{OTN}_c \vdash_1^{k(n)} \neg\forall x(A(x,B) \to B(x)), \neg t \in^n I(A), B(t). \tag{1}$$

Hence by one application of T^∞ to (1), we obtain:

$$\mathrm{OTN}^\infty \vdash_1^\omega \neg\forall x(A(x,B) \to B(x)), \neg t \in I(A), B(t). \tag{2}$$

(2) suffices to verify the lemma by induction on the length of a given $\mathrm{OTN}_c+\mathrm{GID}_s$-derivation. □

5.9. *PROPOSITION (ω–inconsistency)*

For some $A(x)$, we have:
$\mathrm{OTN}^\infty \vdash \neg\forall x A(x)$;
$\mathrm{OTN}^\infty \vdash A(\overline{n})$, for each $n \in \omega$.

PROOF (hint). By the fixed point theorem for operations, there is a term g such that $g\overline{0} = [\overline{0} = \overline{0}]$ and $g(\overline{k+1}) = [T(g\overline{k})]$.
Choose $A(x) := (x \in N \to T(gx))$.

Step 2: Asymmetric Interpretation of OTN^∞ into ON_s^-.

5.10. *Inductive Definition of* $A[m,n]$, *for* $A \in \mathcal{L}_T$.

1. $A[m,n] = A^n$, if A is an atom of the form $Tt, t \prec s, t \preceq s$;
2. $A[m,n] = A^m$, if if A is an atom of the form $\neg Tt, \neg t \prec s, \neg t \preceq s$;
3. $A[m,n] = A$, if A is an atom of the form Nt, $\neg Nt$, $t = s, \neg t = s$;

4. $A[m,n]$ preserves $\wedge, \vee, \forall, \exists$.

5.11. *LEMMA*

1. For each $m, n \in \omega$, $A[m,n] \in \mathcal{L}_{ON}$.
2. If $A \in \mathcal{L}_{ON}$, $A[m,n] = A$.
3. Let $m \leq p \leq q \leq n$: then $\mathrm{ON}^- \vdash \neg A[p,q], A[m,n]$.
4. If A is positive (negative) in $T, \prec, \preceq$, then $A[m,n] = A^n$ ($A[m,n] = A^m$).

Henceforth, we can assume that OTN^∞–derivations are codified by "finite means", so that cut elimination can be defined by a primitive recursive function and it makes sense to apply a number-theoretic function to derivations (identified with some sort of explicit encoding). In particular, we also assume that, whenever a derivation $\mathcal{D}$ ends with a T^∞-rule, then we can read off from the derivation $\mathcal{D}$ a primitive recursive enumeration $\lambda n \in \omega.\mathcal{D}_n$ of its countably many immediate subderivations. For simplicity, let "$\mathcal{D} \vdash_1^\alpha \Gamma$" stand for "$\mathcal{D}$ is an OTN^∞–derivation with length α and cut rule applied to atoms (at most)".

5.12. *THEOREM (Majorization).*
There is a partial recursive function F such that, if $\mathcal{D} \vdash_1^\alpha \Gamma$, then for each $m \in \omega$, $F(\mathcal{D}, m)$ is defined and moreover:

1. $\mathrm{ON}_s^- \vdash m \leq F(\mathcal{D}, m)$;
2. $\mathrm{ON}_s^- \vdash \Gamma[m, F(\mathcal{D}, m)]$.

Hence by embedding, weak cut elimination and majorization, together with lemma 5.11 above:

5.13. *COROLLARY. OTN_c+GID_s is a conservative extension of ON_s^-.*
PROOF of theorem 5.12 (sketch). $F(\mathcal{D}, m)$ is defined by effective transfinite recursion on the height α of $\mathcal{D}$ and the persistence properties of 5.11 are repeatedly applied.
Case 1. Γ is an axiom. Choose $F(\mathcal{D}, m) = m + 1$; here there is some work to be done. Indeed, except for the application of well-known tautology and identity lemmata (for logical axioms), it is routine to check that suitable finite approximations of consistency and WA-axioms hold true in ON_s^- (e.g. $\{\neg T_m t, \neg T_m(\neg t)\}$).
Case 2. Γ is a sequent obtained by means of a logical rule: we simply choose $F(\mathcal{D}, m) = max\{F(\mathcal{D}_i, m) : i \leq k\}$, k being the number of premises of Γ.
Case 3. $\Gamma \equiv \Delta, \neg Tt$ follows by application of T^∞. Hence by IH, for each $n \in \omega$ and each $m \in \omega$:

$\mathrm{ON}_s^- \vdash m \leq F(\mathcal{D}_n, m)$;
$\mathrm{ON}_s^- \vdash \Delta[m, F(\mathcal{D}_n, m)], \neg T^n t$.
Choose $F(\mathcal{D}, m) = F(\mathcal{D}_m, m)$.

Case 4. Class N-induction. Then $F(\mathcal{D},m) = F(\mathcal{D}_0,m)+F(\mathcal{D}_1,m)+F(\mathcal{D}_2,m)$, where $\mathcal{D}_0$, $\mathcal{D}_1$, $\mathcal{D}_2$ are the given immediate subderivations of $\mathcal{D}$. Then apply IH and the derivability in ON_s^- of the sequents:

$\neg Cl_m(t), \neg s \in^k t, s \in^m t$;

$\neg 0 \in^m t, \neg \forall x \in^m t.(x+1) \in^m t, \neg s \in N, s \in^m t$.

Case 5: the final inference is a cut. Here we essentially need that cut formulas are atomic; we define $F(\mathcal{D},m)$ by composing the values assigned to the immediate subderivations and we apply lemma 5.11.

Case 6: the final inference is T-introduction. We let $F(\mathcal{D},m) = F(\mathcal{D}_0,m)+1$.

Case 7: the final inference is $\neg T$-introduction: then $F(\mathcal{D},m) = F(\mathcal{D}_0,m)$ and we argue by induction on m, using IH and definition of T^m. □

5.14. *REMARK.*

$F(\mathcal{D},m)$ cannot be primitive recursive, since OTN_c $+\mathrm{GID}_s$ proves $t : N \to N$, for some closed t which defines the Ackermann function. If we consider derivations of OTN^∞ having a fixed ordinal length below ϵ_0, the function $F(\mathcal{D},m)$ can be chosen total and α-recursive, for some $\alpha < \epsilon_0$.

6. STRENGTHENING THE GROUND APPLICATIVE THEORY WITH A UNIFORMITY PRINCIPLE

6.1.

Let ON be ON^-+ORD (for ON^-, see 4.2), where ORD is the following set of axioms for the constant LEQ (below $x \leq y := LEQxy = \overline{0}$):

1. $\overline{0} \leq \overline{0}$;
2. $x \leq y \wedge y \leq z \wedge z \in N. \to x \leq z$;
3. $x \in N \wedge y \in N. \to x \leq y \vee y \leq x$;
4. $x \leq y \wedge y \leq x \wedge (x \in N \vee y \in N). \to x = y$;
5. $x \leq y \wedge y \in N. \to x \leq y+1$;
6. $x \leq y \wedge y \in N \wedge x \neq y. \to x+1 \leq y$;
7. $x \leq y \wedge y \in N. \to x \in N$.

6.2. *THEOREM.*

ON_s =ON+IND_s has an inner model in PA.

6.2 is proved by a term model construction. Indeed, we add to the well-known reduction rules for the constants K, S, *LEFT*, *RIGHT*, *PRED*, D a new reduction rule for the constant *LEQ*: $LEQ\overline{n}\overline{m} \Rightarrow \overline{0}$, provided $n \leq m$ (where $\Rightarrow$ stands for reduction). Then the Church-Rosser theorem is easily checked for $\Rightarrow$, (provably, say, in PRA), and we can define a closed term model $\mathrm{CTM}^{\leq}$, in PA where N is interpreted as the set of closed terms which are reducible to numerals, while $=$ is reducibility to a common redex. As to the crucial axiom

6.1.7, it holds in $\mathrm{CTM}^{\leq}$ essentially because all proper reductions of $LEQts$ imply that t, s are numerals.

We now consider the problem of finding extensions of ON_s, which do not go beyond PA and yet have a rather rich theory of sets. First of all, the obvious route fails to be viable: if we identify sets as characteristic operations over the whole space, we only get trivial sets. Indeed, define:

$$f : V \to 2 := \forall x(fx = \overline{0} \vee fx = \overline{1}).$$

6.3. *LEMMA (folklore?)*

$\mathrm{ON} \vdash \forall x(fx = \overline{0} \vee fx = \overline{1}) \to \forall x(fx = \overline{0}) \vee \forall x(fx = \overline{1})$.

PROOF. If the conclusion is false, for some z, y: $fz = \overline{1} \wedge fy = \overline{0}$. Using definition by cases on N, there exists a term g such that

$$gx = \begin{cases} z, & \text{if } fx = \overline{0}; \\ y, & \text{if } fx = \overline{1}. \end{cases}$$

Any fixed point e of g leads to contradiction. □

The lemma admits a simple generalization. Let us identify N with the "space of values" and let us say that an operation f converges on a given argument x, if $(fx) \in N$. Then every operation which converges everywhere and is bounded in the space of values, is constant. Formally:

6.4. *PROPOSITION*

$\mathrm{ON}_p \vdash (\forall z \in N)(\forall x(fx \leq z) \to (\exists y \leq z)(\forall x)(fx = y))$.

PROOF (hint). By positive induction on $z \in N$. The case $z = \overline{0}$ follows with axioms ORD. 1, 4, 5, 7. The induction step amounts to apply the lemma 6.3 (but with $\overline{0}$, $\overline{1}$ replaced by suitable distinct elements of N). □

Is it true in some model that that every operation which converges everywhere is constant? Formally, we ask if it is consistent to assume:

$$\forall x(fx \in N) \to (\exists y \in N)(\forall x)(fx = y) \qquad (*)$$

This is not so unreasonable as it appears at a first sight, if one thinks of the range theorem for pure λ–calculus; moreover, Visser shows that any function definable in the open term model of pure λ–calculus which assumes only values in normal form, is constant (see [B]). We now state a principle N-ref (reflection on N), which entails (*) and implies that the space of characteristic operations on N is very rich.

N-ref: $\forall x \exists y \in N.A(x, y) \to \exists z \in N.\forall x.\exists y \leq z.A(x, y)$, where A is N-free.

6.5. *DEFINITION.*

A is N-positive iff A does not contain atoms of the form $\neg(t \in N)$.

A^x := the formula obtained from A by replacing each atom of the form $(t \in N)$ by $(t \leq x)$.

6.6. *PROPOSITION*

1. (Full N-reflection). Let A be N-positive. Then ON+N-Ref $\vdash A \to (\exists x \in N).A^x$.
2. ON_p+N-Ref $\vdash \forall x(fx \in N) \to (\exists y \in N)(\forall x)(fx = y)$.

PROOF. 1. Induction on A: N-ref is applied in the case of $\forall$; the ORD- axioms are applied in the case of atoms of the form $t \in N$.
2. Apply 6.4 and N-Ref. □

$\mathrm{CTM}^{\leq}$ is of no use to prove the consistency of ON_s+N-Ref. Indeed, let EA be the enumeration axiom EA, i.e. $\exists f \forall x \exists y(y \in N \wedge fy = x)$.

6.7. *THEOREM.*

1. $\mathrm{CTM}^{\leq}$ is a countermodel for N-Ref.
2. ON_p+EA+N-Ref is inconsistent

PROOF. 1. We can define an enumeration combinator E such that $\forall x.\exists y \in N.Ey = x$ holds in $\mathrm{CTM}^{\leq}$ and hence $\mathrm{CTM}^{\leq}$ is a model of ON_s+EA. If N-Ref is true in $\mathrm{CTM}^{\leq}$, there is some k such that $\mathrm{CTM}^{\leq} \models \forall x.\exists y \leq \overline{k}.Ey = x$. But this implies that $\mathrm{CTM}^{\leq}$ has finitely many elements: contradiction !
2. We apply N-Ref to EA and we establish an injection of $y+1$ into y, for some $y \in N$. But this conflicts with (a form of) the pigeonhole principle (provable with IND_p). □

There is however a simple model theoretic proof for the consistency of an even stronger system than ON_s+N-ref.

6.8. *DEFINITION.*

1. $\exists^+$–AC(op) is the schema:

$$(\forall x \in N)(\exists y \in N)A(x, y) \to \exists f : N \to N)(\forall x \in N)A(x, fx),$$

   where A is inductively generated from positive atoms by means of $\wedge, \vee, \exists$.
2. OP is the system, which includes COMB, PAIR and the following (primitive recursive) set of axioms:
   (a) $\overline{n} \neq \overline{m}$, if $n \neq m$;
   (b) $\overline{n} \leq \overline{m}$ if $n \leq m$;
   (c) $\mathit{PRED}(\overline{n} + 1) = \overline{n}$
   (d) $D\overline{n}\overline{n}ts = t$;
   (e) $D\overline{n}\overline{m}ts = s$, if $n \neq m$;
   (f) $t \leq \overline{m} \to t \in N^m$, where $t \in N^m := (t = \overline{0} \vee \ldots \vee t = \overline{m})$.

Note that OP has an inner model in PRA (inspect the proof of 6.2).

6.9. *THEOREM.*
ON_s+N-Ref + $\exists^+$–AC(op) is consistent.

The proof is based on the following idea: consider a countable recursively saturated elementary extension $\mathcal{M}$ of the closed term model of OP and interpret N as the subset $N(\mathcal{M})$ containing the denotations of the numerals in $\mathcal{M}$. Recursive saturation implies N-ref (by the main result of [BS]). The choice axiom follows from the selection theorem for r.e. sets and the fact that $\mathrm{CTM}^{\leq}$ is an elementary substructure of $\mathcal{M}$.

6.10. *REMARK.*
The fragment of second order arithmetic, which includes, say, PRA, Σ_1-induction, recursive comprehension, and weak König lemma WKL (see [SI]) has an inner model in OP_p+N-Ref+$\exists^+$–AC(op). Indeed, if $A(f,y)$ is N-free,we have WKL in the form:

$$\forall f : N \to 2.\exists y \in N.A(f,y) \to \exists z \in N.\forall f : N \to 2.\exists y \leq z.A(f,y).$$

This because N occurs positively in the antecedent of the schema above and hence we can apply N-ref. As to the strength of ON_p and ON_s plus N-Ref, we shall prove in the next section that the two systems have the same provably recursive functions as PRA and PA (respectively).

7. ON THE PROOF THEORY OF PARTIAL TRUTH WITH N-REFLECTION

OTN^+ is the theory OTN^-+ORD; so $\mathrm{OTN}^+ \supseteq$ON. We consider the system OTN_p^++GID_p+T-ref, where T-reflection is the inference:

$$\frac{\Gamma, \forall x.\exists y \in N.T(fxy)}{\Gamma, \exists z \in N.\forall x.\exists y \leq z.T(fxy)}$$

T-ref generalizes N-Ref, since $TA \leftrightarrow A$ if A is N-positive. Note also that the conclusion of T-ref is T-positive with logical complexity 5. Call A suitable if A has logical complexity at most 5 and A is T-positive or T-negative (but not both). It is immediate to observe:

7.1. *LEMMA.*
Every OTN_p^++GID_p+T-ref-derivation can be transformed into an OTN_p^++ GID_p+T-ref-derivation with the same conclusion, where cuts occur only on suitable formulas.

7.2. *Asymmetric Interpretation of OTN_p^++GID_p+T-ref into OP*

N is now considered dependent on the verification process. In particular, we let $(t \in N)^n = (t \leq \overline{n})$. So we modify the initial conditions in the definition of $A[m,n]$, for $A \in \mathcal{L}_T$ and we choose:

$(t \in N)[m,n] = (t \in N)^n$ and $(\neg t \in N)[m,n] = (\neg t \in N)^m$.

In the remaining cases, $A[m,n]$ is defined according to 5.10. Clearly the map $A \mapsto A[m,n]$ eliminates every occurrence of N, as $t \leq \overline{k}$ is a formula of $\mathcal{L}_{OP}$. In addition, the asymmetric interpretation of the ON-axioms is provable in OP (e.g. $\neg N^m t, N^{m+1}(t+1)$; $\neg N^0 s$).

7.3. *DEFINITION.*

1. If t is a closed term of $\mathcal{L}_{ON}$, F_t is the partial number-theoretic function defined by t, i.e.

$$F_t(m) = \begin{cases} k, & \text{if } \mathrm{CTM}^{\leq} \models t\overline{m} = \overline{k}; \\ \text{undefined}, & \text{else.} \end{cases}$$

2. If $\mathcal{S}$ is any theory in the language $\mathcal{L}_T$, we define $\mathrm{Rec}(\mathcal{S}) = \{F_t : t \text{ closed and } \mathcal{S} \vdash t : N \to N\}$.

7.4. *THEOREM (Majorization theorem).*

1. Let OTN_p^++T-Ref+$\mathrm{GID}_p \vdash \Gamma$. Then there exists a PR-function F, such that, for every $m \in \omega$, OP$\vdash m \leq F(m)$ and OP$\vdash \Gamma[m, F(m)]$.
2. If t is a closed term such that OTN_p^++T-Ref+$\mathrm{GID}_p \vdash t : N \to N$, then F_t is majorized by a PR-function.

The proof of 7.4 is a finitary version of the previous majorization argument 5.12. Clearly T–ref is trivialized by the asymmetric interpretation, while the case of IND_p corresponds to elimination of Σ_1-induction via primitive recursion in the proof of Parsons's theorem for PA_1 (=PA with number theoretic induction restricted to Σ_1-formulas). The verification of GID_p is similar.

7.5. *COROLLARY (Boundedness).*

1. If $A(x)$ is T-negative and OTN_p^++T-Ref+$\mathrm{GID}_p \vdash \forall x(A(x) \to x \in N)$. Then the set $\{ a : \mathrm{CTM}^{\leq} \models A(a)\}$ is finite.
2. In particular, if OTN_p^++T-Ref+$\mathrm{GID}_p \vdash Cl(t) \wedge \forall x(x \in t \to x \in N)$, then t defines a finite set of natural numbers.

7.5.1 follows by majorization and downward persistence of T-negative formulas; 7.5.2 follows by observing that if we assume classhood of t, $x \in t$ is equivalent to a T-negative condition. By essential use of WA-axioms, 7.5.2 also yields:

7.6. *COROLLARY (Undefinability of N).*
For no T-negative $A(x)$, OTN_p^++T-Ref+$\mathrm{GID}_p \vdash \forall x(A(x) \leftrightarrow x \in N)$.

7.7. *COROLLARY (Closure of OTN_p^++GID_p under T-reflection Rule)*
If OTN_p^++$\mathrm{GID}_p \vdash \forall x \exists y \in N.TA(x,y)$, there is some $k \in \omega$ such that

$$OTN_p^+ + GID_p \vdash \forall x \exists y \leq \overline{k}.TA(x,y).$$

(Proof: by the theorem and observing that OTN_p^++GID_p proves $T^n t \to Tt$.)
If PR is the class of primitive recursive functions, we have (see 7.3):

7.8. *THEOREM. Rec(OTN_p^++T-Ref+GID_p)=PR.*
PROOF. We only consider the non-trivial direction of the statement. The main idea is to adopt PA_1 (PA with number theoretic induction restricted to Σ_1-formulas) as a metatheory.

Step 1: for each OTN_p^++T-Ref+GID_p-derivation $\mathcal{D}$ of Γ, there is a PR-function $G(= G_{\mathcal{D}})$ such that $\mathrm{PA}_1 \vdash$" $\forall m$. OP$\vdash \Gamma[m, G(m)]$ ". Indeed, by inspection, the verification of OP-derivability for $\Gamma[m, G(m)]$ is fully explicit and requires at most Σ_1-induction (see the case of IND_p and of the axioms).

Assume OTN_p^++T-Ref+$\mathrm{GID}_p \vdash t : N \to N$; then by step 1, we obtain:
$\mathrm{PA}_1 \vdash$" $\forall m$. OP$\vdash \bigvee\{t\overline{m} = \overline{k} : k \leq G(m)\}$" (1)

Claim: $\mathrm{PA}_1 \vdash$ "If OP$\vdash \Gamma$ and Γ is a finite set of closed equations, then there is some $(t = s) \in \Gamma$ such that OP$\vdash t = s$".
Hence from (1):
$\mathrm{PA}_1 \vdash$" $\forall m.\exists k \leq G(m)$. OP$\vdash t\overline{m} = \overline{k}$". (2)

By Parsons's theorem (see [HP]), applied to (2), there exists a PR-function $\Pi(= \Pi_{\mathcal{D}})$, such that
$\mathrm{PA}_1 \vdash$" $\forall m.\exists k \leq G(m).\Pi(m)$ is an OP-derivation of $t\overline{m} = \overline{k}$." (3)

Since the predicate $R(m,k) :=$"$\Pi(m)$ is an OP-derivation of $t\overline{m} = \overline{k}$" is primitive recursive, the function $J(m) := \mu k \leq G(m).R(m,k)$ is primitive recursive; but (3) implies $F_t(m) = J(m)$.

As to the claim, we check a slightly more general property. Consider any OP-derivation d of a set $\Gamma = \{A_1, \ldots, A_n\}$, where each $A_i \in \Gamma$ has the form $t = s$ or $\neg(t = s)$. By finitary cut elimination (certainly provable in PA_1), we can assume that cut formulas in d only apply to equations and negated equations. Then PA_1 proves:

"(If $d \vdash \Gamma$, where Γ is a finite set of closed equations and negated equations, then there exists $A \in \Gamma$, such that $CTM^{\leq} \models A$". (4)

(4) can be verified without difficulty by induction on the length of the derivation d. In order to see that the step makes sense in PA_1, again recall that the Church-Rosser theorem can be formalized in PA_1. Moreover, the right-hand-side formula of (4) is obtained by boolean connectives and bounded number quantifiers from the Σ_1-formula, which formalizes the interpretation of $=$ as convertibility (in the sense of the present expanded combinatory logic). But Σ_1-induction is enough to justify induction for such formulas (see [HP]). Once we have (4), the claim follows by the easy check (again in PA_1) that OP is complete for true closed equations of $CTM^{\leq}$. □

7.9. *REMARK.*
The theorem is a sort of elimination of "higher types" in defining number theoretic functions. No matter how complex a term t is (for instance, it may contain terms without normal form, fixed point combinators, etc.), if t is provably total on N with values in N, then t can be replaced by a simpler primitive recursive term.

We conjecture that the theorem is true also if we add $\exists^+$-AC(op).

8. SUMMARY

We conclude with a general picture of the computational content of the theories considered so far. Let $\mathcal{S} \equiv^r \mathcal{T}$ mean that $\mathcal{S}$ and $\mathcal{T}$ have the same provably recursive functions (we leave the definition implicit; the reader can certainly supply the necessary details for himself). For the classical systems ID_1 and $(\Pi^0_1\text{-CA})_{<\alpha}$, we send the reader to [F1].

8.1. *THEOREM ("N total versus N partial").*
"N total":

1. $OTN_c + GID_s \equiv^r ON_s^- \equiv^r PA$.
2. $OTN_p + GID_s \equiv^r ID_1 \equiv^r OTN_s + GID_s$.
3. If $F \in Rec(OTN_p + GID_p)$, F is provably recursive in $(\Pi^0_1\text{–CA})_{<\phi\omega 0}$.
4. $OTN_p \equiv^r (\Pi^0_1\text{–CA})_{<\omega^\omega}$.
5. $OTN_s \equiv^r (\Pi^0_1\text{–CA})_{<\epsilon_0}$.

"N partial":

6. $OTN_p^+ + GID_p + \text{T-Ref} \equiv^r PRA$.
7. $OTN_s^+ + GID_s + \text{T-Ref} \equiv^r PA$.

We comment on the proof. Of course 8.1.1 and 8.1.6 are already given by 5.13 and 7.8.

Ad 2. ID_1 is obviously interpreted into $OTN_s + GID_s$; as to the reverse direction, one can carry out the construction of an inductive model in the Kripke-Platek set theory above natural numbers with full N-induction and full transfinite induction on $\in$. But $KPN \equiv^r ID_1$ by [J].

Ad 3–5: the systems are embedded (possibly via a preliminary partial cut elimination and asymmetric interpretation) into semiformal systems, which are ramified with respet to T, $\preceq$, $\prec$ and are fully predicative. Schütte's cut elimination for predicative systems is then applied (see [F4], [JS2], [C1]).

Ad 7. The proof is an adaptation of 5.12. We get rid of GID_s and IND_s by means of T^∞ -rule and a suitable ω-rule for introducing $(\neg t \in N)$. Then we can prove a partial cut elimination and a majorization theorem; PA and Kreisel's theorem on provably recursive functions of PA are then applied instead of PA_1 and Parsons's theorem (respectively).

Problem. Is $\mathrm{OTN}_p + \mathrm{GID}_p \equiv^r \mathrm{OTN}_p$ or $\mathrm{OTN}_p + \mathrm{GID}_p \equiv^r (\Pi^0_1\text{–CA})_{<\phi\omega 0}$?

As to further research, we wish to mention a final question: is it possible to weaken the ground applicative theory, so that the recursive content of the truth theory, based on it, coincides with the class of PTIME-computable functions? Recently, Strahm [ST] succeeded in developing a theory of operations based on *partial application*, in which the basic type N is replaced by the set of binary strings W, and PTIME is characterized in the style of Ferreira. We investigated a theory of partial truth, based on *total application*, W-reflection (in analogy with N-ref) and a bounded form of W-induction; at present, we have been able to prove only that the functions of type $W \to W$ (provably in the theory) have polynomial growth. A major obstacle for proving a full characterization theorem is that we need to replace PA_1 in the proof of theorem by a system which only proves the well-definedness of PTIME-functions. In particular we must handle the closed term model and the Church-Rosser theorem in a bounded theory. We do not know whether all this can be done.

Università degli Studi di Firenze

REFERENCES

[A] ACZEL, P. The strength of Martin-Löf's intuitionistic type theory with one universe, in: S. Mietissen and J. Väänänen (eds.), *Proceedings of the Symposiums in Mathematical Logic in Oulu* 1974 *and Helsinki* 1975, (Report no.2, Dept. of Philosophy, University of Helsinki), 1977, 1–32.

[AF] ACZEL, P. and S. FEFERMAN. Consistency of the unrestricted abstraction principle using an intensional equivalence operator, in: J.P. Seldin and R. Hindley, (eds.), *To H.B.Curry : Essays on Combinatory Logic, Lambda Calculus and Formalism* (Academic Press, New York 1980), 67–98.

[B] BARENDREGT, H. *The Lambda Calculus: its Syntax and Semantics* (North Holland, Amsterdam 1984).

[BS] BARWISE, J. and J. SCHLIPF. On recursively saturated models of arithmetic, Springer Lecture Notes in Mathematics 498, 1975, 42–55.

[Bu] BURGESS, J. The truth is never simple, *Journal of Symbolic Logic*, 51, 1986, 663–681.

[C1] CANTINI, A. Notes on formal theories of truth, *Zeitschrift für Mathematische Logik und die Grundlagen der Mathematik*, 35, 1989, 97–130.

[C2] CANTINI, A. A theory of formal truth arithmetically equivalent to ID_1, *Journal of Symbolic Logic*, 55, 1990, 244–259.

[C3] CANTINI, A. *Logical Frameworks for Truth and Abstraction* (North-Holland, Amsterdam 1996).
[F1] FEFERMAN, S. Theories of finite type related to mathematical practice, in: J. Barwise (ed.),*Handbook of Mathematical Logic*, (North Holland, Amsterdam 1977), 913–977.
[F2] FEFERMAN, S. Constructive theories of functions and classes, in: D. van Dalen, M. Boffa (eds.), *Logic Colloquium '78* (North Holland, Amsterdam 1979), 159–224.
[F3] FEFERMAN, S. Towards useful type-free theories I, *Journal of Symbolic Logic*, 49, 1984, 75–111.
[F4] FEFERMAN, S. Reflecting on incompleteness, *Journal of Symbolic Logic*, 56, 1991, 1–49.
[FM] FLAGG, R. and J. MYHILL. Implications and analysis in classical Frege structures, *Annals of Pure and Applied Logic*, 34, 1987, 33–85.
[FS] FRIEDMAN, H. and M. SHEARD. An axiomatic approach to self-referential truth, *Annals of Pure and Applied Logic*, 33, 1987, 1–21.
[G] GIRARD, J.Y. A survey of Π^1_2-logic, in: L. Cohen, J. Los, H. Pfeiffer, K.P. Podewski (eds.), *Logic, Methodology and Philosophy of Science VI* (North Holland, Amsterdam 1982) 89–107.
[HP] HAJEK, P. and P. PUDLAK. *Metamathematics of Arithmetic*, Berlin-Heidelberg, Springer 1993
[HZ] HERZBERGER, H. Notes on naive semantics, *Journal of Philosophical Logic*, 11, 1982, 1–60.
[J] JÄGER, G. Zur Beweistheorie der Kripke Platek Mengenlehre über den naturlichen Zahlen, *Archiv fürMathematische Logik und die Grundlagenforschung*, 22, 1982, 121–141.
[JS1] JÄGER, G. and T.A. STRAHM. Totality in applicative theories, *Annals of Pure and Applied Logic*, 74, 1995, 105–120.
[JS2] JÄGER, G. and T.A. STRAHM. Some theories with positive induction of ordinal strength $\phi\omega 0$, preprint, Bern, May 1995
[K] KRIPKE, S. Outline of a theory of truth, *Journal of Philosophy*, 72, 1975, 690–716.
[MC] MC GEE, V. Truth, *Vagueness and Paradox: an essay on the logic of truth*, Hackett Publishing, Indianapolis 1991
[R] REINHARDT, W. Some remarks on extending and interpreting theories with a partial predicate for truth, *Journal of Philosophical Logic* , 15, 1986,219–251
[S] SCHÜTTE, K. *Proof Theory* (Springer, Heidelberg 1977).
[SC] SCOTT, D. Combinators and classes, in: C. Böhm (ed.), *λ-Calculus and Computer Science*, Lecture Notes in Computer Science (Springer, Berlin Heidelberg 1975), 37, 1–26.
[SH] SHEARD, M. A guide to truth predicates in the modern era, *Journal of Symbolic Logic*, 59, 1994, 1032–1054.
[SI] SIMPSON, S.G. Subsystems of Z_2 and reverse mathematics, in: G. Takeuti, *Proof Theory* (North Holland, Amsterdam, 2nd edition 1987), 432–446.
[ST] STRAHM, T. Polynomial time operations in explicit mathematics, preprint, Bern, June 1995.
[T] TARSKI, A. *Logic, Semantics, Metamathematics: Papers from 1923 to 1938* (J. Woodger, ed., Clarendon Press, Oxford 1956).
[V] VISSER, A. Semantics and the liar paradox, in: D. Gabbay and F. Guenthner (eds.), *Handbook of Philosophical Logic* IV (Reidel, Dordrecht 1989), 617–706.

A. JOYAL

FREE LATTICES, COMMUNICATION AND MONEY GAMES

Dedicated to Gian Carlo Rota

ABSTRACT.
Is cost a kind of truth value? What kind of communication is trading? One goal of this work is to show that the theory of free lattices can shed some light on these questions. Another goal is to lay the foundation of a theory of communication based on free lattices and, in further work, on free bicompletions of categories. Here we discuss the game theoretic interpretation of the theory of free lattices. To each term A representing an element of a free lattice we associate a game G_A with two players, σ_A and π_A (symbolically $\sigma_A : G_A : \pi_A$). To any pair (A, B) of such terms we associate a game $\langle G_A, G_B \rangle$ having two components or boards, G_A and G_B, and three players, a mediator σ facing two opposites, σ_A and π_B, one on each board (symbolically $\sigma_A : G_A : \sigma : G_B : \pi_B$). A strategy for the mediator is viewed as a *communication strategy* between the opposites. A communication strategy is complete if it is a winning strategy. We show that the mediator has a complete communication strategy iff $A \leq B$ in the free lattice. The transitivity of the relation $A \leq B$ is then a consequence of the fact that communication strategies compose. We formally extend these results to lattices enriched over a quantale viewed as a domain of generalized logical values. In the special case of the quantale $[0, \infty]$ of (extended) non-negative real numbers we obtain a theory of free metric lattices. The terms representing the elements of a free metric lattice are two person games in which the players each have a stake from which they exchange numerical values, called money. For any pair of such terms we construct a money game $\langle G_A, G_B \rangle$ with three players, a mediator trading with two opposites or sides. The cost of the least expensive winning strategy a mediator can have on $\langle G_A, G_B \rangle$ is a distance $d(A, B)$ defining the metric of the free metric lattice. In the last section we describe certain operations on games that are related to the connectives of linear logic. [1]

[1] Support for this work was provided by the Canadian NSERC-CRSNG

M.L. Dalla Chiara et al. (eds.), Logic and Scientific Methods, 29–68.

§ 0. INTRODUCTION

Our starting point is Philip M. Whitman's theorem on free lattices. Let us recall this theorem. Classically, a lattice is a poset for which every (non-empty) finite subset admits an infimum and a supremum. A lattice can be viewed as an algebraic structure with the operations of binary infimum $x \wedge y$ and of binary supremum $x \vee y$. For any set S there is a free lattice $L(S)$ generated by S. More generally, the *free lattice* generated by a poset P is a lattice $L(P)$ equipped with an order-preserving map $i : P \to L(P)$ satisfying the following universal property: for any order-preserving map $f : P \to L$ with values in a lattice L there is a unique lattice map $f' : L(P) \to L$ such that $f'i = f$.

Theorem. *(Whitman) Let $i : P \to L(P)$ be the free lattice generated by a poset P. Then for any a,b in P and x, y, u and v in $L(P)$ we have:*
(i) $x \wedge y \leq u \vee v$ *iff* $x \wedge y \leq u$ *or* $x \wedge y \leq v$ *or* $x \leq u \vee v$ *or* $y \leq u \vee v$;
(ii) $i(a) \leq u \vee v$ *iff* $i(a) \leq u$ *or* $i(a) \leq v$;
(iii) $x \wedge y \leq i(b)$ *iff* $x \leq i(b)$ *or* $y \leq i(b)$;
(iv) $i(a) \leq i(b)$ *iff* $a \leq b$;
(v) *the lattice $L(P)$ is generated by $i(P)$.*
Moreover, these properties characterize the free lattice generated by P.

The elements of the free lattice $L(P)$ can be formally denoted by the terms defined by the following two clauses: (i) every element of P is a term; (ii) if A and B are terms then so are $(A \wedge B)$ and $(A \vee B)$. The problem of deciding if two terms A and B represent the same element of $L(P)$, or equivalently of deciding if $A \leq B$ in $L(P)$, is the word problem for free lattices. Whitman has shown that a decision procedure can be extracted from his theorem (see [Whitman 1961]). The procedure is rather complex since it involves a systematic exploration of possibilities that are organized logically by a number of disjunctions alternating with a number of conjunctions. We shall discuss a game theoretic interpretation that makes the procedure clearer. It is inspired by Paul Lorenzen's interpretation of intuitionistic logic [Felsher] [Blass]. Another source is John H. Conway's theory of numbers and games [Conway], [Joyal 1977]. To each pair (A, B) of terms we associate a game $\langle G_A, G_B \rangle$ with two players, the defender and the attacker. We show that the relation $A \leq B$ is true iff the defender has a winning strategy. With this interpretation the decision procedure becomes an application of a classical method due to Zermelo for identifying the winning positions in a game [Zermelo]. We introduce another interpretation in which $\langle A, B \rangle$ is viewed as a three person game played on two boards, G_A and G_B, in which a *mediator* σ faces two *opposites*, σ_A and π_B :

$$\sigma_A : G_A : \cdots\cdots \sigma \cdots\cdots : G_B : \pi_B$$

The moves could be questions or answers by the players. The mediator obeys the rules of G_A in his dialogue with σ_A, and the rules of G_B in his dialogue with π_B. The transitivity of the relation $A \leq B$ becomes clear: communication strategies compose:

$$\sigma_A : G_A : \cdots\cdots\sigma_1\cdots\cdots : G_B : \cdots\cdots\sigma_2\cdots\cdots : G_C : \pi_C$$

It is equivalent to a cut elimination theorem for free lattices (see the appendix for a description of the deductive system).

Our conceptual analysis of Whitman's theorem suggests many generalizations. The first, which we develop in this paper, is concerned with extending the theory to $\mathcal{R}$-posets and $\mathcal{R}$-lattices where $\mathcal{R}$ is a quantale viewed as a domain of generalized logical values. The second, which is treated elsewhere, is concerned with bicompletions of categories and with an abstract theory of games and strategies that should model proof theory and programming languages. The terms representing the elements of the free $\mathcal{R}$-lattice generated by an $\mathcal{R}$-poset are formally described in the language of game theory. By using a formal concept of communication strategy we are able to define an element $[A, B] \in \mathcal{R}$ for each pair (A, B) of terms. By proving that communication strategies compose we obtain the inequality

$$[A, B][B, C] \leq [A, C]$$

which is a cut elimination theorem for $\mathcal{R}$-valued logic. When $\mathcal{R}$ is the quantale $[0, \infty]$ of extended non-negative real numbers (with the ordering $\geq$), the $\mathcal{R}$-posets are metric spaces and the $\mathcal{R}$-lattices are metric lattices. In this case the game $\langle G_A, G_B\rangle$ is a *money game* because each player has a stake from which he exchanges numerical values with other players during a play. The communication strategies of the mediator are *trading strategies.* The cost of the least expensive trading strategy is a number $d(A, B)$. The fact that trading strategies compose implies the triangle inequality

$$d(A, C) \leq d(A, B) + d(B, C),$$

which says that eliminating inessential intermediate trading reduces cost.

In the last section we study operations on games corresponding to some connectives of linear logic [Girard]. They are closely related to Andreas Blass' interpretation of these connectives [Blass]. More precisely, for any two games A and B the *tensor product* $A \otimes B$ is a game with two players, σ and π, playing on two boards, A and B, with σ taking the role of σ_A on A and of σ_B on B, and with π taking the role of π_A on A and of π_B on B.

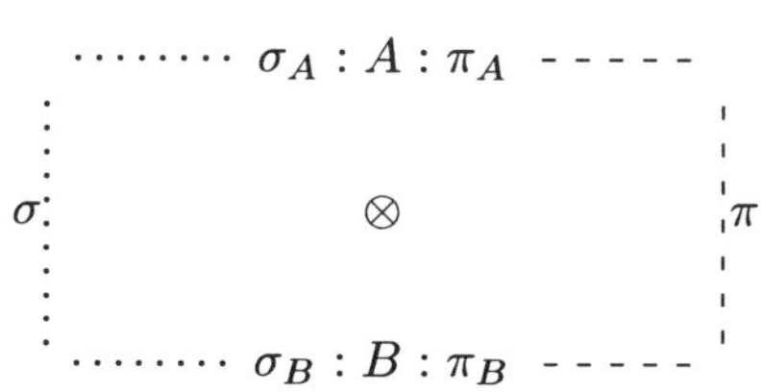

One can imagine that the game $A \otimes B$ has four players divided into two teams, the σ-team $\{\sigma_A, \sigma_B\}$ and the π-team $\{\pi_A, \pi_B\}$. By convention the σ-team has priority on $A \otimes B$, which means that it is the σ-team's turn when it is the turn of *at least one* of its members. There is also a *parallel product* $A \odot B$ which is defined similarly, except the π-team has priority. It is dual to the tensor product, that is, $(A \odot B)^o = A^o \otimes B^o$ where $(-)^o$ denotes the operation of interchanging the roles of σ and π on games.

We illustrate the tensor operations with a small example explaining why loans may reduce cost in trading strategies.

I would like to thank Bill Boshuck, Hongde Hu and Roger Villemaire for their revisions of the manuscript and Paul Libbrecht for the typography.

§ 1. FREE LATTICES AND GAMES

The goal of this section is to introduce the connection between lattice theory and game theory.

1.1 Free Lattices.

We first recall a few concepts from the theory of posets and lattices. Let P be a poset. We shall denote by $\bigvee_{i \in I} x_i$ the supremum of a family $(x_i)_{i \in I}$ of elements of P when it exists. A supremum is said to be *finite* if I is finite. The supremum of x and y is denoted $x \vee y$. If it exists, the bottom element $\bot$ is the supremum of the empty family. Dually, we shall use the notations $\bigwedge_{i \in I} x_i$, $x \wedge y$ and $\top$ for infima and top element.

We shall denote P^o the poset *opposite* to P. By definition, for any $x, y \in P$ we have $x^o \leq y^o$ iff $y \leq x$, where the map $x \mapsto x^o$ is a formal bijection $P \simeq P^o$.

A *lattice* is a poset which admits finite infima and suprema. Our lattices always have top and bottom elements. A map between lattices is a *lattice map* if it preserve finite infima and suprema.

The *free lattice* generated by a poset P is by definition a lattice ΛP equipped with an order preserving map $i : P \to \Lambda P$ satisfying the following universal property: for any order preserving map $f : P \to L$ with values in a lattice L there is a unique lattice map $f' : \Lambda P \to L$ such that $f'i = f$. We shall say that f' is the *lattice map extension* of f.

Definition 1: An element t of a lattice L is σ*-atomic* if for any finite supremum $x = \bigvee_{i \in I} x_i$ the relation $t \leq x$ implies that $t \leq x_i$ for some $i \in I$. Dually, t is π*-atomic* if $t^o \in L^o$ is σ-atomic. An element is *atomic* if it is σ- and π-atomic.

Definition 2: A lattice L is *soft* if for any finite infimum $x = \bigwedge_{i\in I} x_i$ and supremum $y = \bigvee_{j\in J} y_j$ the relation $x \leq y$ implies that either $x \leq y_j$ for some $j \in J$ or that $x_i \leq y$ for some $i \in I$.

Equivalently, a lattice is soft iff (i) $\top$ is σ-atomic, (ii) $\perp$ is π-atomic and (iii) $x \wedge y \leq u \vee v \implies x \wedge y \leq u$ or $x \wedge y \leq v$ or $x \leq u \vee v$ or $y \leq u \vee v$ for any $x, y, u, v \in L$.

We now reformulate Whitman's theorem (for lattices with $\top$ and $\perp$):

Theorem 1. *(Whitman) Let $i : P \to \Lambda P$ be the free lattice generated by a poset P. Then :*
(i) *the lattice ΛP is soft;*
(ii) *for any $x \in P$ the element $i(x) \in \Lambda P$ is atomic ;*
(iii) *for any $x, y \in P$ we have $i(x) \leq i(y)$ iff $x \leq y$;*
(iv) *the lattice ΛP is generated by $i(P)$.*
Moreover, these properties characterize the free lattice generated by P.

There are many proofs of Whitman's theorem. The original proof is really a cut elimination procedure. There is a proof due to Alan Day which uses an ingenious doubling construction (see [Day],[Freese]). In the appendix we show that cut elimination follows from Whitman's theorem. In this section we do not wish to reprove Whitman's theorem but to discuss its meaning. However, in the next section we shall state and prove a generalization to free $\mathcal{R}$-lattices.

Whitman showed that his theorem yields a solution to the word problem for free lattices [Whitman]. Let us describe the decision procedure explicitly. The set $T(P)$ of *terms* is defined by the following clauses:
i) every element of P is a term;
ii) if A and B are terms then so are $(A \wedge B)$ and $(A \vee B)$;
iii) $\perp$ and $\top$ are terms.

There is a unique map $j : T(P) \to \Lambda P$ such that $jx = ix$ if $x \in P$, $j(A \wedge B) = jA \wedge jB$, $j(A \vee B) = jA \vee jB$, $j\perp = \perp$ and $j\top = \top$. Let us write $A \leq B$ for the relation $j(A) \leq j(B)$, and let us put $[A, B] = 1$ if $A \leq B$ is true and $[A, B] = 0$ otherwise. The following proposition shows that $[A, B]$ can be computed by structural induction on the subterms of A and B. We use the operations $\vee$ and $\wedge$ on $\{0, 1\}$.

Proposition 1. *(Decision Procedure)*
The map $[-,-] : T(P) \times T(P) \to \{0, 1\}$ satisfies the following identities:
(1) $[A_1 \vee A_2, B] = [A_1, B] \wedge [A_2, B]$ *and* $[\perp, B] = 1$;
(2) $[A, B_1 \wedge B_2] = [A, B_1] \wedge [A, B_2]$ *and* $[A, \top] = 1$;
(3) $[A_1 \wedge A_2, B_1 \vee B_2] =$
$$[A_1 \wedge A_2, B_1] \vee [A_1 \wedge A_2, B_2] \vee [A_1, B_1 \vee B_2] \vee [A_2, B_1 \vee B_2];$$
(4) $[\top, B_1 \vee B_2] = [\top, B_1] \vee [\top, B_2]$, $[A_1 \wedge A_2, \perp] = [A_1, \perp] \vee [A_2, \perp]$ *and* $[\top, \perp] = 0$;
(5) *If $A \in P$ then* $[A, B_1 \vee B_2] = [A, B_1] \vee [A, B_2]$ *and* $[A, \perp] = 0$;

(6) *If* $B \in P$ *then* $[A_1 \wedge A_2, B] = [A_1, B] \vee [A_2, B]$ *and* $[\top, B] = 0$.
(7) *If* $A, B \in P$ *then* $[A, B] = 1$ *iff* $A \leq B$.

Proof: The identities (1) and (2) express general properties of suprema and infima. The identities (3) and (4) express property (i) in Whitman's theorem. The identities (5) and (6) express property (ii), and identity (7) expresses property (iii). QED

The *polarization* of a term A is an element $\epsilon(A) \in \{0, \sigma, \pi\}$ defined as follows:

$$\epsilon(A) = \begin{cases} \pi & \text{if} \quad A = (A_1 \wedge A_2) \quad \text{or} \quad A = \top \\ \sigma & \text{if} \quad A = (A_1 \vee A_2) \quad \text{or} \quad A = \bot \\ 0 & \text{if} \quad A \in P. \end{cases}$$

The polarization of a pair $(A, B) \in T(P) \times T(P)$ is defined as

$$\epsilon(A, B) = \neg\epsilon(A) \vee \epsilon(B)$$

where $\vee$ denotes the supremum in the poset $\{0 < \sigma < \pi\}$, and where $\neg 0 = 0$, $\neg\sigma = \pi$ and $\neg\pi = \sigma$.

The decision procedure expresses $[A, B]$ as a conjunction or disjunction of simpler elements. It is a conjunction when $\epsilon(A, B) = \pi$ and a disjunction when $\epsilon(A, B) = \sigma$.

1.2 P-games

The idea of seeing proofs as formal strategies for defending a proposition against formal attacks is due to Paul Lorenzen [Felsher]. To a pair $(A, B) \in T(P) \times T(P)$ we shall associate a game $\langle G_A, G_B \rangle$ with two players called σ and π. We then show that $A \leq B$ is true iff σ has a winning strategy on $\langle G_A, G_B \rangle$.

Recall that a *graph* is a pair $G = (G, E)$ where G is a set of *nodes* and $E \subseteq G \times G$ is a set of *edges*. If (x, y) is an edge (symbolically $x \to y$) we shall say that y is a *successor* of x. The set of successors of x will be denoted by $c(x)$. A *path* of length $n \geq 0$ from x to y (symbolically $x \stackrel{*}{\to} y$) is a sequence of edges $x_0 \to x_1 \to \cdots \to x_n$ such that $x_0 = x$ and $x_n = y$. A *cycle* is a path $x \stackrel{*}{\to} x$ of length ≥ 1. A graph is *acyclic* if it has no cycles.

A graph is *well-founded* if it contains no infinite path $x_0 \to x_1 \to \cdots$. A finite acyclic graph is well-founded. When G is well-founded the induction principle is valid: for every $S \subseteq G$ the relation

$$\forall x \in G \;\; c(x) \subseteq S \implies x \in S$$

implies that $S = G$.

A *polarization* on a graph G is a map $\epsilon : G \to \{\sigma, \pi, 0\}$ such that $\epsilon x = 0$ only if x has no children. If $\epsilon x = \sigma$ (resp. $\epsilon x = \pi$, resp. $\epsilon x = 0$) we shall say that x is a *σ-node* (resp. *π-node*, resp. *ground node*). The set of ground nodes of a polarized graph G will be denoted by G_0.

Definition 3: Let P be a poset. A *P-game* is a quadruple $G = (G, x_0, \epsilon, w)$ where G is an acyclic graph, x_0 is a node called the *root*, ϵ is a polarization and $w : G_0 \to P$ is a map called the *valuation.*

There are two players, called σ and π. A *position* of the game is a node of G. A *move* by σ (resp. π), symbolically $x \overset{\sigma}{\to} y$ (resp. $x \overset{\pi}{\to} y$), is an edge $x \to y$ such that $\epsilon x = \sigma$ (resp. $\epsilon x = \pi$). A *play* is a path $x_0 \to x_1 \to \ldots \to x_n$ starting at the root with the move $x_i \to x_{i+1}$ chosen by the player ϵx_i for $0 \leq i < n$. If $\epsilon x_n \neq 0$ the play ends with ϵx_n giving up since it is his turn at x_n. By convention, this player loses the play in this case. If $\epsilon x_n = 0$ the outcome of the play is undefined for general P-games. However, when $P = \{0, 1\}$ the convention is that σ wins iff $w(x_n) = 1$.

Remark: The fact that plays on P-games may have undefined outcomes should not disturb the reader. These games will serve as building blocks for $\{0, 1\}$-games.

Definition 4: Let G be a $\{0, 1\}$-game. A *strategy* (for the σ-player) on G is a subset $S \subseteq G$ such that

(i) for every σ-node $x \in S$ there is at least one move $x \to y$ with $y \in S$,
(ii) for every π-node $x \in S$ and every move $x \to y$ we have $y \in S$,
(iii) $w(x) = 1$ for every ground node $x \in S$.

S is a *winning strategy* if $x_0 \in S$.

By choosing his moves exclusively among the positions of a winning strategy the σ-player will win. A strategy S is *deterministic* if for every σ-node $x \in S$ there is exactly one move $x \to y$ with $y \in S$. Here strategies can be nondeterministic. A strategy S can be contained in a strategy T ($S \subseteq T$). A node $x \in G$ is a *winning position* (for σ) if it belongs to some strategy S for σ. The set of winning positions is the largest strategy. Let $W : G \to \{0, 1\}$ be its characteristic map.

Proposition 2. *(Zermelo) The map $W : G \to \{0, 1\}$ satisfies*

$$W(x) = \begin{cases} w(x) & \text{if} \quad \epsilon x = 0 \\ \bigvee_{u \in c(x)} W(u) & \text{if} \quad \epsilon x = \sigma \\ \bigwedge_{u \in c(x)} W(u) & \text{if} \quad \epsilon x = \pi. \end{cases}$$

It is the unique map satisfying these identities.

We now attach a $\{0, 1\}$-game $\langle G, H \rangle$ to any pair (G, H) of P-games. The game $\langle G, H \rangle$ has two components, G and H, as if it were played on two boards.

We imagine that the players of $\langle G, H\rangle$ are *teams*, the π-team $\{\sigma_G, \pi_H\}$ and the σ-team $\{\pi_G, \sigma_H\}$, each consisting of two players, one for each board:

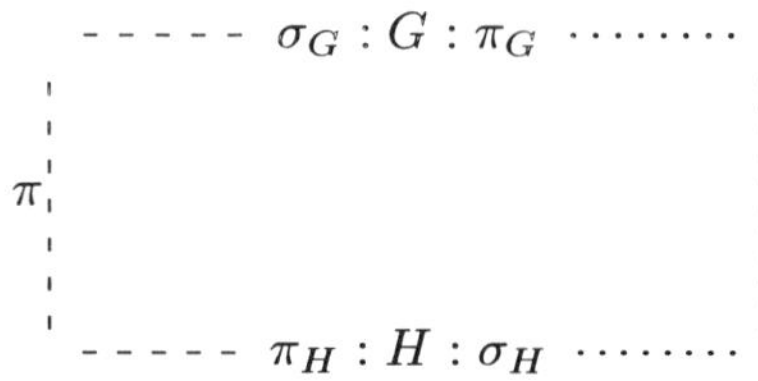

- A position of $\langle G, H\rangle$ is a pair $(x, y) \in G \times H$ of positions. The root of $\langle G, H\rangle$ is the pair (x_0, y_0) of roots.
- The polarization $\epsilon : G \times H \to \{\sigma, \pi, 0\}$ is given by $\epsilon(x, y) = \neg\epsilon x \vee \epsilon y$
- The valuation is such that $w(x, y) = 1$ iff $w(x) \leq w(y)$.
- The moves by π from a position (x, y) are of three kinds:

$$(x,y) \xrightarrow{\pi} \begin{cases} (x,v) & \text{if } y \overset{\pi}{\to} v \\ (u,y) & \text{if } x \overset{\sigma}{\to} u \\ (u,v) & \text{if } x \overset{\sigma}{\to} u \text{ and } y \overset{\pi}{\to} v. \end{cases}$$

- The moves by σ from a position (x, y) are of two kinds:

$$(x,y) \xrightarrow{\sigma} \begin{cases} (x,v) & \text{if } y \overset{\sigma}{\to} v \\ (u,y) & \text{if } x \overset{\pi}{\to} u. \end{cases}$$

Every move can be attributed to one or two team members. For example, the move $(x, y) \to (u, v)$ by the π-team is performed concurrently by σ_G and π_H, and the move $(x, y) \to (u, y)$ by the σ-team is performed by π_G. Notice that the π-team has the exclusive right to perform concurrent moves. This is partly conventional and the rules of $\langle G, H\rangle$ could be modified in various ways, for example by permitting concurrent moves by the σ-team, or on the contrary by forbidding concurrent moves by the π-team. The theory is affected only superficially by the different choices. Our choice is motivated by two goals: one is to ease the interpretation of $\langle G, H\rangle$ as a communication game, and the other is to keep the proofs as simple as possible.

Let P a poset. To each term $A \in T(P)$ we now associate a P-game G_A.
- The positions of G_A are the subterms of A. The root of G_A is A.
- The polarization $\epsilon(U)$ of U was defined in the first section.
- The moves by π are $U_1 \wedge U_2 \to U_1$ and $U_1 \wedge U_2 \to U_2$.
- The moves by σ are $U_1 \vee U_2 \to U_1$ and $U_1 \vee U_2 \to U_2$.
- The valuation of an atomic term U is $U \in P$.

One can imagine that the σ-team of $\langle G_A, G_B\rangle$ is defending the proposition $A \leq B$ and that the π-team is attacking it. For example, when $B = B_1 \wedge B_2$ the opponent can attack $A \leq B$ by choosing $A \leq B_1$ or $A \leq B_2$ and by challenging the defender on his choice. On the other hand, when $A = A_1 \wedge A_2$ and $B = B_1 \vee B_2$, the defender is forced to choose at least one of $A \leq B_1$, $A \leq B_2$, $A_1 \leq B$ and $A_2 \leq B$ to defend.

Proposition 3. *For every $A, B \in T(P)$ the relation $A \leq B$ is true iff σ has a winning strategy on $\langle G_A, G_B \rangle$.*

Proof: Let $W : G_A \times G_B \to \{0, 1\}$ be the characteristic map of the set of winning positions for σ on $\langle G_A, G_B \rangle$. Let us show that $U \leq V$ iff $W(U, V) = 1$. For this it suffices to show that the map $(U, V) \mapsto [U, V]$ satisfies all the identities characterizing W (see Prop. 2). But these identities are those of the decision procedure (see Prop. 1). QED

1.3 Communication strategies

We now interpret $\langle G, H \rangle$ as a *communication game* with three players, a *mediator* σ facing two *opposites* σ_G and π_H:

$$\sigma_G : G : \cdots\cdots \sigma \cdots\cdots : H : \pi_H$$

In the previous description of $\langle G, H \rangle$ we had four players divided into two teams, the π-team $\{\sigma_G, \pi_H\}$ and the σ-team $\{\pi_G, \sigma_H\}$. If we split the π-team into its constituents and hold the σ-team into a single entity, called the mediator, we obtain a three person game. This reorganization can be justified by the following considerations. An inspection of the rules of $\langle G, H \rangle$ reveals that it is σ_G's turn at $(x, y) \in \langle G, H \rangle$ iff it is his turn at $x \in G$. Similarly, it is π_H's turn at $(x, y) \in \langle G, H \rangle$ iff it is his turn at $y \in H$. This means that the members of the π-team may play on their respective boards regardless of the state of the play on the other board. It follows that the π-team can be split. On the contrary, a member of the σ-team cannot adopt such purely local behavior simply because the rules of $\langle G, H \rangle$ forbid him to play on his board when it is the turn of a member of the other team on the other board. Observe that this situation remains even when the σ-team is allowed to perform concurrent moves. We shall treat the σ-team as a single actor σ, called the *mediator*, facing two opposites, σ_G and π_H. In some situations the moves could be questions or answers by the players. It is good to imagine that the mediator is holding a dialogue with the opposites. He obeys the rules of G in his dialogue with σ_G, and the rules of H in his dialogue with π_H. A *communication strategy* is a strategy for the mediator on $\langle G, H \rangle$. It is said to be *complete* if it is a winning strategy. With a complete communication strategy the mediator can be replaced by a mechanism, deterministic or not. In this case a move by an opposite is answered directly by the mechanism or indirectly by the other opposite. Only an opposite can fail to respond, never the mechanism. The communication mechanism could behave like an intelligent medium between the opposites. For other examples of game theoretic paradigms applied to computer science see [A. Nerode & al] and [S. Abramski & al].

Remark 1: A wise mediator should leave nothing to chance and never presume his opposites are not interacting by other means than himself. He should act as though he were facing a single player, the π-team of $\langle G, H\rangle$.

Remark 2: The rules of $\langle G, H\rangle$ could be extended so as to permit extra moves by the mediator. For example, when $\epsilon x = \epsilon y = \pi$ the mediator could be given the right, but without the obligation, of performing $(x, y) \to (u, y)$. The correct formalization of this situation supposes a concept of game in which opponents can move simultaneously.

Remark 3: The mediator wins a play if one of his opposites gives up, or when the play reaches a ground position (x, y) with $w(x) \leq w(y)$. We shall not discuss the meaning of this last condition here. Let us only hint that it is about providing a communication strategy in some other communication game, possibly situated at a different level, and that the process might never stop.

A paradigmatic example of a communication strategy is the *copying strategy*. It is a winning strategy Δ on $\langle G, G\rangle$ for any P-game G. The behaviour of the mediator following Δ is easily described. He plays on two copies of G but in a role opposite on each. On one side he faces σ_G and on the other side he faces π_G:

$$\sigma_G : G : \cdots\cdots \sigma \cdots\cdots : G : \pi_G$$

On his turn, he systematically copies the sequence of moves last performed by his opposites from one side to the other.

We now discuss composition of communication strategies. Let G, H and K be three P-games. Recall that the *composite* of a relation $S \subseteq G \times H$ with a relation $T \subseteq H \times K$ is the relation $S \circ T \subseteq G \times K$ defined as

$$S \circ T = \{(x, z) \in G \times K : \ \exists y \ (x, y) \in S \text{ and } (y, z) \in T\}.$$

Proposition 4. *If S and T are (complete) communication strategies then so is their composite $S \circ T$.*

Proof: Let σ_1, σ_2 and σ denote the mediators on $\langle G, H\rangle$, $\langle H, K\rangle$ and $\langle G, K\rangle$ respectively.

$$\sigma_G : G : \cdots\cdots \sigma_1 \cdots\cdots : H : \cdots\cdots \sigma_2 \cdots\cdots : K : \pi_K$$

Consider a (generalized) game with four players σ_G, σ_1, σ_2 and π_G, whose positions are triples $(x, y, z) \in G \times H \times K$. Each player has a domain on which he moves. For σ_G it is G, for σ_1 it is $\langle G, H\rangle$, for σ_2 it is $\langle H, K\rangle$, and for π_K it is K. A move $(x, y, z) \to (u, v, t)$ on $\langle G, H, K\rangle$ results from a move by one or many players. For example, σ_1 and σ_2 might act concurrently on $\langle G, H, K\rangle$ by moving respectively on G and K (they cannot move concurrently on H because it is never both their turns on H). Inspection shows that if $(x,y,z) \to (u,v,t)$ is a move on $\langle G,H,K\rangle$ then $(x,z) \to (u,t)$ is a move on $\langle G,K\rangle$, unless $(x,z) = (u,t)$. Let us now suppose that σ_1 and σ_2 play according to their winning strategies S and T. Let

$$S \times_H T = \{(x,y,z) : (x,y) \in S \text{ and } (y,z) \in T\}.$$

Starting from a position $(x_0,y_0,z_0) \in S \times_H T$ consider a play $(x_0,y_0,z_0) \overset{*}{\to} (x_n,y_n,z_n)$. By eliminating the repetitions in the sequence (x_0,z_0), (x_1,z_1),..., (x_n,z_n) we obtain a play $(x_0,z_0) \overset{*}{\to} (x_n,z_n)$ on $\langle G,K\rangle$. We claim that σ wins this play. Observe that $(x_k,y_k,z_k) \in S \times_H T \Rightarrow (x_{k+1},y_{k+1},z_{k+1}) \in S \times_H T$ for every $0 \leq k < n$ since σ_1 and σ_2 play according to S and T. Hence $(x_n,y_n,z_n) \in S \times_H T$. Observe that neither σ_1 nor σ_2 have given up the play since they both played according to a winning strategy. Hence the sequence $(x_0,y_0,z_0) \overset{*}{\to} (x_n,y_n,z_n)$ has stopped for one of two reasons: (i) σ_G or π_K has given up; (ii) x_n, y_n and z_n are ground nodes. In the first case σ wins. In the second case we have $w(x_n) \leq w(y_n)$ and $w(y_n) \leq w(z_n)$ since $(x_n,y_n,z_n) \in S \times_H T$. It follows that $w(x_n) \leq w(z_n)$. Finally, if S and T are complete then $(x_0,y_0) \in S$ and $(y_0,z_0) \in T$ hence $(x_0,y_0) \in S \circ T$ and $S \circ T$ is complete. QED

Remark 1: It would be interesting to formalize a concept of generalized game on the basis of the example $\langle G,H,K\rangle$. We could have two teams, the σ-team $\{\sigma_1,\sigma_2\}$ and the π-team $\{\sigma_G,\pi_K\}$. Observe that here opposite teams can move concurrently.

Remark 2: The proposition can be interpreted as a cut elimination theorem for a deductive system which generates the relation $A \leq B$ between the terms of $T(P)$. See the appendix for a description of this deductive system.

In the next section we shall prove a generalization of Whitman's theorem from a generalization of Proposition 4.

§ 2. FREE $\mathcal{R}$-LATTICES AND GAMES

Recall that a *quantale* $\mathcal{R}$ is a partially ordered monoid such that for every $a \in \mathcal{R}$ the maps $x \mapsto ax$ and $x \mapsto xa$ have right adjoints, denoted respectively by $x \mapsto a\backslash x$ and $x \mapsto x/a$ [Rosenthal]. By definition we have

$$xy \leq z \iff y \leq x\backslash z \iff x \leq z/y$$

A quantale can be viewed as a domain of generalized logical values [Lawvere]. In this section $\mathcal{R}$ will denote a fixed quantale which is also a lattice. The unit element will be denoted by 1 and the bottom element by 0. Notice that 1 might differ from the top element $\top$. The *reverse* quantale $\mathcal{R}^r$ is $\mathcal{R}$ equipped with the reverse product $(x,y) \mapsto yx$.

2.1 Free $\mathcal{R}$-lattices

Recall that an *$\mathcal{R}$-preorder* on a set X is a map $[-,-] : X \times X \to \mathcal{R}$ satisfying

$$[x,y][y,z] \leq [x,z] \quad \text{and} \quad 1 \leq [x,x]$$

for every $x, y, z \in X$. The relation $1 \leq [x,y]$ is then an ordinary preorder $x \leq y$ on X. If the antisymmetry condition

$$x \leq y \quad \text{and} \quad y \leq x \quad \Longrightarrow \quad x = y$$

is satisfied the map $[-,-] : X \times X \to \mathcal{R}$ is said to be an *order enriched over* $\mathcal{R}$ and the pair $X = (X, [-,-])$ a *poset enriched over* $\mathcal{R}$, briefly, an *$\mathcal{R}$-poset.*

The *opposite* X^o of an $\mathcal{R}$-poset X is an $\mathcal{R}^r$-poset. By definition, for any $x, y \in X$ we have $[x^o, y^o] = [y, x]$ where the map $x \mapsto x^o$ is a formal bijection $X \simeq X^o$. The quantale $\mathcal{R}$ has the structure of an $\mathcal{R}$-poset with $[x,y] = x\backslash y$, and the structure of an $\mathcal{R}^r$-poset with $[x,y]' = y/x$.

A map $f : X \to Y$ of $\mathcal{R}$-posets is *order preserving* if $[x,y] \leq [fx, fy]$ for every $x, y \in X$. It is an *embedding* if $[x,y] = [fx, fy]$ for every $x, y \in X$.

An element $x \in X$ is the *supremum* $\bigvee_{i \in I} x_i$ (resp. *infimum* $\bigwedge_{i \in I} x_i$) of a finite family $(x_i)_{i \in I}$ of elements of X iff

$$[x,t] = \bigwedge_{i \in I} [x_i, t] \qquad (\text{ resp. } [t,x] = \bigwedge_{i \in I} [t, x_i] \ )$$

for every $t \in X$. We shall consider suprema and infima of finite families only.

The (*tensor*) *product* (resp. (*cotensor*) *product*) of an element $x \in X$ with an element $a \in \mathcal{R}$ is an element $x \cdot a \in X$ (resp. $a \star x \in X$) such that

$$[x \cdot a, t] = a \backslash [x, t] \qquad (\text{ resp. } [t, a \star x] = [t, x]/a \ )$$

for every $t \in X$. When the products exist they are unique and we have $x \cdot a \leq y$ iff $x \leq a \star y$ iff $a \leq [x,y]$. Moreover,

$$(x \cdot a) \cdot b = x \cdot (ab) \quad \text{and} \quad a \star (b \star x) = (ab) \star x.$$

An element $x = \bigvee_{i \in I} x_i \cdot a_i$ is called a *weighted supremum* and the a_i's are called the *weights*. Similarly, $x = \bigwedge_{i \in I} a_i \star x_i$ is called a *weighted infimum.*

Definition 1: An $\mathcal{R}$-poset is (finitely) *complete* (resp. *cocomplete*) if it admits (finite) infima and cotensor products (resp. (finite) suprema and tensor products). An *$\mathcal{R}$-lattice* is an $\mathcal{R}$-poset which is complete and cocomplete.

For any order preserving map $f : X \to Y$ between $\mathcal{R}$-lattices we have the inequalities

$$\bigvee_{i \in I} f(x_i) \cdot a_i \ \leq \ f(\bigvee_{i \in I} x_i \cdot a_i) \quad \text{and} \quad f(\bigwedge_{i \in I} a_i \star x_i) \ \leq \ \bigwedge_{i \in I} a_i \star f(x_i).$$

The map f is said to be *sup-continuous* (resp. *inf-continuous*) if the first (resp. the second) is always an equality. It is *bi-continuous* if both are equalities.

The *free $\mathcal{R}$-lattice* generated by an $\mathcal{R}$-poset X is a pair $(\Lambda X, i)$ where ΛX is an $\mathcal{R}$-lattice and $i : X \to \Lambda X$ an order preserving map satisfying the following universal property: for any $\mathcal{R}$-lattice L and any order preserving map $f : X \to L$ there exists a unique bi-continuous map $f' : \Lambda X \to L$ such that $f'i = f$.

Definition 2: An element $x \in L$ of an $\mathcal{R}$-lattice is *σ-atomic* (resp. *π-atomic*) if the map $[x, -] : L \to \mathcal{R}$ (resp. $[-, x] : L^{op} \to \mathcal{R}$) is sup-continuous. An element is *atomic* if it is σ- and π-atomic.

In particular, we have that $[x, y \cdot a] = [x, y]a$ if x is σ-atomic, and that $[a \star y, x] = a[y, x]$ if x is π-atomic.

Definition 3: An $\mathcal{R}$-lattice L is *soft* if we have

$$[x, y] = \bigvee_{i \in I} a_i[x_i, y] \ \vee \ \bigvee_{j \in J} [x, y_j]b_j$$

for any infimum $x = \bigwedge_{i \in I} a_i \star x_i$ and supremum $y = \bigvee_{j \in J} y_j \cdot b_j$.

The following result is a generalization of Whitman's theorem.

Theorem 1. *Let $i : X \to \Lambda X$ be the free $\mathcal{R}$-lattice generated by an $\mathcal{R}$-poset X. Then we have*

(i) *the lattice ΛX is soft,*
(ii) *for any $x \in X$ the element $i(x) \in \Lambda X$ is atomic,*
(iii) *the map $i : X \to \Lambda X$ is an embedding,*
(iv) *ΛX is the biclosure of $i(X)$.*

Moreover, these properties characterize the free lattice generated by X.

Property (iv) says that ΛX is the closure of $i(X)$ under suprema & tensors and infima & cotensors. This is easily obtained from the universal property. The proof of the theorem will close this section.

2.2 $\mathcal{R}$-games.

A (finite) *$\mathcal{R}$-graph* is a pair (G, θ) where G is a finite set and $\theta : G \times G \to \mathcal{R}$ is a map called the *transition matrix.*

An *edge* of (G, θ) is a pair $(x, y) \in G \times G$ such that $\theta(x, y) \neq 0$. This defines an ordinary graph (G, E) where E is the set of edges. It follows that many concepts applying to ordinary graphs apply equally well to $\mathcal{R}$-graphs. For example, the set $c(x)$ of successors of $x \in G$ is defined to be $\{y \in G : \theta(x, y) \neq 0\}$. An $\mathcal{R}$-graph (G, θ) is *acyclic* if (G, E) is acyclic. A *polarization* on (G, θ) is a polarization on (G, E). The set of ground nodes of a polarized graph is denoted G_0.

Definition 4: Let X be an $\mathcal{R}$-poset. We define an X-*game* to be a quintuple $G = (G, \theta, x_0, \epsilon, w)$ where (G, θ) is a finite acyclic $\mathcal{R}$-graph, $x_0 \in G$ is a node called the *root*, $\epsilon : G \to \{\sigma, \pi, 0\}$ is a polarization and $w : G_0 \to X$ is a map called the *valuation*.

Definition 5: Let $\mathcal{L}$ be an $\mathcal{R}$-lattice and let A be an $\mathcal{L}$-game. We shall say that a map $f : A \to \mathcal{L}$ is a *strategy* if we have:
(i) $f(x) \leq w(x)$ if $\epsilon x = 0$.
(ii) $f(x) \cdot \theta(x, u) \leq f(u)$ for every u if $\epsilon x = \pi$
(iii) $f(x) \leq \bigvee_{u \in c(x)} \ f(u) \cdot \theta(x, u)$ if $\epsilon x = \sigma$

It is easy to see by induction on the nodes of A that there is a largest strategy $W : A \to \mathcal{L}$ and that it is characterized by the following identities:

$$W(x) = \begin{cases} w(x) & \text{if } \epsilon x = 0 \\ \bigwedge_{u \in c(x)} \ \theta(x, u) \star W(u) & \text{if } \epsilon x = \pi \\ \bigvee_{u \in c(x)} \ W(u) \cdot \theta(x, u) & \text{if } \epsilon x = \sigma. \end{cases}$$

Definition 6: Let X be an $\mathcal{R}$-poset and let A and B be two X-games. A *communication strategy* is a map $f : A \times B \to \mathcal{R}$ satisfying the following conditions.
(i) $f(x, y) \leq [w(x), w(y)]$ if $(\epsilon x, \epsilon y) = (0, 0)$.
(ii) $f(x, y)\theta(y, v) \leq f(x, v)$ if $\epsilon y = \pi$.
(iii) $\theta(x, u) f(x, y) \leq f(u, y)$ if $\epsilon x = \sigma$.
(iv) $f(x, y) \leq \bigvee_{v \in c(y)} f(x, v)\theta(y, v)$ if $(\epsilon x, \epsilon y) = (0, \sigma)$.
(v) $f(x, y) \leq \bigvee_{u \in c(x)} \theta(x, u) f(u, y)$ if $(\epsilon x, \epsilon y) = (\pi, 0)$.
(vi) $f(x, y) \leq \bigvee_{u \in c(x)} \theta(x, u) f(u, y) \ \vee \bigvee_{v \in c(y)} f(x, v)\theta(y, v)$
if $(\epsilon x, \epsilon y) = (\pi, \sigma)$.

Proposition 1. *Let X be an $\mathcal{R}$-poset and let A and B be two X-games. There is a largest communication strategy $A \times B \to \mathcal{R}$ and it is the unique map $[-, -] : A \times B \to \mathcal{R}$ satisfying the following identities.*
(i) $[x, y] = [w(x), w(y)]$ *if* $(\epsilon x, \epsilon y) = (0, 0)$.
(ii) $[x, y] = \bigwedge_{v \in c(y)} [x, v]/\theta(y, v)$ *if* $\epsilon y = \pi$.
(iii) $[x, y] = \bigwedge_{u \in c(x)} \theta(x, u) \backslash [u, y]$ *if* $\epsilon x = \sigma$.
(iv) $[x, y] = \bigvee_{v \in c(y)} [x, v]\theta(y, v)$ *if* $(\epsilon x, \epsilon y) = (0, \sigma)$
(v) $[x, y] = \bigvee_{u \in c(x)} \theta(x, u)[u, y]$ *if* $(\epsilon x, \epsilon y) = (\pi, 0)$.
(vi) $[x, y] = \bigvee_{u \in c(x)} \theta(x, u)[u, y] \ \vee \bigvee_{v \in c(y)} [x, v]\theta(y, v)$ *if* $(\epsilon x, \epsilon y) = (\pi, \sigma)$.
Moreover, we have $[x, y]\theta(y, v) \leq [x, v]$ *if* $\epsilon y = \pi$, *and* $[x, y] \geq [x, v]\theta(y, v)$ *if* $\epsilon y = \sigma$.

Proof: A map $[-, -]$ satisfying (i)-(vi) can be defined by induction on the pairs $(x, y) \in A \times B$. However, when $(\epsilon x, \epsilon y) = (\sigma, \pi)$ the cases (ii) and (iii) are overlapping and we have to verify that the values given by (ii) and (iii) agree. By induction hypothesis we can suppose that

$$[u, y] = \bigwedge_{v \in c(y)} [u, v]/\theta(y, v) \quad \text{and} \quad [x, v] = \bigwedge_{u \in c(x)} \theta(x, u)\backslash[u, v]$$

for every $u \in c(x)$ and $v \in c(y)$. But then

$$\begin{aligned} \bigwedge_{u \in c(x)} \theta(x, u)\backslash[u, y] &= \bigwedge_{(u,v) \in c(x) \times c(y)} \theta(x, u)\backslash[u, v]/\theta(y, v) \\ &= \bigwedge_{v \in c(y)} [x, v]/\theta(y, v) \end{aligned}$$

and the two values agree. This proves the existence and uniqueness of a map satisfying (i)-(vi). That it is the largest strategy is easily proved by induction. Let us prove the rest of the proposition. If $\epsilon y = \pi$ then $[x, y] \leq [x, v]/\theta(y, v)$ by identity (ii). Hence $[x, y]\theta(y, v) \leq [x, v]$. Suppose $\epsilon y = \sigma$. We shall prove the inequality $[x, y] \geq [x, v]\theta(y, v)$ by induction on x. This is clear if $\epsilon x = 0$ by identity (iv), and if $\epsilon x = \pi$ by identity (vi). If $\epsilon x = \sigma$ we have

$$\begin{aligned} [x, y] &= \bigwedge_{u \in c(x)} \theta(x, u)\backslash[u, y] \geq \bigwedge_{u \in c(x)} \theta(x, u)\backslash([u, v]\theta(y, v)) \\ &\geq \Big(\bigwedge_{u \in c(x)} \theta(x, u)\backslash[u, v]\Big)\theta(y, v) = [x, v]\theta(y, v) \end{aligned}$$

by identity (iii) and the induction hypothesis. QED

A *matrix* is a map $f : A \times B \to \mathcal{R}$ where A and B are finite sets. Matrices can be composed. The composite of $f : A \times B \to \mathcal{R}$ with $g : B \times C \to \mathcal{R}$ is the matrix $f \circ g : A \times C \to \mathcal{R}$ given by

$$(f \circ g)(x, z) = \bigvee_{y \in B} f(x, y)g(y, z).$$

The *identity matrix* $\delta : A \times A \to \mathcal{R}$ is the usual Kronecker delta:

$$\delta(x, y) = \begin{cases} 1 & \text{if } x = y \\ 0 & \text{otherwise.} \end{cases}$$

Let $\mathcal{L}$ be a cocomplete $\mathcal{R}$-poset. A map $f : A \to \mathcal{L}$ is a kind of vector. Vectors and matrices can be composed. The composite of $f : A \to \mathcal{L}$ with $g : A \times B \to \mathcal{R}$ is the map $f \circ g : B \to \mathcal{L}$ given by

$$(f \circ g)(y) = \bigvee_{x \in A} f(x) \cdot g(x, y).$$

Proposition 2 . *Let $\mathcal{L}$ be an $\mathcal{R}$-lattice, let A and B be two $\mathcal{L}$-games, and let $g : A \times B \to \mathcal{R}$ be a communication strategy. If a map $f : A \to \mathcal{L}$ is a strategy then so is the map $f \circ g : B \to \mathcal{L}$.*

Proof: Let $h = f \circ g$. We shall prove by induction on $y \in B$ that $h(y)$ satisfies the inequalities of Definition 5. Let us suppose that $h(v)$ satisfies the conditions of Definition 5 for every $v \in c(y)$.

(i) If $\epsilon(y) = 0$ we should prove that $h(y) \leq w(y)$. We shall show by induction on $x \in A$ that $f(x) \cdot g(x, y) \leq w(y)$. Let us suppose that $f(u) \cdot g(u, y) \leq w(y)$ for every $u \in c(x)$.
(i.a) If $\epsilon x = 0$ then $f(x) \cdot g(x, y) \leq w(x) \cdot [w(x), w(y)] \leq w(y)$.
(i.b) If $\epsilon x = \pi$ then

$$f(x) \cdot g(x, y) \leq \bigvee_{u \in c(x)} f(x) \cdot \theta(x, u) \cdot g(u, y) \leq \bigvee_{u \in c(x)} f(u) \cdot g(u, y) \leq w(y)$$

by condition (v) of Definition 6, by condition (ii) of Definition 5, and by induction hypothesis.
(i.c) If $\epsilon x = \sigma$ then

$$f(x) \cdot g(x, y) \leq \bigvee_{u \in c(x)} f(u) \cdot \theta(x, u) \cdot g(x, y) \leq \bigvee_{u \in c(x)} f(u) \cdot g(u, y) \leq w(y)$$

by conditions (iii) of Definition 5 and 6, and by induction hypothesis.
(ii) If $\epsilon y = \pi$ we should prove that $h(y) \cdot \theta(y, v) \leq h(v)$ for every $v \in B$. But we have

$$h(y) \cdot \theta(y, v) \leq \bigvee_{x \in A} f(x) \cdot g(x, y) \cdot \theta(y, v) \leq \bigvee_{x \in A} f(x) \cdot g(x, v) = h(v)$$

by condition (ii) of Definition 6.
(iii) If $\epsilon y = \sigma$ we should prove that $h(y) \leq \bigvee_{v \in c(y)} h(v) \cdot \theta(y, v)$. If T denotes the right hand side of this inequality we shall show by induction on $x \in A$ that $f(x) \cdot g(x, y) \leq T$ and the inequality will follow. Let us suppose that $f(u) \cdot g(u, y) \leq T$ for every $u \in c(x)$.
(iii.a) If $\epsilon x = 0$ then

$$f(x) \cdot g(x, y) \leq \bigvee_{v \in c(y)} f(x) \cdot g(x, v) \cdot \theta(y, v) = \bigvee_{v \in c(y)} h(v) \cdot \theta(y, v) = T$$

by condition (iv) of Definition 6.
(iii.b) If $\epsilon x = \sigma$ then

$$f(x) \cdot g(x, y) \leq \bigvee_{u \in c(x)} f(u) \cdot \theta(x, u) \cdot g(x, y) \leq \bigvee_{u \in c(x)} f(u) \cdot g(u, y) \leq T$$

by conditions (iii) of Definition 5 and 6, and by induction hypothesis.
(iii.c) If $\epsilon x = \pi$ then

$$f(x) \cdot g(x, y) \quad \leq \quad \bigvee_{u \in c(x)} f(x) \cdot \theta(x, u) \cdot g(u, y) \vee \bigvee_{v \in c(y)} f(x) \cdot g(x, v) \cdot \theta(y, v)$$

$$\leq \bigvee_{u \in c(x)} f(u) \cdot g(u,y) \vee \bigvee_{v \in c(y)} h(v) \cdot \theta(y,v) \leq T$$

by condition (vi) of Definition 6, by condition (ii) of Definition 5, and by induction hypothesis. QED

Theorem 2. *Let A, B and C be three X-games. If $f : A \times B \to \mathcal{R}$ and $g : B \times C \to \mathcal{R}$ are communication strategies then so is $f \circ g : A \times C \to \mathcal{R}$.*

Proof: If $\epsilon z = \pi$ then we have

$$(f \circ g)(x,z)\theta(z,v) = \bigvee_{y \in B} f(x,y)g(y,z)\theta(z,v) \leq \bigvee_{y \in B} f(x,y)g(y,v) = (f \circ g)(x,v)$$

since g satisfies condition (ii) of Definition 6. This proves that $f \circ g$ satisfies condition (ii). By symmetry, it satisfies condition (iii). Let us prove that $f \circ g$ satisfies conditions (i) and (iv). If $\epsilon x = 0$ we shall use a game B' having the same structure as B but with a different valuation $w' : B_0 \to \mathcal{R}$ given by $w'(y) = [w(x), w(y)]$. Similarly, let C' be the game with a valuation given by $w'(z) = [w(x), w(z)]$. Then g is a communication strategy between B' and C' since $g(y,z) \leq [w(y), w(z)] \leq [w'(y), w'(z)]$ for any $(y,z) \in B_0 \times C_0$. Moreover, the map $f(x,-) : B \to \mathcal{R}$ is a strategy on B'. If we apply Proposition 2 it follows that $f \circ g(x,-)$ is a strategy on C'. From condition (i) of Definition 5 it follows that $(f \circ g)(x,z) \leq w'(z) = [w(x), w(z)]$ for every ground node. This proves that $f \circ g$ satisfies condition (i) of Definition 6. Also, from condition (iii) of Definition 5 we obtain that $f \circ g$ satisfies condition (iv) of Definition 6. By symmetry (with f taking the role of g and vice-versa), it also satisfies condition (v). Let us prove that $f \circ g$ satisfies condition (vi). Suppose $(\epsilon x, \epsilon z) = (\pi, \sigma)$. It suffices to show that

$$f(x,y)g(y,z) \leq \bigvee_{u \in c(x)} \theta(x,u)(f \circ g)(u,z) \vee \bigvee_{w \in c(z)} (f \circ g)(x,w)\theta(z,w)$$

for every $y \in B$. We shall prove this by induction on $y \in B$. Let T be the right hand side of this inequality. By induction hypothesis, we may assume that $f(x,v)g(v,z) \leq T$ for every $v \in c(y)$.

Suppose $\epsilon y = 0$. In this case

$$f(x,y)g(y,z) \leq \bigvee_{u \in c(x)} \theta(x,u)f(u,y)g(y,z) \leq \bigvee_{u \in c(x)} \theta(x,u)(f \circ g)(u,z) \leq T$$

by condition (v) for f.

Now suppose $\epsilon y = \sigma$. In this case

$$f(x,y)g(y,z) \quad \leq \quad \bigvee_{u \in c(x)} \theta(x,u)f(u,y)g(y,z) \vee \bigvee_{v \in c(y)} f(x,v)\theta(y,v)g(y,z)$$

$$\leq T \vee \bigvee_{v \in c(y)} f(x,v)g(v,z) \leq T$$

by condition (vi) for f, by condition (iii) for g, and by induction hypothesis. The case $\epsilon y = \pi$ is symmetrical. QED

On any X-game A we shall use the following matrices $[\pi]$ and $[\sigma]$:

$$[\pi](x,y) = \begin{cases} \theta(x,y) & \text{if } \epsilon x = \pi \\ 0 & \text{otherwise;} \end{cases} \qquad [\sigma](x,y) = \begin{cases} \theta(y,x) & \text{if } \epsilon y = \sigma \\ 0 & \text{otherwise.} \end{cases}$$

We have $\theta = [\pi] \vee [\sigma]^t$ where $[\sigma]^t$ is the transpose of $[\sigma]$. By definition $[\pi](x,y) = 0$ unless $\epsilon x = \pi$ and $[\sigma](x,y) = 0$ unless $\epsilon y = \sigma$. Thus $[\sigma] \circ [\pi] = 0$.

The *power* f^n of a matrix $f : A \times A \to \mathcal{R}$ is defined for every natural number n. We have $f^0 = \delta$ and $f^{n+1} = f \circ f^n = f^n \circ f$. A matrix f is *nilpotent* if $f^n = 0$ for sufficiently large n (here 0 is the matrix with null entries). As an example, for any X-game A the matrices $[\pi], [\sigma] : A \times A \to \mathcal{R}$ are nilpotent since the graph A is acyclic. The *transitive closure* f^* of a nilpotent matrix is obtained as a finite supremum $f^* = \bigvee_{n=1}^{N} f^n$ for sufficiently large N.

Consider the matrix $\Delta = \delta \vee [\pi]^* \vee [\sigma]^*$.

Proposition 3. *The matrix $\Delta : A \times A \to \mathcal{R}$ is a communication strategy.*

Proof: Let us verify that Δ satisfies the conditions (i)-(vi) of Definition 6. From the relation $[\sigma] \circ [\pi] = 0$ it follows that $[\sigma]^* \circ [\pi] = 0$ and therefore

$$\Delta \circ [\pi] = [\pi] \vee [\pi]^* \circ [\pi] \vee [\sigma]^* \circ [\pi] \leq [\pi]^* \leq \Delta.$$

This proves condition (ii) of Definition 6. Condition (iii) is proved similarly. Let us prove condition (iv). If $\epsilon x = 0$ and $\epsilon y = \sigma$ we have $\Delta(x,y) = [\sigma]^*(x,y)$ since in this case $[\pi]^*(x,y) = 0$ and $\delta(x,y) = 0$. But

$$\Delta \circ [\sigma] = [\pi]^* \circ [\sigma] \vee [\sigma]^* \circ [\sigma] \vee [\sigma] \geq [\sigma]^*.$$

Therefore $\Delta(x,y) \leq \Delta \circ [\sigma](x,y)$ and this proves condition (iv). Condition (v) is proved similarly. Let us prove condition (vi). If $\epsilon x = \pi$ and $\epsilon y = \sigma$ we have $\Delta(x,y) = [\sigma]^*(x,y) \vee [\pi]^*(x,y)$ since in this case $\delta(x,y) = 0$. But we have

$$[\sigma]^* \vee [\pi]^* \leq \Delta \circ [\sigma] \vee [\pi] \circ \Delta$$

Therefore $\Delta(x,y) \leq (\Delta \circ [\sigma])(x,y) \vee ([\pi] \circ \Delta)(x,y)$ and this proves condition (vi). It remains to verify condition (i). If $\epsilon x = \epsilon y = 0$ then $\Delta(x,y) = \delta(x,y)$ since in this case $[\sigma]^*(x,y) = [\pi]^*(x,y) = 0$. Therefore $\Delta(x,y) \leq [w(x), w(y)]$ since $1 \leq [w(x), w(x)]$. This proves condition (i). QED

In Proposition 1 we have defined the largest strategy $[-,-] : A \times A \to \mathcal{R}$.

Corollary. *For any $x, y \in A$ we have $\theta(x,y) \leq [x,y]$ if $\epsilon x = \pi$ and $\theta(y,x) \leq [x,y]$ if $\epsilon y = \sigma$. Moreover, $1 \leq [x,x]$ for any $x \in A$.*

Proof: We have $\Delta(x,y) \leq [x,y]$ for any $x, y \in A$ since Δ is a communication strategy and $[-,-]$ is the largest such. Hence $[\pi](x,y) \leq \Delta(x,y) \leq [x,y]$. This proves that $\theta(x,y) \leq [x,y]$ if $\epsilon x = \pi$. The rest is proved similarly. QED

For any pair of X-games A and B let us put $[A,B] = [a_0, b_0]$ where a_0 and b_0 are the roots of A and B respectively.

Proposition 4. *(Cut Elimination) Let A, B and C be three X-games. We have*

$$1 \leq [A,A] \quad \textit{and} \quad [A,B][B,C] \leq [A,C].$$

Proof: The first inequality follows from the Corollary. Let us show that $[x,y][y,z] \leq [x,z]$ for any $(x,y,z) \in A \times B \times C$. According to Theorem 2 the map $h : A \times C \to \mathcal{R}$ given by

$$h(x,z) = \bigvee_{y \in B} [x,y][y,z]$$

is a communication strategy. Hence $h(x,z) \leq [x,z]$ for any $(x,z) \in A \times C$ by the maximality of $[-,-]$. Thus $[x,y][y,z] \leq [x,z]$ for any $(x,y,z) \in A \times B \times C$. QED

2.3 The Construction

We now give a construction of the free $\mathcal{R}$-lattice generated by an $\mathcal{R}$-poset X. Let us denote by $\mathcal{G}(X)$ the collection of X-games. It follows from Proposition 4 that the relation $1 \leq [A,B]$, denoted $A \leq B$, is reflexive and transitive. Hence the relation $A \leq B$ and $B \leq A$ is an equivalence $A \equiv B$. Let ΛX be the quotient of $\mathcal{G}(X)$ by this equivalence relation and let $q : \mathcal{G}(X) \to \Lambda X$ be the quotient map. Then ΛX becomes an $\mathcal{R}$-poset with

$$[qA, qB] = [A,B]$$

For any $x \in X$ let $I(x)$ be the game with a single node x, of polarization 0 and of valuation x. Clearly, $[I(x), I(y)] = [x,y]$ for any $x, y \in X$. By composing I with q we obtain an embedding $i : X \to \Lambda X$. We shall prove that ΛX is an $\mathcal{R}$-lattice. For any finite family $(A_i : i \in I)$ of X-games and elements $(a_i : i \in I)$ of $\mathcal{R}$ the *σ-suspension*

$$A = \bigvee_{i \in I} A_i \cdot a_i$$

is the X-game constructed as follows. The set of nodes of A is the disjoint union

$$A = \{\star\} + \sum_{i \in I} A_i$$

where $\star$ is the root of A. The transition matrix $\theta : A \times A \to \mathcal{R}$ is defined as follows:

$$\theta(x, y) = \begin{cases} \theta(x, y) & \text{if } x, y \in A_i \text{ for some } i \in I \\ a_i & \text{if } x = \star \text{ and } y \text{ is the root of } A_i \\ 0 & \text{otherwise.} \end{cases}$$

The polarization on A is defined by the polarization on the A_i's and we have $\epsilon\star = \sigma$ by convention. The valuation on the ground nodes of A is defined by the valuation on the ground nodes of the A_i's. The π-*suspension*

$$A = \bigwedge_{i \in I} a_i \star A_i$$

has the same description as the σ-suspension except that the polarization of $\star$ is defined to be π.

Theorem 3. *The $\mathcal{R}$-poset ΛX equipped with the map $i : X \to \mathcal{G}(X) \to \Lambda X$ is the free $\mathcal{R}$-lattice generated by X.*

Proof: According to properties (ii) and (iii) in Prop.1 we have

$$[\bigvee_{i \in I} A_i \cdot a_i\ , B] = \bigwedge_{i \in I} a_i \backslash [A_i\ , B] \quad \text{and} \quad [B,\ \bigvee_{i \in I} a_i \star A_i] = \bigwedge_{i \in I} [B,\ A_i]/a_i$$

for any X-game B. This shows that ΛX is an $\mathcal{R}$-lattice. To prove that ΛX is free we shall use the following results.

Let $f : X \to Y$ be an order preserving map between $\mathcal{R}$-posets. For any X-game A let A_f be the Y-game having the same structure as A but with a valuation w_f given by $w_f(x) = f(w(x))$ for $x \in A_0$.

Proposition 5. *For any pair of X-games A and B we have*

$$[A, B] \leq [A_f, B_f].$$

Proof: Let $[-, -]_f : A_f \times B_f \to \mathcal{R}$ be the largest communication strategy between A_f and B_f. The result will follow if we show that $[x, y] \leq [x, y]_f$ for every $(x, y) \in A \times B$. For this let us verify that the map $[-, -] : A \times B \to \mathcal{R}$, when viewed as a map $[-, -] : A_f \times B_f \to \mathcal{R}$, is a communication strategy between A_f and B_f. The conditions in Definition 6 are obviously satisfied except possibly condition (i). But if $\epsilon x = \epsilon y = 0$ we have

$$[x, y] = [w(x), w(y)] \leq [f(w(x)), f(w(y))] = [w_f(x), w_f(y)].$$

By the maximality of $[-, -]_f$ we have $[x, y] \leq [x, y]_f$ for all $(x, y) \in A \times B$.
QED

Let $\mathcal{L}$ be an $\mathcal{R}$-lattice. On any $\mathcal{L}$-game A there is a largest strategy $W : A \to \mathcal{L}$. Let us put $W(A) = W(a_0)$ where a_0 is the root of A.

Proposition 6. *For any pair of $\mathcal{L}$-games A and B we have*

$$[A, B] \leq [W(A), W(B)].$$

Proof: Let us prove that $[x, y] \leq [W(x), W(y)]$, or equivalently, that $W(x) \cdot [x, y] \leq W(y)$, for every $(x, y) \in A \times B$. According to Proposition 2 the map $K : B \to \mathcal{L}$ given by

$$K(y) = \bigvee_{x \in A} W(x) \cdot [x, y]$$

is a strategy on B. It follows that $K(y) \leq W(y)$ for every $y \in B$ since W is the largest strategy. QED

We can now complete the proof of Theorem 3. Let $f : X \to \mathcal{L}$ be an order preserving map with values in some $\mathcal{R}$-lattice $\mathcal{L}$. For any X-game A let A_f be the $\mathcal{L}$-game having the same structure as A but with a valuation w_f given by $w_f(x) = f(w(x))$ for ground nodes $x \in A$. Let us set $f'(A) = W(A_f) \in \mathcal{L}$. Then according to Propositions 5 and 6 we have

$$[A, B] \leq [A_f, B_f] \leq [W(A_f), W(B_f)] = [f'(A), f'(B)]$$

for any $A, B \in \mathcal{G}(X)$. Therefore the relation $A \equiv B$ implies the equality $f'(A) = f'(B)$. This defines an order preserving map $f' : \Lambda X \to \mathcal{L}$. It follows directly from the definition of f' that

$$f'(\bigvee_{i \in I} A_i \cdot a_i) = \bigvee_{i \in I} f'(A_i) \cdot a_i \quad \text{and} \quad f'(\bigwedge_{i \in I} a_i \star A_i) = \bigwedge_{i \in I} a_i \star f'(A_i).$$

Moreover, $f'(i(x)) = f(x)$ for every $x \in X$. The uniqueness of f' is left to the reader. QED

We can now prove the generalization of Whitman's theorem to $\mathcal{R}$-lattices (Theorem 1). To prove properties (i), (ii) and (iii) we use the explicit construction of ΛX obtained from Theorem 3. To prove (i) it suffices to show that

$$[A, B] = \bigvee_{i \in I} a_i [A_i, B] \ \vee \ \bigvee_{j \in J} [A, B_j] b_j$$

for any π-suspension $A = \bigwedge_{i \in I} a_i \star A_i$ and σ-suspension $B = \bigvee_{j \in J} B_j \cdot b_j$. But this a direct consequence of identity (vi) in Proposition 1. Properties (ii) and (iii) of Theorem 1 are similarly obtained. The last part of Theorem 1 is a consequence of the following.

Proposition 7. *Let $j : X \to \mathcal{L}$ be an order preserving map with values in an $\mathcal{R}$-lattice and suppose that: (i) $\mathcal{L}$ is soft ; (ii) $\forall x \in X$ $j(x)$ is atomic, (iii) $j : X \to \mathcal{L}$ is an embedding. Then the unique bi-continuous map $k : \Lambda X \to \mathcal{L}$ satisfying $ki = j$ is an embedding.*

Proof: We shall use the fact that $i(X)$ generates ΛX as an $\mathcal{R}$-lattice. For every element x of ΛX consider the set $E(x)$ of elements $y \in \Lambda X$ satisfying $[kx, ky] = [x, y]$, or equivalently, $[kx, ky] \leq [x, y]$. $E(x)$ is closed under (weighted) infima since they are preserved by the map $[x, -]$. Next, let us verify that $E(ix) = \Lambda X$ for every $x \in X$. The map $[kix, k(-)] = [jx, k(-)]$ preserves suprema since k is sup-continuous and jx is σ-atomic. It follows that $E(ix)$ is closed under suprema and that it is a sublattice of ΛX. Moreover, we have $[kix, kiy] = [jx, jy] = [x, y]$ since j is an embedding. It follows that $i(X) \subset E(ix)$, and therefore that $E(ix) = \Lambda X$ since $i(X)$ generates ΛX. A dual argument shows that $E(x)$ contains $i(X)$ for every $x \in \Lambda X$. Let E be the set of elements $x \in \Lambda X$ satisfying $E(x) = \Lambda X$. We shall prove that $E = \Lambda X$ by proving that it is a sublattice of ΛX. Clearly E is closed under suprema since they are preserved by $[-, x]$. We saw that E contains $i(X)$. It remains to verify that E is closed under infima. Let $x = \bigwedge_{\alpha \in A} u_\alpha \star x_\alpha$ be an infimum where $x_\alpha \in E$ for every $\alpha \in A$. Let us show that $E(x) = \Lambda X$. We saw before that $E(x)$ is closed under infima and that it contains $i(X)$. To prove that $E(x) = \Lambda X$ it suffices to verify that $E(x)$ is closed under suprema. Let $y = \bigvee_{\beta \in B} y_\beta \cdot v_\beta$ be a supremum where $y_\beta \in E(x)$ for every $\beta \in B$. Then we have

$$[kx\ ,ky] = \bigvee_{\alpha \in A} u_\alpha [kx_\alpha\ ,ky] \ \vee \ \bigvee_{\beta \in B} [kx\ ,ky_\beta] v_\beta$$

since L is soft. We have that $[kx_\alpha, ky] \leq [x_\alpha, y]$, since $x_\alpha \in E$, and that $[kx, ky_\beta] \leq [x, y_\beta]$, since $y_\beta \in E(x)$. Therefore

$$[kx\ ,ky] \leq \bigvee_{\alpha \in A} u_\alpha [x_\alpha\ ,y] \ \vee \ \bigvee_{\beta \in B} [x\ ,y_\beta] v_\beta \ \leq [x, y].$$

This completes the proof that $E = \Lambda X$, it follows that k is an embedding. QED

§ 3. FREE METRIC LATTICES AND MONEY GAMES

The goal of this section is to illustrate the theory of free $\mathcal{R}$-lattices developed in the previous section in the case where $\mathcal{R}$ is the interval $[0, 1]$ of real numbers equipped with the operation of multiplication. We shall use the interval $[0, \infty]$ instead because $[0, \infty]$-games have a natural interpretation as money games. There is an isomorphism $l : [0, 1] \to [0, \infty]$ transforming multiplication into addition, given by $l(x) = -\log(x)$. As the isomorphism l is order reversing we need to use the opposite ordering on $[0, \infty]$. This can be done by systematically writing $x \geq y$ for the relation $y \leq x$ between the elements of $[0, \infty]$. We wish to view $[0, 1]$ and $[0, \infty]$ as domains of generalized logical values [Lawvere]. The idea of using numbers to represent logical values goes back to George Boole. Nowadays subsets and relations are commonly represented by their characteristic maps with values in $\{0,1\}$. The so-called fuzzy subsets of a set X are defined by maps $X \to [0,1]$ (but we do not adhere to the fuzzy doctrine). It is natural to view arbitrary maps $X \to [0,\infty]$ as defining generalized subsets. The characteristic map $\tau_S : X \to [0,\infty]$ of a subset $S \subseteq X$ would then be defined as $\tau_S = l \circ \chi_S$, where $\chi_S : X \to \{0,1\}$ is the usual characteristic map and $l(x) = -\log(x)$. It follows that

$$\tau_S(x) = \begin{cases} 0 & \text{if } x \in S \\ \infty & \text{if } x \notin S. \end{cases}$$

This represents true by 0 and false by ∞! Such a radical departure from common usage requires some discussion.

We propose to view the elements of $[0,\infty]$ as representing economic values rather than ordinary truth values. A map $f : X \to [0,\infty]$ might represent the cost of doing a certain task, for example, of executing certain algorithm for an input $x \in X$, or of extracting a certain information, or of manufacturing a certain object. When the execution cost is too large the task is considered impossible from an economic standpoint, but the concepts of impossibility and falsity are almost interchangeable from a logical standpoint. Hence ∞ represents an impossibility. On the other hand, the truth as represented by current knowledge is, or should be, free.

The reader will surely perceive the limits of our interpretation. For example, real cost varies through time and this is not reflected in our model. We could try to improve the model with a more sophisticated quantale, but we refrain from modelling reality in this fashion.

3.1 Metric Spaces and Lattices.

Let us explicitly describe the quantale structure on $[0,\infty]$. For any $a \in [0,\infty]$ the map $x \mapsto a + x$ has a right adjoint $x \mapsto x \dot{-} a$:

$$a + x \geq y \iff x \geq y \dot{-} a.$$

It follows that $x \dot{-} (y+z) = (x \dot{-} y) \dot{-} z$ and $(x \dot{-} y) + (y \dot{-} z) \geq x \dot{-} z$ for any $x, y, z \in [0,\infty]$.

A *pre-metric* on a set X is a map $d : X \times X \to [0,\infty]$ such that

$$d(x,y) + d(y,z) \geq d(x,z) \quad \text{and} \quad d(x,x) = 0$$

for every $x, y, z \in X$. The number $d(x,y)$ might represent the cost of transforming an object x into an object y, or of moving from a position x to position y. The symmetry $d(x,y) = d(y,x)$ is not generally desirable. For example, more work is needed to move from the bottom to the top of a mountain than from the top to the bottom for which no work is required. The relation $d(x,y) = 0$ is a pre-order $x \geq y$ on X (x is *higher* than y).

A pre-metric d is a *metric* if moreover,

$$d(x,y)=0 \text{ and } d(y,x)=0 \implies x=y$$

for every $x,y \in X$. A *metric space* is a pair $X=(X,d)$ where X is a set and d is a metric on X. For example, $[0,\infty]$ is a metric space with $d(x,y)=y \dot{-} x$. Observe that $d(x,y)=0$ iff $x \geq y$.

The *opposite* X^o of a metric space X is the metric space (X^o,d) where $X \simeq X^o$ is a formal bijection and $d(x^o,y^o)=d(y,x)$ for any $x,y \in X$.

A map $f: X \to Y$ between metric spaces is a *contraction* if $d(x,y) \geq d(fx,fy)$ for every $x,y \in X$, it is an *embedding* if $d(x,y)=d(fx,fy)$ for every $x,y \in X$.

An element $x \in X$ of a metric space is the *infimum* (resp. *supremum*) of a family $(x_i : i \in I)$ of elements of X if we have

$$d(x,t)=\bigvee_{i\in I} d(x_i,t) \qquad (\text{ resp. } d(t,x)=\bigvee_{i\in I} d(t,x_i))$$

for every $t \in X$. An infimum (resp. supremum) is *finite* if I is finite (beware that this terminology might be confusing since $\infty \in [0,\infty]$ is the (finite) infimum of the empty set).

We define the *raising* (resp. *lowering*) of $x \in X$ by $a \in [0,\infty]$ to be an element $a+x \in X$ (resp. $x \dot{-} a \in X$) such that

$$d(a+x,t)=d(x,t) \dot{-} a \qquad (\text{ resp. } d(t,x \dot{-} a)=d(t,x) \dot{-} a\ )$$

for every $t \in X$. We shall say that X admits raising operators (resp. lowering operators) if $a+x$ (resp. $x \dot{-} a$) exists for any $x \in X$ and $a \in [0,\infty]$. An element $x=\bigwedge_{i\in I} a_i + x_i$ is called a *weighted infimum*. Similarly, an element $x=\bigvee_{i\in I} x_i \dot{-} a_i$ is called a *weighted supremum*.

Definition 1: A *metric lattice** is a metric space which admits finite suprema and infima as well as raising and lowering operators.

It is natural to consider metric lattices with *countable* infima and suprema. We have imposed a finiteness condition only for the sake of uniformity of the present paper.

For any contraction $f: X \to Y$ between metric lattices we have the inequalities

$$\bigwedge_{i\in I} a_i + f(x_i) \geq f\Big(\bigwedge_{i\in I} a_i + x_i\Big) \quad \text{and} \quad f\Big(\bigvee_{i\in I} x_i \dot{-} a_i\Big) \geq \bigvee_{i\in I} f(x_i) \dot{-} a_i.$$

A contraction f is said to be *inf-continuous* (resp. *sup-continuous*), if the first (resp. the second) inequality is an equality. A map f is *bi-continuous* if it is sup- and inf-continuous.

---

*our concept differs from the concept of metric lattice described in [Birkhoff]

The *free* metric lattice generated by a metric space X is a pair $(\Lambda X, i)$ where ΛX is a metric lattice and $i : X \to \Lambda X$ is a contraction having the following universal property: for any metric lattice L and any contraction $f : X \to L$ there exists one and only one bi-continuous map $f' : \Lambda X \to L$ such that $f'i = f$.

3.2 Money games

Money games arise in the description of free metric lattices.

A *cost matrix* is a map $f : A \times B \to [0, \infty]$. The composite of f with $g : B \times C \to [0, \infty]$ is the matrix $f \circ g : A \times C \to [0, \infty]$ defined by

$$(f \circ g)(x, z) = \bigwedge_{y \in B} f(x, y) + g(y, z).$$

The characteristic map $\delta : A \times A \to [0, \infty]$ of the diagonal is given by

$$\delta(x, y) = \begin{cases} 0 & \text{if } x = y \\ \infty & \text{otherwise.} \end{cases}$$

A *metric graph* G is a pair (G, θ) where G is a set (of nodes) and where $\theta : G \times G \to [0, \infty]$ is a map called the *penalty map*. An *edge* of G is a pair $(x, y) \in G \times G$ of finite penalty, that is, such that $\theta(x, y) < \infty$. The set of *successors* of $x \in G$ is $c(x) = \{y \in G : \theta(x, y) < \infty\}$. A node x is *terminal* if $c(x) = \emptyset$. The concepts of acyclic metric graph, polarization, σ-node, π-node and ground node are defined as in section 1. The set of ground nodes is denoted G_0.

Definition 2: Let X be a metric space. An X-*game* is a quintuple $G = (G, \theta, x_0, \epsilon, w)$ where (G, θ) is a finite acyclic metric graph, x_0 is a node called the root, $\epsilon : G \to \{\sigma, \pi, 0\}$ is a polarization and $w : G_0 \to X$ is a map called the valuation.

When $X = [0, \infty]$ we shall say that G is a *money game*. We now describe the plays on money games. There are two players σ and π. A *state* is a triple (u, x, v) where $x \in G$ is a *position* and $u, v \in [0, \infty)$ are the *stakes* of σ and π respectively. Every play starts in a state (u_0, x_0, v_0) where x_0 is the root and where u_0 and v_0 are the initial stakes of σ and π respectively. It is σ's turn at (u, x, v) if $\epsilon x = \sigma$, and it is π's turn if $\epsilon x = \pi$. On his turn, a player moving from position x to a position x' must transfer an amount $t \geq \theta(x, x')$ from his stake to the stake of the other player. More precisely, if $(u, x, v) \to (u', x', v')$ is a move by σ then $u' = u - t$ and $v' = v + t$ for some $u \geq t \geq \theta(x, x')$, if it is a move by π then $u' = u + t$ and $v' = v - t$ for some $v \geq t \geq \theta(x, x')$.

By convention, the σ-player *wins* when his opponent gives up or when the play reaches a ground state (u, x, v) with $u \geq w(x)$, that is, when his stake is above a certain level.

Remark: One might argue that a rational player will systematically choose to give up when his position in a money game become unfavourable. This would be an objection to our interpretation, but it is based on the assumption that cashing the penalty is the only motivation for playing. This need not be the case because money games can model situations where the positions have a value of their own. For example, a position might represent an answer to a question. A flow of money in one direction is almost surely connected with a flow of values in the opposite direction. The motivation for playing a game could be broad, as when the game is part of a larger game (and this happens in our theory). It is perfectly reasonable to support a limited loss at a certain level to attain a better position at another level.

Definition 3: A *strategy* for the σ-player on a money game G is a map $f : G \to [0, \infty]$ such that

(i) for each σ-node x there is an $x' \in G$ such that $f(x) \geq \theta(x, x') + f(x')$,
(ii) for every π-node x and every $x' \in G$ we have $\theta(x, x') + f(x) \geq f(x')$,
(iii) for every ground node x we have $f(x) \geq w(x)$.

If f is a strategy the σ-player can win with any initial stake $u_0 \geq f(x_0)$ because he can choose his moves so as to keep the value of his stake at least the value of f throughout the play. More precisely, if $\epsilon x = \sigma$ at a state (u, x, v) with $u \geq f(x)$ there is a node x' with $f(x) \geq \theta(x, x') + f(x')$. If σ chooses the move $(u, x, v) \to (u - t, x', v + t)$ with $t = \theta(x, x')$ then we have $u - t \geq f(x')$. If $\epsilon x = \pi$ and $u \geq f(x)$ then for any move $(u, x, v) \to (u', x', v')$ we have $u' \geq \theta(x, x') + u \geq \theta(x, x') + f(x) \geq f(x')$. Finally, if x is a ground node and $u \geq f(x)$ then $u \geq w(x)$ since $f(x) \geq w(x)$ and the σ player wins.

The infimum of two strategies is a strategy. There is a smallest or least expensive strategy which we call the *level map* (for σ).

Proposition 1. *Let G a money game. The level map $W : G \to [0, \infty]$ is the unique map such that*

$$W(x) = \begin{cases} w(x) & \text{if } \epsilon x = 0 \\ \bigwedge_{u \in c(x)} W(u) + \theta(x, u) & \text{if } \epsilon x = \sigma \\ \bigvee_{u \in c(x)} W(u) \dot{-} \theta(x, u) & \text{if } \epsilon x = \pi. \end{cases}$$

The *height* of a money game G is defined to be $W(x_0)$ where $W : G \to [0, \infty]$ is the level map and x_0 is the root.

To a pair of X-games A and B we associate a *communication game* $\langle A, B\rangle$ with three players, a *mediator* σ facing two *opposites* σ_A and π_B on two boards, A and B:

$$\sigma_A : A : \cdots\cdots \sigma \cdots\cdots : B : \pi_B$$

A *state* is a quintuple (t, x, u, y, v) where $(x, y) \in A \times B$ is a position and $t, u, v \in [0, \infty)$ are finite numbers representing the stakes of σ_A, σ and π_B respectively. The players σ_A and π_B can be assembled into a team, the π-team, but we shall keep their stakes separate. The root is the pair (a_0, b_0) of roots.

- The polarization $\epsilon : A \times B \to \{\sigma, \pi, 0\}$ is given by $\epsilon(x, y) = \neg\epsilon x \vee \epsilon y$.
- The valuation $w : A_0 \times B_0 \to [0, \infty]$ is given by $w(x, y) = d(w(x), w(y))$.
- The moves by the mediator are of two kinds:
  (i) a move $(u, a, x, b, v) \to (u + t, a', x - t, b, v)$
  where $(u, a, x) \to (u + t, a', x - t)$ is a move by π_A on A;

  (ii) a move $(u, a, x, b, v) \to (u, a, x - s, b', v + s)$
  where $(x, b, v) \to (x - s, b', v + s)$ is a move by σ_B on B.
- The moves by the π-team are of three kinds:

  (i) a move by σ_A: $(u, a, x, b, v) \to (u - t, a', x + t, b, v)$
  where $(u, a, x) \to (u - t, a', x + t)$ is a move by σ_A on A;

  (ii) a move by π_B: $(u, a, x, b, v) \to (u, a, x + s, b', v - s)$
  where $(x, b, v) \to (x + s, b', v - s)$ is a move by π_B on B;

  (iii) a concurrent move by σ_A and π_B: $(u, a, x, b, v) \to (u - t, a', x + t + s, b', v - s)$ where $(u, a, x) \to (u - t, a', x + t)$ is a move by σ_A on A and $(x, b, v) \to (x + s, b', v - s)$ is a move by π_B on B.

A play is won by the mediator if one of his opposites gives up or if the play ends in a state (u, a, x, b, v) where a and b are ground nodes and $x \geq d(w(a), w(b))$; the latter means that his stake should remain sufficiently high.

By combining the stakes of σ_A and π_B we obtain a two person money game $\langle A, B \rangle$ with the mediator facing the π-team.

Definition 4: The *distance* $d(A, B)$ between two X-games A and B is the height of the money game $\langle A, B \rangle$.

Proposition 2. *For any three X-games A, B and C we have $d(A, B) + d(B, C) \geq d(A, C)$ and $d(A, A) = 0$.*

Let $\mathcal{G}(X)$ be the collection of X-games. There is an equivalence relation $\equiv$ on $\mathcal{G}(X)$ defined by

$$A \equiv B \iff d(A, B) = d(B, A) = 0.$$

The quotient of $\mathcal{G}(X)$ by this equivalence relation is a metric space ΛX. There is a canonical map $i : X \to \mathcal{G}(X) \to \Lambda X$, where for any $x \in X$ the game $i(x)$ has a single node x, with $\epsilon x = 0$ and $w(x) = x$. Clearly, $d(i(x), i(y)) = d(x, y)$ for any $x, y \in X$.

Proposition 3. *Let X be a metric space. The metric space ΛX equipped with the map $i : X \to \mathcal{G}(X) \to \Lambda X$ is the free metric lattice generated by X.*

§ 4. OPERATIONS

In this section we introduce operations on games and lattices related to the connectives of linear logic. We need to suppose that the quantale $\mathcal{R}$ is commutative. For $x, y \in \mathcal{R}$, $[x, y]$ will denote the element $x\backslash y = y/x$. The opposite X^o of an $\mathcal{R}$-poset X is an $\mathcal{R}$-poset with $[x^o, y^o] = [y, x]$. The product $X \times Y$ of two $\mathcal{R}$-posets is an $\mathcal{R}$-poset with

$$[(u, v), (x, y)] = [u, x]\ [v, y].$$

4.1 Operations on games

As in §2, for any $\mathcal{R}$-poset X we shall denote by $\mathcal{G}(X)$ the collection of X-games.

The *opposite* of a game $A \in \mathcal{G}(X)$ is a game $A^o \in \mathcal{G}(X^o)$ obtained by interchanging the role of σ and π. More precisely, let $x \mapsto x^o$ be a formal bijection $A \simeq A^o$. The root of A^o is a^o where a is the root of A; the transition matrix of A^o is given by $\theta(x^o, y^o) = \theta(x, y)$; the polarization $\epsilon : A^o \to \{\sigma, 0, \pi\}$ is given by $\epsilon x^o = \neg \epsilon x$; the valuation is given by $w(x^o) = w(x)^o$ for any $x \in A_0$. It follows directly from the definition that we have $[A^o, B^o] = [B, A]$ for any $A, B \in \mathcal{G}(X)$. The map $A \mapsto A^o$ induces the duality isomorphism $(\Lambda X)^o \simeq \Lambda X^o$.

The *(external) parallel product* of $A \in \mathcal{G}(X)$ with $B \in \mathcal{G}(Y)$ is the game $A \odot B \in \mathcal{G}(X \times Y)$ in which σ and π play in parallel on A and B and π has the priority. One can imagine that there are four players divided into two teams: the σ-team $\{\sigma_A, \sigma_B\}$ and the π-team $\{\pi_A, \pi_B\}$.

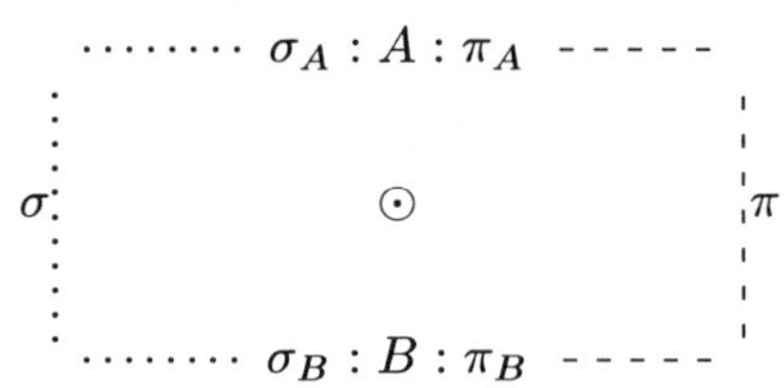

By convention, it is the π-team's turn when it is the turn of at *least one* of its members. Here is the precise definition:

(i) A position of $A \odot B$ is a pair $(x, y) \in A \times B$ of positions, the root is the pair of roots.

(ii) The transition matrix $\theta : (A \times B) \times (A \times B) \to \mathcal{R}$ is given by

$$\theta(x, y, u, v) = \delta(x, u)\theta(y, v) \vee \theta(x, u)\delta(y, v) \vee \theta(x, u)\theta(y, v).$$

(iii) The polarization $\epsilon : A \times B \to \{\sigma, \pi, 0\}$ is given by $\epsilon(x, y) = \epsilon x \vee \epsilon y$.
(iv) The valuation $w : A_0 \times B_0 \to X \times Y$ is given by $w(x, y) = (w(x), w(y))$.

The *(external) tensor product* $A \otimes B \in \mathcal{G}(X \times Y)$ has the same definition except that the σ-team now has priority. The polarisation of $A \otimes B$ is given by the formula $\epsilon(x, y) = \neg(\neg\epsilon x \vee \neg\epsilon y)$. By duality, we have

$$(A \otimes B)^o = A^o \odot B^o \quad \text{and} \quad (A \odot B)^o = A^o \otimes B^o.$$

There are obvious associativity isomorphisms

$$(A \odot B) \odot C \simeq A \odot (B \odot C), \qquad (A \otimes B) \otimes C \simeq A \otimes (B \otimes C)$$

for $A \in \mathcal{G}(X)$, $B \in \mathcal{G}(Y)$ and $C \in \mathcal{G}(Z)$.

Remark: There is a similarity between the mediator game $\langle A, B\rangle$ and the game $A^o \odot B$. One difference is the valuation, which is in $\mathcal{R}$ for the first, and in $X^o \times Y$ for the second. Another difference is that the members of the σ-team may move concurrently on $A^o \odot B$. This is not an essential difference: we have $A \odot' B \equiv A \odot B$ in $\mathcal{G}(X \times Y)$ where $A \odot' B$ is obtained from $A \odot B$ by forbidding concurrent moves by either team.

Example: We shall illustrate the operations $\otimes$ and $\odot$ in the case $\mathcal{R} = [0, \infty]$ with the activities of lending and borrowing. Let $B(u)$ be the game having three positions $0, 1, 2$, with 2 as root, with polarization $\epsilon(0) = 0$, $\epsilon(1) = \sigma$ and $\epsilon(2) = \pi$, with penalty $\theta(2, 1) = \theta(1, 0) = u$ ($\theta(x, y) = \infty$ otherwise), and with valuation $w(0) = 0$. It is a $\{0\}$-game where $\{0\}$ is the singleton metric space. The plays on $B(u)$ are deterministic (unless a player decides to gives up). The player π opens the play by transferring an amount u from his stake to σ. Then it is σ's turn and this player moves by transferring u back to π, ending the play. This is the prototype of the game of lending and borrowing (without interest). The game becomes more interesting when combined with other games. Let $S, T \in \mathcal{G}(X)$. Then $T \odot B(u) \in \mathcal{G}(X)$ since $X \times \{0\} \simeq X$. The game $\langle S, T \odot B(u)\rangle$ is similar to $\langle S, T\rangle$ except for one important difference. On $\langle S, T \odot B(u)\rangle$ the mediator will not move until he has received an amount u from an opposite moving on $B(u)$. From this point on he can play as if on $\langle S, T\rangle$, but his stake is higher, he is in a better position to win. However, he will have to pay back u before the end of the play unless he gives up, in which case he loses, or an opposite gives up, in which case he wins (it is not his fault!). Replacing T by $T \odot B(u)$ reduces the cost of trading, that is, $d(S, T) \geq d(S, T \odot B(u))$. Equivalently, we could replace S by $S \otimes B(u)^o$. The inequality $d(S, T) \geq d(S, T \odot B(u))$ is valid for any S and T only because $i(0) \geq B(u)$ in $\Lambda\{0\}$, that is, borrowing is free. A more realistic model would use a game $B(u, v)$ where σ borrows u and pays back $v > u$. Depending on the particular nature of S and T one can have $d(S, T) > d(S, T \odot B(u, v))$.

Theorem 1. *For any $A, C \in \mathcal{G}(X)$ and $B, D \in \mathcal{G}(Y)$ we have*

$$[A,C][B,D] \leq [A \odot B, C \odot D] \quad \text{and} \quad [A,C][B,D] \leq [A \otimes B, C \otimes D].$$

Proof: We only indicate the idea of the proof. The second inequality follows from the first by duality. It suffices to prove $[B, D] \leq [A \odot B, A \odot D]$ because $[A, C] \leq [A \odot D, C \odot D]$ follows by symmetry and then

$$[A,C][B,D] \leq [A \odot B, A \odot D][A \odot D, C \odot D] \leq [A \odot B, C \odot D].$$

The inequality $[B, D] \leq [A \odot B, A \odot D]$ is obtained by proving $(A \odot B) \cdot [B, D] \leq A \odot D$. For any $x \in A$ let $(A \mid x)$ be the game having the same structure as A except for its root, which is set to be x. The map $x \mapsto q(A \mid x)$ is the largest strategy on A considered as a ΛX-game. For any $(x, y) \in A \times B$ let us put $x \odot y = q(A \odot B \mid (x, y))$. Then the map $(x, y) \mapsto x \odot y$ is the largest strategy on $A \odot B$ considered as a $\Lambda(X \times Y)$-game. Similarly, we have a largest strategy $(x, z) \mapsto x \odot z$ on $A \odot C$ considered as a $\Lambda(X \times Z)$-game. The inequality $(A \odot B) \cdot [B, D] \leq A \odot D]$ will follow if we show that

$$\bigvee_{y \in B} (x \odot y) \cdot [y, z] \leq x \odot z$$

for any $(x, z) \in A \times C$. For this it suffices to show that the map $h : A \times D \to \Lambda(X \times Z)$ defined by the left hand side is a strategy since the right hand side is the largest strategy. This follows from an easy modification of Theorem 2 §2. QED

4.2 Operations on free lattices

It follows from Theorem 1 that the operations $\odot$ and $\otimes$ on games induce corresponding operations on free lattices:

$$\odot, \otimes : \Lambda X \times \Lambda Y \to \Lambda(X \times Y).$$

Both operations are associative as witnessed by the associativity isomorphisms. They can be characterized by a few properties. We shall use these operations to define various structures on free lattices. For example we prove that $\Lambda \mathbf{1}$, the free $\mathcal{R}$-lattice generated by a singleton, is a quantale and that any ΛX is enriched over $\Lambda \mathbf{1}$.

Let X, Y and Z be complete $\mathcal{R}$-posets. Recall that an order preserving map $f : X \times Y \to Z$ is *separately inf-continuous* if the maps $f(x, -) : Y \to Z$ and $f(-, y) : X \to Z$ are inf-continuous for any $x \in X$ and $y \in Y$.

Definition 1: Let X, Y and Z be cocomplete $\mathcal{R}$-posets. An order preserving map $f : X \times Y \to Z$ is *soft* if

$$f(x, y) = \bigvee_{i \in I} f(x_i, y) \cdot a_i \;\vee\; \bigvee_{j \in J} f(x, y_j) \cdot b_j$$

for any $x = \bigvee_{i\in I} x_i \cdot a_i$ and $y = \bigvee_{j\in J} y_j \cdot b_j$.

The concept of soft map in $n \geq 1$ variables is defined similarly. In one variable it is a sup-continuous map. For $n = 3$ the condition is that

$$f(x,y,z) = \bigvee_{i\in I} f(x_i,y,z)\cdot a_i \;\vee\; \bigvee_{j\in J} f(x,y_j,z)\cdot b_j \;\vee\; \bigvee_{j\in J} f(x,y,z_k)\cdot c_k$$

for any $x = \bigvee_{i\in I} x_i \cdot a_i$, $y = \bigvee_{j\in J} y_j \cdot b_j$ and $z = \bigvee_{k\in K} z_k \cdot c_k$.

There is a dual concept of cosoft map.

Definition 2: Let X and Y be $\mathcal{R}$-posets and let $\mathcal{L}$ be an $\mathcal{R}$-lattice. An order preserving map $f : \Lambda X \times \Lambda Y \to \mathcal{L}$ is *responsive* if the following conditions are satisfied:

(i) f is separately inf-continuous,

(ii) f is soft and the maps $f(ix,-) : \Lambda Y \to \mathcal{L}$ and $f(-,iy) : \Lambda X \to \mathcal{L}$ are sup-continuous for any $x \in X$ and $y \in Y$.

For example, according to Theorem 1 §2, the map $[-,-] : \Lambda X^o \times \Lambda X \to \mathcal{R}$ is responsive.

The concept of responsive map of any number $n \geq 1$ of variables is easily defined. In one variable it is a bi-continuous map. For $n = 3$ the conditions are that (i) f is separately inf-continuous, (ii) the maps $f(-,-,-)$, $f(ix,-,-)$, $f(-,iy,-)$, $f(-,-,iz)$, $f(ix,iy,-)$, $f(-,iy,iz)$, $f(ix,-,iz)$ are soft for any $x \in X$, $y \in Y$ and $z \in Z$.

There is a dual concept of coresponsive map.

Proposition 2. *The parallel product $\odot : \Lambda X \times \Lambda Y \to \Lambda(X \times Y)$ is responsive. Dually, the tensor product $(x,y) \mapsto x \otimes y$ is coresponsive.*

Proof: Let $A \in \mathcal{G}(X)$ and $B \in \mathcal{G}(X)$. As in the proof of Theorem 1 let us put $x \odot y = q((A \odot B) \mid (x,y))$ for $(x,y) \in A \times B$. The proposition will follow if we show that

(i) $x \odot y = \; (w(x), w(y))$ if $(\epsilon x, \epsilon y) = (0,0)$,

(ii) $x \odot y = \bigwedge_{v\in c(y)} \theta(y,v) \star (x \odot v)$ if $\epsilon y = \pi$,

(iii) $x \odot y = \bigwedge_{u\in c(x)} \theta(x,u) \star (u \odot y)$ if $\epsilon x = \pi$,

(iv) $x \odot y = \bigvee_{v\in c(y)} (x \odot v) \cdot \theta(y,v)$ if $(\epsilon x, \epsilon y) = (0,\sigma)$,

(v) $x \odot y = \bigvee_{u\in c(x)} (u \odot y) \cdot \theta(x,u)$ if $(\epsilon x, \epsilon y) = (\sigma,0)$,

(vi) $x \odot y = \bigvee_{u\in c(x)} (u \odot y) \cdot \theta(x,u) \vee \bigvee_{v\in c(y)} (x \odot v) \cdot \theta(y,v)$ if $(\epsilon x, \epsilon y) = (\sigma,\sigma)$.

This is easily proved by induction on $(x,y) \in A \times B$. QED

Let X_j $(1 \leq j \leq n)$ be a finite sequence of $\mathcal{R}$-posets. The n-fold parallel product

$$\odot : \prod_{j=1}^{n} \Lambda X_j \to \Lambda(\prod_{j=1}^{n} X_j)$$

is responsive. This follows from Proposition 2 since it is a composite of two-fold products.

Theorem 2. *Let $\mathcal{L}$ be an $\mathcal{R}$-lattice. Any order preserving map $f : \prod_{j=1}^n X_i \to \mathcal{L}$ has a unique responsive (resp. coresponsive) extension $f' : \prod_{j=1}^n \Lambda X_j \to \mathcal{L}$.*

Proof: Let $l : \Lambda(\prod_{j=1}^n X_i) \to \mathcal{L}$ be a bi-continuous extension of f. The map f' given by $f'(x_1, \ldots, x_n) = l(x_1 \odot \cdots \odot x_n)$ is responsive since the parallel product is responsive and l is bi-continuous. It is also an extension of f since

$$f'(ix_1, \ldots, ix_n) = l(ix_1 \odot \cdots \odot ix_n) = l(i(x_1, \ldots, ix_n)) = f(x_1, \ldots, ix_n).$$

The uniqueness of f' is a consequence of the following lemma.

Let us put $Y = \prod_{j=1}^n \Lambda X_j$ and let ∂Y be the subset of Y whose elements are the n-tuples $(y_j : 1 \leq j \leq n)$ having at least one component y_j in the image of $i : X_j \to \Lambda X_j$.

Lemma. *Let $f, g : Y \to \mathcal{L}$ be order preserving maps such that $f(y) \leq g(y)$ for every $y \in \partial Y$. If f is soft and g is separately inf-continuous then $f(y) \leq g(y)$ for every $y \in Y$.*

Proof: If $y = (y_j : 1 \leq j \leq n)$ is an element of Y let us choose a representative $A_j \in \mathcal{G}(X_j)$ of y_j for each $1 \leq j \leq n$. Let A be the the parallel product of the games A_j $(1 \leq j \leq n)$. Let $V : A \to Y$ be the map given by $V(x_1, \ldots, x_n) = (Wx_1, \ldots, Wx_n))$ where $Wx_i = q(A_1 \mid x_1)$ as the proof of Theorem 1. Then $y = V(a)$ where a is the root of A. We shall prove by induction on $x \in A$ that $fV(x) \leq gV(x)$. By hypothesis, the inequality $fV(x) \leq gV(x)$ is true when at least one component x_j of x is a ground node since $V(x) \in \partial Y$ in this case. Let us make the induction hypothesis that $fV(u) \leq gV(u)$ for every $u \in c(x)$. If $\epsilon x = \pi$ then at least one component x_j of x is a π-node. By symmetry we can suppose that $j = 1$ since the order of the variables is irrelevant. Let us put $x = (x_1, z)$. We have $W_1(x_1) = \bigwedge_{u \in c(x_1)} \theta(x_1, u) \star W_1(u)$ since x_1 is a π-node. Therefore

$$fV(x_1, z) \leq \bigwedge_{u \in c(x_1)} \theta(x_1, u) \star fV(u, z) \leq \bigwedge_{u \in c(x_1)} \theta(x_1, u) \star gV(u, z) = gV(x_1, z)$$

since $f(-, z)$ is order preserving, since $fV(u, z) \leq gV(u, z)$ for every $u \in c(x_1)$ by induction hypothesis, and since g is separately inf-continuous. It remains to consider the case where $\epsilon x = \sigma$. We can suppose that no component of x is a ground node. In this case $\epsilon x_j = \sigma$ for every $1 \leq j \leq n$. Then $W_j(x_j) = \bigvee_{u_j \in c(x_j)} W_j(u_j) \cdot \theta(x_j, u_j)$ for every $1 \leq j \leq n$ since x_j is a σ-node. If we put $f_j(u_j) = fV(x_1, \ldots, u_j, \ldots, x_n)$ and $g_j(u_j) = gV(x_1, \ldots, u_j, \ldots, x_n)$ then

$$fV(x) = \bigvee_{1 \leq j \leq n} \bigvee_{u_j \in c(x_j)} f_j(u_j) \cdot \theta(x_j, u_j)$$

$$\leq \bigvee_{1\leq j\leq n} \bigvee_{u_j\in c(x_j)} g_j(u_j)\cdot\theta(x_j,u_j) \leq gV(x)$$

since f is soft, since $f_j(u_j) \leq g_j(u_j)$ for every $u_j \in c(x_j)$ and $1 \leq j \leq n$ by induction hypothesis, and since g is order preserving. QED

We can now finish the proof of Theorem 2. For every $0 \leq k \leq n$ let $Y(k)$ be the subset of Y whose elements are the n-tuples $(y_j : 1 \leq j \leq n)$ having at least $n-k$ components y_j belonging to the image of $i : X_j \to \Lambda X_j$. We have $Y(0) \subset \cdots \subset Y(n) = Y$. Let $g, h : Y \to \mathcal{L}$ be two responsive extensions of f. By hypothesis we have $f \mid Y(0) = g \mid Y(0)$. Let us prove that $f \mid Y(k-1) \leq g \mid Y(k-1) \Rightarrow f \mid Y(k) \leq g \mid Y(k)$ for every $0 < k \leq n$. Assuming that $f \mid Y(k-1) \leq g \mid Y(k-1)$ let us prove that $f(x_1,\ldots,x_k,ix_{k+1},\ldots,ix_n) \leq g(x_1,\ldots,x_k,ix_{k+1},\ldots,ix_n)$ for every $(x_1,\ldots,x_k) \in \Lambda(\prod_{j=1}^k X_j)$ and every $(x_{k+1},\ldots,x_n) \in \prod_{j=k+1}^n \Lambda X_j$. The first map is soft in the first k variables since f is responsive, the second is separately inf-continuous since g is responsive, and the lemma applies since $f \mid Y(k-1) \leq g \mid Y(k-1)$. Thus the conclusion is valid for any subset of k variables, and therefore $f \mid Y(k) \leq g \mid Y(k)$. It follows by induction that $f \mid Y(n) \leq g \mid Y(n)$. We have proved that $f \leq g$. The same argument proves that $g \leq f$. QED

Proposition 3. *For any $x \in \Lambda X$ the map $x \otimes (-) : \Lambda Y \to \Lambda(X \times Y)$ has a right adjoint $z \mapsto x\backslash z$. Moreover, $(x,z) \mapsto x\backslash z$ is responsive as a map $\Lambda X^o \times \Lambda(X\times Y) \to \Lambda Y$ and $i(x)\backslash i(u,y) = iy\cdot[x,u]$ for any $x, u \in \Lambda X$ and $y \in \Lambda Y$.*

Proof: Let $K : \Lambda X^o \times \Lambda(X\times Y) \to \Lambda Y$ be the responsive extension of the map $X^o \times (X \times Y) \to \Lambda Y$ given by $(x^o, u, y) \mapsto iy\cdot[x,u]$. Let us prove that

$$[x\otimes y, z] = [y, K(x^o, z)]$$

for any $x \in \Lambda X$, $y \in \Lambda Y$ and $z \in \Lambda(X\times Y)$. The two sides of this equality determine responsive maps of three variables, so by Theorem 2 it suffices to verify the equality

$$\begin{aligned}[ix\otimes iy, i(u,v)] &= [i(x,y), i(u,v)] = [(x,y),(u,v)] = [x,u][y,v] \\ &= [x,u][iy,iv] = [iy, iv\cdot[x,u]] = [iy, K(ix^o, i(u,v))].\end{aligned}$$

QED

By symmetry, for any $y \in \Lambda Y$ the map $(-)\otimes y : \Lambda X \to \Lambda(X\times Y)$ has a right adjoint $z \mapsto z/y$.

Proposition 4. *For any $x \in \Lambda X$, $y \in \Lambda(X\times Y)$ and $z \in \Lambda Z$ we have $(x\backslash y)\odot z = x\backslash(y\odot z)$.*

Proof: Both sides of the equality determine responsive maps $\Lambda X \times \Lambda(X \times Y) \times \Lambda Z \to \Lambda(Y\times Z)$. Moreover, for any $x, u \in X$, $y \in Y$ and $z \in Z$

$$(ix\backslash i(u,y))\odot iz = i(y)\cdot[x,u]\odot iz = (i(y)\odot iz)\cdot[x,u]$$

$$= \quad i(y,z)\cdot[x,u] = i(x)\backslash i(u,y,z) = i(x)\backslash(i(u,y)\odot iz),$$

and the result follows from Theorem 2. QED

Corollary. *For any $x \in \Lambda X$, $y \in \Lambda Y$ and $z \in \Lambda Z$ we have the Grishin inequalities*

$$x \otimes (y \odot z) \leq (x \otimes y) \odot z \quad \text{and} \quad (x \odot y) \otimes z \leq x \odot (y \otimes z).$$

Proof: By adjointness we have $y \leq x\backslash(x \otimes y)$. It follows that $y \odot z \leq (x\backslash(x \otimes y))\odot z$, and hence $y\odot z \leq x\backslash((x\otimes y)\odot z)$. By adjointness we obtain $x\otimes(y\odot z) \leq (x \otimes y) \odot z$. The other inequality is symmetrical. QED

When X is finite the *diagonal* $\Delta_X \in \Lambda(X \times X^o)$ can be defined:

$$\Delta_X = \bigvee_{x\in X} i(x, x^o) = \bigvee_{x\in X} x \otimes x^o.$$

Lemma. *Let $f : X \times Y \to Z$ be a soft map between cocomplete $\mathcal{R}$-posets. The set of element $x \in X$ for which the map $f(x,-) : Y \to Z$ is sup-continuous is closed under (weighted) suprema.*

Proof: Let $x = \bigvee_{i\in I} x_i \cdot a_i$ and suppose that $f(x_i,-) : Y \to Z$ is sup-continuous for every $i \in I$. If $y = \bigvee_{j\in J} y_j \cdot b_j$ then

$$f(x,y) = \bigvee_{i\in I} f(x_i,y)\cdot a_i \;\vee\; \bigvee_{j\in J} f(x,y_j)\cdot b_j$$

since f is soft. If we show that $f(x_i,y)\cdot a_i \leq \bigvee_{j\in J} f(x,y_j)\cdot b_j$ for every $i \in I$ the result will follow. But

$$f(x_i,y)\cdot a_i = \bigvee_{j\in J} f(x_i,y_j)\cdot b_j\cdot a_i = \bigvee_{j\in J} f(x_i,y_j)\cdot a_i\cdot b_j \leq \bigvee_{j\in J} f(x,y_j)\cdot b_j$$

since $f(x_i,-)$ is sup-continuous and $f(x_i,y_j)\cdot a_i \leq f(x,y_j)$ for every $i \in I$ and $j \in J$. QED

Proposition 5. *For any $x, y \in \Lambda X$ we have*

$$x^o = x\backslash\Delta_X \quad \text{and} \quad [x,y] = [x \otimes y^o, \Delta_X] = [\Delta_X^o, x^o \odot y].$$

Proof: Let us see that the map $d : \Lambda X^o \to \Lambda X^o$ given by $d(x^o) = x\backslash\Delta_X$ is bi-continuous. According to Proposition 4, $(x^o, z) \mapsto x\backslash z$ is a responsive map $\Lambda X^o \times \Lambda(X \times X^o) \to \Lambda X^o$. In particular, d is inf-continuous. Moreover, $x^o \mapsto x\backslash i(u,u^o)$ is sup-continuous for every $u \in X$. By the lemma, $x^o \mapsto x\backslash\Delta_X$

is sup-continuous since $\Delta_X = \bigvee_{u\in X} i(u,u^o)$. The equality $x^o = x\backslash\Delta_X$ will follow if we show that $i(x)^o = i(x)\backslash\Delta_X$ for every $x \in X$. But

$$i(x)\backslash\Delta_X = \bigvee_{u\in X} i(x)\backslash i(u,u^o) = \bigvee_{u\in X} i(u^o)\cdot[x,u] = i(x^o)$$

since $i(x)\backslash(-)$ is sup-continuous and $i(u^o)\cdot[x,u] \le i(x^o)$ for every $u \in X$. It follows that $[x\otimes y^o, \Delta_X] = [y^o, x\backslash\Delta_X] = [y^o, x^o] = [x,y]$. The equality $[x\otimes y^o, \Delta_X] = [\Delta_X^o, x^o \odot y]$ follows by duality. QED

We now explain how certain structures can be extended to free $\mathcal{R}$-lattices. Recall that an $\mathcal{R}$-monoid M is an $\mathcal{R}$-poset equipped with a monoid structure such that the product $(x,y)\mapsto x\bullet y$ is order preserving, that is, such that $[x,u][y,v] \le [x\bullet y, u\bullet v]$ for every $x,y,u,v \in M$. An $\mathcal{R}$-quantale is an $\mathcal{R}$-monoid which admits two binary operations $(x,y)\mapsto x/y$ and $(x,y)\mapsto y\backslash x$ satisfying the identities

$$[x\bullet y, z] = [y, x\backslash z] = [y, z/x].$$

An element $d \in M$ is a *dualizing* element if we have $d/(x\backslash d) = x = (d/x)\backslash d$ for every $x \in M$.

Theorem 3. *Let M be an $\mathcal{R}$-monoid. The coresponsive extension of the product on M defines an $\mathcal{R}$-monoid structure on ΛM. Moreover, ΛM is a quantale if M is, and $id \in \Lambda M$ is a dualizing element if $d \in M$ is.*

Proof: Let $m : \Lambda M\times\Lambda M \to \Lambda M$ be the coresponsive extension of the map $(x,y)\to i(x\bullet y)$. By definition $m(ix,iy) = i(x\bullet y)$ for every $(x,y)\in M\times M$. It is easy to see that $(x,y,z)\mapsto m(x,m(y,z))$ and $(x,y,z)\mapsto m(x,m(y,z))$ are both coresponsive as maps $(\Lambda M)^3 \to \Lambda M$. For any $x,y,z\in M$ we have

$$\begin{aligned} m(ix,m(iy,iz)) &= m(ix,i(y\bullet z)) = i(x\bullet(y\bullet z)) \\ &= i((x\bullet y)\bullet z) = m(i(x\bullet y),iz) = m(m(ix,iy),iz) \end{aligned}$$

Associativity follows by the uniqueness of coresponsive extensions (Theorem 2). If $1\in M$ is a unit then $i(1)\in\Lambda M$ is a unit for m by a similar argument. This proves that ΛM is an $\mathcal{R}$-monoid. For any $x,y\in\Lambda M$ let us write $x\bullet y$ instead of $m(x,y)$. Let us now suppose that M is an $\mathcal{R}$-quantale. Let $K : \Lambda M^o\times\Lambda M\to\Lambda M$ be the responsive extension of the map $(x^o,y)\to i(x\backslash y)$. By definition we have $K(ix^o,iy) = i(x\backslash y)$ for any $x,y\in M$. If we use the fact that $[-,-] : \Lambda M^o\times\Lambda M\to\mathcal{R}$ is a responsive map, it is easy to see that $(x,y,z)\mapsto[x,K(y^o,z)]$ and $(x,y,z)\mapsto[x\bullet y,z]$ are responsive maps. Moreover,

$$[ix,K(iy^o,iz)] = [ix,i(y\backslash z)] = [x,y\backslash z][x\bullet y,z] = [(i(x\bullet y),iz] = [ix\bullet iy,iz]$$

for any $x,y,z\in M$. It follows from Theorem 2 that $[x,K(y^o,z)] = [x\bullet y,z]$ for any $x,y,z\in\Lambda M$. For any $x,y\in\Lambda M$ let us write $x\backslash y$ instead of $K(x^o,y)$.

By symmetry, the responsive extension of the map $(x^o, y) \to i(y/x)$ defines an operation y/x on ΛM such that $[x \bullet y, z] = [x, z/y]$ for any $x, y, z \in \Lambda M$. This proves that ΛM is an $\mathcal{R}$-quantale. Finally, assume that $d \in M$ is a dualizing element towards showing that $id/(x\backslash id) = x = (id/x)\backslash id$ for every $x \in \Lambda M$. Both sides define bi-continuous maps of x, and $id/(ix\backslash id) = i(d/(x\backslash d)) = ix = i((d/x)\backslash d) = (id/ix)\backslash id$ for every $x \in M$, so the result follows by the uniqueness of bi-continuous extensions. QED

Let $\mathbf{1}$ be the singleton $\mathcal{R}$-poset. By definition, $\mathbf{1}$ has a single node that will be denoted 1, and $[1, 1] = 1 \in \mathcal{R}$. Trivially, $\mathbf{1}$ has the structure of a quantale. It follows that $\Lambda\mathbf{1}$ is a quantale when equipped with the tensor product $\otimes : \Lambda\mathbf{1} \times \Lambda\mathbf{1} \to \Lambda\mathbf{1}$. Moreover, $1 = i(1)$ is a dualizing element. Consider the map $j : \mathcal{R} \to \Lambda\mathbf{1}$ given by $j(a) = 1 \cdot a$ for $a \in \mathcal{R}$. For any $a, b \in \mathcal{R}$ we have $j(a) \otimes j(b) = (1 \cdot a) \otimes (1 \cdot b) = 1 \cdot (ab) = j(ab)$ since $\otimes$ is separately sup-continuous. An $\mathcal{R}$-poset X becomes a $\Lambda\mathbf{1}$-poset with $j[-,-] : X \times X \to \Lambda\mathbf{1}$. Let $[[-,-]] : \Lambda X^o \times \Lambda X \to \Lambda\mathbf{1}$ be the responsive extension of $j[-,-] : X^o \times X \to \Lambda\mathbf{1}$. It follows from Proposition 4 that $[[x, y]] = x\backslash y$ for any $x, y \in \Lambda X$.

Proposition 6. *The map $[[-,-]] : \Lambda X^o \times \Lambda X \to \Lambda\mathbf{1}$ is an enrichment of ΛX over $\Lambda\mathbf{1}$, and ΛX is bicomplete as a $\Lambda\mathbf{1}$-poset. For any $x, y \in \Lambda X$ we have $[[x^o, y^o]] = [[y, x]]$. Moreover, for any $\mathcal{R}$-poset Y we have $[[x \otimes y, z]] = [[y, x\backslash z]]$ for any $x \in \Lambda X$, $y \in \Lambda Y$ and $z \in \Lambda(X \times Y)$.*

Proof: According to Proposition 4, for any $x \in \Lambda X$ the map $y \mapsto [[x, y]]$ is right adjoint to the map $z \mapsto x \otimes z$. Hence $x \otimes [[x, y]] \leq y$ for any $x, y \in \Lambda X$. It follows that $x \otimes [[x, y]] \otimes [[y, z]] \leq y \otimes [[y, z]] \leq z$, or equivalently, $[[x, y]] \otimes [[y, z]] \leq [[x, z]]$, for any $x, y, z \in \Lambda X$. Moreover, $1 \leq [[x, x]]$ since $x \otimes 1 = x \leq x$. This shows that $[[-,-]]$ is a $\Lambda\mathbf{1}$-preorder on ΛX. By adjointness, $[1, [[x, y]]] = [x \otimes 1, y] = [x, y]$. Thus $1 \leq [[x, y]]$ iff $x \leq y$. This shows that $[[-,-]]$ defines the given order on ΛX. Let us prove the last statement. It follows from Proposition 4 that $(x \otimes y)\backslash z = y\backslash(x\backslash z)$ for any $x \in \Lambda X$, $y \in \Lambda Y$ and $z \in \Lambda(X \times Y)$, therefore $[[x \otimes y, z]] = [[y, x\backslash z]]$. Let us prove that $[[x^o, y^o]] = [[y, x]]$ for any $x, y \in \Lambda X$. Both sides of the equality define responsive maps $\Lambda X \times \Lambda Y^o \to \Lambda\mathbf{1}$ and we have $[[ix^o, iy^o]] = j[x^o, y^o] = j[y, x] = [[iy, ix]]$ for any $x \in X$ and $y \in Y$. The result follows. Clearly ΛX admits tensor products by any $a \in \Lambda\mathbf{1}$ since $[[a, [[x, y]]]] = [[a, x\backslash y]] = [[x \otimes a, y]]$ for any $x, y \in \Lambda X$. By duality, it admits cotensor products: $a \star x = (x^o \otimes a)^o = x \odot a^o$.
QED

Corollary. *For any $u, x \in \Lambda X$ and $v, y \in \Lambda Y$ we have*

$$[[u, x]] \otimes [[v, y]] \leq [[u \otimes v, x \otimes y]] \quad \text{and} \quad [[u, x]] \otimes [[v, y]] \leq [[u \odot v, x \odot y]].$$

Proof: We have $u \otimes v \otimes [[u, x]] \otimes [[v, y]] = u \otimes [[u, x]] \otimes v \otimes [[v, y]] \leq x \otimes y$. Thus $[[u, x]] \otimes [[v, y]] \leq [[u \otimes v, x \otimes y]]$ by adjointness. This proves the first inequality. The second follows by duality:

$$[[u,x]] \otimes [[v,y]] = [[x^o,u^o]] \otimes [[y^o,v^o]] \leq [[x^o \otimes y^o, u^o \otimes v^o]] = [[u \odot v, x \odot y]].$$

QED

Remark: The map $j : \mathcal{R} \to \Lambda\mathbf{1}$ given by $j(x) = 1 \cdot x$ has a right adjoint $\gamma : \Lambda\mathbf{1} \to \mathcal{R}$ given by $x \mapsto [1,x]$. The map γ is bi-continuous since $[-,-]$ is responsive and $[1,-] = [i1,-]$. Moreover, $\gamma j(x) = x$ for any $x \in \mathcal{R}$. We saw that j preserves multiplication and that $\gamma[[x,y]] = [x,y]$ for any $x,y \in \Lambda\mathbf{1}$.

Example: For any $a \in \Lambda\mathbf{1}$ we have $1 \leq [[a,a]]$. If $a = ju$ for some $u \in \mathcal{R}$ then $[[a,a]] = u\star(1\cdot u)$. When $\mathcal{R}$ is the interval $[0,\infty]$ the element $[[ju,ju]]$ can be represented by $B(u)$, the game of lending and borrowing u of the example before Theorem 1. Similarly, in this example $B(u,v) = [[ju,jv]]$.

§ 5. APPENDIX

In this appendix we show that cut elimination for free lattices follows from Whitman's Decision Procedure (Prop.1 §1).

Let P be a poset. Let $T(P)$ be the set of terms as defined in Section 1. The relation $A \leq B$ between terms can be generated by a deductive system, in which case it is usually written $A \vdash B$. Consider the deductive system $\mathcal{D}(P)$ whose axioms and deduction rules are the following: *The axioms are:*

i) (identity) $\qquad A \vdash A$

ii) (top and bottom) $\qquad A \vdash \top, \qquad \bot \vdash A$

iii) for every $x \leq y$ in P $\qquad x \vdash y$

The deduction rules are:

i) (cut rule)

$$\frac{A \vdash B, \quad B \vdash C}{A \vdash C}$$

ii) (weakening rules)

$$\frac{A \vdash C}{A \wedge B \vdash C}, \qquad \frac{B \vdash C}{A \wedge B \vdash C}, \qquad \frac{A \vdash B}{A \vdash B \vee C}, \qquad \frac{A \vdash C}{A \vdash B \vee C}$$

iii) (introduction rules)

$$\frac{A \vdash B, \quad A \vdash C}{A \vdash B \wedge C}, \qquad \frac{A \vdash C, \quad B \vdash C}{A \vee B \vdash C}.$$

We leave to the reader the verification that the relation $A \vdash B$ defined by these rules is equivalent to the relation $A \leq B$.

Theorem. *(cut elimination) The cut rule and the identity axiom can be removed from the deductive system $\mathcal{D}(P)$ without weakening it.*

Proof: Let $\mathcal{D}'(P)$ be the deductive system obtained from $\mathcal{D}(P)$ by removing the cut rule and the identity axiom. Let us put $[A,B]' = 1$ if the relation $A \vdash B$ can be proved from $\mathcal{D}'(P)$ alone and $[A,B]' = 0$ otherwise. Obviously, $[A,B] \geq [A,B]'$ for every $A, B \in T(P)$. We shall prove that $[A,B]' \geq [A,B]$. Let us first observe that the inequalities $\geq$ associated to the equalities (1) to (6) of the Decision Procedure follow from the rules of $\mathcal{D}'(P)$ alone. More precisely, we have:

(1) $[A_1 \vee A_2, B]' \geq [A_1, B]' \wedge [A_2, B]'$ and $[\bot, B]' \geq 1$;
(2) $[A, B_1 \wedge B_2]' \geq [A, B_1]' \wedge [A, B_2]'$ and $[A, \top]' \geq 1$;
(3) $[A_1 \wedge A_2, B_1 \vee B_2]' \geq$
$[A_1 \wedge A_2, B_1]' \vee [A_1 \wedge A_2, B_2]' \vee [A_1, B_1 \vee B_2]' \vee [A_2, B_1 \vee B_2]'$;
(4) $[\top, B_1 \vee B_2]' \geq [\top, B_1]' \vee [\top, B_2]'$, $[A_1 \wedge A_2, \bot]' \geq [A_1, \bot]' \vee [A_2, \bot]'$ and $[\top, \bot]' \geq 0$;
(5) If $A \in P$ then $[A, B_1 \vee B_2]' \geq [A, B_1]' \vee [A, B_2]'$ and $[A, \bot]' \geq 0$;
(6) If $B \in P$ then $[A_1 \wedge A_2, B]' \geq [A_1, B]' \vee [A_2, B]'$ and $[\top, B]' \geq 0$.

For example, the inequality $[A_1 \vee A_2, B]' \geq [A_1, B]' \wedge [A_2, B]'$ follows from the introduction rule

$$\frac{A_1 \vdash B, \quad A_1 \vdash B}{A_1 \vee A_2 \vdash B}$$

and the inequality $[\bot, B]' \geq 1$ follows from axiom (ii). As another example, if $A = A_1 \wedge A_2$ and $B = B_1 \vee B_2$, the inequality $[A,B]' \geq [A, B_1]' \vee [A, B_2]' \vee [A_1, B]' \vee [A_2, B]'$ follows from the weakening rules:

$$\frac{A_1 \vdash B}{A_1 \wedge A_2 \vdash B}, \qquad \frac{A_2 \vdash B}{A_1 \wedge A_2 \vdash B}, \qquad \frac{A \vdash B_1}{A \vdash B_1 \vee B_2}, \qquad \frac{A \vdash B_2}{A \vdash B_1 \vee B_2}.$$

Let us now prove the inequality $[A,B]' \geq [A,B]$ by structural induction on the pair (A,B). We shall consider three cases depending on the polarization of (A,B). If $\epsilon(A,B) = \pi$ then by the result above and the decision procedure (Prop. 1 §1) we have

$$[A,B]' \geq \bigwedge_{i \in I} [A_i, B_i]' \quad \text{and} \quad [A,B] = \bigwedge_{i \in I} [A_i, B_i]$$

where A_i and B_i are subterms of A and B. It follows that $[A,B]' \geq [A,B]$ since $[A_i, B_i]' \geq [A_i, B_i]$ for every $i \in I$ by induction hypothesis. If $\epsilon(A,B) = \sigma$ then

$$[A,B]' \geq \bigvee_{i \in I} [A_i, B_i]' \quad \text{and} \quad [A,B] = \bigvee_{i \in I} [A_i, B_i]$$

where A_i and B_i are subterms of A and B. Again, we have $[A,B]' \geq [A,B]$ since by induction hypothesis, $[A_i, B_i]' \geq [A_i, B_i]$ for every $i \in I$. Finally, if $\epsilon(A,B) = 0$ then $[A,B]' \geq [A,B]$ since according to the decision procedure the relation $[A, B] = 1$ implies $A \leq B$, and therefore $[A, B]' = 1$, by axiom (iii). QED

§6. EPILOGUE

Why are free lattices so potent in modelling commmunication? One explanation is that being lattices they can model long alternations of disjunctions and conjunctions, and being free, they can capture many aspects of the behavioral meaning that is associated to such alternation. They produce invariants for games in much the same way that boolean algebras model logical propositions. But they are weak in modeling communication strategies, as their boolean counterpart are weak in modeling proofs. We propose to use free bicomplete categories instead of lattices to capture more of communication strategies. A fundamental problem of computation theory is to find a reasonnable answer the following question: when are two programs equivalent? The problem should be given both a semantic and a formal solution. The concept of communication is fundamental in the practice of computing and it should be basic in the theory. This is done with the Calculus of Communicating System of Robin Milner and with Jean-Yves Girard's Linear Logic. We have shown that free lattices and free bicomplete categories [Joyal] can model communication and linear logic in a very natural way. We believe that they have a role in the emergence of semantics that will eventually inspire both the theory and the practice of computing.

Université du Quebec à Montréal

§ 7. REFERENCES

Abramsky S. and Jagadeesan R., *Games and Full Completion for Multiplicative Linear Logic*, The Jour. of Symbolic Logic, Vol. 59, No 2, 1994, pp. 543-574.

Birkhoff G., *Lattice Theory*, AMS Colloquium Publication Vol XXV.

Blass A., *A Game Semantic for Linear Logic*, Ann. Pure & App. Logic, 56 (1992), pp.183-220.

Conway J.H, *On Numbers and Games*, Academic Press, London.

Day A., *Doubling constructions in lattice theory*, Canad. J. Math. 44 (1992). 252-269.

Girard, J.-Y., *Linear Logic*, Theoretical Computer Science 50, (1987) pp. 1-102.

Felsher W., *Dialogues, Strategies, and Intuitionistic Provability*, Ann. Pure & App. Logic, 28 (1985). pp. 217-254.

Freese R. & al *Free Lattices*, AMS Math. Surveys and Monographs, Volume 42, 1995.

Grishin, V. N. *Predicate and set-theoretic calculi based on logic without contractions* , Math. USSR. Izvestja 18 ,1982, (translation), 41-59.

Joyal A., *Remarques sur la théorie des jeux à deux personnes*, La Gazette des Sciences Mathématiques du Québec- Vol I, Numéro 4, April 1977 et Rapports de recherche #253, UQAM Canada

Joyal A., *Free Bicomplete Categories*, Math. Rep., Acad. Sci. Canada - Vol. XVII, No. 5, October 1995, p. 219-225.

Joyal A., *Free Bicompletion of Enriched Categories*, Math. Rep., Acad. Sci. Canada - Vol. XVII, No. 5, October 1995, pp. 213-218.

Joyal A., *The Structure of Free Bicomplete Categories*, Rapport de recherche #252, UQAM Canada, April 1996.

Lambek J. *From Categorical Grammar to Bilinear Logic*, in Substructural Logics, Schroeder-Heister & Dosen ed., Oxford Science Publications, 1993.

Lawvere F. W., *Metric Spaces, Generalized Logic and Closed Categories*, Rend. Sem. Mat. Fis. Milano, XLIII, 1973, pp. 135-66.

Milner, A. J. R. G., *Communication and concurrency*, Prentice Hall, 1989

Nerode A., Yakhnis A., Yakhnis V., *Concurrent Programs as Strategies in Games*, in Logic from Computer Science, (V. Moschovakis, ed.), Springer-Verlag, 1992

Rosenthal K. *Quantales and their Applications*, Pitman Research Notes in Math. Vol. 234, Longman Scientific and Technical, 1990.

Whitman P. M., *Free Lattices*, Annals of Mathematics, Vol. 42, No 1, 1941. 325-330.

Whitman P. M., *Status of the Word Problems for Lattices* , Proc. Sympos. Pure Math. Vol. II, pp. 17-21, AMS, 1961.

Zermelo E., *Über eine Anwendung der Mengenlehre auf die Theorie des Schachspiels.* Proc. 5th Int. Cong. Math., Cambridge, Camb. Univ. Press, 1913. Vol II, pp. 501-504.

JAN KRAJÍČEK*

ON METHODS FOR PROVING LOWER BOUNDS IN PROPOSITIONAL LOGIC

ABSTRACT.

This paper is based on my lecture [26]. It examines the problem of proving non-trivial lower bounds for the length of proofs in propositional logic from the perspective of methods available rather than surveying known partial results (i.e., lower bounds for weaker proof systems). We discuss neither motivations for proving lower bounds for propositional logic nor relations to other problems in logic or complexity theory. The reader is referred to [20] for the background information (as well as for all details missing in this paper). The paper is aimed at curious non-specialists. The style of our exposition is accordingly informal at places and we do not burden the text (especially in the introduction) with exhausting references not directly related to our main objective. The reader starving for details can find them, together with all original references, in [20] (see also expository articles [25, 32]).

INTRODUCTION

The language of propositional calculus contains constants 0 and 1 (false and true), connectives $\neg$, $\vee$ and $\wedge$ with their usual meaning, and atoms $p_0, p_1, \ldots$.

A propositional calculus is given by a finite number of axiom schemes (like $A \vee \neg A$) and schematic inference rules (like modus ponens: from A and $\neg A \vee B$ infer B). The only requirement is that the axioms and rules are sound and implicationally complete (i.e., if B is a tautological consequence of $A_1, \ldots A_k$ then B is provable from $A_1, \ldots, A_k$). Calculi of this type are called *Frege systems* after [9]. Fix one of them and call it F.

The complexity of a proof (tacitly a proof in F) is measured by two important functions: the number of proof steps and the size, i.e., the total number of symbols in the proof. These functions yield two measures of complexity of a tautology τ, namely $k_F(\tau)$ and $s_F(\tau)$, the minimal number of steps in a proof of τ and the minimal size of a proof of τ respectively. Important observation is that if we augment F by the substitution rule (allowing to infer any substitution instance of an already proved formula in one step) then these two measures become essentially equal. The same effect has the extension rule (allowing to abbreviate formulas in a proof). None of these rules is schematic in the same sense as is modus ponens.

Now we are in a position to state the main problem.[1]

*Partially supported by the *US - Czechoslovak Science and Technology Program* grant # 93025, and by the grant #119107 of the *AV ČR*.

[1]Experts among readers should note now, and keep in mind throughout the paper, that we are concerned with the measure k_F rather than with s_F.

M.L. Dalla Chiara et al. (eds.), Logic and Scientific Methods, 69–83.

Problem: Show that there is no polynomial $p(x)$ such that for a Frege system F it holds $k_F(\tau) \leq p(n)$, for every n and every tautology τ of size at most n.

The set of tautologies $TAUT$ is $coNP$-complete and a proof of τ with k steps can be coded by a string of size $O(k + |\tau|)$ (this is non-trivial, cf. [20, Lemmas 4.5.3 and 4.5.7]). Hence a Frege system can be thought of as a non-deterministic acceptor of $TAUT$ and the existence of a polynomial bounding the number of steps in proofs would imply that $TAUT \in NP$. Thus, unless the generally accepted conjecture that $NP \neq coNP$ fails, indeed no such polynomial $p(x)$ exists.

The problem is quite robust as it is stated. It does not depend on a particular language (any complete language is good) or on particular axioms schemes and rules. We may add even the substitution and the extension rules. It also does not depend on a particular format of proofs we require (tree-like or sequence-like). In fact, it also does not matter whether we confine to Frege systems or whether we use Gentzen's sequent calculus (with cut) or natural deduction, or extended resolution. The phrase *does not matter* means that proofs in one system can be translated into proofs in another system with just a polynomial increase in the number of steps.

No non-trivial lower bounds for $k_F(\tau)$ are known at present. For some provably less efficient systems superpolynomial lower bounds are known. These systems include, in particular, resolution, constant-depth subsystems of a Frege system and their extensions by some counting principles, and cutting planes proof system.

The methods used in the proofs of these lower bounds include counting arguments, an adaptation of the method of random restrictions from Boolean complexity (that can be cast also as forcing in non-standard models of bounded arithmetic), and effective interpolation theorems. We shall discuss these methods later.

Finally we should mention the tautology that was used in one or other form in all lower bounds mentioned earlier. It is the *pigeonhole principle* PHP_n^m, $m > n$,

$$\bigwedge_{i<m} \bigvee_{j<n} p_{ij} \rightarrow \bigvee_{i_1<i_2<m} \bigvee_{j<n} (p_{i_1 j} \wedge p_{i_2 j})$$

formalizing that no map from $\{0, \ldots, m-1\}$ to $\{0, \ldots, n-1\}$ can be injective. In retrospect we can see that it was a crucial step taken by [9] to put this tautology forward as a candidate for being hard (i.e., without short proofs) for usual proof systems. One particular difficulty in approaching the problem stated earlier is that we lack such good candidates for F.

In next three sections we describe and compare three universal frameworks for lower bounds to k_F. In section 4 we consider some related mathematical structures. Section 5 discusses limitations of particular methods used so far for subsystems of F. In section 6 we discuss new candidates of hard tautologies.

1. PARTIAL BOOLEAN VALUATIONS

A partial Boolean algebra $\mathcal{B}$ is a set together with distinguished constants $0_\mathcal{B}$, $1_\mathcal{B}$ and partial functions $\neg, \wedge, \vee$ such that any instance $t = s$ of any identity of some fixed equational axiomatization (e.g., [14]) of the theory of Boolean algebras, in which both terms t, s are defined, is actually valid in $\mathcal{B}$. The following notion was defined in [23].

Let Γ be a finite set of formulas closed under subformulas. A *partial Boolean valuation* of Γ is a map

$$\nu : \Gamma \to \mathcal{B}$$

of Γ into a partial Boolean algebra $\mathcal{B}$ such that

1. $\nu(0) = 0_\mathcal{B}$ and $\nu(1) = 1_\mathcal{B}$, whenever 0 and 1 are in Γ
2. $\neg\nu(A)$ is defined and equal to $\nu(\neg A)$, whenever $A, \neg A$ are in Γ
3. $\nu(A) \wedge \nu(B)$ is defined and equal to $\nu(A \wedge B)$, if all $A, B, A \wedge B$ are in Γ. Analogously for $\vee$.

Theorem 1.1 ([23]) *Let τ be a tautology and let Γ range over sets of formulas containing τ and closed under subformulas.*

(a) Assume that every set Γ with $\leq k$ formulas admits a partial Boolean valuation $\nu : \Gamma \to \mathcal{B}$ in which $\nu(\tau) \neq 1_\mathcal{B}$. Then:

$$k_F(\tau) > \epsilon \cdot k$$

(b) Assume that $k_F(\tau) > k^c$. Then every set Γ of $\leq k$ formulas admits a partial Boolean valuation ν in which $\nu(\tau) \neq 1_\mathcal{B}$.

Constants ϵ, c depend only on the particular formulation of the Frege system.

Part (a) says that partial Boolean valuations can be used to prove a lower bound to $k_F(\tau)$. Part (b) complements this by showing that at least a lower bound of the form $k_F(\tau)^{\Omega(1)}$ can be proved, in principle, by a construction of suitable partial Boolean valuations.

Let us illustrate the method on an example. Consider the tautology $(\neg)^{(2k)}(1)$ $2k$ negation signs in front of 1. It was proved in [21] that $k_F((\neg)^{(2k)}(1)) = \Omega(k)$. We show this by constructing a suitable partial Boolean valuation for every set Γ of at most $2k$ formulas and containing $(\neg)^{(2k)}(1)$. As $\Gamma \leq 2k$, Γ does not contain some formula $(\neg)^{(i)}(1)$, for $i < 2k$. Define map ν_i in two stages:

1. For $j < i$ give to $(\neg)^{(j)}(1)$ the value $1_\mathcal{B}$ if j is even and the value $0_\mathcal{B}$ if j is odd.
   For $j \geq i$ give to $(\neg)^{(j)}(1)$ the value $0_\mathcal{B}$ if j is even and the value $1_\mathcal{B}$ if j is odd.

2. For any formula $A(\overline{p})$ evaluate first atoms $\overline{p}$ by $0_{\mathcal{B}}$, maximal subformulas of the form $(\neg)^{(j)}(1)$ according to 1., and then evaluate A using ordinary truth-tables of $\neg, \vee, \wedge$.

It is easy to see that this defines a partial Boolean valuation of Γ in $\{0_{\mathcal{B}}, 1_{\mathcal{B}}\}$ in which $\nu_i((\neg)^{(2k)}(1)) = 0_{\mathcal{B}}$. Hence a lower bound $k_F((\neg)^{(2k)}(1)) = \Omega(k)$ follows by Theorem 1.1.

Admittedly the tautology $(\neg)^{(2k)}(1)$ is rather primitive. Sadly enough, this is essentially the best lower bound for any $k_F(\tau)$ known at present.

2. LOCAL BOOLEAN VALUATIONS

Another form of a Boolean-type valuation was used in [28] and formalized into a universal framework in [8, 32].

Let Γ be a set of formulas. A *local Boolean valuation* is a map η assigning to every subset $\Delta \subseteq \Gamma$ of at most c formulas an ordinary total Boolean algebra $\mathcal{B}_\Delta$ and a map $\eta_\Delta : \Delta \to \mathcal{B}_\Delta$ preserving all connectives for formulas in Δ satisfying the following compatibility condition. Whenever $\Delta_1 \subseteq \Delta_2 \subseteq \Gamma$ are two sets of size at most c then there is a homomorphism ρ from $\mathcal{B}_{\Delta_1}$ into $\mathcal{B}_{\Delta_2}$ such that $\rho(\eta_{\Delta_1}(A)) = \eta_{\Delta_2}(A)$ holds for all $A \in \Delta_1$. The constant c depends only on the particular Frege system F.

It was proved in [20, Chpt.13] that any set Γ with a local Boolean valuation in which $\eta_{\{\tau\}}(\tau) \neq 1_{\mathcal{B}_{\{\tau\}}}$ admits a partial Boolean valuation $\nu : \Gamma \to \mathcal{B}$ in which $\nu(\tau) \neq 1_{\mathcal{B}}$ (assuming that c is large enough). This shows that if we can find a local Boolean valuation with this property for any Γ of size $\leq k$, then $k_F(\tau) = \Omega(k)$.

In the other direction consider the following construction. From Γ form a new set $\Gamma' \supseteq \Gamma$ as follows. Let $\mathcal{B}_c$ be the finite total Boolean algebra (of size 2^{2^c}) freely generated by $b_1, \ldots, b_c$, let $t_1, t_2, \ldots$ be 2^{2^c} terms (formed from $\overline{b}$) expressing all elements of $\mathcal{B}_c$, and let $s_1, s_2, \ldots$ be all terms occurring in some fixed derivations (from the identities defining Boolean algebras) of all identities of the form $\neg t_i = t_j$, $t_i \vee t_j = t_k$ and $t_i \wedge t_j = t_k$ valid in $\mathcal{B}_c$.

Given set Δ of at most c formulas $A_1, \ldots, A_d$, $d \leq c$, form the set Δ^+ of all formulas $s_1(b_1/A_1, \ldots, b_d/A_d, b_{d+1}/1, \ldots, b_c/1), \ldots$. Define $\Gamma' \supseteq \Gamma$ to be the minimal set closed under subformulas and containing $\bigcup_{\Delta \subseteq \Gamma, |\Delta| \leq c} \Delta^+$. Note that $|\Gamma'| = |\Gamma|^{O(1)}$.

Obviously, a partial Boolean valuation ν of Γ' yields a local Boolean valuation of Γ as the images of Δ^+ in ν are total Boolean algebras. Hence part (b) of Theorem 1.1 implies part (b) of the next theorem.

Theorem 2.1 *Let τ be a tautology and let Γ range over sets of formulas containing τ and closed under subformulas.*

(a) [8, 32] Assume that every set Γ with $\leq k$ formulas admits a local Boolean valuation η in which $\eta_{\{\tau\}}(\tau) \neq 1_{\mathcal{B}_{\{\tau\}}}$. Then $k_F(\tau) = \Omega(k)$.

(b) On the other hand, any set Γ, $\Gamma \leq k_F(\tau)^\epsilon$, *admits a local Boolean valuation* η *in which* $\eta_{\{\tau\}}(\tau) \neq 1_{\mathcal{B}_{\{\tau\}}}$.

Constant ϵ *depends only on the particular formulation of the Frege system.*

3. A PROVER - ADVERSARY GAME

The following game played by two players, *Prover* and *Adversary*, was defined in [8, 32]. In every round Prover asks about the truth-value of a formula and Adversary replies with a value. When aiming at proving τ, Prover asks in the first round about the truth-value of τ and Adversary is obliged to reply with 0. The game ends when the answers of Adversary contain an elementary contradiction, i.e., 0 gets value 1 or vice versa, or formulas $A, \neg A$ get the same value, or formulas $A, B, A \wedge B$ (resp. $A \vee B$) get values contradicting the truth-tables of $\wedge$ (resp. of $\vee$). The following theorem was proved in [8, 32] by induction on t and $k_F(\tau)$.

Theorem 3.1 ([8, 32]) *Let* t *be the minimal number such that there is a Prover forcing the end of the game on* τ *in* t *rounds against any Adversary. Then* t *is proportional to* $\log_2(k_F(\tau))$.

The theorem can be also seen as a corollary (in fact, an equivalent version) of the fact that every proof can be rewritten into a form of a balanced tree with only a polynomial increase in the number of steps.

Let us illustrate the method on the same example as in section 1 (following [8]). Adversary may use partial Boolean valuations ν_i defined in section 1 as follows. In every round s, being asked about a formula A, he takes the collection X_s of all ν_i such that all his answers in rounds $1, \ldots, s-1$ were consistent with ν_i. Then he answers 1 if at least half of valuations from X_s give to A value $1_{\mathcal{B}}$, otherwise he answers 0. Clearly $|X_s| \geq \frac{2k}{2^{s-1}}$ and Prover cannot force an elementary contradiction if $|X_s| \geq 2$. Hence Adversary using this strategy survives at least $\log_2(2k)$ rounds with any prover.

This construction of a strategy for Adversary from partial Boolean valuations can be generalized to any τ. Fix t such that $k_F(\tau) > 2^{O(t)}$, the constant in the exponent to be determined later. For any prover P take the binary tree of depth $t-1$ corresponding to all possible plays P can have against some adversary. Put Γ_P to be the smallest set closed under subformulas and containing all $2^t - 1$ conjunctions of answers of an adversary (determining a path in the tree) in first $i = 1, \ldots, t$ rounds. In particular, $\neg\tau \in \Gamma_P$. A crucial technical fact is that if there would be a prover P forcing the end of a game in t rounds against any adversary then, in fact, there would be such prover yielding the set Γ_P of size $O(2^t + |\tau|)$, and such that it is a priori sufficient to restrict to finitely many of such provers. This follows from Theorem 3.1 and [20, Lemmas 4.4.4 and 4.4.6]. Consider only such provers P.

Assume that $k_F(\tau) > |\Gamma_P|^c$, c the constant from Theorem 1.1. Then there is a partial Boolean valuation $\nu_P : \Gamma_P \to \mathcal{B}_P$ in which $\nu_P(\tau) \neq 1_{\mathcal{B}_P}$, i.e.,

$\nu_P(\neg\tau) \neq 0_{\mathcal{B}_P}$. Put X_1 to be the set of all these valuations. At round s Adversary has the set X_s of those ν_P for which $\nu_P(\psi_1 \wedge \ldots \wedge \psi_{s-1}) \neq 0_{\mathcal{B}_P}$, where $\psi_1(= \neg\tau), \ldots, \psi_{s-1}$ are his answers in the first $s-1$ rounds. Enquired about the truth-value of ψ he computes the ν_P-values of the conjunctions $\psi_1 \wedge \ldots \wedge \psi_{s-1} \wedge \psi$ and $\psi_1 \wedge \ldots \wedge \psi_{s-1} \wedge \neg\psi$ (both are in Γ_P) and answers in such a way that the new conjunction has still the ν_P-value different from $O_{\mathcal{B}_P}$.

4. RELATED STRUCTURES

We shall digress at this point and consider four different mathematical structures related to the methods described earlier. In particular, we shall discuss non-standard models of bounded arithmetic, the approximation method from Boolean complexity, logical structures arising in quantum mechanics, and partial algebraic structures.

4.1. *Non-standard models of bounded arithmetic*

Let N be a countable non-standard model of a sufficiently strong fragment of true arithmetic, say of Peano arithmetic. Let $\tau \in N$ be a non-standard propositional formula of size $n = |\tau|$, built from non-standardly many atoms $p_0, \ldots, p_m$. Assume that N thinks that τ is a tautology and that $k_F(\tau) > n^\ell$, all standard ℓ. By compactness argument this is a consistent situation if k_F cannot be bounded by a polynomial. Consider the following construction.

Let $M \subseteq_e N$ be the initial part of N consisting of numbers with at most n^ℓ bits, $\ell < \omega$. Let $Flas^M$ be all elements of M coding in N and M a formula built from atoms $p_0, \ldots, p_m$. Assume that we find a map $\nu : Flas^M \to \mathcal{B}$, where $\mathcal{B}$ is a total Boolean algebra, and such that $\nu(\tau) \neq 1_{\mathcal{B}}$.

Take an ultrafilter $\mathcal{U} \subseteq \mathcal{B}$ such that $\nu(\tau) \notin \mathcal{U}$ and define the map

$$\eta : A \in Flas^M \to \nu(\tau)/\mathcal{U} \in \{0, 1\}$$

In particular, $\eta(\tau) = 0$ and, in a sense, we may think of the tuple $\eta(p_0), \ldots, \eta(p_m)$ as of a truth-assignment satisfying $\neg\tau$. A problem with this interpretation is that we cannot define the assignment neither in N (as we do not assume $\mathcal{U}$ to be definable in N) nor in the standard model (as both τ and the assignment are nonstandard). This suggest to consider the set of all tuples of bits $\langle \eta(A_1), \ldots, \eta(A_t) \rangle$ formed from all tuples $A_1, \ldots, A_t$ of elements of $Flas^M$ that are coded in M. These tuples of bits can be identified via dyadic encoding with numbers. The set of these numbers M^* is, in fact, a structure extending M and with a suitable choice of $\mathcal{U}$ it is a model of bounded arithmetic theory S^1_2, see [23]. Moreover, in M^* τ is not a tautology anymore.

On the other hand, having an extension M' of M in which $\neg\tau$ is satisfied by a truth-assignment $\overline{a} \in \{0,1\}^{m+1}$ allows to evaluate in M' any $A \in Flas^M$ under $\overline{a}$, obtaining a Boolean valuation in which τ gets value 0.

We note that the valuation ν of $Flas^M$ is really only a partial Boolean valuation from the point of view of N, as $Flas^M$ is not closed even under all conjunctions (in N).

In this way Boolean valuations of sets of propositional formulas correspond to extensions of models of bounded arithmetic. The particular theory S^1_2 appears as we consider Frege systems; other proof system correspond to other theories of bounded arithmetic.

4.2. *The approximation method*

[36] formalized [34, 35] into a universal framework for proving lower bounds on the size of general circuits. A particular presentation is as follows, cf. [36, 16].

Let $f : \{0,1\}^n \to \{0,1\}$ be a Boolean function about which we want to prove that it cannot be computed by a small circuit. Let C be a Boolean circuit (over the base $0, 1, \neg, \vee, \wedge$) with inputs $x_1, \ldots, x_n$. Every node (i.e., a subcircuit) y of C determines a function C_y of $x_1, \ldots, x_n$ computed by the subcircuit y. Take the set $X := f^{(-1)}(0)$ and assign to every node y a subset $\|y\|$ of X of those inputs accepted by C_y. We may think of the map $\|\ldots\|$ as a map from C into the Boolean algebra $\exp(X)$ of the subsets of X preserving the connectives.

We assign to nodes of C a value in $\{0,1\}$ as follows. Pick $U \subseteq \exp(X)$ and assign to y the value $\eta_U(y) := 1$ if $\|y\| \in U$ and $\eta_U(y) := 0$ otherwise. Assume that C computes f. Then $\eta_U(C) = 0$ iff $\emptyset \notin U$. U defines a vector $w_U \in \{0,1\}^n$ whose i^{th} coordinate is $\eta_U(x_i)$. Thus if we find U not containing $\emptyset$ such that $f(w_U) = 1$ and η_U preserves the connectives in C we, in fact, demonstrate that $C(w_U) \neq f(w_U)$ and so C does not compute f. This method is universal in the sense that such U exists whenever the size of C is at most k^ϵ, k being the circuit-size of f (and ϵ an independent constant), cf. [36, 16]).

Note that the map η_U is only a partial Boolean valuation unless U is an ultrafilter. Thus partial Boolean valuations are inevitable as on the finite Boolean algebra $\exp(X)$ the only ultrafilters are principal and thus always $w_U \in X$.

For the approximation method in the particular form as in [34, 35, 38] a partial Boolean valuation of the subcircuits of C can be described directly. Let y^* be the approximating function assigned to y and let E be the set of all inputs that are erroneously computed by y^* at some y. Then define the value of y to be the collection of all subcircuits z such that $y^* = z^*$ holds outside the set E. By the definition of E this map preserves all connectives of C. The point then is to show that C^* and f do not agree outside E.

Now assume that f is the characteristic function of an NP-predicate. Form a formula τ of the form

$$\alpha(x_1, \ldots, x_n, z_1, \ldots, z_{n^{O(1)}}) \to \beta(x_1, \ldots, x_n, y_1, \ldots, y_m)$$

where α is a formula formalizing that $f(x_1, \ldots, x_n) = 1$ as witnessed by $\overline{z}$, and β formalizes that circuit C defined by $\overline{y}$ accepts $\overline{x}$. Hence τ is a tautology iff C accepts all inputs accepted by f. Now assume also that we have a partial Boolean valuation ν of the set of subformulas of τ in which $\nu(\tau) = 0_{\mathcal{B}}$. Hence

$$\nu(\alpha(x_1, \ldots, x_n, z_1, \ldots, z_{n^{O(1)}})) = \alpha(\nu(x_1), \ldots, \nu(x_n), \nu(z_1), \ldots, \nu(z_{n^{O(1)}})) = 1_{\mathcal{B}}$$

and

$$\nu(\beta(x_1, \ldots, x_n, y_1, \ldots, y_m)) = \beta(\nu(x_1), \ldots, \nu(x_n), \nu(y_1), \ldots, \nu(y_m)) = 0_{\mathcal{B}}$$

so we may think of the tuple $\nu(x_1), \ldots, \nu(x_n)$ as an input from $\mathcal{B}^n$ accepted by f but not by C. In fact, the model-extension construction from the previous section can made the intuition precise (and, in particular, yield a new 0-1 input in M^*).

4.3. *Logical structures arising in quantum mechanics*

Important object arising in quantum mechanics is the lattice of closed linear subspaces of a Hilbert space. This object is called also a lattice of (quantum mechanical) propositions, see [5, 15]. It is not a Boolean algebra.

The sublattices of the lattice of propositions that are Boolean algebras play an important role in defining compatible propositions; propositions are compatible if the sublattice they generate is a Boolean algebra. The sublattices that are complete Boolean algebras correspond to reference frames of observations, see [42]. Another way of approaching a logic in which not all propositions are compatible is by formalizing the informal notion of logic with partial propositions, see [39, 18]. These were studied in a connection with the hidden variables problem, see [17, 19].

A third logical object arising in quantum mechanics are formal dialogs, used as a form of semantic for quantum logic (see [30, 40]).

These three structures, the lattice of propositions, reference frames and formal dialogs, appear to be analogous to partial Boolean valuations, local Boolean valuations and the Prover-Adversary game. For the latter structures sections 1-3 explain why they occur simultaneously, while for the former structures this seems to be a matter of discussion, see [3].

While we are at pointing informal relations we may as well add one more. Note that the construction from section 4.1 is somewhat analogous to a construction of a Quantum set theory, see [41]. This is a theory constructed as

Boolean-valued set theory with formulas taking their truth-values in the lattice of propositions defined earlier rather than in a complete Boolean algebra.

4.4. *Partial algebraic structures*

A partial algebraic structure is a set together with a collection of partial operations. A variety of algebraic structures of a given signature axiomatized by identities T (call the variety also T) gives rise to an associated class T^p of partial algebras in the same signature. Namely, a partial algebra is in T^p iff instances $t = s$ of all identities of T that are actually defined in the partial algebra are also valid in it. (Occasionally few more technical conditions are added, see [10, 11].) An interesting problem associated with T^p is the embeddability problem: given a finite partial structure from T^p decide whether it can be embedded into a total (possible infinite) structure from T. Important theorem of [10] says that the embeddability problem is decidable iff the word problem associated to T is decidable. It follows that if all partial algebras from T^p can be embedded into total ones then the word problem for T is decidable. It is instructive to compare the following proof-sketch of this corollary with the proof of Theorem 1.1 in [23]. Let $\overline{a}$ be generators of an algebra $\mathcal{A}$ from T, $r_i(\overline{a}) = s_i(\overline{a})$ its finite presentation and $w_1(\overline{a}), w_2(\overline{a})$ two words. For every subterm t occurring in one of r_i, s_i, w_1, w_2 introduce new unknown b_t and rewrite the identities $r_i = s_i$ into the set of indentities the form $b_i = b_j$ and $f(b_{i_1}, \ldots, b_{i_k}) = b_j$. From these identities derive, using T, all possible identities of the same form, possible identifying new pairs $b_i = b_j$. This process yields a partial algebra $\mathcal{A}^p$ from T^p. If $b_{w_1} = b_{w_2}$ in $\mathcal{A}^p$ then $w_1 = w_2$ in $\mathcal{A}$. Otherwise $\mathcal{A}^p$ can be embedded into a total algebra $\mathcal{A}'$ and hence $w_1 \neq w_2$ in its subalgebra generated by the image of $\mathcal{A}^p$. Hence $w_1 \neq w_2$ in $\mathcal{A}$ as well.

Partial Boolean algebras arising in section 1 can be never embedded in a total one. This is because every tautology must get value 1 in every total Boolean algebra (otherwise a suitable ultrafilter would define a truth-assignment not satisfying the tautology). It is an interesting problem whether we can get any quantitative information about the structure of partial Boolean valuations. Specifically, assume that $\nu : \Gamma \to \mathcal{B}$ is a partial Boolean valuation in which a tautology gets a value different from $1_\mathcal{B}$, and let $|\mathcal{B}| = \ell$. The tables of operations $\neg, \vee, \wedge$ of $\mathcal{B}$ have $2\ell^2 + \ell$ potential entries.

Can we show that, say, at most 99% of entries of the tables of a partial Boolean algebra are filled? This would hold if the fact that more than 99% of the entries are filled would imply that $\mathcal{B}$ is embeddable in a total Boolean algebra (something similar holds for partial groups, see [29]).

5. LIMITATIONS OF PARTICULAR METHODS

In this section we want to examine why methods used for lower bounds for subsystems of a Frege system mentioned in the introduction do not work for F as well. First we should frankly admit that the word *methods* is not appropriate. This is because we are able to prove lower bounds only for very few tautologies and we are definitely not able to determine the minimal number of steps needed in a proof of an arbitrary tautology.

Secondly there seems to be a popular opinion that a lower bound for a proof system is essentially a corollary of a Boolean complexity lower bound for the class of formulas occurring in proofs in the particular proof system. This is, unfortunately or fortunately, not the case. (A sceptical reader may try to adapt [35, 38] to a lower bound for tautologies $Count_{q,1}$ - defined below - in a natural proof system working with constant-depth formulas with counting-mod-p gates, p, q different primes, see [20, 4] for a definition of such a system.) One reason perhaps is that formulas or circuits represent deterministic computations while proof systems are non-deterministic.

A counting argument used originally by [12] to prove a lower bound for resolution works as follows. There is a large set X of truth assignments (to atoms of PHP_n^{n+1} - see the introduction) with the following property. Any refutation of the clauses representing $\neg PHP_n^{n+1}$ determines a map from X to the clauses occurring in the refutation such that the preimage of any clause is small. As X is large, the number of clauses must be large too.

There is no a priori reason why similar approach cannot work for F. Indeed, [13] succeeded in reproving an exponential lower bounds for monotone circuits ([34, 2]) in this way. However, the structure of F-proofs appears to be much more complex than that of resolution refutations (the difference between the two proof systems may be also documented by a big difference between bounded arithmetic theories corresponding to them, see [20]) to allow a direct construction of a suitable map.

The method of random restrictions in Boolean complexity works as follows. Given a constant-depth and not too big circuit one shows (by a probabilistic argument) that there is an assignment of 0 and 1 to some inputs simplifying the circuit substantially. This means that the Boolean function computed by the restricted circuit on the unassigned inputs can be, in fact, computed by a circuit of a very simple form. Then one shows that the function the original circuit supposedly computed does not have this property.

In lower bounds for constant-depth subsystems of a Frege system a related approach is used (e.g., [1, 22, 28, 31]). First partially evaluate atoms so that all formulas in a fixed and not too big proof of a tautology (e.g., PHP_n^{n+1}) get simplified. Then argue that there cannot be a proof of the restricted tautology involving only simple formulas. As the restrictions must not simplify the tautology too much they must be of a special form. In the case of the formula PHP_n^{n+1} the restrictions correspond to partial injective maps from $\{0,\ldots,n\}$ into $\{0,\ldots,n-1\}$. The second step is then achieved in [28] by a form of a local Boolean valuation (of the set of simplified formulas occurring in the restricted proof), in [22, 31] it is an argument in the spirit of the Prover-Adversary game.

A principal reason why this does not work for F is that a general formula or a circuit do not simplify much after a restriction (but see [20, Chpt.13]). We may at least try to learn a lesson of how partial Boolean valuations implicitly occurring in proofs of lower bounds for PHP_n^{n+1} are constructed and used, and think about how to extend such an approach to F. The reader may find this in detail in [20, Chpt.12] and here we add only few general observations (we shall assume a familiarity with [28, 23] for the next paragraph).

Important objects in [28] are complete systems. A disjunction of maps (map-conjuctions precisely) from a complete system is shortly provably equivalent to 1, assuming $\neg PHP_n^{n+1}$ as an axiom. *Shortly provably equivalent* means that we take a set Γ of not too many formulas and any proof may use only formulas from Γ, cf. [23]. Thus, assuming $\neg PHP_n^{n+1}$, complete systems are just new disjunctive normal forms of 1. In terms of a partial Boolean valuation $\nu : \Gamma \to \mathcal{B}$ this means that ν-values of maps from a complete system form a partition of unity in $\mathcal{B}$. To restrict the proof means to augment $\neg PHP_n^{n+1}$ by a new axiom, a map-conjunction. Under this new axiom all formulas in Γ have (shortly provably equivalent) new disjunctive normal forms. In proving this a crucial step is an application of a switching lemma. In the language of ν this corresponds to refining a subset of $\mathcal{B}$ into an antichain.

Let us now turn to the method of effective interpolation. It was first explicitly proposed in [22] and put to work in [37, 24], implicitly in [6], and in [33]. (This development is not much covered in [20] so we add references.)

We say that a proof system admits effective interpolation if every implication

$$A(\overline{p},\overline{q}) \to B(\overline{p},\overline{r})$$

provable in k steps has an interpolant $I(\overline{p})$ whose circuit-size is $k^{O(1)}$. An effective monotone interpolation asserts the same for implications without negative occurrences of p_i in A and B, and requires $I(\overline{p})$ to define a Boolean function computable by a monotone circuit of size $k^{O(1)}$. These effective versions of interpolation are valid for resolution, cut-free sequent calculus, a version of the depth 2 subsystem of the sequent calculus, linear equational logic and cutting planes (see the references above).

The idea of using effective interpolation for lower bounds is as follows. Assume that a proof system P admits effective (monotone) interpolation and that $A_n(\overline{p},\overline{q}) \to B_n(\overline{p},\overline{r})$ is a family of implications with atoms $\overline{p} = p_1,\ldots,p_n$ and of size $n^{O(1)}$ that do not have (monotone) interpolants of polynomial size. Then the implications necessarily do not have P-proofs with $n^{O(1)}$ steps.

It is an important open problem to decide which proof systems do admit effective interpolation. There are some limitations to effective monotone interpolation (see [24]) showing that the known positive results are almost best possible. However, for the non-monotone version similar statement is open. [27] show that unless a standard cryptographical conjecture about the security of the RSA encryption scheme fails, Frege systems do not admit effective interpolation.

For constant-depth subsystems we do not have similar result but we may note that a theorem of [4] implies that a version of effective interpolation valid for proof systems mentioned earlier does not hold for extensions of constant-depth systems by modular counting principles. This version implies, in particular, that if formulas A and B have no common atoms and $A \vee B$ is provable in k steps then one of A, B is provable in $k^{O(1)}$ steps.

Consider formulas $\neg Count^n_{q,i}(\overline{r})$ formalizing that the atoms $\overline{r}$ define a q-partition of n, where $n \equiv i \pmod q$. In particular, there is one atom r_e for each q-element subset e of $\{0, \ldots, n-1\}$ and the formula formalizes the fact that $\{e \mid r_e = 1\}$ is not a total partition of $\{0, \ldots, n-1\}$. Define:

$$A := Count^{6a+1}_{2,1}(\overline{r}) \quad \text{and} \quad B := Count^{6b+1}_{3,1}(\overline{s})$$

From $\overline{r}$ and $\overline{s}$ we can define constant-depth formulas $\overline{C}$ such that $\neg A \wedge \neg B$ imply by a short constant-depth proof

$$\neg Count^{(6a+1)(6b+1)}_{6,1}(\overline{C})$$

(think of the rectangle $(6a+1) \times (6b+1)$ partitioned into rectangles of size 2×3 defined by $\overline{r}, \overline{s}$). Hence a constant-depth Frege system augmented by instances of $Count_{6,1}$ proves $A \vee B$ by a poly-size proof. However, by the main theorem (Thm 1.2) of [4] none of $Count_{2,1}$ or $Count_{3,1}$ admits poly-size proofs from $Count_{6,1}$.

It is not known whether $min(k_F(A), k_F(B)) = k_F(A \vee B)^{O(1)}$ for A, B without common atoms. The reader may contemplate difficulties encountered while trying to prove such an inequality using one of Theorems 1.1, 2.1 or 3.1.

6. HARD TAUTOLOGIES ?

It is apparently rather difficult to come up with tautologies τ that would be reasonable candidates to have large $k_F(\tau)$. A heuristic reason for that is that Frege systems can count and hence combinatorial principles derivable by counting arguments (like the pigeonhole principle) cannot be hard for F (even when measuring the complexity of proofs by $s_F(\tau)$ rather than by $k_F(\tau)$, see [7]).

In [27] the following tautologies τ_p were defined. It is not known whether they have Frege proofs with polynomially many steps.

Fix p a prime with n bits. The tautology τ_p is formed from atoms $x_1, \ldots, x_n$, $y_1, \ldots, y_n$ and $z_1, \ldots, z_{n^{O(1)}}$. Atoms $\overline{x}, \overline{y}$ represent bits of two integers and $\overline{z}$ stand for bits of the table of a computation of the product $\overline{x} \cdot \overline{y}$. The formula τ_p asserts that if both $\overline{x}, \overline{y}$ are bigger than 1 and $\overline{z}$ is indeed a computation of $\overline{x} \cdot \overline{y}$ then the output of the computation (as encoded in $\overline{z}$) is not equal to p.

Currently no lower bounds for these tautologies are known in any proof system. On the other hand, we also do not know an infinite set of primes p for which there would be Frege proofs with polynomially many steps. It would be very interesting to prove a superpolynomial lower bound to $k_F(\tau_p)$ even under some unproved but plausible conjecture from computational complexity theory.

Mathematical Institute of the Academy of Sciences
Prague

ACKNOWLEDGEMENTS

I thank P. Pudlák and A. A. Razborov for comments on the draft of this paper.

REFERENCES

1. M. Ajtai: The complexity of the pigeonhole principle, in: Proc. IEEE 29th Annual *Symp. on Foundation of Computer Science*, (1988), pp. 346-355.
2. A. E. Andreev : On a method for obtaining lower bounds for the complexity of individual monotone functions (in Russian), *Doklady AN SSSR*, **282(5)**, (1985), pp.1033-1037.
3. *Current issues in quantum logic*, Eds. E.G.Beltrametti and B.van Fraassen, Plenum Press, New York-London, Ettore Majorana International Science Series, Vol.8, Proc. of the Workshop on Quantum Logic held Dec.2-9, 1979 in Erice, Sicily. (1981).
4. P. Beame, R. Impagliazzo, J. Krajíček, T. Pitassi and P.Pudlák : Lower bounds on Hilbert's Nullstellensatz and propositional proofs, *Proceedings of the London Mathematical Society*, to appear.
5. G. Birkhoff and J. von Neumann: The logic of quantum mechanics, *Annals of Mathematics*, **37**, (1936), pp.823-843.
   Reprinted in *The logico-algebraic approach to quantum mechanics*, Vol.I., Ed. C.A.Hooker, D.Reidel Publ.Co., Dordrecht-Boston, (1975).
6. M. L. Bonet, T. Pitassi, and R. Raz : Lower bounds for cutting planes proofs with small coefficients, preprint, (1994).
7. S. Buss : The propositional pigeonhole principle has polynomial size Frege proofs, *J. Symbolic Logic* **52**, (1987), pp. 916-927.
8. S. Buss and P. Pudlák: How to lie without being (easily) convicted and the length of proofs in propositional calculus, in: Proceedings of the meeting *Computer Science Logic*, Kazimierz 1994, Eds. L.Pacholski and J.Tiuryn, LNCS **933**, Springer-Verlag, (1995), pp.151-162.
9. S.A. Cook and A.R. Reckhow: The relative efficiency of propositional proof systems, *J. Symbolic Logic*, **44(1)**, (1979), pp. 36-50.
10. T. Evans: Word problems, *Bulletin of the AMS*, **84(5)**, (1978), pp.789-802.
11. G. Gratzer : *Universal algebra* (2nd ed.), Springer - Verlag, (1979), 581 p.

12. A. Haken: The intractability of resolution, *Theoretical Computer Science*, **39**, (1985), pp. 297-308.
13. ——— : Counting bottlenecks to show monotone $P \neq NP$, in: Proc. IEEE 36th Annual *Symp. of Foundations of Computer Science*, (1995), pp. 36–40.
14. E.V. Huntington: Sets of independent postulates for the algebra of logic, *Transaction of the AMS*, **5**, (1904), pp.288-309.
15. J.M. Jauch: *Foundations of quantum mechanics*, Addison-Wesley, Reading-Menlo Park-London-Don Mills, Series in Advanced Physics, (1968).
16. M. Karchmer : On proving lower bounds for circuit size, in : *Proc. Structure in Complexity*, 8th Annual Conference, *IEEE* Computer Science Press, (1993), pp. 112-119.
17. S. Kochen and E. Specker: Logical structures arising in quantum theory, in: *The theory of models*, Eds. J. Addison, L. Henkin and A. Tarski, North-Holland, Amsterdam, (1956).
Reprinted in *The logico-algebraic approach to quantum mechanics*, Vol.I., Ed. C.A.Hooker, D.Reidel Publ.Co., Dordrecht-Boston, (1975).
18. ——— : The calculus of partial propositional functions, in: *Logic, Methodology and Philosophy Science*, Ed. Y. Bar-Hillel, North-Holland, Amsterdam, (1965).
Reprinted in *The logico-algebraic approach to quantum mechanics*, Vol.I., Ed. C.A.Hooker, D.Reidel Publ.Co., Dordrecht-Boston, (1975).
19. ——— : The problem of hidden variables in quantum mechanics, *J. of Mathematics and Mechanics*, **17**, (1967), pp.59-67.
Reprinted in *The logico-algebraic approach to quantum mechanics*, Vol.I., Ed. C.A.Hooker, D.Reidel Publ.Co., Dordrecht-Boston, (1975).
20. J. Krajíček : *Bounded arithmetic, propositional logic and complexity theory*, Cambridge University Press, Cambridge-New York-Melbourne, (1995).
21. ——— : Speed-up for propositional Frege systems via generalizations of proofs, *Commentationes Mathematicae Universitas Carolinae*, **30(1)**, (1989), pp.137-140.
22. ——— : Lower bounds to the size of constant-depth propositional proofs, *J. of Symbolic Logic*, **59(1)**, (1994), pp.73-86.
23. ——— : On Frege and extended Frege proof systems, in: *Feasible Mathematics II*, Eds. P. Clote and J. Remmel, Birkhauser, (1995), pp. 284-319.
24. ——— : Interpolation theorems, lower bounds for proof systems, and independence results for bounded arithmetic, *J. of Symbolic Logic*, (1995), to appear.
25. ——— : A fundamental problem of mathematical logic, *Annals of Kurt Gödel Society*, Springer Verlag, (1995), to appear.
26. ——— : Valuations of Boolean formulae in partial algebras, in: Volume of Abstracts of the Tenth International Congress *Logic, Methodology and Philosophy of Science*, International Union of History and Philosophy of Science, Florence (August 19-25, 1995), (1995), pp.6-7.
27. J. Krajíček and P. Pudlák: Some consequences of cryptographical conjectures for S_2^1 and EF, Proc. of the meeting *Logic and Computational Complexity*, (Indianapolis, October 1994), Ed. D. Leivant, Lecture Notes in Computer Science, Vol. 960, Springer-Verlag, (1995), pp. 210–220.
28. J. Krajíček, P. Pudlák and A. Woods: An exponential lower bound to the size of bounded depth Frege proofs of the pigeonhole principle, *Random Structures and Algorithms*, **7(1)**, (1995), pp.15-39.
29. A. Lewenberg : On elementary pairs of O-minimal structures, PhD.Thesis, University of Illinois at Champaign-Urbana, (1995).
30. P. Mittelstaedt : *Quantum logic*, D. Reidel Publ.Co., Dordrecht, Synthese Library, Vol. 126, (1978).
31. T. Pitassi, P. Beame and R. Impagliazzo: Exponential lower bounds for the pigeonhole principle, *Computational Complexity*, **3**, (1993), pp.97-308.
32. P. Pudlák : The lengths of proofs, in: *Handbook of Proof Theory*, Ed. S. Buss, to appear.

33. ——— : Lower bounds for resolution and cutting planes proofs and monotone computations, preprint, (1995).
34. A. A. Razborov : Lower bounds on the monotone complexity of some Boolean functions, *Soviet Mathem. Doklady*, **31**, (1985), pp.354-357.
35. ——— : Lower bounds on the size of bounded depth networks over a complete basis with logical addition, *Matem. Zametki*, **41(4)**, (1987), pp.598-607.
36. ——— : On the method of approximations, in: *Proc. 21th Annual ACM Symp. on Theory of Computing*, (1989), pp. 168-176. ACM Press.
37. ——— : Unprovability of lower bounds on the circuit size in certain fragments of bounded arithmetic, *Izvestiya of the R.A.N.*, **59(1)**, (1995), pp.201-224.
38. R. Smolensky : Algebraic methods in the theory of lower bounds for Boolean circuit complexity, in: *Proc. 19th Ann. ACM Symp. on Th. of Computing*, (1987), pp. 77-82.
39. E. P. Specker: The logic of propositions which are not simultaneously decidable, *Dialectica*, **14**, (1960), pp.239-246.
Reprinted in *The logico-algebraic approach to quantum mechanics*, Vol.I., Ed. C.A.Hooker, D.Reidel Publ.Co., Dordrecht-Boston, (1975).
40. E.-W. Stachow: Completeness of quantum logic, *J. Phil. Logic*, **5**, (1976), pp.237-280.
41. G. Takeuti: Quantum set theory, in: [3], pp.303-322.
42. ——— : Quantum logic and quantization, in; *Proc. Int. Symp. Foundations of Quantum Mechanics*, Tokyo, (1983), pp.256-260.

VLADIMIR YU. SAZONOV

ON BOUNDED SET THEORY*

Dedicated to the memory of Professor Robin Gandy

ABSTRACT.
We consider some Bounded Set Theories (BST), which are analogues to Bounded Arithmetic. Corresponding provably-recursive operations over sets are characterized in terms of explicit definability and PTIME- or LOGSPACE-computability. We also present some conservativity results and describe a relation between BST, possibly with Anti-Foundation Axiom, and a Logic of Inductive Definitions (LID) and Finite Model Theory.

1. INTRODUCTION

The goal of this paper is to outline the present state of affairs in investigating various versions of *Bounded Set Theory* (BST). These are analogues of S. Cook's theory PV [8] of Bounded Primitive Recursive Functions of A. Cobham [7] over natural numbers and to Bounded Arithmetic (cf. e.g. [5, 33, 30]). Also, they are similar to the predicative systems of S. Feferman [13] and G. Jäger [24], however essentially more weak, even weaker than the Primitive Recursive Set Theory of R. Jensen and C. Karp [26]. BST's reasonably extend the fragment KP_0 of *Kripke-Platek Set Theory* KP (cf. [3]) obtained by omitting $\in$-*Induction* (or *Foundation*) *Schema*. Note, that KP_0 is equivalent to R. Gandy's Basic Set Theory BST_0 [15] plus Δ-Collection principle. We also consider BST as a PTIME- or LOGSPACE-computability theory for operations over the universe of hereditarily-finite sets **HF**. This could be called Δ-programming approach, in comparison with somewhat different, however related approaches to Σ-programming [17], also based on KP and [3], and to the programming languages SETL [9, 10] and STARSET [16, 47].

There are three closely related aspects of BST:

Definability and computability of operations over sets. This subject is closely related to descriptive complexity theory or finite model theory for a Logic of Inductive Definitions (LID) or First Order Logic with the construct of Transitive Closure ($\mathsf{FO}^{\otimes}$) (cf. [18, 19, 20, 21, 22, 23, 28, 32, 50], etc.).

Possibility of applications in computer science, because the hereditarily-finite sets form a very important high-level data structure. For example, they may be used to represent the state of a "nested" data base (cf. [40]). However, in this paper we will not touch on this aspect explicitly.

*Supported by RFBR (project 93-011-16016).

M.L. Dalla Chiara et al. (eds.), Logic and Scientific Methods, 85–103.

Proof theory. We investigate provably-recursive operations of BST's and obtain some conservativity results. In particular, KP_0 proves to be conservative over the only extensionality axiom with respect to Δ_0-formulas, and a theorem of G. Jäger [24] is obtained by using a general technique. Also, some set-theoretic extension of a version of Peano Arithmetic $\mathsf{PA}_\Box$ over a finite row of natural numbers $\{0, 1, \ldots, \Box - 1, \Box\}$ is considered.

The principal feature of any version of BST is the use of mainly *bounded quantifiers* $\forall x \in a$ and $\exists x \in a$ and other constructs which makes it a *theory of computability* (over sets) *with bounded resources*. For example, the powerset operation is *not* considered as "bounded". Analogously, quantifiers $\forall x \subseteq a$ and $\exists x \subseteq a$ are bounded too roughly in comparison with $\forall x \in a$ and $\exists x \in a$. The latter correspond to *sharply bounded* quantifiers in Bounded Arithmetic.

We introduce the main formalisms in Sections 2 and 3. Then for hereditarily-finite sets the notions of PTIME- and LOGSPACE-computability are considered in Section 4.1, 4.2 and 4.3. An anti-wellfounded universe and connections with finite model theory are discussed in Sections 4.4 and 4.5. We present proof-theoretic results in Sections 3 and 5 (which may be read after Section 2).

There are also some digressions from the technical considerations of the paper devoted to the related informal aspect of *feasibility*.

2. SOME BOUNDED SET THEORIES

In the following we implicitly assume that any theory T is naturally presented as Gentzen's sequent calculus LK extended by a set of *initial* sequents closed under term substitutions (besides the logical LK-axioms of the form $\varphi \to \varphi$). Then let $T\restriction\mathcal{F}$ denote the restriction of axioms and rules of the calculus T to sequents involving only formulas from some class $\mathcal{F}$.

Usually set theory is formulated on the base of primitives $\in$ and $=$ with set variables as the only terms. We consider set-theoretic languages with a richer class of terms, because we are interested in explicitly definable operations over sets so that they will also coincide with *provably-recursive* ones. Moreover, as in the case of Bounded Arithmetic, such operations may constitute, in an appropriate sense, a computational complexity class of low level.

We call the languages of BST's by a generic name Δ, possibly mentioning the corresponding set theory in square brackets or the like, and define one of them (in fact, $\Delta[\mathsf{BSTC}]$) by the clauses

$$\begin{array}{rcl} \Delta\text{-}\mathit{formulas} & ::= & a \in b \,|\, a = b \,|\, \varphi \& \psi \,|\, \varphi \vee \psi \,|\, \neg\varphi \,|\, \forall x \in a.\varphi \,|\, \exists x \in a.\varphi \\ \Delta\text{-}\mathit{terms} & ::= & \text{set-variables } p, q, x, y, \ldots \,|\, \{a, b\} \,|\, \cup\{t(x) : x \in a \& \varphi(x)\} \\ & & |\, \mathsf{the\text{-}least}\, p.(p = \{x \in a : \varphi(x, p)\}) \,|\, \mathsf{TC}(a) \,|\, \mathsf{C}(a, b) \end{array}$$

where a, b, t are Δ-terms and φ, ψ are Δ-formulas, set-variables x, p are not free in a, all occurrences of p in φ of the-least construct, called also (Δ-)Rec, satisfy a syntactic condition of *positivity*. Formally, *these occurrences must be only of the form '- $\in p$', not inside of any complex subterm of* φ *and not in the* *scope of a negation sign.* This condition on p guarantees that φ is monotonic in p relative to $\subseteq$ (where $u \subseteq v \rightleftharpoons \forall x \in u.(x \in v)$) and, therefore, the least required set $p \subseteq a$ may be postulated to exist. Rec is used to define *recursively* a subset p of a given set a. Only *bounded* quantifiers are allowed in Δ. Other constructs in Δ also may be comprehended as reasonably "bounded".

The above clauses, except for the last three constructs Rec, TC and C, define the language of *basic* = *rudimentary* operations of R.O.Gandy [15] and R.B.Jensen [25], respectively. We take it as a sublanguage of any other version of a Δ-language (which could be denoted as $\Delta[\mathsf{BST}_0]$ or $\Delta[\mathsf{KP}_0]$; cf. below). This basic language extended by TC is denoted by Δ^*. Formulas of a Δ-language involving only atomic Δ-terms are just A.Lévy's Δ_0-formulas. Let $^*t$ abbreviate $\mathsf{TC}(t)$. By taking $s \in {}^*t$, or $s \in^* t$ as a unique relation (the transitive closure of $\in$) we may define the class of formulas Δ_0^*, like Δ_0, as the class of those formulas in the language $\{\in, \in^*\}$ which contain only bounded quantifiers $\forall x \in^{(*)} y$ and $\exists x \in^{(*)} y$ and only atomic terms.

Sometimes we will need also the closure of Δ-formulas under $\&, \vee, \neg$ and *unbounded* $\forall$ and $\exists$ (bearing in mind the ordinary definition of the bounded quantifiers of Δ via unbounded ones). The equality predicate may be removed from the language by taking $u = v \rightleftharpoons u \subseteq v \& v \subseteq u$.

All considered constructs have a natural semantics in any reasonable universe of sets $\mathcal{V}$, say, for Zermelo-Fraenkel set theory ZF or in the universe **HF** of *hereditarily-finite sets* and are described by corresponding axioms:

Pair: $t \in \{a, b\} \Leftrightarrow t = a \vee t = b$,

Δ-Union: $s \in \bigcup\{t(x) : x \in a \ \&\ \varphi(x)\} \Leftrightarrow \exists x \in a(\varphi(x) \ \&\ s \in t(x))$,

Recursive Δ-Separation: If $p_0 = \text{the-least}\, p.(p = \{x \in a : \varphi(x, p)\})$ then

$$p_0 = \{x \in a : \varphi(x, p_0)\} \ \&\ (\{x \in a : \varphi(x, q)\} \subseteq q \Rightarrow p_0 \subseteq q).$$

where a, b, t, s, q are any Δ-terms and φ is any Δ-formula. *Transitive closure* $\mathsf{TC}(a)$ denotes the least transitive set containing the set a as a subset.

TC: $x \subseteq \mathsf{TC}(x)$, $\mathsf{Tran}(\mathsf{TC}(x))$, $x \subseteq y \Rightarrow \mathsf{TC}(x) \subseteq \mathsf{TC}(y)$ and

$$\mathsf{Tran}(y) \Rightarrow \mathsf{TC}(y) \subseteq y.$$

where $\mathsf{Tran}(y) \rightleftharpoons \forall u \in y(u \subseteq y)$. The *Collapsing* operation $\mathsf{C}(g, v)$ of A. Mostowski assigns a set to each vertex v of a *well-founded* graph g (considered as a set of ordered pairs[1] $\langle u, v\rangle \rightleftharpoons \{\{u\}, \{u, v\}\}$) according to the identity

C : $\mathsf{Well\text{-}Founded}(g) \Rightarrow \mathsf{C}(g, v) = \{\mathsf{C}(g, u) : \langle u, v\rangle \in g\}$.

where well-foundedness of a graph is naturally Δ-definable by using Rec.

Any version of Bounded Set Theory BST is defined by a corresponding version of the Δ-language and *non-logical axioms*, which are Δ-formulas describing constructs of the Δ-language like those given above, and also the axioms

Extensionality: $a \subseteq b \ \&\ b \subseteq a \Rightarrow a = b \Rightarrow (\varphi(a) \Leftrightarrow \varphi(b))$, $\varphi \in \Delta$,

[1]Edges $u \to v$ in the graphs are directed as in [3, 37], just conversely to [2].

and, possibly, the axiom which itself is not a Δ-formula

$$\Delta\text{-Collection:} \quad \forall x \in a \exists y \varphi(x,y) \Rightarrow \exists z \forall x \in a \exists y \in z \varphi(x,y)$$

with z a new variable. Additionally, we may include for any $\Delta_{(0)}$-formulas $\forall x \in t\varphi(x)$, Φ and Ψ with no free occurrences of the variable y

$$\in\text{-}\Delta_{(0)}\text{-Ind (Foundation) rule:} \quad \frac{\forall x \in y\varphi(x), \Phi \to \Psi, \varphi(y)}{\Phi \to \Psi, \forall x \in t\varphi(x)}$$

or any other reasonable rules of inference which contain only Δ-formulas.

Let BSTC be the version of BST corresponding to the Δ-language presented exactly by the above clauses. Gandy's Basic Set Theory, which we call BST_0, is formulated in the basic (rudimentary) fragment of the above Δ-language by taking only axioms Ext, Pair and Δ-Union. The theory KP_0 (whose language is traditionally $\{\in, =\}$; cf. [3]) may be equivalently reformulated as BST_0 + Δ-Collection. The extension of KP_0 by the-least (respectively, by the TC) construct with corresponding axiom is called KPR_0 (respectively, KP_0^*). Let $\mathsf{KPR}_0^* \rightleftharpoons \mathsf{KPR}_0 \cup \mathsf{KP}_0^*$. Moreover, $\mathsf{KP}_0^\otimes$ is the extension of KP_0^* by the "horizontal" transitive closure construct $r^\otimes$ where r is any term whose value is a set of ordered pairs, i.e. a graph, and $\otimes$ gives the transitive closure of this graph.

Collapsing may be replaced by the *decoration* operation $\mathsf{D}(g,v)$ of P. Aczel (cf. [2, 4]) which satisfies the same axiom as C, except for being applicable to *arbitrary* (possibly non-wellfounded) graphs g. The existence and uniqueness of such an operation is known as the *Anti-Foundation Axiom* AFA and can be formulated as a Δ-axiom (in terms of D). The corresponding version of BST (without any $\in$-induction, to avoid a contradiction with AFA) is called BSTA.

Finally, we will need also the simplest set theory EXT in the language $\{\in\}$, containing only the Extensionality axiom (with φ in Δ_0), and its conservative extension EXT^* (with φ in Δ_0^*) in the language $\{\in, \in^*\}$ by the following axioms and inference rule:

$$\forall v \in^* w \forall u \in v(u \in^* w),$$

$$u \in^* v \leftrightarrow u \in v \vee \exists x \in v(u \in^* x),$$

$$\frac{u \in v \to \psi(u,v) \quad \psi(u,v), \psi(v,w) \to \psi(u,w)}{x \in^* y \to \psi(x,y)}, \quad \psi \text{ in } \Delta_0^*.$$

3. LOGIC OF INDUCTIVE DEFINITIONS

A "purely" logical analog of KPR_0 is the *Logic of Inductive Definitions* LID defined below. A version of the corresponding *Language of Inductive Definitions* (sometimes called as FO+LFP) was used by Y. Moschovakis [29] in his approach to generalized computability. Arithmetical versions of LID are known rather as systems like ID_i considered by S. Feferman [12], G. Jäger [24] and others. Also P. Aczel [1] asked about a coherent conceptual doctrine making inductive definitions the main notion.

We define LID as *many-sorted* First-Order Logic with predicate variables and with the (positive) Least Fixed Point construct

$$[\textsf{the-least}\ P.\forall\bar{x}.(P(\bar{x}) \Leftrightarrow \varphi(\bar{x},\bar{y},P,\bar{Q}))]$$

abbreviated below as P_φ $(= P_\varphi(\text{-}) = P_\varphi(\text{-},\bar{y},\bar{Q}))$. Here, $\bar{x},\bar{y}$ and $P,\bar{Q}$ are, respectively, all free individual and predicate variables of φ and the predicate variable P is required to have only *positive* occurrences in φ. This new predicate construct may participate in other LID-formulas with any nesting except that its predicate variable P *is not allowed to appear in the corresponding body* $\varphi(\bar{x},P)$ *inside any other such construct which is a part of this body* (essentially the same as in the Δ-language). No second-order quantification is allowed, however: n-ary predicates are considered as comprising a *second-order sort.*

LID as a *logic* extends ordinary First-Order Logic with equality by the following axiom schemes (for φ and P_φ as above):

$$\forall\bar{x}(\varphi(\bar{x},P_\varphi)\Rightarrow P_\varphi(\bar{x})) \quad\text{and}\quad \forall\bar{x}(\varphi(\bar{x},P)\Rightarrow P(\bar{x})) \Rightarrow \forall\bar{x}(P_\varphi(\bar{x})\Rightarrow P(\bar{x})).$$

When φ does not depend on P these axioms give rise to the ordinary comprehension axiom for first-order formulas. It is also assumed that variable P in the second axiom may be replaced by any predicate construct P_ψ.

Let '$\models_{\mathcal{V}} \varphi$' abbreviate a set-theoretic formula $\forall M \in \mathcal{V}\tilde{\forall}.(\text{'}M \models \varphi\text{'})$ where M is a first-order structure and $\tilde{\forall}$ denotes universal quantification over all free individual and predicate variables of φ ranging eventually over the intended universe of sets $\mathcal{V} = \{x : x = x\}$. This gives a natural embedding of LID in KPR_0 and, therefore, also in BSTC and BSTA which is, in fact, *conservative.*

Theorem 1 (cf. [43] and partially [39])

$$\begin{aligned}\mathsf{LID} \vdash \varphi \quad \textit{iff} \quad \mathsf{KPR}_0 \vdash \text{'}\models_{\mathcal{V}} \varphi\text{'} \quad &\textit{iff} \quad \mathsf{BSTC} \vdash \text{'}\models_{\mathcal{V}} \varphi\text{'} \\ &\textit{iff} \quad \mathsf{BSTA} \vdash \text{'}\models_{\mathcal{V}} \varphi\text{'}\end{aligned}$$

This theorem is based, among other things, on Theorem 15 (b) below.

In [21, 28, 50] the following theorem is proved (in fact, for variations of the language of LID; cf. also essentially the same result in terms of recursive functions in finite domains [18, 19, 32, 33]).

Theorem 2 *The expressive power of the language* LID *in finite linearly ordered structures exactly corresponds to* PTIME*-computability.*

A weaker version of LID is $\mathsf{FO}^\otimes$, first-order logic with the predicate forming construct $[\lambda\bar{x}\lambda\bar{y}.\varphi(\bar{x},\bar{y},\bar{z})]^\otimes$ which denotes the transitive closure of the relation $\lambda\bar{x}\lambda\bar{y}.\varphi(\bar{x},\bar{y},\bar{z})$ of two lists of variables $\bar{x}$ and $\bar{y}$ of the same length (and sorts) and which may participate in any other formula as a predicate (both positively and negatively!). Then it was proved by N. Immerman [22, 23]

Theorem 3 *Over finite linearly ordered structures* $\mathsf{FO}^\otimes \sim$ NLOGSPACE.[2]

This implies the following very unexpected

---

[2]In [22, 23] the notation TC is used essentially for $\otimes$.

Corollary 4 [23, 48] NLOGSPACE = co-NLOGSPACE.

4. DEFINABILITY AND COMPUTABILITY IN BST

4.1. PTIME-*computability over* **HF**

The key notion for our approach to computability over **HF** is the following definition of PTIME-*computability* of any operation $f : \mathbf{HF}^n \to \mathbf{HF}$. Given any encoding $\nu : \text{Codes} \to \mathbf{HF}$ of **HF**-sets, we say that f is in $\mathcal{P}_\nu$ if (for the case $n = 1$) the following diagram

$$\begin{array}{ccc} \mathbf{HF} & \xrightarrow{f} & \mathbf{HF} \\ {\scriptstyle \nu}\uparrow & & \uparrow{\scriptstyle \nu} \\ \text{Codes} & \xrightarrow{F} & \text{Codes} \end{array}$$

commutes for some PTIME-computable transformation F. A general theory of computability over abstract domains relative to some encoding is developed in [11] as *numbering theory* with Codes = **N** (Natural Numbers).

There is some ambiguity in considering Natural Numbers as Codes from the point of view of Complexity Theory. Usually the ordinary decimal or, essentially equivalent, binary notation system is adopted for numbers. Actually, this means that we take Codes = $\{0,1\}^*$, the set of all finite *binary* strings. The last decision seems quite natural from the point of view of both abstract and real computers. However, *identifying* numbers with strings in a finite alphabet of cardinality > 1 appears *not sufficiently correct* by the following reasons. It is quite natural to consider numbers, first of all, as a tool for counting ("one, two, three, ...") some really existing things (stones, stars, houses, peoples, bits in a computer memory, etc.) or even some abstract entities in a framework of "Feasible Applied Mathematics". We may also consider any *concrete* (say, written on a sheet of paper) binary string of length one thousand as something existing in the "Nature". But which integer quantity does it *really* denote? Which computer can ever count "all" strings of the length one thousand? Who knows what this "all" means? "All" strings which have been or which could be written? What does it mean "could"? Finally, why *potential feasibility* is usually considered (with rare exceptions, e.g. [27, 49, 51]) as posing no essential foundational problems in comparison with the *actual infinity*?

If we are looking for a realistic complexity theory or a "Feasible Mathematics" we must rather conceptually identify **N** with $\{1\}^*$, the set of finite *unary* strings, or with a "set" of sufficiently "short" binary strings. (Cf. also the corresponding discussion and some formalizations of feasibility and constructibility concepts for finite objects in [30, 33, 38, 42] and in the end of this paper.) There are *many* different kinds of finite objects such as unary and binary strings, hereditarily-finite sets, etc., whose interrelations must be investigated from the point of view of feasibility of these objects and also of corresponding computations.

There are quite different natural encodings ν of **HF** giving rise to different notions of $\mathcal{P}_\nu$-computability. For uniformity and definiteness we could fix Codes = $\{0,1\}^*$ and vary only ν. However conceptually it is useful to allow different versions of Codes. It seems that the most natural example of Codes is the class $\mathcal{AG}$ of all finite *acyclic pointed graphs* (apg), i.e. graphs G with a distinguished point (vertex) p in each. (Sometimes we consider several points distinguished.) Then let $\gamma : \mathcal{AG} \to \mathbf{HF}$ be the corresponding *graph encoding*, i.e. essentially the collapsing operation C considered above. The following theorem gives a purely set-theoretic characterization of $\mathcal{P} \rightleftharpoons \mathcal{P}_\gamma$ (over **HF**). Note, that in the case of the graph encoding γ polynomial time computability may be considered just relative to the number of vertices (or, equivalently, relative to the length of a natural binary code) of a graph.

Theorem 5 [35, 37, 40] *The class of Δ-definable operations of the theory* BSTC *in the universe of hereditarily-finite sets coincides with $\mathcal{P} = \mathcal{P}_\gamma$:*

$$\Delta[\mathsf{BSTC}] \sim_{\mathrm{HF}} \mathcal{P}\,.$$

Another natural *tree encoding* of **HF** is the restriction τ of graph encoding γ to the subclass $\mathcal{T} \subseteq \mathcal{AG}$ of finite trees. Well-formed bracket expressions like $\{\{\}\{\{\}\}\}$ give an essentially equivalent encoding β.

Note, that any finite ordinal n requires 2^{n+1} braces. It follows that the ordinals like 'one hundred' essentially *cannot exist* (say, in a computer memory) according to this encoding.

Consider also the best-known bijective encoding (Ackerman) Ack : $\mathbf{N} \to \mathbf{HF}$:

$$\mathrm{Ack}(2^{n_1} + 2^{n_2} + \ldots + 2^{n_k}) = \{\mathrm{Ack}(n_1), \mathrm{Ack}(n_2), \ldots, \mathrm{Ack}(n_k)\}$$

for $n_1 > n_2 > \ldots > n_k \geq 0$, $k \geq 0$. Note that a very simple singleton operation $x \mapsto \{x\}$ corresponds to the arithmetical exponential $k \mapsto 2^k$ with respect to this encoding of **HF**, which is surely untractable. It follows that we rather must consider other encodings like the above γ and τ (or β).

Ackernman's bijection defines a *canonical linear ordering* $<_{\mathrm{HF}}$ on **HF**

$$\mathrm{Ack}(m) <_{\mathrm{HF}} \mathrm{Ack}(n) \rightleftharpoons m <_{\mathrm{N}} n$$

where $<_{\mathrm{N}}$ is the ordinary linear ordering of the set of natural numbers **N**. This ordering is also uniquely defined by the axiom

Canonical Order $<_{\mathrm{HF}}$:

$$x < y \Leftrightarrow \exists u \in y \setminus x(\{v \in x : u < v\} = \{v \in y : u < v\})\,.$$

Moreover, $<_{\mathrm{HF}}$ is definable in the Δ-language of KPR_0^* and this fact is essentially used in the proof of Theorem 5 (cf. also Theorem 2).

While it is clear that the encoding Ack is rather bad from the point of view of feasible computability, we do not know which encoding is the best one; we even do not know what "best" should mean here. One possible mathematical approach to this problem is via polytime reducibility for encodings. We say that ν_1 is *polytime reducible* to ν_2, $\nu_1 \leq_{\mathrm{P}} \nu_2$, if for some polytime transformer of codes F we have $\nu_1(c) = \nu_2(F(c))$ (i.e. the corresponding triangle diagram commutes). Also, let $\nu_1 <_{\mathrm{P}} \nu_2 \rightleftharpoons \nu_1 \leq_{\mathrm{P}} \nu_2 \& \nu_2 \not\leq_{\mathrm{P}} \nu_1$. Then we have

$$\begin{array}{ccccccc} & & & & \gamma & & \\ & & & <_{\mathrm{P}} & & <_{\mathrm{P}} & \\ \mathrm{Ack} & <_{\mathrm{P}} & \tau & & & & \gamma' \\ & & & <_{\mathrm{P}} & & <_{\mathrm{P}} & \\ & & & & \tau' & & \end{array}$$

with no other reducibility relations $\leq_{\mathrm{P}}$, except those explicitly shown. Here, encodings γ' and τ' are obtained from γ and τ by allowing numerical labels on the edges of finite graphs and trees, say, in decimal notation, so that any edge $u \overset{n}{\to} v$ actually denotes a chain of edges $u \to \ldots \to v$ with n *new* intermediate vertices. These somewhat artificial encodings are just more economical versions of γ and τ introduced here only to illustrate that $\leq_{\mathrm{P}}$ is non-trivial.

An analog of Theorem 5 holds for a wide class of encodings defined in the next paragraph. It contains all the above encodings, except Ack.

Any four-tuple of mappings ν, i^R : Codes $\to$ **HF** and ν^R, i : **HF** $\to$ Codes is called a P-*regular coding system* if the following five conditions are satisfied:

1. The predicate $\nu(c) \in \nu(d)$ is PTIME-computable on $c, d \in$ Codes.
2. Two PTIME-computable mappings $c \mapsto c_1, \ldots, c_k$ and $c_1, \ldots, c_k \mapsto c$ (*splitting* and *gathering*) between codes and lists of codes of arbitrary length exist such that both satisfy the equality $\nu(c) = \{\nu(c_1), \ldots, \nu(c_k)\}$.
3. $\nu\nu^R = 1_{\mathrm{HF}}$ and $ii^R = 1_{\mathrm{Codes}}$. (The mappings ν^R and i^R give a *canonical* code of any **HF**-set and, respectively, representation of any code in **HF**.)
4. The compositions $\nu^R i^R$, $i\nu$ and $\nu^R\nu$: Codes $\to$ Codes are PTIME-computable.
5. Given any PTIME-computable transformer F : Codes $\to$ Codes there exists corresponding $\Delta[\mathsf{KPR}_0]$-definable operation f : **HF** $\to$ **HF** (representing F in Δ) such that $Fi = if$.

Note, that i (satisfying (5)) may be taken as a mapping identifying a reasonable set-theoretic representations of Codes with the actual Codes. For example, in the case of Codes $= \{0,1\}^*$, any binary string is naturally represented in **HF** as a finite linear ordered set with some distinguished subset.

Let Δ_ν be a version of the Δ-language defined in Section 2 with C and TC replaced by operations $\tilde{\nu} \rightleftharpoons \nu i$ and $\tilde{\nu}^R \rightleftharpoons i^R\nu^R$: **HF** $\to$ **HF**, respectively.

Theorem 6 [41, 46] *For all* P*-regular encodings* ν, θ*:*

(a) $\Delta_\nu \sim_{\mathrm{HF}} \mathcal{P}_\nu$*, i.e.* Δ_ν*-language represents exactly* $\mathcal{P}_\nu$*-operations in* **HF**.

(b) $\nu \not\leq_{\mathrm{P}} \theta$ *implies* $\mathcal{P}_\theta \not\subseteq \mathcal{P}_\nu$ *and* $\mathcal{P}_\nu \not\subseteq \mathcal{P}_\theta$ *with counter-examples* $\tilde{\theta}^R$ *and* $\tilde{\nu}$*, respectively; in fact,* $\nu \leq_{\mathrm{P}} \theta \Leftrightarrow \tilde{\nu} \in \mathcal{P}_\theta \Leftrightarrow \tilde{\theta}^R \in \mathcal{P}_\nu$.

Note, that the *graph encoding* γ *is the largest* (the best?) *one in the preorder relation* $\leq_{\mathrm{P}}$ *among those* P*-regular encodings* ν *for which* $\mathsf{TC} \in \mathcal{P}_\nu$.

4.2. *The Case of* NLOGSPACE

The case of (nondeterministic) computability in logarithmic space (abbreviated as (N)LOGSPACE) over **HF** proves to be more intricate.[3] First, we have to consider instead of $\mathcal{AG}$ the class of *extensional acyclic graphs* $\mathcal{EAG}$ (which are isomorphic to transitive sets in **HF**). *Extensionality* of any graph G means just the ordinary extensionality property with respect to the *edge relation* $u \in_G v \rightleftharpoons u \to_G v$. In the case of $\mathcal{AG}$'s it would be necessary to identify in (N)LOGSPACE those vertices of a graph which denote the same

[3]The results of Sections 4.2 and 4.3 were obtained in cooperation with A. Lisitsa [44, 45].

set under collapsing. However, this is hardly possible because it was shown in [9] that this is a PTIME-complete problem and, therefore, this would imply (N)LOGSPACE = PTIME. Remember, that only *non-proper* inclusions LOGSPACE $\subseteq$ NLOGSPACE $\subseteq$ PTIME are known.

Moreover, Theorem 3 suggests taking Codes = $\mathcal{EAG}_<$, the class of $\mathcal{EAG}$'s with a given *explicitly* canonical linear order on their vertices (which is induced by $<_{\mathrm{HF}}$ or defined uniquely by the axiom for $<_{\mathrm{HF}}$).

Unfortunately, we cannot consider just NLOGSPACE-computable transformers $\mathcal{EAG}_< \to \mathcal{EAG}_<$ to get a reasonable definition of the corresponding notion of computable operations over **HF**-sets. The problem consists in fulfilling natural requirements that the class of such (in general, many-place) operations must be *closed under substitutions* and that a corresponding version of the Δ-language must have an *effective syntax*. After a number of intermediate, only partially satisfactory steps this was achieved in [45] by defining a class IC-NLOGSPACE of NLOGSPACE-computable transformers $\mathcal{EAG}_< \to \mathcal{EAG}_<$ which are additionally (NLOGSPACE-) *inflationary* and *coherent*.

Here 'inflationary' means the following: Given any input $\mathcal{EAG}_<$-graph G, not only the corresponding output $\mathcal{EAG}_<$-graph G' should be NLOGSPACE-computable, but also an *auxiliary* $\in$-isomorphism of G onto some transitive subgraph[4] of G'. Moreover, *only* distinguished vertices of graph G' may depend on distinguished vertices of graph G.

'Coherent' means the existence of one more construction: Let there be given any two input $\mathcal{EAG}_<$-graphs G_1 and G_2 with the canonical linear *preorder* relation $\leq_{G_1+G_2}$ on the disjoint union $G_1 + G_2$.[5] Then the analogous linear preorder relation $\leq_{G'_1+G'_2}$ on $G'_1 + G'_2$ is required to be also NLOGSPACE-computable from the given two-sorted structure $\langle G_1 + G_2; \leq_{G_1+G_2}\rangle$. Again, *no* dependence of $\leq_{G'_1+G'_2}$ on the distinguished vertices of G_1 and G_2 is allowed.

As for the case of PTIME, the class of graph transformers IC-NLOGSPACE defines a class of corresponding computable operations over **HF** which we denote by $\mathcal{NL}$. This class may serve as a reasonable candidate for the notion of NLOGSPACE-computable operations because the above IC-restriction can be proved to be *inevitable* in some precise sense. Evidently, $\mathcal{NL} \subseteq \mathcal{P}$.

Note, that it is unclear whether any given IC-NLOGSPACE-transformer does compute an operation in $\mathcal{NL}$ (actually in terms of corresponding commutative square diagram) and whether appropriate transformers for *all* these computable operations may be effectively described.

[4]A *transitive subgraph* of a graph is a full subgraph such that for all its vertices v, if there is an edge $u \to v$ then u is also in this subgraph.

[5]This relation is induced by $<_{\mathrm{HF}}$ restricted to the (ordinary) union of corresponding transitive sets $T[G_1] \cup T[G_2]$ in **HF**. It is a preorder because some vertices in G_1 must be identified with those in G_2 if they denote the same sets in **HF**. Also it would be insufficient to use in this definition corresponding equivalence relations $\equiv_{G_1+G_2}$ and $\equiv_{G'_1+G'_2}$ instead of $\leq_{G'_1+G'_2}$ and $\leq_{G'_1+G'_2}$.

Nevertheless, we will show in Theorem 8 that some Δ-description of $\mathcal{NL}$ has an effective syntax by using the following proposition. Any transformer $F : \mathcal{EAG}_< \to \mathcal{EAG}_<$ in IC-NLOGSPACE may be considered as an operation over **HF**-sets representing $\mathcal{EAG}_<$'s. The class of such operations $\mathbf{HF} \to \mathbf{HF}$ is also denoted as $\mathcal{NL}_0$. (Essentially, $\mathcal{NL}_0$ coincides with IC-NLOGSPACE.) This class evidently has a natural effective enumeration via any effective enumeration of NLOGSPACE Turing Machines. Then we can prove a technical

Proposition 7 $\mathcal{NL}_0 \subseteq \mathcal{NL}$.

Let $\Delta^*_<$ be Δ^* extended by the predicate $<$ for the canonical linear order on **HF**. Then we have an analog of Theorem 5 for the case of NLOGSPACE.

Theorem 8 $\mathcal{NL} \sim_{\mathrm{HF}} \Delta^{\otimes}_< + \mathcal{NL}_0$.

It is interesting to strengthen this Δ-description of $\mathcal{NL}$ by substituting for $\mathcal{NL}_0$ some finite number of $\mathcal{NL}$-operations or some *schemes* (like $\{s(x) : x \in a \,\&\varphi(x)\}$) so that, moreover, all of them could be *mutually axiomatized* in a natural way as in the case of our original Δ-language.

The following proposition confirms that there is some concordance in our approaches to PTIME- and NLOGSPACE-computability over **HF**.

Proposition 9

$$\Delta[\mathsf{BSTC}] \sim_{\mathrm{HF}} \Delta^{\otimes}_< + \mathcal{NL}_0 \textit{ iff } \mathcal{P} = \mathcal{NL} \textit{ iff } \mathrm{PTIME} = \mathrm{NLOGSPACE}.$$

There are $\mathcal{NL}$-operations over **HF** which are not $\Delta^{\otimes}_<$-definable. It is an interesting exercise to show that such examples are *addition* and *multiplication* operations on finite ordinals and also the operation $\mathrm{rank}(x) = k$, giving an ordinal k of the same rank (= nesting) as the set x.

4.3. *The Case of* DLOGSPACE

The above theorem on $\mathcal{NL}$ may be adopted for the case of deterministic LOGSPACE-computability over **HF**. First, we have to replace the "horizontal" transitive closure construct $\otimes$ by its *deterministic* version $\odot$

$$r^{\odot} \rightleftharpoons D(r)^{\otimes} \quad \text{where} \quad D(r) \rightleftharpoons \{\langle x,y\rangle \in r : \exists! z(\langle x,z\rangle \in r)\}\,.$$

It was also considered in [22] for the description of DLOGSPACE, analogous to that of NLOGSPACE in Theorem 3, in terms of $\mathsf{FO}^{\odot}$-definability in linear ordered finite structures.[6] Further, it is necessary to let Codes $= \mathcal{EAG}^*_<$ where $*$ designates that $\mathcal{EAG}$'s are taken together with an *explicitly* given transitive closure predicate $\in^*_G$, as well as the main graph relation $\in_G$ (i.e. $\to_G$). We need this because the construct $\odot$ can hardly express the "vertical" transitive closure $\in^*_G$ which is used, in turn, for demonstrating DLOGSPACE-computability of the operation $\mathsf{TC} : \mathbf{HF} \to \mathbf{HF}$.

[6] [22] uses the denotation DTC essentially for the present $\odot$.

As above, we can define a corresponding class IC-DLOGSPACE of transformers $\mathcal{EAG}^*_< \to \mathcal{EAG}^*_<$ with its **HF**-version $\mathcal{DL}_0$ and also the class of **HF**-operations $\mathcal{DL} \supseteq \mathcal{DL}_0$ computable by these transformers and prove

Theorem 10 $\mathcal{DL} \sim_{\mathbf{HF}} \Delta^{\circledcirc}_< + \mathcal{DL}_0$.

4.4. *Anti-Foundation Axiom*

Let $\mathcal{V}$ be any universe for KP_0. We construct a corresponding anti-wellfounded universe $\mathcal{G}$ (in fact, its quotient $\mathcal{G}/_{=_0}$). Following [2, 14], define (in terms of KP_0) a Δ_0-class $\mathcal{G} \subseteq \mathcal{V}$ of all pointed graphs (pg's) together with two Σ-relations over this class: $\in_0$ and a *congruence* $=_0$. A class $R \subseteq \mathcal{V}$ is called a *bisimulation relation* between graphs G and G', shortly $\mathrm{Bis}(R, G, G')$, if R is a class of pairs and always

$$bRb' \text{ implies } (\forall x \to_G b \exists y \to_{G'} b'.\ xRy)\ \&\ (\forall y \to_{G'} b' \exists x \to_G b.\ xRy)\,.$$

Let $\sim_{G,G'}$ be the *largest* such R. Then define

$$\langle G, a\rangle =_0 \langle G', a'\rangle \rightleftharpoons a \sim_{G,G'} a' \ \ (\text{iff } \ \exists R \in \mathcal{V}.(\mathrm{Bis}(R, G, G')\& aRa')) \ \text{ and}$$

$$\langle G, a\rangle \in_0 \langle G', a'\rangle \rightleftharpoons \exists b' \to_{G'} a'.\langle G, a\rangle =_0 \langle G', b'\rangle\,.$$

Theorem 11 **([14])** $\mathsf{KP}_0 + \mathsf{Power} \vdash$ '$\langle \mathcal{G}, =_0, \in_0\rangle \models \mathsf{KP}_0 + \mathsf{Power} + \mathsf{AFA}$'.

It was essentially used in the proof of this theorem that Σ-relations $=_0$ and $\in_0$ become Δ in the presence of Powerset, which seems a too strong axiom from various points of view. For example, the corresponding operation is non-tractable and gives rise to *Kalmar elementary* rather than to *polytime* computability.

Our first aim is to show that an analogous effect may be reached for BSTA by using, instead of Powerset, the Recursive Δ-Separation axiom.

Note, that the largest bisimulation relation $\sim_{GG'} \in \mathcal{V}$ between any two graphs (sets of pairs) $G, G' \in \mathcal{V}$ is expressible as a $\Delta[\mathsf{KPR}_0]$-term of two variables G, G' (by using the-least construct). Therefore $=_0$ and $\in_0$ defined with the help of $\sim_{GG'} \in \mathcal{V}$ are also in $\Delta[\mathsf{KPR}_0]$. Evidently, $=_0$ is an equivalence relation which is *congruent* with respect to $\in_0$ and other Δ-constructs. This allows one to prove, by a lengthy induction argument

Theorem 12 $\mathsf{KPR}_0 \vdash$ '$\langle \mathcal{G}, =_0, \in_0\rangle \models \mathsf{BSTA}$'.

Here we take formally '$\langle \mathcal{G}, =_0, \in_0\rangle \models \varphi(\bar{x})$' $\rightleftharpoons \forall \bar{x} \in \mathcal{G}.[\![\varphi]\!](\bar{x})$ where $[\![\varphi]\!]$ is a properly defined ($\Delta[\mathsf{KPR}_0]$-) translation of any ($\Delta[\mathsf{BSTA}]$-) formula φ which will give the intended meaning of φ in the universe $\langle \mathcal{G}, =_0, \in_0\rangle$. Simultaneously, we must define (by a structural induction) for any $\Delta[\mathsf{BSTA}]$-term $t = t(\bar{x})$ a corresponding $\Delta[\mathsf{KPR}_0]$-term $[\![t]\!]$. In particular, we must have $\mathsf{KPR}_0 \vdash \forall \bar{x} \in \mathcal{G}.([\![t]\!](\bar{x}) \in \mathcal{G})$. Let us define the point-changing operation for any pg $\langle G, x\rangle$ by $\langle G, x\rangle * y \rightleftharpoons \langle G, y\rangle$ and also $\mathsf{elements}\langle G, x\rangle \rightleftharpoons \{y \in |G| : y \to_G x\}$. Then let (only for several simple cases; a full definition of $[\![\text{-}]\!]$ is given in [43])

$$[\![a \in b]\!] \rightleftharpoons [\![a]\!] \in_0 [\![b]\!]; \quad [\![a = b]\!] \rightleftharpoons [\![a]\!] =_0 [\![b]\!];$$

$$[\![\varphi \,\&\, \psi]\!] \rightleftharpoons [\![\varphi]\!] \,\&\, [\![\psi]\!]; \quad [\![\varphi \vee \psi]\!] \rightleftharpoons [\![\varphi]\!] \vee [\![\psi]\!]; \quad [\![\neg\varphi]\!] \rightleftharpoons \neg[\![\varphi]\!];$$
$$[\![\forall x \in a.\varphi(x)]\!] = [\![\forall x(x \in a \Rightarrow \varphi(x))]\!] \rightleftharpoons \forall x \in \mathsf{elements}[\![a]\!].[\![\varphi]\!]([\![a]\!] * x);$$
$$[\![\exists x \in a.\varphi(x)]\!] = [\![\exists x(x \in a \& \varphi(x))]\!] \rightleftharpoons \exists x \in \mathsf{elements}[\![a]\!].[\![\varphi]\!]([\![a]\!] * x);$$
$$[\![\forall x\varphi(x)]\!] \rightleftharpoons \forall x \in \mathcal{G}.[\![\varphi]\!](x), \quad \text{if } \varphi(x) \text{ has not the form } x \in a \Rightarrow \psi(x);$$
$$[\![\exists x\varphi(x)]\!] \rightleftharpoons \exists x \in \mathcal{G}.[\![\varphi]\!](x), \quad \text{if } \varphi(x) \text{ has not the form } x \in a \& \psi(x);$$
$$[\![\text{set-variable}]\!] \rightleftharpoons \text{the same variable};$$

etc., for all term constructs of $\Delta[\mathsf{BSTA}]$.

There exists an analogous $\langle\!\langle\text{-}\rangle\!\rangle$-translation of $\Delta[\mathsf{BSTA}]$-formulas (and terms) to LID-formulas so that

Theorem 13 $\mathsf{LID} \vdash \langle\!\langle$*all* Δ*-theorems of* $\mathsf{BSTA}\rangle\!\rangle$.

Note, that BSTA is conservative over $\mathsf{BSTA}\restriction\Delta$ by Theorem 15 (b) below. Also we can characterize in terms of LID the operations and predicates definable in $\Delta[\mathsf{BSTA}]$ over the universe $\langle \mathcal{G}, =_0, \in_0 \rangle /_{=_0}$, $\mathcal{G} \subseteq \mathcal{V}$, and its version **HFA** based on $\mathcal{V} = \mathbf{HF}$.

Theorem 14 (a) *The* Δ*-definable operations and predicates over* $\langle \mathcal{G}, =_0, \in_0 \rangle$ *in* BSTA *coincide, up to* $=_0$*, with those operations and predicates definable in the language* LID*, for which* $=_0$ *is a congruence relation.*

(b) *Each such operation over* **HFA** *is polytime computable with respect to the natural graph encoding based on* D.

One direction of (a) is based just on the $\langle\!\langle\text{-}\rangle\!\rangle$-translation of the $\Delta[\mathsf{BSTA}]$-language. Conversely, any LID-definable operation $g' = F(g)$ in $\langle \mathcal{G}, =_0, \in_0 \rangle$ over pg's (as sorts for LID) which preserves $=_0$ may be expressed in $\Delta[\mathsf{BSTA}]$ as a composition of the corresponding $\Delta[\mathsf{BSTA}]$-term $\lambda g \in \mathcal{G}.t_F(g)$ (as the second and main factor of the composition) directly imitating the given LID-definition of F up to an isomorphism of the output graph g', with $\lambda x \in \mathcal{V}.\langle \in \restriction \mathsf{TC}(\{x\}), x \rangle$ (as the first factor) and with D (as the third factor).

In contrast to Theorem 5, the converse to (b) is an *open question.* For an affirmative answer, it would be sufficient to define in $\Delta[\mathsf{BSTA}]$ (as in $\Delta[\mathsf{KPR}_0^*]$ for **HF**), if possible at all, *any* linear order on (each set of) the universe **HFA**.

4.5. *Relations to Finite Model Theory*

Computability and definability over **HF**- (or **HFA**-) sets were discussed in this paper in terms of finite graph transformers. This leads to the following interesting questions in finite model theory (cf. also [45]).

1. *Is the canonical linear ordering* $<$ *on each* $G \in \mathcal{EAG}_{(\prec)}$ *uniformly definable in* $\mathsf{FO}^{\otimes}$*?* Here '$(\prec)$' designates that graphs are considered either with or without *arbitrary* given linear order $\prec$ on their vertices.

2. *Is any linear order $\prec$ on $\mathcal{EAG}$'s uniformly definable in* $\mathsf{FO}^{\otimes}$*? And for the case of* LID *and arbitrary strongly extensional graphs* $(\mathcal{SEG})^7$*, possibly with cycles?* If 'YES', then the following question also has affirmative answer.
3. *Do the classes of* $\mathsf{FO}^{\otimes}$*-definable and* NLOGSPACE*-computable global predicates over* $\mathcal{EAG}$*'s coincide? And for the case of* LID *and arbitrary* $\mathcal{SEG}$*'s?*
4. *Is the canonical preordering* $\leq_G$ *on* $\mathcal{AG}$*'s definable in* $\mathsf{FO}^{\otimes}$*? What about the even weaker corresponding relation* $\equiv_G$ *identifying vertices of a graph* G *in* $\mathcal{AG}$*?* (For the case of $\mathcal{AG}_{\prec}$'s this is equivalent to 'PTIME $\stackrel{?}{=}$ NLOGSPACE').
5. *Does any* $\mathsf{FO}^{\otimes}$*-transformer* $\mathcal{EAG}_{<} \to \mathcal{EAG}_{<}$ *belong to the class* IC-$\mathsf{FO}^{\otimes}$*?* ((IC-)$\mathsf{FO}^{\otimes}$ coincides here with (IC-)NLOGSPACE.)

5. A PROOF THEORY FOR BST

Besides the proof-theoretic Theorems 1, 12 and 13 there are the following results.

Theorem 15 (a) *Provably-total* Σ*-definable operations of any* BST *coincide with corresponding* Δ*-definable ones.*

(b) *Any* BST *is conservative over its fragment* $\mathsf{BST}\upharpoonright\Delta$ *(without collection).*

In particular, the provably-total Σ-definable operations of KP_0 coincide with the basic = rudimentary ones. This theorem follows from the following generalized form of Harrop's property for some logical calculus L with the term constructs $\{t_1, t_2\}$ and $\cup\{t(x) : x \in s\}$ ([36]; cf. also [33], p.573):

$$[\mathsf{L} + \Delta\mathsf{Collection}] \vdash \exists x\varphi \quad \text{implies} \quad \mathsf{L} \vdash \exists x \in t\varphi$$

for some term t (which has no free occurrences of x). Here the relation $x \in y$ is read as 'x is bounded by y' and satisfies, with the above constructs, very weak natural axioms so that $\in$ may be interpreted either as a membership relation or a linear order or as $\subseteq$, etc. The proof is based on a cut-elimination and on interpretation of sequents $\Gamma \to \Pi$ in any proof containing only Σ-formulas [3] by the sequents of the form $\Gamma^v \to \Pi^{t(v)}$ for a new variable v and an appropriate term $t(v)$ bounding unbounded quantifiers $\exists$ in Γ and Π, respectively. (An analogous result was proved by R. Parikh [30] model-theoretically for a bounded arithmetical theory containing no collection principle; cf. also [5, 6, 31].)

Let now Δ be just the *basic language* $\Delta[\mathsf{BST}_0] = \Delta[\mathsf{KP}_0]$. Consider a *reduction relation* $\rhd$ based on the following rules for formulas in $\Delta^* = \Delta[\mathsf{KP}_0^*]$:

$s \in t \rhd \exists x \in t(s = x)$ if s is not an atomic term ('$\in t$' may be '$\in^* t$'),

$s \in \{t_1, t_2\} \rhd s = t_1 \vee s = t_2$,

[7] i.e. graphs having no different bisimilar vertices [2]

$$s \in \cup\{t(x) : x \in a \& \varphi(x)\} \rhd \exists x \in a(\varphi(x) \& s \in t(x)),$$
$$\exists y \in \{t_1, t_2\}\psi(y) \rhd \psi(t_1) \vee \psi(t_2),$$
$$\exists y \in \cup\{t(x) : x \in a \& \varphi(x)\}\psi(y) \rhd \exists x \in a(\varphi(x) \& \exists y \in t(x)\psi(y)),$$
$$s \in^* t \rhd s \in t \vee \exists x \in t(s \in^* x) \quad \text{for non-atomic } t,$$
$$\exists x \in^* t\psi(x) \rhd \exists x \in t\psi(x) \vee \exists x \in t \exists y \in^* x\psi(y) \quad \text{for non-atomic } t,$$

and analogously for $\forall$. (Formulas are considered up to renaming closed variables without any collisions.) Evidently, $\varphi \rhd \psi$ implies $\mathsf{KP}_0^* \vdash \varphi \leftrightarrow \psi$.

Theorem 16 (cf. [37]) *The reduction relation* $\rhd$ *satisfies the Church-Rosser and the normalizing properties.*

Therefore, any $\Delta^{(*)}$-formula φ is reducible to a unique $\Delta_0^{(*)}$-formula φ^0, a *normal form* of φ ($\varphi^0 = \varphi$ for $\varphi \in \Delta_0^{(*)}$). A version of this result on equivalence of any $\varphi \in \Delta^{(*)}$ to some $\varphi^0 \in \Delta_0^{(*)}$ and with the reducibility notion used only implicitly was proved by R. Gandy [15]. The corresponding *translation*

$$\Delta^* \to \Delta_0^*, \quad \varphi \mapsto \varphi^0$$

also serves as an *interpretation* of the formal system $\mathsf{KP}_0^* \restriction \Delta^*$ in $\mathsf{EXT}^* \restriction \Delta_0^*$. This means that each inference rule (in particular, axiom) of the former system becomes an admissible rule of the latter under this translation. This is proved by a complicated proof-theoretic induction argument. Therefore, it follows from Theorem 15 (b) and from the conservativity of EXT^* over EXT that

Theorem 17 *The theory* KP_0^* *conservatively extends* $\mathsf{EXT}^* \restriction \Delta_0^*$ *and* $\mathsf{EXT} \restriction \Delta_0$.

More generally, we may also allow urelements (i.e. atomic objects in the universe which are not sets and from which sets may be constructed step-by-step; some complication is involved in the corresponding reducibility notion). We write KPU and EXTU instead of KP and EXT in this case. Also individual constant symbols may be allowed, say, for the ordinal ω or for the set U of urelements with arbitrary additional closed Δ_0^*-axioms $\mathcal{A}_0$. Let

$$\Delta^*\mathsf{KPU} \rightleftharpoons \Delta^*\mathsf{KPU}(\mathcal{A}_0) \rightleftharpoons \mathsf{KPU}_0^* + \in\text{-}\Delta^*\text{-}\mathsf{Ind} + \mathcal{A}_0 \quad \text{and}$$
$$\Delta_0^*\mathsf{EXTU} \rightleftharpoons \Delta_0^*\mathsf{EXTU}(\mathcal{A}_0) \rightleftharpoons \mathsf{EXTU}^* + \in\text{-}\Delta_0^*\text{-}\mathsf{Ind} + \mathcal{A}_0 \,.$$

Further, let $\mathcal{R}$ and $\mathcal{R}^0$, $\mathcal{R}_0$ be classes of Δ^*- and, respectively, Δ_0^*-rules of inference (in particular, axioms), with $\mathcal{R}^0$ the translation of $\mathcal{R}$ under $(\text{-})^0$, such that (i) $\mathcal{R}$ is closed under substitutions of Δ^*-terms, (ii) $\mathcal{R}_0 \subseteq \mathcal{R}$, up to deducibility in $\Delta^*\mathsf{KPU}$, and (iii) Δ_0^*-rules $\mathcal{R}^0$ are admissible in $[\Delta_0^*\mathsf{EXTU} + \mathcal{R}_0] \restriction \Delta_0^*$. For example, we may include in $\mathcal{R}_{(0)}$ the axiom 'U is a set' and the schema of set-theoretic *Induction over* U

$$\mathsf{IU}_{(0)}: \quad \varphi(0) \ \& \ \forall nn' \in U(\varphi(n) \& \langle n, n' \rangle \in S \Rightarrow \varphi(n')) \Rightarrow \forall n \in U\varphi(n)$$

with $\varphi \in \Delta_{(0)}^*$ and the constants $0 \in U$ for "zero" and $S \subseteq U^2$ for any reasonable (say, as in $\mathsf{PA}_\square$ below) "successor" relation. Then we have (cf. [37] and a forthcoming paper of the author with full proofs)

Theorem 18 *The translation $\varphi \mapsto \varphi^0$, φ in Δ^*, is an interpretation of the calculus $[\Delta^*\mathsf{KPU} + \mathcal{R}]\restriction\Delta^*$ in $[\Delta_0^*\mathsf{EXTU} + \mathcal{R}_0]\restriction\Delta_0^*$ and, therefore,*

$$[\Delta^*\mathsf{KPU} + \mathcal{R}]\restriction\Delta^* \vdash \varphi \quad \textit{iff} \quad [\Delta_0^*\mathsf{EXTU} + \mathcal{R}_0]\restriction\Delta_0^* \vdash \varphi^0\,,$$

so that $\Delta^*\mathsf{KPU} + \mathcal{R}$ conservatively extends $[\Delta_0^*\mathsf{EXTU} + \mathcal{R}_0]\restriction\Delta_0^*$.

The same results hold with '$*$' and/or $\in$-Ind rules omitted. It follows also

Theorem 19 *Any model E_0 of $\Delta_0^{(*)}\mathsf{EXTU} + \mathcal{R}_0$ is a transitive class of corresponding $\Delta^*(E_0)$-termal model E of $[\Delta^*\mathsf{KPU} + \mathcal{R}]\restriction\Delta^*$.*

By Theorem 15 (b) the model E can be further embedded (not necessary as a transitive class) into some model E' of full $\Delta^*\mathsf{KPU} + \mathcal{R}$ satisfying the same Δ^*-formulas. We may also obtain e.g. the following extension of a result by G. Jäger's [24] to the case of an *arbitrary* first-order theory T in a relational language $\{R_1, R_2, \ldots; c_1, c_2, \ldots\}$ with $0 \rightleftharpoons c_1$ and $S \rightleftharpoons R_1 \subseteq U^2$. (Our proof is based on Theorems 18, 19 and 15 (b).) Generalizing [24], define two *covers* of T as the following set theories (of a universe of sets over a model of T)

$$T_c \rightleftharpoons \Delta_0^*\mathsf{EXTU} + \text{`}U \text{ and all } R_i \subseteq U^{k_i} \text{ are sets and } c_j \in U\text{'} + T^{\mathcal{U}} \ (+\,\mathsf{IU}_0),$$

$$T^c \rightleftharpoons \Delta^*\mathsf{KPU} + \text{`}U \text{ and all } R_i \subseteq U^{k_i} \text{ are sets and } c_j \in U\text{'} + T^{\mathcal{U}} \ (+\,\mathsf{IU})\,.$$

Here $\varphi^{\mathcal{U}}$ is relativization of any formula φ in the language of T to the first-order structure $\mathcal{U} \rightleftharpoons \langle U; R_1, R_2, \ldots; c_1, c_2, \ldots\rangle$ whose components U, R_i are considered as sets in the universe for T^c or T_c. We include IU (IU_0) in T^c (T_c) if T contains the corresponding full first-order *Induction Schema*

$$\mathsf{IND}: \quad \varphi(0)\ \&\ \forall nn'(\varphi(n) \& S(n,n') \Rightarrow \varphi(n')) \Rightarrow \forall n \varphi(n).$$

Moreover, let $\mathcal{A}_0$ in $\Delta_0^*\mathsf{EXTU}$ be such that $T \vdash \varphi$ iff $T_c \vdash \varphi^{\mathcal{U}}$ for all φ in the language of T. E.g., this holds for empty $\mathcal{A}_0$ or for $\mathcal{A}_0 \rightleftharpoons$ 'ω is the first limit ordinal', an interesting possibility not considered in [24]. Then

Theorem 20 $\quad T^c \vdash \varphi^{\mathcal{U}} \quad$ *iff* $\quad T \vdash \varphi$.

In particular, we may take T to be *Peano Arithmetic* PA, say, in the language containing predicates for graphs of *all* partial recursive function(al)s with axioms corresponding to recursive, i.e. least fixed point equations between them (and possibly with second-order parameters in recursive equations and induction axiom; but, the arithmetic is still *first-order*). Remember, that PA is often formulated, for convenience, in the language of all primitive recursive functions. The above version of PA may involve basic algorithmically *undecidable* predicates. However, our present interest is in the following modification of such PA where this small defect disappears.

As a new application of Theorem 20 let $T \rightleftharpoons \mathsf{PA}_\Box^{(\infty)}$ be the same as PA and in the same language as above, except it has one new constant $\Box$ for the *largest natural number* and respectively changed axioms for the successor relation, e.g. $S(\Box, \Box)$ (plus, in the case of ∞, the axioms $\Box \neq 0, \neq 1, \neq 2, \ldots$). In fact, Theorem 20 must be specially *adapted* to the case of $\mathsf{PA}_\Box$ with R_i second-order predicates representing $\Box$-recursive functionals which need *not* be sets. Fortunately, they may be included in Δ and $\mathsf{PA}^c_\Box$ as *operations over sets*.

Note, that $\mathsf{PA}_\Box$ is actually a $\Box$-*bounded arithmetic* of $\Box$-*recursive* or, equivalently, *polynomial time computable* global functions (relative to some second-order parameters) over the segment $U_\Box = \{0, 1, \ldots, \Box - 1, \Box\}$ with $\Box$ any *indefinite*, formally finite and possibly non-standard natural number; cf. [32, 33, 38]. This statement is analogous to the formulation of Theorem 2, however, with some axiomatic ($\Box$-Peano Arithmetic) flavour.

We may let $\Box = 17$ or one million, etc. and consider this number as some occasional *resource bound* which may be fixed in some concrete application. Of course, it is desirable that a theory should be quite general and not oriented, say, toward some specific computer "hardware".

On the other hand, we may take $\Box$ as some "non-feasible" and "concrete" (from the point of view of Classical or Intuitionistic Mathematics) natural number, say, $\Box = 2^{1000}$. We can write this axiom rather shortly as $\log_2 \Box = 1000$ by a formula of $\sim$ one thousand of symbols. (A shorter axiom $\log_2 \log_2 \Box = 10$ corresponds to $\Box = 2^{1024}$.) In some *non-traditional*, though very natural sense this axiom is consistent even with $\mathsf{PA}^\infty_\Box$ if we assume that all (formulas in) formal proofs should be "physically" written on a sheet of paper or in a computer memory. We can even imagine very informally a (second-order) model $\mathcal{U}_\Box$ for the resulting theory, even though no universe for ZF has such a model as a legal object.

Therefore, we have a first-order theory of "recursive" or "constructive" *finite* functions and functionals (say, over "non-standard" finite row of numbers $\{0, 1, \ldots, 2^{1000}\}$). Moreover, $\Box$-recursive functions (and numbers, as 0-place functions) of any model $\mathcal{U}_\Box \models \mathsf{PA}_\Box$ also constitute a submodel for $\mathsf{PA}_\Box$ because first-order quantifiers may be eliminated in $\mathsf{PA}_\Box$. It is a very interesting question related to the problem 'P=NP?' whether non-$\Box$-recursive or non-$\Box$-constructive finite functions are inevitable in $\mathsf{PA}^\infty_\Box$ with added full *second-order* induction and comprehension axioms. (Cf. also [33, 38, 42].) Of course, here is both some evident analogy with Kolmogorov's complexity theory for finite objects and also a difference.

By Theorem 20 we have that the admissible cover $\mathsf{PA}^c_\Box$, as a set theory for a universe $\mathcal{V}_\Box$ which even may contain the ordinal ω, is conservative over $\mathsf{PA}_\Box$. The same holds for PA or $\mathsf{PA}^\infty_\Box$. Therefore it also serves as a bounded set theory corresponding to polytime computability, say, over sets of natural numbers $0, 1, \ldots, \Box - 1, \Box$ (urelements). Due to the presence of Δ-Collection and Δ-Separation (cf. Δ-Union) axioms in set theory $\mathsf{PA}^c_{(\Box)}$ we also have

Corollary 21 $\mathsf{PA}_{(\Box)} + (\Delta^1_1\text{-}\mathsf{CA})_{(\Box)}$ *conservatively extends (first-order)* $\mathsf{PA}_{(\Box)}$.

Here $(\Delta^1_1\text{-}\mathsf{CA})_{(\Box)}$ denotes the *Comprehension Axiom*

$$\forall \bar{x}(\varphi_1(\bar{x}) \Leftrightarrow \neg\varphi_2(\bar{x})) \Rightarrow \exists R \forall \bar{x}(R(\bar{x}) \Leftrightarrow \varphi_1(\bar{x}))$$

for any two Σ^1_1-formulas $\varphi_i(\bar{x}) \rightleftharpoons \exists P \psi_i(\bar{x}, P)$ (with ψ_i first-order and $P \subseteq U^k$).

Program Systems Institute of Russian Academy of Sciences
Pereslavl-Zalessky

ACKNOWLEDGEMENTS

The author is very grateful for encouragement and support to Ju.L. Ershov, S. Feferman, G.E. Mints, Y.N. Moschovakis, R. Parikh, M.A. Taitslin and B.A. Trakhtenbrot.

REFERENCES

1. Aczel, P.: 1977, 'Introduction to the theory of inductive definitions', in: Barwise, J. ed., *Handbook of Mathematical Logic*, North-Holland, Amsterdam.
2. Aczel, P.: 1988, *Non-well-founded sets*, CSLI Lecture Notes **Vol. 14**, Stanford, 1988.
3. Barwise, J.: 1975, *Admissible sets and structures*, Berlin, Springer.
4. Barwise, J. and Moss, L.S.: 1995, *Vicious Circles: On the Mathematics of Non-wellfounded Phenomena*, manuscript of a book.
5. Buss, S.R.: 1986, *Bounded Arithmetic*, Bibliopolis, Napoli.
6. Buss, S.R.: 1987, 'A conservation result concerning bounded theories and the collection axiom', *Proceedings of the American Mathematical Soiciety*, **100**, pp. 709–716.
7. Cobham, A.: 1964, 'The intrinsic computational difficulty of functions', in: Bar-Hillel, Y., ed. *Logic, Methodology and Philosophy of Science II*, Jerusalem, North-Holland, pp. 24–30.
8. Cook, S.: 1975, 'Feasible constructive proofs and the propositional calculus', *7th ACM Symp. on Theoty of Computing*, pp. 83–97.
9. Dahlhaus, E.: 1987, 'Is SETL a suitable language for parallel programming? – a theoretical approach', in: Börger, E. *at al.* eds., *CSL'87, LNCS* **Vol. 329**, pp. 56–63.
10. Dahlhaus, E. and Makowsky, J.: 1986, 'The Choice of programming Primitives in SETL-like Languages', in: *ESOP'86, LNCS* **Vol. 213**, pp. 160–172.
11. Ershov, Ju.L.: 1977, *Numberings theory*, "Nauka", Glavnaja redakcija phisiko-matematicheskoi literatury, Moskva (in Russian).
12. Feferman, S.: 1970, 'Formal theories for transfinite iterations of generalized inductive definitions and some subsystems of analysis', in: Kino, A. *et al.* eds., *Intuitionism and Proof Theory*, Amsterdam, North-Holland, pp. 303–326.
13. Feferman, S.: 1974, 'Predicatively reducible systems of set theory', in: *Proc. Symp. in Pure Math.* **Vol. 13, Part II**, pp. 11–32.
14. Fernando, T.: 'A Primitive recursive set theory and AFA: on the logical complexity of the largest bisimulation', Report CS-R9213 ISSN 0169-118XCWI P.O.Box 4079, 1009 AB Amsterdam, Netherlands.
15. Gandy, R.O.: 1974, 'Set-theoretic functions for elementary syntax', in: *Proc. Symp. in Pure Math.* **Vol. 13, Part II**, pp. 103–126.
16. Gilula, M.M. 1994, *The Set Model For Database and Information Systems*, Addison Wesley.
17. Goncharov, S.S., Ershov, Ju.L. and Sviridenko, D.I.: 1986, 'Semantic programming', Kugler H.J., ed. *Information Processing'86*, Elsevier, North Holland, pp. 1093–1100.
18. Gurevich, Y.: 1983, 'Algebras of feasible functions', in: *Proc. 24th IEEE Conf. on Foundations of Computer Science*, pp. 210–214.
19. Gurevich, Y.: 1984, 'Towards logic tailored for computational complexity', in: *Computation and Proof Theory, LNM* **Vol. 1104**, Springer, Berlin, pp. 175–216.
20. Gurevich, Y. and Shelah, S.: 1986, 'Fixed point extensions of first-order logic', *Ann. Pure Appl. Logic* **32**, pp. 265–280.
21. Immerman, N.: 1982, 'Relational queries computable in polynomial time', in: *Proc. 14th. ACM Symp. on Theory of Computing*, pp. 147–152; cf. also *Inform. and Control* **68** (1986), pp. 86–104.
22. Immerman, N.: 1987, 'Languages which captures complexity classes', *SIAM J. Comput.*, **Vol. 16**, 4, pp. 760–778
23. Immerman, N.: 1989, 'Descriptive and computational complexity', *Proceedings of Symposia in Applied Mathematics* **Vol. 38**, pp. 75–91.
24. Jäger, G.: 1986, *Theories for admissible sets. A unifying approach to proof theory*, Studies in Proof Theory, LN **Vol. 2**, Bibliopolis.
25. Jensen, R.B.: 1972, 'The fine structure of the constructible hierarchy', *Ann. Math. Logic* **4**, pp. 229-308.
26. Jensen, R.B. and Karp, C.: 1971, 'Primitive recursive set functions', in: *Proc. Sympos. Pure Math.* **Vol. 13, Part I**, pp. 143–176.

27. Kolmogorov, A.N.: 1979, Automata and life (in Russian), *Kibernetika – neogranichennye vozmozhnosti i vozmozhnye ogranichenija. Itogi razvitija.* Moskwa, Nauka, pp. 10–29.
28. Livchak, A.B.: 1982, 'Languages of polynomial queries', in: *Raschet i optimizacija teplotehnicheskih ob'ektov s pomosh'ju EVM* , Sverdlovsk, p. 41 (in Russian).
29. Moschovakis, Y.N.: 1974, *Elementary Induction on Abstract Structures,* Amsterdam, North-Holland.
30. Parikh, R.: 1971, 'Existence and feasibility in arithmetic', *Journal of Symbolic Logic* **36**, (3), pp. 494–508.
31. Paris, J.B.: 1980, 'Some conservation results for fragments of arithmetic', *Model Theory and Arithmetic, LNM* **Vol. 890**, Springer, Berlin, pp. 251–262.
32. Sazonov, V.Yu.: 1980, 'Polynomial computability and recursivity in finite domains', *Elektronische Informationsverarbeitung und Kybernetik* **Vol. 16**, no. 7, pp. 319–323.
33. Sazonov, V.Yu.: 1980, 'A logical approach to the problem "P=NP?"', in: *Proc., Math. Found. of Comput. Sci., LNCS* **Vol. 88**, Springer. New York, pp. 562–575. (An important correction to this paper is given in [34], P. 490)
34. Sazonov, V.Yu.: 1981, 'On existence of complete predicate calculus in metamathematics without exponentiation', in: *LNCS* **Vol. 118** Springer, New York, pp. 483–490.
35. Sazonov, V.Yu.: 1985, 'Bounded set theory and polynomial computability', in: *All Union Conf. on Applied Logic*, Novosibirsk, pp. 188–191 (In Russian).
36. Sazonov, V.Yu.: 1985, 'Collection principle and existetial quantifier' (In Russian), in: *Logiko-matematicheskie problemy MOZ., Vychislitel'nye sistemy* **Vol. 107**, Novosibirsk, pp. 30–39. Cf. also English translation in: *AMS Transl.* (2) **Vol. 142** (1989) pp. 1–8.
37. Sazonov, V.Yu.: 1987, 'Bounded set theory, polynomial computability and Δ-programming', in: *Application aspects of mathematical logic, Computing systems* **122**, pp. 110–132 (In Russian). Cf. also a short English version of this paper in: *LNCS* **Vol. 278**, Springer, pp. 391–397.
38. Sazonov, V.Yu.: 1989, 'An equivalence between polynomial constructivity of Markov's principle and the equality P=NP' (in Russian), in: *Trudy instituta matematiki, Sibirskoje otdelenie akademii nauk SSSR, Matematicheskaja logika i algoritmicheskije problemy*, Novosibirsk, "Nauka", Sibirskoje otdelenije, pp. 138–165. (See also English translation in *Siberian Advances in Mathematics*, Allerton Press Inc., 1991, **Vol. 1**, N4, pp.92–121.)
39. Sazonov, V.Yu.: 1991, 'Bounded set theory and inductive definability', (Abstracts of Logic Colloquium'90), *JSL*, **Vol. 56**, no. 3, pp. 1141–1142.
40. Sazonov, V.Yu.: 1993, 'Hereditarily-finite sets, data bases and polynomial-time computability', *Theoretical Computer Science* **Vol. 119**, Elsevier, pp. 187–214.
41. Sazonov, V.Yu.: 1993, 'On polynomial regular codings of hereditarily finite sets', *Bulletin of Symbolic Logic* (Abstracts of European Summer Meeting of the Association for Symbolic Logic, Keel, 1993).
42. Sazonov, V.Yu.: 1995, 'On feasible numbers', in: Leivant, D. ed., *Logic and Computational Complexity, LNCS* **Vol. 960**, Springer, pp. 30-51.
43. Sazonov, V.Yu.: 1995, 'A bounded set theory with anti-foundation axiom and inductive definability', in: *Computer Science Logic – CSL'94, LNCS* **Vol. 933**, Springer, pp. 527–541.
44. Sazonov, V.Yu. and Lisitsa, A.P.: 1995, 'Δ-languages for sets and sub-PTIME graph transformers', in: *Database Theory – ICDT'95, LNCS* **Vol. 893**, Springer, pp. 125–138.
45. Sazonov, V.Yu. and Lisitsa, A.P.: 1995, 'Δ-languages for sets and LOGSPACE computable graph transformers', *manuscript of a forthcoming paper*.
46. Sazonov, V.Yu. and Leontjev, A.V.: 1992, 'On coding of hereditarily-finite sets, polynomial-time computability and Δ-expressibility', in: *Proc. Conf. on Applied Logic, Novosibirsk, May 1992, Computing systems* **Vol. 146** pp. 195-198 (In English).

47. Stolboushkin, A.P.: 1990, 'On the expressive power of program schemas with sets', in: *Proc. 5th Jerusalem Conf. on Information Technology*, JCIT, Jerusalem, Israel, Oct., 1990, Los Alamitos, CA: IEEE Computer Soc. Press.
48. Szelepcsényi, R.: 1987, 'The method of forcing for nondeterministic automata', *Bull. European Association Theor. Comp. Sci.* (Oct. 1987), pp. 96–100.
49. Troelstra, A.S. and van Dalen, D.: 1988, *Constructivism in Mathematics. An introduction*, **Vol. I, II**, North-Holland, Amsterdam.
50. Vardi, M.Y.: 1982, 'The complexity of relational query languages', in: *Proc. of the 14th. ACM Symp. on Theory of Computing*, pp. 137–146.
51. Yesenin-Volpin, A.S.: 1959, 'Analysis of the potential feasibility', in: *Logicheskije issledovanija*, Moskwa, AN SSSR, pp. 218–262. (In Russian).

2
Model Theory, Set Theory and Formal Systems

PHOKION G. KOLAITIS

INFINITARY LOGIC IN FINITE MODEL THEORY

ABSTRACT.
The infinitary logic $\mathcal{L}^{\omega}_{\infty\omega}$ consists of all formulas of $\mathcal{L}_{\infty\omega}$ with a finite number of variables. During the past several years, the study of $\mathcal{L}^{\omega}_{\infty\omega}$ has occupied a prominent place in finite model theory. We present an overview of results concerning the interaction of $\mathcal{L}^{\omega}_{\infty\omega}$ with least fixed-point logic and first-order implicit definability on finite structures.

1. INTRODUCTION

Finite model theory can be succinctly described as the study of logics on classes of finite structures. As a general rule, *first-order* logic has severely limited expressive power on classes of finite structures. For this reason, the emphasis has been on powerful extensions of first-order logic, such as *fragments* of *second-order* logic, logics with *fixed-point operators*, *infinitary* logics, and logics with *generalized quantifiers*. The class $\mathcal{F}$ of all finite structures over a given relational vocabulary σ and the class $\mathcal{O}$ of all *ordered* finite structures over σ are the two primary terrains on which the aforementioned extensions of first-order logic have been investigated. Moreover, numerous restricted classes of finite structures of combinatorial or algorithmic interest have been considered, including graphs of bounded degree, triangle-free graphs, trees, and various classes of equivalence relations.

There are three main research areas in finite model theory: *expressive power of logics*, *descriptive complexity theory*, and *logic and random structures*. The first area is the study of *uniform definability* in extensions of first-order logic on various classes of finite structures; its aim is to compare and classify logics in terms of their expressive power on finite structures. Descriptive complexity theory is concerned with the connections between uniform definability on finite structures and computational complexity. The prototypical result in this area is Fagin's [12] theorem, which asserts that on finite structures "$\Sigma^1_1 = \mathrm{NP}$", that is to say, a class of finite structures is definable by an existential-second order sentence if and only if it is accepted by a non-deterministic polynomial-time bounded Turing machine. The area of logic and random structures focuses on the asymptotic probabilities of properties expressible in a logic with respect to a given measure on finite structures. Of particular significance is the discovery of *zero-one laws* for certain logics, which means that the asymptotic probability (with respect to a given measure) of every property expressible in the logic under consideration exists and is either 0 or 1. This area started with the zero-one law for first-order logic with respect to the uniform measure on the class of all finite structures over a relational vocabulary, a result due to Glebskii et al. [15] and, independently, Fagin [14].

M.L. Dalla Chiara et al. (eds.), Logic and Scientific Methods, 107–125.

In this paper, we present an overview of certain recent advances in the first two areas of finite model theory. After introducing the basic notions, we cover the fundamentals of least fixed-point logic and implicit definability on finite structures. We then bring into center stage the infinitary logic $\mathcal{L}^{\omega}_{\infty\omega}$, which consists of all formulas of $\mathcal{L}_{\infty\omega}$ with a finite number of variables. We illustrate some of the uses and applications of $\mathcal{L}^{\omega}_{\infty\omega}$ to finite model theory by developing its interaction with least fixed-point logic and implicit definability. Space limitations prevent us from including detailed proofs of the main results. Nevertheless, we give the definitions of all relevant concepts and, whenever possible, describe the key ideas or techniques behind the proofs. Our goal will be achieved if this paper succeeds in serving as an introduction and guide to some of the recent developments in finite model theory.

2. BASIC CONCEPTS

A *vocabulary* is a set $\sigma = \{R'_1, \ldots, R'_m\}$ of relation symbols of specified arities. In the sequel, we will assume a fixed vocabulary σ with at least one non-unary relation symbol and will concentrate on the class $\mathcal{F}$ of all finite structures $\mathbf{A} = (A, R_1, \ldots, R_m)$ over σ. In most of the examples the vocabulary σ consists of a single binary relation symbol, which means that a structure over σ is a *graph* $G = (V, E)$. The following concept plays a key role in both finite model theory and database theory.

Definition 2.1 (Chandra and Harel [6])

- A *k-ary query*, $k \geq 1$, is a function Q on $\mathcal{F}$ such that
  (1) for every $\mathbf{A} \in \mathcal{F}$ the value $Q(\mathbf{A})$ is a k-ary relation on A;
  (2) Q is *preserved under isomorphisms*, which means that if $h : \mathbf{A} \rightarrow \mathbf{B}$ is an isomorphism, then $Q(\mathbf{B}) = h(Q(\mathbf{A}))$.
- A *Boolean query* is a function $Q : \mathcal{F} \rightarrow \{0, 1\}$ that is preserved under isomorphisms.

The canonical example of a binary query on graphs is the *transitive closure* query TC, which asks: "Given a graph $\mathbf{G} = (V, E)$ and two nodes a and b, is there a path from a to b?". More formally, TC is the function

$$\mathbf{G} \longmapsto TC(\mathbf{G}) = \{(a, b) \in V \times V : \text{there is a path from } a \text{ to } b\}.$$

Similarly, the question "Given a graph $\mathbf{G}$, is it 3-colorable?" gives rise to the 3-*colorability* Boolean query Q such that

$$Q(\mathbf{G}) = \begin{cases} 1, & \text{if } \mathbf{G} \text{ is 3-colorable,} \\ 0, & \text{otherwise.} \end{cases}$$

Note that every Boolean query can be identified with a class of finite structures closed under isomorphisms, which in the above example is the class of all 3-colorable graphs.

By their very nature, queries are semantical objects. ¿From the standpoint of a given logic, one is interested in determining which queries are expressible by formulas of that logic. The following definition makes precise what it means for a query to be uniformly definable in a logic.

Definition 2.2 Let $\mathcal{L}$ be a logic.

- A k-ary query Q is *$\mathcal{L}$-definable* if there is a formula $\varphi(x_1, \ldots, x_k)$ of $\mathcal{L}$ such that for every structure $\mathbf{A}$ in $\mathcal{F}$
$$Q(\mathbf{A}) = \{(a_1, \ldots, a_k) : \mathbf{A} \models \varphi(x_1/a_1, \ldots, x_k/a_k)\}.$$
- A Boolean query Q is *$\mathcal{L}$-definable* if there is a sentence ψ of $\mathcal{L}$ such that for every structure $\mathbf{A}$ in $\mathcal{F}$
$$Q(\mathbf{A}) = 1 \quad \text{if and only if} \quad \mathbf{A} \models \psi.$$
- $\mathcal{L}[\mathcal{F}]$ denotes the class of all $\mathcal{L}$-definable queries.

In effect, $\mathcal{L}[\mathcal{F}]$ measures the expressive power of a logic $\mathcal{L}$ on finite structures and serves as the basis for comparing $\mathcal{L}$ to other logics.

Definition 2.3 Let $\mathcal{L}$ and $\mathcal{L}'$ be two logics.

- $\mathcal{L}'$ is *at least as expressive as $\mathcal{L}$ on $\mathcal{F}$* if $\mathcal{L}[\mathcal{F}] \subseteq \mathcal{L}'[\mathcal{F}]$, which means that every $\mathcal{L}$-definable query is also $\mathcal{L}'$-definable.
- $\mathcal{L}'$ *strictly more expressive than $\mathcal{L}$ on $\mathcal{F}$* if $\mathcal{L}[\mathcal{F}] \subset \mathcal{L}'[\mathcal{F}]$, which means that $\mathcal{L}'$ is at least as expressive as $\mathcal{L}$ and, moreover, there exist queries that are $\mathcal{L}'$-definable but not $\mathcal{L}$-definable.

Remark 2.4 Let $\mathcal{C}$ be an arbitrary class of structures (finite or infinite) over σ closed under isomorphisms. All preceding definitions relativize to $\mathcal{C}$ in a natural way, so that they give rise to the concepts of *k-ary query on* C, *Boolean query on* C, and *$\mathcal{L}$-definable query on* C. We write $\mathcal{L}[\mathcal{C}]$ for the class of all $\mathcal{L}$-definable queries on $\mathcal{C}$. This extended framework is flexible enough to account for the following two important cases:

- *Uniform definability* on the class $\mathcal{S}$ of *all* structures (finite and infinite) over the vocabulary σ.
- *Local definability* on a fixed infinite structures of mathematical significance, such as the structure $\mathbf{N} = (N, +, *)$ of *arithmetic*.

Let us now return to finite structures and examine the expressive power of first-order logic. Using Ehrenfeucht-Fraïssé games, it is possible to show that the *transitive closure* query is *not* first-order definable (see [13]). The same method can also be used to establish that essentially no algorithmically interesting property of finite graphs is expressible in first-order logic. In particular, *connectivity*, *acyclicity*, *planarity*, *k-colorability* (for each fixed $k \geq 2$), and other similar Boolean queries are not first-order definable on finite graphs. This state of affairs attests that first-order logic does not embody a *recursion* or *iteration* mechanism on classes of finite structures. Therefore, it is natural to ask: how can the limitations of first-order logic be overcome?

What has to be added to first-order logic to increase its expressive power on finite structures? At first, one may consider moving to second-order logic. Note, however, that *transitive closure* and all other queries encountered earlier are Σ_1^1-definable on finite graphs. Moreover, Fagin's theorem [12] asserts that every NP-computable query is Σ_1^1-definable. Thus, even the syntactically simplest fragments of second-order logic can express queries that, in all likelihood, are computationally intractable. These findings suggest that the original question should be modified as follows: what has to be added to first-order logic to increase its expressive power on finite structures, while maintaining computational tractability? This modified question was addressed fairly early by researchers in database theory, who were interested in developing powerful, yet computationally feasible, query languages for the relational data model. Aho and Ullman [3] proposed to add recursion to first-order logic by augmenting it with suitable *fixed-point operators*. Their proposal proved to be particularly influential and resulted to an in-depth study of several logics with fixed-point operators, including *datalog*, *stratified datalog*, *least fixed-point* logic, and *inflationary fixed-point* logic (see the books by Ullman [32], and Abiteboul, Hull and Vianu [1]). The most prominent among these extensions is least fixed-point logic, which makes the topic of the next section.

3. LEAST FIXED-POINT LOGIC ON FINITE STRUCTURES

Let $\varphi(x_1, \ldots, x_k, S)$ be a first-order formula over an expanded vocabulary $\sigma \cup \{S\}$, where S is a k-ary relation symbol not in σ. Let $\mathbf{A}$ be a structure over σ and let $\mathcal{P}(A^k)$ be the power set of A^k, that is, the set of all k-ary relations on the universe A of $\mathbf{A}$. Then the formula φ gives rise to an operator $\Phi : \mathcal{P}(A^k) \rightarrow \mathcal{P}(A^k)$, where

$$\Phi(P) = \{(a_1, \ldots, a_k) : \mathbf{A} \models \varphi(x_1/a_1, \ldots, x_k/a_k, S/P)\}.$$

If φ is a *positive*[1] in S formula, then Φ is a *monotone* operator, which means that $\Phi(P_1) \subseteq \Phi(P_2)$ whenever $P_1 \subseteq P_2$. Therefore, by the Knaster-Tarski Theorem, Φ has a *least fixed-point*, that is, there is a smallest k-ary relation P^* such that $\Phi(P^*) = P^*$. We let $\varphi^\infty(x_1, \ldots, x_k)$ denote the least fixed-point of Φ and call it the *least fixed-point* of φ.

The proof of the Knaster-Tarski theorem actually shows that the least fixed-point of a monotone operator can be obtained by iterating the operator in a *bottom-up* manner. This basic fact also implies that evaluating least fixed-points of positive first-order formulas is a computationally tractable task.

Proposition 3.1 *If $\varphi(x_1, \ldots, x_k, S)$ is a positive first-order formula, then the least fixed-point $\varphi^\infty(x_1, \ldots, x_k)$ is computable on finite structures in time bounded by a polynomial in the size of the structure.*

---

[1]A first-order formula φ is *positive* in a relation symbol S if every occurrence of S in φ is within an even number of negations.

Proof: Define the *stages* φ^m, $m \geq 0$, of φ by the induction:

$$\varphi^0 = \emptyset, \quad \varphi^{m+1} = \Phi(\varphi^m).$$

Then, since Φ is monotone,

$$\varphi^0 \subseteq \varphi^1 \subseteq \ldots \subseteq \varphi^m \subseteq \varphi^{m+1} \subseteq \ldots.$$

Moreover, for every finite structure $\mathbf{A}$ there is an integer $m_0 \leq |A|^k$ such that $\mathbf{A} \models \varphi^{m_0} = \varphi^{m_0+1}$. It follows that

$$\mathbf{A} \models \varphi^\infty = \bigvee_{m=0}^{\infty} \varphi^m = \varphi^{m_0}. \blacksquare$$

Definition 3.2 *Least Fixed-Point Logic* LFP is the extension of first-order logic that is closed under the operations of first-order logic and the following *least fixed-point rule*: if $\varphi(x_1, \ldots, x_k, S)$ is a first-order formula that is positive in S, then $\varphi^\infty(x_1, \ldots, x_k)$ is also a formula of LFP.

The following examples illustrate the expressive power of LFP and manifest that it is a proper extension of first-order logic on finite structures.

Example 3.3 *Transitive Closure* and *Connectivity*

Let $\varphi(x, y, S)$ be the formula $E(x, y) \vee (\exists z)(E(x, z) \wedge S(z, y))$. Then

$$\begin{aligned} \varphi^m(x,y) &\equiv \text{ there is a path of length at most } m \text{ from } x \text{ to } y \\ \varphi^\infty(x,y) &\equiv \text{ there is a path from } x \text{ to } y. \end{aligned}$$

Thus, $\varphi^\infty(x, y)$ is an LFP-formula defining the *transitive closure* query TC on graphs $\mathbf{G} = (V, E)$. It follows that *connectivity* is also LFP-definable, since

$$\mathbf{G} \text{ is connected} \iff \mathbf{G} \models (\forall x)(\forall y)\varphi^\infty(x, y).$$

Example 3.4 *Acyclicity*

Let $\psi(x, S)$ be the formula $(\forall y)(E(y, x) \rightarrow S(y))$. Then

$$\begin{aligned} \psi^1(x) &\equiv \text{ indegree}(x) = 0 \\ \psi^2(x) &\equiv (\forall y)(E(y,x) \rightarrow \text{indegree}(y) = 0) \\ &\vdots \\ \psi^\infty(x) &\equiv \text{ no path through } x \text{ leads to a cycle.} \end{aligned}$$

It follows that *acyclicity* is LFP-definable, since for every finite graph $\mathbf{G}$

$$\mathbf{G} \text{ is acyclic} \iff \mathbf{G} \models (\forall x)\psi^\infty(x).$$

Example 3.5 *Path Systems*

A *path system* is a finite structure $\mathbf{S} = (F, A, R)$ such that A is a unary relation on F, and R is a ternary relation on F. Such a structure can be viewed as consisting of a finite set F of *formulas* in some formal system, a set of *axioms* A, and an *inference rule* R, for example, *resolution*. In this case, one is interested in computing the *theorems* of $\mathbf{S}$, that is, the smallest set of formulas containing A and closed applications of R. Therefore, we have the unary query T such that $T(\mathbf{S}) = \{t \in F : t \text{ is a theorem of } \mathbf{S}\}$. It is clear that T is definable by the least fixed-point $\psi^\infty(x)$ of the formula

$$\psi(x, S) \equiv A(x) \vee (\exists y)(\exists z)(S(y) \wedge S(z) \wedge R(x, y, z)).$$

This example is important, because, as established by Cook [8], T is a query that is complete for polynomial time[2], that is to say, every polynomial-time computable problem has a logarithmic-space reduction to T.

Although LFP can express queries that are complete for polynomial time, it can not express every polynomial-time computable query. Indeed, Chandra and Harel [6] showed that *counting* queries, such as *parity*, are not LFP-definable, where *parity* is the Boolean query that takes value 1 on a structure if and only if the universe of the structure has an even number of elements. The intuitive reason that LFP can not express *counting* queries is that no formula of LFP can tell apart two elements of an arbitrary finite structure.

In summary, we have the following picture for the expressive power of LFP:

$$\text{FO}[\mathcal{F}] \subset \text{LFP}[\mathcal{F}] \subset \text{PTIME}[\mathcal{F}],$$

which means that least fixed-point logic is strictly more expressive than first-order logic, but strictly less expressive than polynomial-time computability on the class $\mathcal{F}$ of all finite structures over a given vocabulary σ. As we are about to see, however, the picture changes on classes of *ordered* finite structures.

Definition 3.6 Let $\sigma = \{R'_1, \ldots, R'_m, \prec\}$ be a vocabulary such that $\prec$ is a binary relation symbol.

- An *ordered* finite structure over the vocabulary σ is a finite structure $\mathbf{A} = (A, R_1, \ldots, R_m, <)$ such that $<$ is a *linear order* on A.
- $\mathcal{O}$ is the class of all ordered finite structures over σ.

The next result, which is due to Immerman [19] and, independently, Vardi [34], reveals that on classes of ordered finite structures fixed-point logic captures polynomial-time computability.

Theorem 3.7 $\text{LFP}[\mathcal{O}] = \text{PTIME}[\mathcal{O}]$.

Proof: (*Hint*) Let M be a deterministic Turing machine and let $\mathbf{A}$ be an ordered structure with n elements. For every $k \geq 1$, the first n^k tape cells of M and the first n^k time units can be encoded by k-tuples of the structure using the *lexicographic order* induced by $\prec$. Then the first n^k steps of the computation of M can be simulated by an LFP-formula that describes how to obtain the *next* configuration from the *current* one. ∎

[2] As a matter of fact, this was the first query shown to be complete for polynomial time.

Remark 3.8 The investigation of fixed-point logic on finite structures was sparked by Aho and Ullman's [3] proposal to consider extensions of first-order logic with fixed-point operators. In particular, Chandra and Harel [6] embarked on a systematic study of least fixed-point logic on finite structures as a database query language. Nevertheless, least fixed-point logic had been introduced much earlier by logicians under the name *inductive definability* and studied extensively on fixed infinite structures. The origins of this line of work go back to Kleene [21] and Spector [30], who investigated inductive definability on the structure $\mathbf{N} = (N, +, *)$ of arithmetic. An in-depth investigation of inductive definability on arbitrary fixed infinite structures is contained in Moschovakis' monograph [26]. It should be noted, however, that there is an important technical difference between the definition of least fixed-point, as presented here, and the definition of inductive definability. Specifically, according to Definition 3.2, LFP-formulas are closed under negations, which implies that the class of LFP-definable queries is closed under complements. On the other hand, in classical inductive definability no applications of negation are allowed on top of least fixed-points of positive first-order formulas. Immerman [19] showed that these two different formalisms give rise to the same class of queries on finite structures, that is to say, complements of least fixed-points of positive first-order formulas are definable on finite structures as projections of least fixed-points of positive first-order formulas. Finiteness is important in this result, since, for example, the inductively definable relations on $\mathbf{N} = (N, +, *)$ are not closed under complements.

4. IMPLICIT DEFINABILITY ON FINITE STRUCTURES

Up to this point, we have considered queries that are *explicitly* definable in a logic. All the same, queries can be *implicitly* definable in a logic.

Definition 4.1 Let $\mathcal{L}$ be a logic, $\mathcal{C}$ a class of structures over σ closed under isomorphisms, and Q a query on $\mathcal{C}$. A sentence $\psi(T)$ of $\mathcal{L}$ over an expanded vocabulary $\sigma \cup \{T\}$ defines the query Q *implicitly* if for every structure $\mathbf{A}$ in $\mathcal{C}$, the value $Q(\mathbf{A})$ is the *only* relation P on $\mathbf{A}$ such that $\mathbf{A} \models \psi(T/P)$.

Implicit definability in first-order logic is a classical notion. Indeed, *Beth's definability theorem* [5] asserts that first-order implicit definability coincides with first-order explicit definability on the class $\mathcal{S}$ of all structures (finite and infinite) over a vocabulary σ. Note that Beth's theorem extends to *partial queries*, that is, queries that are defined on some (but not necessarily all) structures in $\mathcal{S}$. In this paper, a query Q on a class $\mathcal{C}$ of structures over σ is, by definition, a *total* function on $\mathcal{C}$.

The concept of implicit definability of *single* queries extends to implicit definability of *finite sequences* of queries. For our purposes here, it is important to recall that Beth's theorem extends to sequences of queries on the class $\mathcal{S}$ of all structures over a vocabulary σ (see [7]).

Theorem 4.2 *Let $(Q_1, \ldots, Q_m)$ be a sequence of queries on the class $\mathcal{S}$ of all structures over σ. Then the following are equivalent.*

- *$(Q_1, \ldots, Q_m)$ is implicitly definable on $\mathcal{S}$ by some first-order sentence $\psi(T_1, \ldots, T_m)$ over $\sigma \cup \{T_1, \ldots, T_m\}$.*
- *Each query Q_i, $1 \leq i \leq m$, is first-order definable on $\mathcal{S}$.*

Gurevich [16] demonstrated that most classical results of model theory *fail* in the context of finite structure. We now sketch his counterexample witnessing the failure of Beth's definability theorem in this context. Let $\sigma = \{\prec\}$ be a vocabulary consisting of a single binary symbol $\prec$, and let T be a unary relation symbol. It is not hard to construct a first-order sentence $\psi(T)$ over $\sigma \cup \{T\}$ such that for every finite structure $\mathbf{A} = (A, <)$ over σ and every unary relation P on A the following two statements are equivalent:

- $\mathbf{A} \models \psi(T/P)$
- ($<$ is a linear order and $P = \mathit{Even}$) or ($<$ is not a linear order and $P = \emptyset$), where *Even* is the set of all even elements of $<$.

Thus, $\psi(T)$ defines implicitly a query Q on the class $\mathcal{F}$ of all finite structures over σ. This query, however, is *not* first-order definable on $\mathcal{F}$, because it is is easy to see that *Even* is *not* first-order definable on finite linear orders.

What does the failure of Beth's definability theorem on finite structures signify? Two interpretations of this phenomenon are possible.

Negative Interpretation: View this failure as yet another manifestation that essentially all *nice* results of classical model theory *fail* in the context of finite structures. Give up on implicit definability on finite structures.

Positive Interpretation: View this failure as an *invitation* to study the expressive power of first-order implicit definability in the context of finite structures. Compare first-order implicit definability to other extensions of first-order logic on finite structures.

By adhering to the positive interpretation, in [22] we investigated the expressive power of first-order implicit definability on finite structures and compared it to least fixed-point logic.

Theorem 4.3 [22] *Let Q be a query on the class $\mathcal{F}$ of all finite structures over σ. If Q is* LFP*-definable, then there is an* LFP*-definable query Q^* such that the pair (Q, Q^*) is first-order implicitly definable on the class $\mathcal{F}$ of all finite structures over σ.*

Proof: (*Hint*) Assume that Q is definable by the least fixed-point $\varphi^\infty(\mathbf{x})$ of some positive first-order formula $\varphi(\mathbf{x}, S)$. Let Q^* be the *stage comparison query* $\prec^*_\varphi$ associated with φ, where

$$\mathbf{a} \prec^*_\varphi \mathbf{b} \iff \mathbf{a} \text{ enters } \varphi^\infty \text{ at an earlier stage than } \mathbf{b}.$$

The stage comparison theorem guarantees that $\prec^*_\varphi$ is LFP-definable [26]. We claim that the pair $(\varphi^\infty, \prec^*_\varphi)$ is implicitly definable on $\mathcal{F}$ by some first-order sentence $\psi(T, T^*)$ over the vocabulary $\sigma \cup \{T, T^*\}$ that describes properties of $\prec_{\varphi}^{*}$ and the relationship between φ^{∞} and $\prec_{\varphi}^{*}$. Indeed, it can be shown that $\psi(T, T^{*})$ is the conjunction $\psi_1(T, T^{*}) \wedge \psi_2(T, T^{*}) \wedge \psi_3(T, T^{*})$, where

$$\begin{aligned}
\psi_1(T, T^{*}) &\equiv T^{*} \text{ is a partial order on } T \\
\psi_2(T, T^{*}) &\equiv (\forall \mathbf{x})(\varphi(\mathbf{x}, T) \rightarrow T(\mathbf{x})) \\
&\qquad \wedge (\forall \mathbf{x})(T(\mathbf{x}) \rightarrow \varphi(x, \{\mathbf{w} : T^{*}(\mathbf{w}, \mathbf{x})\})) \\
\psi_3(T, T^{*}) &\equiv (\forall \mathbf{x})(\forall \mathbf{y})(T^{*}(\mathbf{y}, \mathbf{x}) \rightarrow \neg\varphi(\mathbf{x}, \{\mathbf{w} : T^{*}(\mathbf{w}, \mathbf{y})\})) \quad \blacksquare
\end{aligned}$$

Remark 4.4 We illustrate both the preceding Theorem 4.3 and the concept of the stage comparison query by revisiting one of our earlier examples.

Let $\varphi(x, y, S)$ be the formula $E(x, y) \vee (\exists z)(E(x, z) \wedge S(z, y))$, which, as seen in Example 3.3, defines the *transitive closure* query on graphs. It is easy to check that $\prec_{\varphi}^{*}$ is the *distance* query D on graphs, that is to say, $\mathbf{G} \models (a, b) \prec_{\varphi}^{*} (c, d)$ if and only if the length of the shortest path from a to b in $\mathbf{G}$ is less than the length of any path from c to d in $\mathbf{G}$ (there may be no path from c to d). Theorem 4.3 asserts that the pair (TC, D) is implicitly definable by some first-order sentence $\psi(T, T^{*})$ on the class of all finite graphs. Note that neither TC nor D is a first-order definable query on the class of all (finite and infinite) graphs; therefore, by Beth's theorem, neither of these queries is first-order implicitly definable on the class of all graphs.

The reader may wonder why the proof of Theorem 4.3 breaks down on the class of all (finite and infinite) graphs. The answer is that $\psi_1(T, T^{*})$ would have to be replaced by a sentence asserting that T^{*} is a *well-founded* partial order on T; for better or for worse, no such first-order sentence exists.

Remark 4.5 It should be pointed out that the use of implicitly definable *pairs* of queries is indispensable in Theorem 4.3. Indeed, it is not hard to verify that the *transitive closure* query TC is *not* implicitly definable by any first-order sentence $\psi(T)$ on the class of all finite graphs [22]. Therefore, once again we have encountered a phenomenon that is special to finite structures, since Theorem 4.2 implies that on the class of all structures first-order implicitly definable *pairs* of queries are just as expressive as first-order implicitly definable *single* queries.

Definition 4.6 If $\mathcal{C}$ is a class of structures over σ closed under isomorphisms, then IMP[$\mathcal{C}$] is the class of all queries on $\mathcal{C}$ that are members of a first-order implicitly definable sequence of queries on $\mathcal{C}$.

In other words, IMP[$\mathcal{C}$] consists of all queries Q on $\mathcal{C}$ for which there is an implicitly definable sequence $(Q_1, \ldots, Q_m)$ of queries on $\mathcal{C}$ such that $Q = Q_1$. Moreover, one can verify that it is enough to consider implicitly definable pairs of queries, that is, $m = 2$ suffices.

The results presented so far yield the following picture for least fixed-point logic and first-order implicit definability on the class $\mathcal{F}$ of all finite structures:

$$\text{FO}[\mathcal{F}] \subset \text{LFP}[\mathcal{F}] \subseteq \text{IMP}[\mathcal{F}].$$

To complete the picture, however, we need to find the answer to the question: does first-order implicit definability have *strictly* higher expressive power than least fixed-point logic on $\mathcal{F}$? This question was raised in [22], where it was conjectured that it has an affirmative answer.

Conjecture 4.7 LFP[$\mathcal{F}$] $\subset$ IMP[$\mathcal{F}$].

The time has now come to address a methodological issue, namely, what are the tools used to delineate the expressive power of least fixed-point logic on finite structures? In particular, how does one show that a given query is *not* LFP-definable? On finite structures, logics with fixed-point operators can be viewed as fragments of *infinitary logics* with *finitely many* variables. In turn, the expressive power of such infinitary logics can be analyzed using certain *pebbble games.* These two important facts give rise to a fruitful methodology for analyzing the expressive power of LFP and other similar logics. The next section contains the fundamentals of this methodology.

5. THE INFINITARY LOGIC $\mathcal{L}^{\omega}_{\infty\omega}$

The infinitary logic $\mathcal{L}_{\infty\omega}$ is the extension of first-order logic with arbitrary disjunctions and conjunctions: if Ψ is a set of $\mathcal{L}_{\infty\omega}$-formulas, then $\bigvee \Psi$ and $\bigwedge \Psi$ are $\mathcal{L}_{\infty\omega}$-formulas. Since the isomorphism type of every finite structure is first-order definable, it is easy to see that every query on the class $\mathcal{F}$ of all finite structures is $\mathcal{L}_{\infty\omega}$-definable. As a result, $\mathcal{L}_{\infty\omega}$ is too powerful on finite structures to make any meaningful distinctions. Note, however, that an $\mathcal{L}_{\infty\omega}$-formula may have an infinite number of distinct variables. By treating the number of variables as a resource and requring it to be finite, we can obtain fragments of $\mathcal{L}_{\infty\omega}$ that turn out to be of relevance in finite model theory. It is worth pointing out that these fragments were introduced by Barwise [4] to solve a problem in inductive definability on infinite structures.

Definition 5.1 (Barwise [4])

- $\mathcal{L}^k_{\infty\omega}$ is the collection of all $\mathcal{L}_{\infty\omega}$-formulas with at most k distinct variables, where k is a fixed positive integer.
- $\mathcal{L}^{\omega}_{\infty\omega} = \bigcup_{k\geq 1} \mathcal{L}^k_{\infty\omega}$ is the collection of all $\mathcal{L}_{\infty\omega}$-formulas with finitely many variables.

Notation: $\mathcal{L}^k_{\omega\omega}$ is the collection of all first-order formulas with k variables.

Example 5.2 *Cardinalities of Linear Orders*

On arbitrary structures, the statement *there are at least m distinct elements*, where m is a positive integer, can *not* be expressed by any first-order sentence with fewer than m variables. In contrast, on linear orders for every $m \geq 1$ there is an $\mathcal{L}^2_{\omega\omega}$-sentence θ_m asserting that there are at least m distinct elements. The idea is to reuse variables. For example,

$$\theta_4 \equiv (\exists x)(\exists y)[y < x \wedge (\exists x)(x < y \wedge (\exists y)(y < x))]$$

It follows that on the class of finite linear orders $\mathbf{L}_n$, $n \geq 1$, the infinitary logic $\mathcal{L}^2_{\infty\omega}$ can express *arbitrary* cardinalities, since for every set S of integers

$$n \in S \iff \mathbf{L}_n \models \bigvee_{m \in S} (\theta_m \wedge \neg\theta_{m+1}).$$

Example 5.3 *Transitive Closure* and *Connectivity* revisited
Let $p_m(x, y)$ be the statement *there is a path of length m from x to y*. This statement is expressible by a first-order formula with $m+1$ variables, namely,

$$(\exists z_1) \dots (\exists z_{m-1})(E(x, z_1) \wedge \dots \wedge E(z_{m-1}, y)).$$

At first sight every first-order formula expressing $p_m(x, y)$ seems to require $m+1$ distinct varialbes. Nonetheless, it turns out that just three variables suffice. Once again, the idea is to reuse variables; indeed, each $p_m(x, y)$ is definable by the formula $\psi_m(x, y)$ of $\mathcal{L}^3_{\omega\omega}$, where

$$\begin{aligned} \psi_1(x, y) &\equiv E(x, y) \\ \psi_m(x, y) &\equiv (\exists z)[E(x, z) \wedge (\exists x)(z = x \wedge \psi_{m-1}(x, y))]. \end{aligned}$$

Therefore, the $\mathcal{L}^3_{\infty\omega}$-formula $(\forall x)(\forall y)(\bigvee_{m \geq 1} \psi^m(x, y))$ defines *connectivity*.

The preceding Example 5.3 is an instance of Theorem 5.4 below asserting that $\mathcal{L}^\omega_{\infty\omega}$ subsumes LFP on the class $\mathcal{F}$ of all finite structures. A proof of this result, which essentially goes back to Rubin [28], can be found in [24].

Theorem 5.4 LFP$[\mathcal{F}] \subset \mathcal{L}^\omega_{\infty\omega}[\mathcal{F}]$.

Remark 5.5 It should be noted that $\mathcal{L}^\omega_{\infty\omega}$ does *not* subsume LFP on the class $\mathcal{S}$ of all (finite and infinite) structures. To see this, let $\psi(x, S)$ be the formula $(\forall y)(E(y, x) \rightarrow S(y))$. As demonstrated in Example 3.4, on the class of all finite graphs the LFP-sentence $(\forall x)\psi^\infty(x)$ defines the *acyclicity* query. A moment's reflection shows that on the class of all (finite and infinite) graphs this sentence asserts that the relation E is *well-founded.* López-Escobar [25], however, showed that *well-foundedness* is *not* definable in $\mathcal{L}_{\infty\omega}$. Intuitively, Theorem 5.4 fails on the class $\mathcal{S}$ of all structures, because we can not take an infinitary disjunction over all ordinal numbers. Nevertheless, $\mathcal{L}^\omega_{\infty\omega}$ is at least as expressive as LFP on every class $\mathcal{C}$ of structures whose cardinalities are bounded by some fixed cardinal number.

$\mathcal{L}^\omega_{\infty\omega}$ can be used as a tool in studying the expressive power of LFP on finite structures. In particular, lower bounds for definability in $\mathcal{L}^\omega_{\infty\omega}$ yield similar negative results for definability in LFP. The main advantage of $\mathcal{L}^\omega_{\infty\omega}$ over LFP is that for every $k \geq 1$ the relation of $\mathcal{L}^k_{\infty\omega}$*-equivalence* between structures has a clean characterization in terms of combinatorial *k-pebble games*. As a matter of fact, this characterization holds for structures of arbitrary cardinalities. For this reason, in the next two definitions the structures under consideration need not be finite.

Definition 5.6 Let k be a positive integer and $\mathbf{A}$, $\mathbf{B}$ two structures.

We say that $\mathbf{A}$ is $\mathcal{L}^k_{\infty\omega}$*-equivalent* to $\mathbf{B}$, and write $\mathbf{A} \equiv^k_{\infty\omega} \mathbf{B}$, if $\mathbf{A}$ and $\mathbf{B}$ satisfy the same sentences of $\mathcal{L}^k_{\infty\omega}$.

We say that $\mathbf{A}$ is $\mathcal{L}^k_{\omega\omega}$*-equivalent* to $\mathbf{B}$, and write $\mathbf{A} \equiv^k_{\omega\omega} \mathbf{B}$, if $\mathbf{A}$ and $\mathbf{B}$ satisfy the same sentences of $\mathcal{L}^k_{\omega\omega}$.

Definition 5.7 Let k be a positive integer and $\mathbf{A}$, $\mathbf{B}$ two structures.

The *k-pebble game on* $\mathbf{A}$ *and* $\mathbf{B}$ is played by two players, called[3] *Spoiler* and *Duplicator*. Each player uses k distinct pebbles: the *Spoiler*'s pebbles are $p_1, \ldots, p_k$ and the *Duplicator*'s are $q_1, \ldots, q_k$. During a round of the game, the *Spoiler* places a pebble p_i on an element of one of the two structures or removes a pebble p_i from such an element; the *Duplicator* tries to duplicate the move on the other structure by placing the corresponding pebble q_i on an element of the other structure or by removing it from such an element. So, a snapshot of the game produces the picture

$$\begin{array}{lcccccc} \mathbf{A}: & a_1 & a_2 & \ldots & a_l & & \\ & \updownarrow & \updownarrow & \cdots & \updownarrow & & \\ \mathbf{B}: & b_1 & b_2 & \ldots & b_l & & 1 \leq l \leq k, \end{array}$$

where for every $i \leq l$ the pair (p_i, q_i) of pebbles is placed on the pair (a_i, b_i) or on the pair (b_i, a_i) of elements from $\mathbf{A}$ and $\mathbf{B}$. The *Spoiler* wins if at some point the mapping

$$a_1 \mapsto b_1, a_2 \mapsto b_2, \ldots, a_l \mapsto b_l$$

is *not* a partial isomorphism of $\mathbf{A}$ to $\mathbf{B}$. The *Duplicator* wins if the above never happens, that is, (s)he can continue playing "forever" in such a way that a partial isomorphism is always maintained.

The above description of the k-pebble game can be formalized by defining the concept of a *winning strategy* for the *Duplicator* in terms of a family of partial isomorphisms with the *back and forth property up to k* (see [4]).

Example 5.8 *Paths of different length*

Let $\mathbf{P}_m$ be a graph that is a directed path of length m, where $m \geq 2$. It is easy to see that, whenever $m \neq n$, the *Spoiler* wins the 2-pebble game on $\mathbf{P}_m$ and $\mathbf{P}_n$. This should be contrasted with the r-move Ehrenfeucht–Fraïssé game for first-order logic. Indeed, the *Duplicator* wins the r-move Ehrenfeucht–Fraïssé game on $\mathbf{P}_m$ and $\mathbf{P}_n$ for all m and n that are sufficiently larger than r.

Example 5.9 *Cliques of different size*

Let $\mathbf{K}_m$ be the *complete* graph with m nodes, where $m \geq 2$. Then for every $k \geq 1$ and every $m, n \geq k$, the *Duplicator* wins the k-pebble game on $\mathbf{K}_m$ and $\mathbf{K}_n$. In particular, the *Duplicator* wins the k-pebble game on $\mathbf{K}_k$ and $\mathbf{K}_{k+1}$.

[3]The original names of the two players were *Player I* and *Player II*. The more descriptive names *Spoiler* and *Duplicator* were coined by Joel Spencer a number of years ago.

Barwise [4] showed that $\mathcal{L}^k_{\infty\omega}$-equivalence can be characterized in terms of the existence of a family of partial isomorphisms with the back and forth property up to k. Although Barwise [4] did not consider games explicitly, in effect his result states that k-pebble games characterize $\mathcal{L}^k_{\infty\omega}$-equivalence on arbitrary (finite or infinite) structures. Immerman [20] studied k-pebble games and showed that they characterize $\mathcal{L}^k_{\omega\omega}$-equivalence on finite structures. In this way, we have arrived at the promised result.

Theorem 5.10 *Let k be a positive integer and* $\mathbf{A}$, $\mathbf{B}$ *be two structures. Then the following are equivalent.*

- $\mathbf{A} \equiv^k_{\infty\omega} \mathbf{B}$.
- *The Duplicator wins the k-pebble game on* $\mathbf{A}$ *and* $\mathbf{B}$.

If both $\mathbf{A}$ *and* $\mathbf{B}$ *are finite structures, then the above are also equivalent to*

- $\mathbf{A} \equiv^k_{\omega\omega} \mathbf{B}$.

Corollary 5.11 Method for studying definability in $\mathcal{L}^k_{\infty\omega}$

Let $\mathcal{C}$ be a class of structures closed under isomorphisms and Q a Boolean query on $\mathcal{C}$. To show that Q is not *$\mathcal{L}^k_{\infty\omega}$-definable on $\mathcal{C}$, it suffices to show that there exist structures* $\mathbf{A}$ *and* $\mathbf{B}$ *in $\mathcal{C}$ such that*

- $\mathbf{A} \models Q$ *and* $\mathbf{B} \not\models Q$
- *The* Duplicator *wins the k-pebble game on* $\mathbf{A}$ *and* $\mathbf{B}$.

Moreover, this method is also complete, that is, if Q is not $\mathcal{L}^k_{\infty\omega}$-definable on $\mathcal{C}$, then such structures $\mathbf{A}$ *and* $\mathbf{B}$ *are guaranteed to exist.*

As mentioned in Section 3, Chandra and Harel [6] showed that *parity* is not LFP-definable. We can now give a one-line proof of a stronger result.

Corollary 5.12 *Parity is not $\mathcal{L}^\omega_{\infty\omega}$-definable on finite structures.*

Proof: As seen in Example 5.9, for every $k \geq 1$ we have that $\mathbf{K}_k \equiv^k_{\infty\omega} \mathbf{K}_{k+1}$.

6. THE INTERACTION OF $\mathcal{L}^\omega_{\infty\omega}[\mathcal{F}]$ WITH LFP[$\mathcal{F}$] AND IMP[$\mathcal{F}$]

Let us now return to Conjecture 4.7, which stated that

$$\text{LFP}[\mathcal{F}] \subset \text{IMP}[\mathcal{F}].$$

This conjecture was finally settled by establishing the sharper separation

$$\text{IMP}[\mathcal{F}] \cap \mathcal{L}^\omega_{\infty\omega}[\mathcal{F}] \subset \text{IMP}[\mathcal{F}].$$

In words, this means that there is a query Q on the class $\mathcal{F}$ of all finite structures that is first-order implicitly definable, but *not* $\mathcal{L}^\omega_{\infty\omega}$-definable. Since $\text{LFP}[\mathcal{F}] \subseteq \text{IMP}[\mathcal{F}] \cap \mathcal{L}^\omega_{\infty\omega}[\mathcal{F}]$, it follows that $\text{LFP}[\mathcal{F}] \subset \text{IMP}[\mathcal{F}]$.

The resolution to Conjecture 4.7 was obtained as a byproduct of research in a different area of finite model theory that has focused on the question: when does least fixed-point logic capture polynomial-time computability?

Problem 6.1 Characterize or identify all classes of finite structures $\mathcal{C}$ such that LFP[$\mathcal{C}$] = PTIME[$\mathcal{C}$].

A first milestone would be to identify all classes $\mathcal{C}$ of finite structures for which there is an LFP-formula that defines a *linear order* on every member of $\mathcal{C}$. Indeed, the earlier Theorem 3.7 of Immerman [19] and Vardi [34] implies that if $\mathcal{C}$ is such a class, then LFP[$\mathcal{C}$] = PTIME[$\mathcal{C}$].

Assume that $\mathcal{C}$ is a class of finite structures such that a formula of some logic defines a linear order on every member of $\mathcal{C}$. Then every member of $\mathcal{C}$ is a *rigid* structure, that is, the identity function is the only automorphism of that structure. For this reason, an extensive study of the definability of linear order on classes of rigid finite structures has been carried out in recent years, including work by Stolboushkin [31], Dawar [9], Hella, Kolaitis and Luosto [18], Gurevich and Shelah [17], and Seth [29]. In particular, Dawar [9] conjectured that *first-order axiomatizability* is a sufficient condition for the existence of an LFP-definable linear order on a class of rigid finite structures. Naturally, a class $\mathcal{C}$ of finite structures is *first-order axiomatizable* if there is a first-order sentence θ such that for every finite structure $\mathbf{A}$ we have that $\mathbf{A} \in \mathcal{C}$ if and only if $\mathbf{A} \models \theta$.

Conjecture 6.2 *Let $\mathcal{C}$ be a class of rigid finite structures. If $\mathcal{C}$ is first-order axiomatizable, then there is a formula of* LFP *that defines a linear order on every member of $\mathcal{C}$.*

Initially, the empirical evidence was in favor of this attractive conjecture; Gurevich and Shelah [17], however, refuted it in a decisive manner.

Theorem 6.3 [17] *There is a class $\mathcal{K}$ of rigid finite structures, called odd multipedes, with the following properties:*

- *$\mathcal{K}$ is first-order axiomatizable.*
- *No formula of $\mathcal{L}^{\omega}_{\infty\omega}$ defines a linear order on $\mathcal{K}$.*

The proof of Theorem 6.3 involves a technically difficult probabilistic construction of *odd multipedes* with special properties. Then k-pebble games are used to establish that $\mathcal{L}^{\omega}_{\infty\omega}$ can not define a linear order on $\mathcal{K}$. Nonetheless, as pointed out by Dawar, Hella and Kolaitis [10], first-order implicit definability can achieve this.

Proposition 6.4 *There is a sentence of first-order logic that defines implicitly a linear order on every member of the class $\mathcal{K}$ of odd multipedes.*

By combining Theorem 6.3 and Proposition 6.4, we obtain an affirmative answer to Conjecture 4.7 concerning the containment of LFP[$\mathcal{F}$] in IMP[$\mathcal{F}$].

Corollary 6.5 [10] *There is a query on the class $\mathcal{F}$ of all finite structures that is first-order implicitly definable, but not $\mathcal{L}^{\omega}_{\infty\omega}$-definable. As a result,* LFP[$\mathcal{F}$] *is properly contained in* IMP[$\mathcal{F}$].

Remark 6.6 Although our chief interest is in finite model theory, it is instructive to digress to two classical contexts for a moment and recall how first-order logic, least fixpoint logic, and first-order implicit definability compare there. In this comparison, we will also use Δ^1_1-definability, that is to say, we will consider queries that are definable both by some existential second-order formula (Σ^1_1) and some universal second-order formula (Π^1_1). The contexts we have in mind are the class $\mathcal{S}$ of all (finite and infinite) structures and the structure $\mathbf{N} = (N, +, *)$ of arithmetic. The point of this exercise is to demonstrate that the state of affairs in finite model theory differs dramatically from that in classical model theory and in hyperarithmetic theory.

- On the class $\mathcal{S}$ of all structures

$$\mathrm{FO}[\mathcal{S}] = \mathrm{IMP}[\mathcal{S}] = \Delta^1_1[\mathcal{S}] \subset \mathrm{LFP}[\mathcal{S}].$$

Beth's definability theorem and Craig's interpolation theorem yield the two equalities. Since the *connectivity* query is LFP-definable but not first-order definable, it follows that $\mathrm{FO}[\mathcal{S}]$ is properly contained in $\mathrm{LFP}[\mathcal{S}]$.

- On the structure $\mathbf{N} = (N, +, *)$ of arithmetic

$$\mathrm{FO}[\mathbf{N}] \subset \mathrm{IMP}[\mathbf{N}] = \Delta^1_1[\mathbf{N}] \subset \mathrm{LFP}[\mathbf{N}].$$

Several results in hyperarithmetic theory must be used to derive the above relationships (see [26]).

- On the class $\mathcal{F}$ of all finite structures

$$\mathrm{FO}[\mathcal{F}] \subset \mathrm{LFP}[\mathcal{F}] \subset \mathrm{IMP}[\mathcal{F}] \subset \Delta^1_1[\mathcal{F}].$$

The first two proper containments were established earlier. The last one is a consequence of Fagin's theorem [12] that $\Sigma^1_1[\mathcal{F}] = \mathrm{NP}[\mathcal{F}]$ and the fact that the *parity* query is not in $\mathrm{IMP}[\mathcal{F}]$.

In the sequel, we re-examine the relationship between least fixed-point logic and first-order implicit definability on finite structures. Recall that the containment of $\mathrm{LFP}[\mathcal{F}]$ in $\mathrm{IMP}[\mathcal{F}]$ was a consequence of Theorem 4.3, which asserted that for every LFP-definable query Q on $\mathcal{F}$ there is an LFP-definable query Q^* on $\mathcal{F}$ such that the pair (Q, Q^*) is first-order implicitly definable on $\mathcal{F}$. Consequently, every LFP-definable query Q on $\mathcal{F}$ is a member of a first-order implicitly definable pair (Q_1, Q_2) of queries such that both Q_1 and Q_2 are $\mathcal{L}^\omega_{\infty\omega}$-definable. This observation was the motivation for introducing in [10] the following refinement of $\mathrm{IMP}[\mathcal{F}]$.

Definition 6.7 $\mathrm{IMP}[\mathcal{F}]|\mathcal{L}^\omega_{\infty\omega}$ is the class of all queries Q on $\mathcal{F}$ for which there is an implicitly definable sequence $(Q_1, \ldots, Q_m)$ of queries on $\mathcal{F}$ such that $Q = Q_1$ and each Q_i, $1 \le i \le m$, is $\mathcal{L}^\omega_{\infty\omega}$-definable.

We call $\mathrm{IMP}[\mathcal{F}]|\mathcal{L}^\omega_{\infty\omega}$ the *restriction of* $\mathrm{IMP}[\mathcal{F}]$ *to* $\mathcal{L}^\omega_{\infty\omega}$.

With the class $\mathrm{IMP}|\mathcal{L}^{\omega}_{\infty\omega}$ at hand, Theorem 4.3 and Corollary 6.5 can be combined as follows.

Theorem 6.8 $\mathrm{LFP}[\mathcal{F}] \subseteq \mathrm{IMP}[\mathcal{F}]|\mathcal{L}^{\omega}_{\infty\omega} \subset \mathrm{IMP}[\mathcal{F}]$.

It is now natural to ask:

- What is the exact relationship between $\mathrm{LFP}[\mathcal{F}]$ and $\mathrm{IMP}[\mathcal{F}]|\mathcal{L}^{\omega}_{\infty\omega}$?
- Is $\mathrm{LFP}[\mathcal{F}]$ properly contained in $\mathrm{IMP}[\mathcal{F}]|\mathcal{L}^{\omega}_{\infty\omega}$?

Rather unexpectedly, it turns out that at this point finite model theory meets complexity theory. Indeed, we will show that separating $\mathrm{LFP}[\mathcal{F}]$ from $\mathrm{IMP}[\mathcal{F}]|\mathcal{L}^{\omega}_{\infty\omega}$ is equivalent to resolving an outstanding open problem in computational complexity. Admittedly, this result does not answer the above questions; nevertheless, it strongly suggests that separating $\mathrm{LFP}[\mathcal{F}]$ from $\mathrm{IMP}[\mathcal{F}]|\mathcal{L}^{\omega}_{\infty\omega}$ is a much bigger challenge than separating it from $\mathrm{IMP}[\mathcal{F}]$.

To develop the link with computational complexity, we need to bring in the class UP of NP-problems accepted by unambiguous Turing machines.

Definition 6.9 (Valiant [33])

- An *unambiguous Turing Machine* is a non-deterministic Turing machine such that for every input there is *at most one* accepting computation.
- UP is the class of all problems accepted by polynomial-time bounded unambiguous Turing machines.
- coUP is the class of all problems whose complement is in UP.

The following containments are more or less obvious from the definitions:

$$\mathrm{UP} \subseteq \mathrm{NP} \quad \text{and} \quad \mathrm{PTIME} \subseteq \mathrm{UP} \cap \mathrm{coUP} \subseteq \mathrm{NP} \cap \mathrm{coNP}.$$

In computational complexity, the predominant belief is that the above containments are proper (see [27]); moreover, strong evidence in support of this belief has been provided. Thus far, however, it has *not* been substantiated with a proof, despite numerous concerted efforts during the past twenty years. Therefore, the questions

$$\mathrm{UP} \stackrel{?}{=} \mathrm{NP} \quad \text{and} \quad \mathrm{PTIME} \stackrel{?}{=} \mathrm{UP} \cap \mathrm{coUP}$$

remain outstanding open problems in computational complexity. Note that UP and UP ∩ coUP are complexity classes of relevance to cryptography. In particular, the separation of PTIME from UP∩coUP is known to be equivalent to the existence of *easy to compute*, but *hard to invert* functions (see [27]).

The next result, due to Dawar, Hella and Kolaitis [10], pinpoints the difficulty of separating $\mathrm{LFP}[\mathcal{F}]$ from $\mathrm{IMP}[\mathcal{F}]|\mathcal{L}^{\omega}_{\infty\omega}$.

Theorem 6.10 *The following are equivalent.*

- $\mathrm{LFP}[\mathcal{F}] = \mathrm{IMP}[\mathcal{F}]|\mathcal{L}^{\omega}_{\infty\omega}$.
- $\mathrm{PTIME} = \mathrm{UP} \cap \mathrm{coUP}$.

Proof: (*Hint*) Let Q be a query in the class $\mathrm{IMP}[\mathcal{F}]|\mathcal{L}^{\omega}_{\infty\omega}$. It follows that Q is $\mathcal{L}^{\omega}_{\infty\omega}$-definable and there is an $\mathcal{L}^{\omega}_{\infty\omega}$-definable query Q^* such that the pair (Q, Q^*) is first-order implicitly definable on $\mathcal{F}$. In particular, both Q and Q^* are unions of $\mathcal{L}^{k}_{\infty\omega}$-equivalence classes for some $k \geq 1$. Abiteboul and Vianu [2] proved that for each $k \geq 1$ there is an LFP-definable preorder on the $\mathcal{L}^{k}_{\infty\omega}$-*types* of elements of finite structures. This important technical result makes it possible to translate Q and Q^* to queries on a class of ordered finite structures. Then we can use the following two theorems of descriptive complexity about the class $\mathcal{O}$ of all ordered finite structures: first, the aforementioned result in Immerman [19] and Vardi[34] that $\mathrm{LFP}[\mathcal{O}] = \mathrm{PTIME}[\mathcal{O}]$; second, a result in Kolaitis [22] that $\mathrm{IMP}[\mathcal{O}] = \mathrm{UP} \cap \mathrm{coUP}[\mathcal{O}]$. ∎

It is clear that $\mathrm{IMP}[\mathcal{F}]|\mathcal{L}^{\omega}_{\infty\omega} \subseteq \mathrm{IMP}[\mathcal{F}] \cap \mathcal{L}^{\omega}_{\infty\omega}[\mathcal{F}]$ and so, by combining several earlier results, we have the following picture:

$$\mathrm{LFP}[\mathcal{F}] \subseteq \mathrm{IMP}[\mathcal{F}]|\mathcal{L}^{\omega}_{\infty\omega} \subseteq \mathrm{IMP}[\mathcal{F}] \cap \mathcal{L}^{\omega}_{\infty\omega}[\mathcal{F}] \subset \mathrm{IMP}[\mathcal{F}].$$

The exact relationship between $\mathrm{LFP}[\mathcal{F}]$ and $\mathrm{IMP}[\mathcal{F})\cap\mathcal{L}^{\omega}_{\infty\omega}[\mathcal{F}]$ was determined in Dawar *et al.* [10].

Theorem 6.11 $\mathrm{LFP}[\mathcal{F}] \subset \mathrm{IMP}[\mathcal{F}] \cap \mathcal{L}^{\omega}_{\infty\omega}[\mathcal{F}]$.

Also in [10], it was conjectured that $\mathrm{IMP}[\mathcal{F}]|\mathcal{L}^{\omega}_{\infty\omega}$ is properly contained in $\mathrm{IMP}[\mathcal{F}] \cap \mathcal{L}^{\omega}_{\infty\omega}[\mathcal{F}]$. This conjecture does *not* appear to be equivalent to an open problem in computational complexity; it is conceivable that one can settle it using machinery that is currently available in finite model theory.

7. CONCLUDING REMARKS

The study of the infinitary logic $\mathcal{L}^{\omega}_{\infty\omega}$ with finitely many variables has occupied a central place in finite model theory during the past several years. One of the reasons for its preeminence is that $\mathcal{L}^{\omega}_{\infty\omega}$ is the "right" infinitary logic in which LFP and other logics with fixed-point operators can be embedded and studied. In this paper, we demonstrated that $\mathcal{L}^{\omega}_{\infty\omega}$ has manifold interactions with least fixed-point logic and implicit definability on finite structures. Specifically, we saw that $\mathcal{L}^{\omega}_{\infty\omega}$ is used in an essential way to establish that $\mathrm{IMP}[\mathcal{F}]$ contains $\mathrm{LFP}[\mathcal{F}]$ properly. Furthermore, we found out that the restriction $\mathrm{IMP}[\mathcal{F}]|\mathcal{L}^{\omega}_{\infty\omega}$ of $\mathrm{IMP}[\mathcal{F}]$ to $\mathcal{L}^{\omega}_{\infty\omega}$ forms a class of queries on finite structures that in many respects captures the complexity class UP ∩ coUP. In particular, separating $\mathrm{LFP}[\mathcal{F}]$ from $\mathrm{IMP}(\mathcal{F})|\mathcal{L}^{\omega}_{\infty\omega}$ amounts to separating PTIME from UP ∩ coUP, which would be a major breakthrough in computational complexity.

We conclude by pointing out that there are many distinctive features of $\mathcal{L}^{\omega}_{\infty\omega}$ that have not been covered here, such as certain deeper aspects of the intimate relationship of $\mathcal{L}^{\omega}_{\infty\omega}$ with least fixed-point logic [2, 9, 11, 23], and the zero-one law for $\mathcal{L}^{\omega}_{\infty\omega}$ with respect to the uniform measure [24].

Computer and Information Sciences
University of California
Santa Cruz

ACKNOWLEDGEMENTS

During the preparation of this paper, the author was supported by NSF Grants CCR-9307758 and INT-9024681.

REFERENCES

1. S. Abiteboul, R. Hull, and V. Vianu. *Foundations of Databases.* Addison-Wesley, 1995.
2. S. Abiteboul and V. Vianu. Computing with first-order logic. *Journal of Computer and System Sciences*, 50(2):309–335, April 1995. Special Issue: Selections from 23rd Annual ACM Symposium on Theory of Computing, New Orleans, LA, USA, 6-8 May 1991.
3. A. V. Aho and J. D. Ullman. Universality of data retrieval languages. In *Proc. 6th ACM Symp. on Principles of Programming Languages*, pages 110–117, 1979.
4. J. Barwise. On Moschovakis closure ordinals. *Journal of Symbolic Logic*, 42:292–296, 1977.
5. E. W. Beth. On Padoa's method in the theory of definition. *Indag. Math.*, 15:330–339, 1953.
6. A. Chandra and D. Harel. Structure and complexity of relational queries. *Journal of Computer and System Sciences*, 25:99–128, 1982.
7. C. C. Chang and H. J. Keisler. *Model Theory.* North Holland, 3rd edition, 1990.
8. S. A. Cook. An observation of time-storage trade-off. *Journal of Computer and System Sciences*, 9:308–316, 1974.
9. A. Dawar. *Feasible computation through model theory.* PhD thesis, University of Pennsylvania, Philadelphia, 1993.
10. A. Dawar, L. Hella, and Ph. G. Kolaitis. Implicit definability and infinitary logic in finite model theory. In *Proceedings of 22nd International Colloquium on Automata, Languages, and Programming. ICALP 95*, pages 624–635, Springer-Verlag, 1995.
11. A. Dawar, S. Lindell, and S. Weinstein. Infinitary logic and inductive definability over finite structures. *Information and Computation*, 119(2):160–75, June 1995.
12. R. Fagin. Generalized first–order spectra and polynomial–time recognizable sets. In R. M. Karp, editor, *Complexity of Computation, SIAM-AMS Proceedings, Vol. 7*, pages 43–73, 1974.
13. R. Fagin. Monadic generalized spectra. *Zeitschrift für Mathematische Logik und Grundlagen der Mathematik*, 21:89–96, 1975.
14. R. Fagin. Probabilities on finite models. *Journal of Symbolic Logic*, 41:50–58, 1976.
15. Y. V. Glebskii, D. I. Kogan, M. I. Liogonki, and V. A. Talanov. Range and degree of realizability of formulas in the restricted predicate calculus. *Cybernetics*, 5:142–154, 1969.
16. Y. Gurevich. Toward logic tailored for computational complexity. In M. M. Richter et al., editor, *Computation and Proof Theory, Lecture Notes in Mathematics 1104*, pages 175–216, Springer-Verlag, 1984.
17. Y. Gurevich and S. Shelah. On finite rigid structures. August 1994. Preprint. To appear in the *Journal of Symbolic Logic.*
18. L. Hella, Ph. G. Kolaitis, and K. Luosto. How to define a linear order on finite models. In *Proc. 9th IEEE Symp. on Logic in Computer Science*, pages 40–49, 1994.

19. N. Immerman. Relational queries computable in polynomial time. *Information and Control*, 68:86–104, 1986.
20. N. Immerman. Upper and lower bounds for first-order expressibility. *Journal of Computer and System Sciences*, 25:76–98, 1982.
21. S.C. Kleene. On the forms of the predicates in the theory of constructive ordinals (second paper). *American Journal of Mathematics*, 77:405–428, 1955.
22. Ph. G. Kolaitis. Implicit definability on finite structures and unambiguous computations. In *Proc. 5th IEEE Symp. on Logic in Computer Science*, pages 168–180, 1990.
23. Ph. G. Kolaitis and M. Y. Vardi. Fixpoint logic vs. infinitary logic in finite-model theory. In *Proc. 7th IEEE Symp. on Logic in Computer Science*, pages 46–57, 1992.
24. Ph. G. Kolaitis and M. Y. Vardi. Infinitary logic and 0-1 laws. *Information and Computation*, 98:258–294, 1992.
25. E.G.K. López-Escobar. On defining well-orderings. *Fundamenta Mathematicae*, 59:13–21, 1966.
26. Y. N. Moschovakis. *Elementary Induction on Abstract Structures.* North Holland, 1974.
27. C.H. Papadimitriou. *Computational Complexity.* Addison-Wesley, 1994.
28. A. Rubin. *Free algebras in von Neumann-Bernays-Gödel set theory and positive elementary induction in reasonable structures.* PhD thesis, California Institute of Technology, 1975.
29. A. Seth. When do fixed point logics capture complexity classes. In *Proc. 10th IEEE Symp. on Logic in Computer Science*, pages 353–363, 1995.
30. C. Spector. Inductively defined sets of natural numbers. In *Infinitistic Methods*, pages 97–102, Pergamon Press, New York, 1961.
31. A. Stolboushkin. Axiomatizable classes of finite models and definability of linear order. In *Proc. 7th IEEE Symp. on Logic in Computer Science*, pages 64–70, 1992.
32. J. D. Ullman. *Database and Knowledge-Base Systems, Volumes I and II.* Computer Science Press, 1989.
33. L. Valiant. Relative complexity of checking and evaluating. *Information Processing Letters*, 5:20–23, 1976.
34. M. Y. Vardi. The complexity of relational query languages. In *Proc. 14th ACM Symp. on Theory of Computing*, pages 137–146, 1982.

P.D. LINCOLN, A. SCEDROV, AND N. SHANKAR

DECISION PROBLEMS FOR SECOND-ORDER LINEAR LOGIC

ABSTRACT.

The decision problem is studied for fragments of second-order linear logic without modalities. It is shown that the structural rules of Contraction and Weakening may be simulated by second-order propositional quantifiers and the multiplicative connectives. Among the consequences are the undecidability of the intuitionistic second-order fragment of propositional multiplicative linear logic and the undecidability of multiplicative linear logic with first-order and second-order quantifiers. Several open problems related to second-order linear quantification are stated.

1. INTRODUCTION

Much of the expressive power and plasticity of linear logic, which was introduced in Girard [7], may be traced to the absence of structural rules of Contraction and to some extent Weakening. These rules are reintroduced into linear logic in a controlled fashion by the logical rules for modalities (or: exponentials), which allow, for instance, intuitionistic implication or function type $A \Rightarrow B$ to be expressed as $!A \multimap B$. Without the use of modalities, however, any reintroduction of the structural rules appears unlikely because of a phenomenon of *linearity*, namely, a formula occurrence may be the principal formula of an inference rule at most once in a (cut-free) proof. This phenomenon still holds in second-order linear logic. It is therefore surprising to discover that Contraction and Weakening can indeed be simulated by the multiplicative connectives using second-order propositional quantification.

Perhaps the most interesting consequences of our simulation of the structural rules concern an intuitionistic version of linear logic, presented by two-sided sequents with one consequent formula. We show that even the weakest of these fragments is undecidable in the presence of second-order propositional quantification. Our simulation of the structural rules also applies to the classical or multiple-conclusion framework for linear logic, but it yields undecidability only in the presence of both first-order and second-order quantifiers. The undecidability results in the classical propositional framework for linear logic have been subsequently obtained in Lafont [18] and Lafont and Scedrov [19].

Several overviews and expositions of linear logic are available in the literature: Girard [9], Scedrov [29, 30], and Troelstra [32]. In referring to linear logic fragments, let M stand for multiplicatives, A for additives, E for exponentials (or: modalities), 1 for first-order quantifiers, 2 for second-order quantifiers, and I for intuitionistic version of linear logic fragments. Thus IMLL2 denotes

M.L. Dalla Chiara et al. (eds.), Logic and Scientific Methods, 127–143.

the multiplicative fragment of second-order propositional intuitionistic linear logic. Furthermore, let LJ2 denote second-order intuitionistic propositional logic and let LK12 denote second-order classical predicate logic. The reader should recall that LK12 is undecidable in the presence of at least binary predicate variables. The undecidability of LJ2 was shown in Löb [26] and in Gabbay[5].

Decision problems for propositional (quantifier-free) linear logic were first studied in our previous work with J. Mitchell [21], where it was shown that full propositional linear logic is undecidable and that MALL is PSPACE-complete. The main problems left open in [21] were the NP-completeness of MLL, the decidability of MELL, and the decidability of various fragments of propositional linear logic without exponentials but extended with second-order propositional quantifiers.

The decision problem for MELL is still open. The NP-completeness of MLL has been obtained by Kanovich [14]. Lincoln and Winkler [24] have established that the provability of multiplicative propositional sentences built on constants, not literals, is already NP-complete. MALL1 with function symbols is NEXPTIME-complete: the hardness has been obtained by Lincoln and Scedrov [22] and the membership, and hence completeness, by Lincoln and Shankar [23]. Here we show the undecidability of IMLL2, IMALL2, MLL12, and MALL12. Subsequently, the undecidability of MALL2 has been shown by Lafont [18] and the undecidability of MLL2 by Lafont and Scedrov [19].

Let us point out that the undecidability of full second-order propositional linear logic was known: it follows from the undecidability of LJ2 by Girard's translation [7], which involves exponentials. Our translation below provides an alternative argument.

Fragments of intuitionistic linear logic have been studied in regard to several aspects of computer science. For instance, Gehlot and Gunter [6] have investigated relationship to Petri nets. Girard, Scedrov, and P. Scott [11] and more recently Girard [8] have studied the computational complexity aspects and in particular the versions of linear logic related to polynomial time. Functional programming aspects have been considered in, for instance, Abramsky [1], Lincoln and Mitchell [20], Benton *et al.* [3], and Chirimar, Gunter, and Riecke [4]. Logic programming aspects of intuitionistic linear logic have been investigated in Hodas and Miller [13]. Logic programming aspects of the classical or multiple-conclusion framework for linear logic have been studied in Miller [27, 28] and Lincoln and Shankar [23]. While the nature of our results is foundational, we believe that the techniques described here contribute to the understanding of the role of linear logic in describing control structure of second-order logic programs and perhaps in the study of compilation techniques of polymorphically typed functional languages as well. Furthermore, our result addresses the issue of replacing copy and delete features by second-order polymorphism. A preliminary version of this work has appeared in [25].

2. LOGICAL FRAMEWORK

Gentzen-style sequent calculus is the formal logical framework throughout the paper. For our purposes it is convenient to consider sequents of the form $\Gamma \vdash \Delta$ where Γ, Δ are finite multisets of formulas. In intuitionistic versions of sequent calculi one considers only those sequents for which Δ consists of a single formula. Let us refer to such sequents $\Gamma \vdash A$ as intuitionistic sequents. Note that in standard presentations of sequent calculi, sequents are often built from sets of formulas, where we use multisets here. For second-order predicate calculus we assume a countably infinite set of individual variables, a nonempty set of arities, and for each arity a countably infinite set of predicate variables and a countably infinite set of predicate atoms of that arity. Atomic formulas (other than constants) are of the form $X(x_1, \ldots, x_n)$ where $x_1, \ldots, x_n$ are individual variables and X is a predicate symbol (variable or atom) of arity n. For second-order propositional calculus we assume a countably infinite set of propositional variables (*i.e.*, predicate variables of arity zero) and a countably infinite set of propositional atoms. Atomic formulas are propositional variables or atoms.

The following notational conventions are observed throughout this paper: H denotes atomic formulas other than constants, G and G' denote predicate or propositional atoms (not variables or constants), X denotes propositional or predicate variables (not atoms or constants), A, B, D, and E denote formulas, and $\Gamma, \Delta, \Sigma, \Theta, \Xi$ denote finite multisets of formulas. In the rules $\forall\mathbf{R}$ and $\exists\mathbf{L}$ the propositional atom G must be chosen so that it does not occur in the conclusion of the rule application. Predicate calculi also include the analogous first-order versions of the quantifier rules with a similar restriction on first-order $\forall\mathbf{R}$ and $\exists\mathbf{L}$.

A presentation of inference rules for second-order classical propositional calculus, LK2, is given in Figure 1. This presentation of LK2 differs somewhat from some of the other expositions (say, in [31]) in several ways. For instance, negation is here treated as defined by $\neg True = False$, $\neg False = True$, $\neg\neg H = H$, and by so that $\neg\neg A = A$ for all formulas A and the negation rules are derived rules. Furthermore, the structural rules of contraction and weakening are not implicit in the logical rules. In this latter aspect our presentation is similar to the presentation in [10]. It is proof-theoretic folklore that the presentations are equivalent in the sense that the set of provable sequents is the same. Similar remarks apply to the intuitionistic calculus as well. In particular, let us consider the presentation of second-order intuitionistic propositional calculus, LJ2, in which the formulas are built from propositional variables and the constant $True$ by the connectives $\wedge$ and $\Rightarrow$ and by the universal second-order propositional quantifier $\forall$. The sequents of LJ2 are intuitionistic sequents of such formulas (*i.e.*, exactly one formula appears on the right-hand side.) The inference rules of LJ2 are the applicable rules of LK2, see Figure 2. It is proof-theoretic folklore that other intuitionistic connectives

$$\mathbf{I+}\quad \frac{}{H \vdash H} \qquad\qquad \frac{}{\neg H \vdash \neg H}\quad \mathbf{I\text{-}}$$

$$\mathbf{IL}\quad \frac{}{H, \neg H \vdash} \qquad\qquad \frac{}{\vdash H, \neg H}\quad \mathbf{IR}$$

$$\mathbf{CL}\quad \frac{\Gamma, A, A \vdash \Delta}{\Gamma, A \vdash \Delta} \qquad\qquad \frac{\Gamma \vdash A, A, \Delta}{\Gamma \vdash A, \Delta}\quad \mathbf{CR}$$

$$\mathbf{WL}\quad \frac{\Gamma \vdash \Delta}{\Gamma, A \vdash \Delta} \qquad\qquad \frac{\Gamma \vdash \Delta}{\Gamma \vdash A, \Delta}\quad \mathbf{WR}$$

$$\wedge\mathbf{L}\quad \frac{\Gamma, A, B \vdash \Delta}{\Gamma, (A \wedge B) \vdash \Delta} \qquad\qquad \frac{\Sigma \vdash A, \Theta \qquad \Gamma \vdash B, \Delta}{\Sigma, \Gamma \vdash (A \wedge B), \Theta, \Delta}\quad \wedge\mathbf{R}$$

$$\Rightarrow\mathbf{L}\quad \frac{\Sigma \vdash A, \Theta \qquad \Gamma, B \vdash \Delta}{\Sigma, \Gamma, (A \Rightarrow B) \vdash \Theta, \Delta} \qquad\qquad \frac{\Gamma, A \vdash B, \Delta}{\Gamma \vdash (A \Rightarrow B), \Delta}\quad \Rightarrow\mathbf{R}$$

$$\mathbf{FL}\quad \frac{}{False \vdash} \qquad\qquad \frac{}{\vdash True}\quad \mathbf{TR}$$

$$\forall\mathbf{L}\quad \frac{\Gamma, A[B/X] \vdash \Delta}{\Gamma, \forall X.A \vdash \Delta} \qquad\qquad \frac{\Gamma \vdash A[G/X], \Delta}{\Gamma \vdash \forall X.A, \Delta}\quad \forall\mathbf{R}$$

$$\exists\mathbf{L}\quad \frac{\Gamma, A[G/X] \vdash \Delta}{\Gamma, \exists X.A \vdash \Delta} \qquad\qquad \frac{\Gamma \vdash A[B/X], \Delta}{\Gamma \vdash \exists X.A, \Delta}\quad \exists\mathbf{R}$$

$$\frac{\Sigma \vdash A, \Theta \qquad A, \Gamma \vdash \Delta}{\Sigma, \Gamma \vdash \Theta, \Delta}\quad \mathbf{Cut}$$

Fig. 1. Rules for LK2

and the existential second-order propositional quantifier are definable in this presentation.

A presentation of the inference rules for second-order linear propositional calculus without modalities, MALL2, is given in Figure 4. The English names for the rules given in Figure 4 are identities, tensor, linear implication, plus, with, bottom, one, zero, top, universal, existential, and cut, respectively. $\otimes$, $\multimap$, and $^\perp$ are *multiplicative* connectives; 1 and $\perp$ are *multiplicative* propositional constants. $\oplus$ and & are *additive* connectives; 0 and $\top$ are *additive* propositional constants. Second-order multiplicative linear propositional calculus, MLL2, may be seen as that part of MALL2 that mentions only multiplicatives and second-order quantifiers. For notational convenience, it is often assumed that $\multimap$ and $\otimes$ associate to the right, and that $\otimes$ has higher precedence than $\multimap$.

$$\frac{}{H \vdash H}\ \mathbf{I}$$

$$\frac{\Gamma, A, A \vdash B}{\Gamma, A \vdash B}\ \mathbf{C}$$

$$\frac{\Gamma \vdash B}{\Gamma, A \vdash B}\ \mathbf{W}$$

$$\frac{}{\vdash True}\ \mathbf{T}$$

$$\wedge\mathbf{L}\ \frac{\Gamma, A, B \vdash C}{\Gamma, (A \wedge B) \vdash C} \qquad \frac{\Sigma \vdash A \qquad \Gamma \vdash B}{\Sigma, \Gamma \vdash (A \wedge B)}\ \wedge\mathbf{R}$$

$$\Rightarrow\!\mathbf{L}\ \frac{\Sigma \vdash A \qquad \Gamma, B \vdash C}{\Sigma, \Gamma, (A \Rightarrow B) \vdash C} \qquad \frac{\Gamma, A \vdash B}{\Gamma \vdash (A \Rightarrow B)}\ \Rightarrow\!\mathbf{R}$$

$$\forall\mathbf{L}\ \frac{\Gamma, A[B/X] \vdash C}{\Gamma, \forall X.A \vdash C} \qquad \frac{\Gamma \vdash A[G/X]}{\Gamma \vdash \forall X.A}\ \forall\mathbf{R}$$

$$\frac{\Sigma \vdash A \qquad A, \Gamma \vdash B}{\Sigma, \Gamma \vdash B}\ \mathbf{Cut}$$

Fig. 2. Rules for LJ2

$$\frac{}{H \vdash H}\ \mathbf{I}$$

$$\otimes\mathbf{L}\ \frac{\Gamma, A, B \vdash C}{\Gamma, (A \otimes B) \vdash C} \qquad \frac{\Sigma \vdash A \qquad \Gamma \vdash B}{\Sigma, \Gamma \vdash (A \otimes B)}\ \otimes\mathbf{R}$$

$$\multimap\mathbf{L}\ \frac{\Sigma \vdash A \qquad \Gamma, B \vdash C}{\Sigma, \Gamma, (A \multimap B) \vdash C} \qquad \frac{\Gamma, A \vdash B}{\Gamma \vdash (A \multimap B)}\ \multimap\mathbf{R}$$

$$\mathbf{1L}\ \frac{\Gamma \vdash A}{\Gamma, 1 \vdash A} \qquad \frac{}{\vdash 1}\ \mathbf{1R}$$

$$\forall\mathbf{L}\ \frac{\Gamma, A[B/X] \vdash C}{\Gamma, \forall X.A \vdash C} \qquad \frac{\Gamma \vdash A[G/X]}{\Gamma \vdash \forall X.A}\ \forall\mathbf{R}$$

$$\frac{\Sigma \vdash A \qquad A, \Gamma \vdash B}{\Sigma, \Gamma \vdash B}\ \mathbf{Cut}$$

Fig. 3. Rules for IMLL2

$$\textbf{I+}\quad \frac{}{H \vdash H} \qquad\qquad \frac{}{H^{\perp} \vdash H^{\perp}}\quad \textbf{I-}$$

$$\textbf{IL}\quad \frac{}{H, H^{\perp} \vdash} \qquad\qquad \frac{}{\vdash H, H^{\perp}}\quad \textbf{IR}$$

$$\otimes\textbf{L}\quad \frac{\Gamma, A, B \vdash \Delta}{\Gamma, (A \otimes B) \vdash \Delta} \qquad\qquad \frac{\Sigma \vdash A, \Theta \qquad \Gamma \vdash B, \Delta}{\Sigma, \Gamma \vdash (A \otimes B), \Theta, \Delta}\quad \otimes\textbf{R}$$

$$\multimap\textbf{L}\quad \frac{\Sigma \vdash A, \Theta \qquad \Gamma, B \vdash \Delta}{\Sigma, \Gamma, (A \multimap B) \vdash \Theta, \Delta} \qquad\qquad \frac{\Gamma, A \vdash B, \Delta}{\Gamma \vdash (A \multimap B), \Delta}\quad \multimap\textbf{R}$$

$$\oplus\textbf{L}\quad \frac{\Gamma, A \vdash \Delta \qquad \Gamma, B \vdash \Delta}{\Gamma, (A \oplus B) \vdash \Delta} \qquad\qquad \frac{\Gamma \vdash A, \Delta \qquad \Gamma \vdash B, \Delta}{\Gamma \vdash (A \,\&\, B), \Delta}\quad \&\,\textbf{R}$$

$$\&\,\textbf{L1}\quad \frac{\Gamma, A \vdash \Delta}{\Gamma, (A \,\&\, B) \vdash \Delta} \qquad\qquad \frac{\Gamma \vdash A, \Delta}{\Gamma \vdash (A \oplus B), \Delta}\quad \oplus\textbf{R1}$$

$$\&\,\textbf{L2}\quad \frac{\Gamma, B \vdash \Delta}{\Gamma, (A \,\&\, B) \vdash \Delta} \qquad\qquad \frac{\Gamma \vdash B, \Delta}{\Gamma \vdash (A \oplus B), \Delta}\quad \oplus\textbf{R2}$$

$$\perp\textbf{L}\quad \frac{}{\perp \vdash} \qquad\qquad \frac{\Gamma \vdash \Delta}{\Gamma \vdash \perp, \Delta}\quad \perp\textbf{R}$$

$$\textbf{1L}\quad \frac{\Gamma \vdash \Delta}{\Gamma, 1 \vdash \Delta} \qquad\qquad \frac{}{\vdash 1}\quad \textbf{1R}$$

$$\textbf{0L}\quad \frac{}{\Gamma, 0 \vdash \Delta} \qquad\qquad \frac{}{\Gamma \vdash \top, \Delta}\quad \top\textbf{R}$$

$$\forall\textbf{L}\quad \frac{\Gamma, A[B/X] \vdash \Delta}{\Gamma, \forall X.A \vdash \Delta} \qquad\qquad \frac{\Gamma \vdash A[G/X], \Delta}{\Gamma \vdash \forall X.A, \Delta}\quad \forall\textbf{R}$$

$$\exists\textbf{L}\quad \frac{\Gamma, A[G/X] \vdash \Delta}{\Gamma, \exists X.A \vdash \Delta} \qquad\qquad \frac{\Gamma \vdash A[B/X], \Delta}{\Gamma \vdash \exists X.A, \Delta}\quad \exists\textbf{R}$$

$$\frac{\Sigma \vdash A, \Theta \qquad A, \Gamma \vdash \Delta}{\Sigma, \Gamma \vdash \Theta, \Delta}\quad \textbf{Cut}$$

Fig. 4. Rules for MALL2

The reader will note that linear negation may be defined by recursion on the structure of formulas: $(H^{\perp})^{\perp}$ is H, $(A \otimes B)^{\perp}$ is $A \multimap B^{\perp}$, $(A \multimap B)^{\perp}$ is $A \otimes B^{\perp}$, $(A \,\&\, B)^{\perp}$ is $A^{\perp} \oplus B^{\perp}$, $(A \oplus B)^{\perp}$ is $A^{\perp} \,\&\, B^{\perp}$, $(\forall X.A)^{\perp}$ is $\exists X.A^{\perp}$, $(\exists X.A)^{\perp}$ is $\forall X.A^{\perp}$, $1^{\perp}$ is $\perp$, $\perp^{\perp}$ is 1, $\top^{\perp}$ is 0, and $0^{\perp}$ is $\top$. If Γ consists of $A_1, \ldots, A_n$, then $\Gamma^{\perp}$ denotes $A_1^{\perp}, \ldots, A_n^{\perp}$.

Although the identity rules, *i.e.,* the first four rules in Figure 4 are restricted to atomic formulas H, the analogous identity rules for arbitrary formula A instead of H are derivable. Negation rules are derivable as well, namely from $\Gamma, A \vdash \Delta$ one can infer $\Gamma \vdash A^{\perp}, \Delta$, and from $\Gamma \vdash A, \Delta$ one can infer $\Gamma, A^{\perp} \vdash \Delta$. The reader will also observe that the linear logic connective *par* may be defined by letting $A ⅋ B$ be $A^{\perp} \multimap B$ and that the *par* rule [7] is derivable. In fact, MALL2 is equivalent to the second-order propositional sequent calculus in [7] without exponentials $!, ?$. More precisely, a sequent $\Gamma \vdash \Delta$ is provable in MALL2 (without cut) iff the sequent $\vdash \Gamma^{\perp}, \Delta$ is provable (without cut, respectively) in the second-order propositional sequent calculus in [7], excluding exponentials. Similarly, MLL2 is equivalent to the second-order propositional sequent calculus in [7] restricted to multiplicatives and quantifiers.

The formulas of second-order intuitionistic linear propositional calculus without modalities, IMALL2, are built from propositional variables and constants $1, \top$ by the linear connectives and by the universal second-order propositional quantifier $\forall$. The sequents considered are intuitionistic sequents of such formulas. Second-order intuitionistic multiplicative linear propositional calculus, IMLL2, is that part of IMALL2 that mentions only multiplicatives and the universal quantifier. The inference rules of IMLL2 are given in Figure 3. The inference rules of IMALL2 are the applicable rules of MALL2 given in Figure 4, *i.e.*, the rules of IMLL2 given in Figure 3 together with the following rules from Figure 4: $\oplus$**L**, **& L** (all with Δ consisting of a single formula), and $\oplus$**R** **& R**, and $\top$**R** (all with Δ empty).

The cut-elimination property holds for all calculi considered here (see [7, 10] for cut-elimination property in second-order calculi). In particular, if a sequent is provable, then it is provable without using the **Cut** rule.

An *analysis* of a sequent is the process of applying a rule to a sequent matching the conclusion of the rule in order to generate the corresponding premises. The *principal formula* of the rule is then said to be *analyzed* by the reduction. The occurrence of an instance of a rule in a proof is said to be an *inference*. The proper subformulas of a principal formula of a rule that appear in the premises of the rule are called the *side formulas*. A sequent calculus proof is represented as a tree rooted at its conclusion sequent at the bottom and with the leaves at the top. Given this orientation, the notion of a rule occurring above or below another rule should be clear.

3. REPRESENTING INTUITIONISTIC LOGIC

We first define the IMLL2 formulas C and W as follows:

$$C \triangleq \forall X.X \multimap (X \otimes X)$$

$$W \triangleq \forall X.X \multimap 1$$

The idea is that C represents Contraction and W represents Weakening. We represent LJ2 in IMLL2 using the formulas C and W.

In order to simplify our constructions, we first develop some derived rules of inference in IMLL2 for sequents containing the formulas C, C, C, W in the left-hand side. Let us begin by demonstrating that C actually embodies the Contraction rule.

Lemma 3.1 *If $C, C, \Sigma, B, B \vdash A$ is provable in* IMLL2, *then $C, C, \Sigma, B \vdash A$ is provable in* IMLL2.

Proof. The following proof fragment shows that Contraction on the left-hand side of a sequent is a derived rule of inference, "Contract", for sequents with at least two copies of C. That is, reading this proof fragment from the bottom up, if one has one copy of the formula B in a sequent, then one can effectively copy it and have two copies of the formula B. Recall that the sequent $D \vdash D$ is provable in IMLL2 for any formula D.

$$\dfrac{\dfrac{\dfrac{\dfrac{}{B\vdash B}\quad\dfrac{}{C\vdash C}}{C,B\vdash B\otimes C}{\otimes R} \qquad \dfrac{\dfrac{\dfrac{C,C,\Sigma,B,B\vdash A}{B,C,B\otimes C,\Sigma\vdash A}{\otimes L}}{B\otimes C,B\otimes C,\Sigma\vdash A}{\otimes L}}{(B\otimes C)\otimes(B\otimes C),\Sigma\vdash A}{\otimes L}}{(B\otimes C)\multimap(B\otimes C)\otimes(B\otimes C),C,\Sigma,B\vdash A}{\multimap L}}{C,C,\Sigma,B\vdash A}{\forall L}$$

■

In the presence of Contraction, W embodies the Weakening rule.

Lemma 3.2 *If $C, C, W, \Sigma \vdash A$ is provable in* IMLL2, *then $C, C, W, \Sigma, B \vdash A$ is provable in* IMLL2.

Proof. The rule of inference "Weaken" can be derived as follows:

$$\dfrac{\dfrac{\dfrac{\dfrac{}{B\vdash B}\qquad\dfrac{C,C,W,\Sigma\vdash A}{1,W,C,C,\Sigma\vdash A}{1L}}{B\multimap 1,W,C,C,\Sigma,B\vdash A}{\multimap L}}{C,C,W,W,\Sigma,B\vdash A}{\forall L}}{C,C,W,\Sigma,B\vdash A}{\text{Contract}} \qquad \text{Weaken}$$

■

Let us proceed to the main result of this section. The representation of LJ2 in IMLL2 for any LJ2 multiset of formulas Σ and formula A is given by:

$$
\begin{array}{rcl}
[True] & \triangleq & 1 \\
[H] & \triangleq & H \\
[A \Rightarrow B] & \triangleq & [A] \multimap [B] \\
[A \wedge B] & \triangleq & [A] \otimes [B] \\
[\forall X.A] & \triangleq & \forall X.[A] \\
[A_1, \ldots, A_n] & \triangleq & [A_1], \ldots, [A_n] \\
[\Sigma \vdash A] & \triangleq & C, C, C, W, [\Sigma] \vdash [A]
\end{array}
$$

Lemma 3.3 *If* $\Sigma \vdash A$ *is provable in* LJ2, *then* $[\Sigma \vdash A]$ *is provable in* IMLL2.

Proof. This is shown by induction on the size of an LJ2 proof. We will construct an IMLL2 proof from a given LJ2 proof.

The LJ2 inference rules Contraction and Weakening are derived rules of inference in the class of IMLL2 sequents we consider, as shown above.

The LJ2 inference rule identity can be simulated in IMLL2 as follows:

$$
\dfrac{\dfrac{\dfrac{\overline{C \vdash C} \quad \overline{C \vdash C} \quad \overline{C \vdash C}}{C, C, C \vdash C \otimes C \otimes C}\,{\otimes R} \qquad \dfrac{\overline{H \vdash H}\,{I}}{1, H \vdash H}\,{1L}}{((C \otimes C \otimes C) \multimap 1), C, C, C, H \vdash H}\,{\multimap L}}{C, C, C, W, H \vdash H}\,{\forall L}
$$

The LJ2 inference rules of ∀L and ∀R coincide with the IMLL2 inference rules of the same name.

The LJ2 inference rules of ⇒R, and ∧L coincide with the IMLL2 inference rules ⊸R and ⊗L.

The LJ2 inference rules of ⇒L, and ∧R, and Cut all have two hypotheses. These rules coincide with ⊸L, ⊗R, and Cut in IMLL2, but the C, C, C, W component of the conclusion sequent must be duplicated into both hypotheses in order to apply the inductive hypothesis. This is straightforward since our encoding provides three copies of the formula C. As the derived rule "Contract" makes use of two copies of the formula C, we need a third copy of this formula to be the object of Contraction. For example, the ⊸ L rule appears as follows:

$$
\dfrac{\dfrac{\dfrac{\dfrac{\dfrac{C, C, C, W, \Sigma \vdash B \qquad C, C, C, W, \Gamma, D \vdash A}{C, C, C, C, C, C, W, W, \Sigma, \Gamma, B \multimap D \vdash A}\,{\multimap L}}{C, C, C, C, C, C, W, \Sigma, \Gamma, B \multimap D \vdash A}\,\text{Contract}}{C, C, C, C, C, W, \Sigma, \Gamma, B \multimap D \vdash A}\,\text{Contract}}{C, C, C, C, W, \Sigma, \Gamma, B \multimap D \vdash A}\,\text{Contract}}{C, C, C, W, \Sigma, \Gamma, B \multimap D \vdash A}\,\text{Contract}
$$

The other binary rules are similar. ∎

$$
\dfrac{\vdots \atop \vdash W' \qquad \dfrac{\vdots \atop \vdash C' \qquad \dfrac{\vdots \atop \vdash C' \qquad \dfrac{\vdots \atop \vdash C' \qquad \vdots \atop C', C', C', W', \Sigma \vdash A}{C', C', W', \Sigma \vdash A}\,\text{Cut}}{C', W', \Sigma \vdash A}\,\text{Cut}}{W', \Sigma \vdash A}\,\text{Cut}}{\Sigma \vdash A}\,\text{Cut}
$$

Fig. 5. Faithfulness of the encoding

The converse of Lemma 3.3 is straightforward and actually one easily obtains a slightly stronger fact about the fragment IMALL2.

Lemma 3.4 *If* $[\Sigma] \vdash [A]$ *is provable in* IMALL2, *then* $\Sigma \vdash A$ *is provable in* LJ2.

Proof. This is shown by a straightforward induction on size of an IMALL2 proof. Essentially, IMALL2 proof rules are easily derivable from their counterparts in LJ2, since the latter blurs the distinction between multiplicatives and additives. ■

The IMLL2 formulas C and W are the images of provable LJ2 formulas $C' = \forall X.(X \Rightarrow X \wedge X)$ and $W' = \forall X.(X \Rightarrow True)$. That is, $[C'] = C$, where $\vdash C'$ is provable in LJ2, and $[W'] = W$, where $\vdash W'$ is provable in LJ2. One then obtains the following result.

Lemma 3.5 *If* $C, C, C, W, [\Sigma] \vdash [A]$ *is provable in* IMALL2, *then* $\Sigma \vdash A$ *is provable in* LJ2.

Proof. By Lemma 3.4, given a proof of $C, C, C, W, [\Sigma] \vdash [A]$ in IMALL2, one can obtain a proof of $C', C', C', W', \Sigma \vdash A$ in LJ2. By Cut and the above mentioned proofs of $\vdash C'$ and $\vdash W'$, one can construct the LJ2 proof of $\Sigma \vdash A$ given in Figure 5. ■

Theorem 3.6 IMLL2 *and* IMALL2 *are undecidable.*

Proof. By Lemma 3.5 and Lemma 3.3, the provability of LJ2 sequents can be decided using IMLL2 or using IMALL2. However, LJ2 is undecidable [26, 5] and therefore both IMLL2 and IMALL2 are undecidable. ■

3.1. *Discussion*

The conservativity of IMALL2 over IMLL2, an open problem, is not needed for Theorem 3.6. Moreover, when the initially constructed proof of

$$C', C', C', W', \Sigma \vdash A$$

in LJ2 contains no uses of Contraction or Weakening (for instance, when it arises from an IMLL2 proof), the final proof uses Contraction in only three places (the proofs of $\vdash C'$) and Weakening in only one place (the proof of $\vdash W'$). The cut-elimination process essentially distributes the uses of Contraction and Weakening throughout the proof to all the places where C and W were used in essential ways in the IMLL2 proof. If one considers a cut-free LJ2 proof translated to IMLL2 and back again using the above two lemmas, cut elimination on the result can be made to produce exactly the original proof. Thus there is a great deal of efficiency in this translation.

It is curious that we require three copies of the formula C in the translated sequent. As it happens, there is an alternate encoding:

$$T \triangleq \forall X.X \multimap (X \otimes X \otimes X)$$

$$[\Sigma \vdash A] \quad \triangleq \quad T, T, W, [\Sigma] \vdash [A]$$

The formula T (for triplicate) is similar to C, but produces three copies of the target formula. The sequent can then be encoded using only two copies of the formula T and one of W. The Contraction and Weakening rules are derived rules of inference for sequents of the above form, and the proof of faithfulness proceeds along the same lines as for the above encoding.

Our encoding uses three occurrences of a contraction formula (C) that produces two copies, or two occurrences of a formula (T) that produces three copies. An encoding with only two copies of a contraction formula that produces two copies of a given formula is sufficient for those cases in which there is exactly one main branch of their proof. However, in our encoding we must follow an intuitionistic proof that may require Contraction rule along many branches. It is the need to copy the contraction formula itself in order to be able to perform Contraction rule along all branches of a proof that motivates using three copies of the formula C in our encoding.

3.2. *Limitations*

Our encoding is quite efficient at the levels of formula and proof, but is somewhat problematic at the level of cut normalization (lambda reduction). The standard sequent calculus cut-elimination procedure is extremely fast (polynomial) for IMLL and IMLL2 proofs. Cut normalization in propositional or second-order intuitionistic logic is equivalent to lambda reduction in the simply or polymorphically typed lambda calculus, which has much greater computational complexity. Thus there is no possibility for the standard cut-elimination procedures to correspond nicely. An alternative is to create a non-standard cut-elimination algorithm that packages up the sequence of proof rules associated with each application of derived rules Contract or Weaken and introduce special rules for cut elimination through these packages. If one were able to

$$\dfrac{\begin{array}{c}\vdots\\ \Gamma\vdash A\end{array} \qquad \dfrac{\dfrac{\dfrac{}{A\vdash A}I \quad \dfrac{}{A\vdash A}I}{A,A\vdash A\wedge A}\wedge R}{A\vdash A\wedge A}C}{\Gamma\vdash A\wedge A}Cut$$

Fig. 6. Cut on a contracted formula in LJ2

$$\begin{array}{c}\dfrac{\dfrac{\begin{array}{c}\vdots\\ \Gamma\vdash A\end{array} \qquad \begin{array}{c}\vdots\\ \Gamma\vdash A\end{array}}{\Gamma,\Gamma\vdash A\wedge A}\wedge R}{\vdots}C\\ \dfrac{}{\Gamma\vdash A\wedge A}C\end{array}$$

Fig. 7. The result of standard LJ2 cut-elimination on the proof in Figure 6

reliably synchronize the cut-elimination procedure in IMLL2 with the cut-elimination procedure in LJ2, then the two would correspond nicely, although some additional bookkeeping would be necessary in the IMLL2 proof. However, it is impossible to determine from an IMLL2 proof which steps correspond to derived rules Contract or Weaken and which correspond to LJ2 steps. Unfortunately, the result of cut normalization then becomes ill-defined, as there are multiple normal forms for some proofs.

For example, consider the LJ2 proof displayed in Figure 6, which involves cut of a contracted formula. This proof is mapped to IMLL2 proof shown in Figure 8.

In LJ2 there is only one possible result of cut elimination, displayed in Figure 7. In IMLL2, if one treats derived rule Contract as a single unit with special handling during cut-elimination process, then one obtains a proof analogous to the LJ2 cut-free proof:

$$\begin{array}{c}\dfrac{\dfrac{\begin{array}{c}\vdots\\ C,C,C,W,[\Gamma]\vdash[A]\end{array} \qquad \begin{array}{c}\vdots\\ C,C,C,W,[\Gamma]\vdash[A]\end{array}}{C,C,C,C,C,C,W,W,[\Gamma],[\Gamma]\vdash[A]\otimes[A]}\otimes R}{\vdots}\text{Contract}\\ \dfrac{}{C,C,C,W,[\Gamma]\vdash[A]\otimes[A]}\text{Contract}\end{array}$$

But if the standard IMLL2 cut-elimination process is applied to the original IMLL2 proof given in Figure 8, where the sequence of rule applications that make up derived rule Contract are treated individually, the cut-elimination process is short-circuited and the result is the cut-free proof shown in Figure 9.

$$\dfrac{\dfrac{\begin{matrix}\vdots\\ C,C,C,W,[\Gamma]\vdash[A]\end{matrix} \qquad \dfrac{\dfrac{\dfrac{}{C,C,C,W,[A]\vdash[A]}{\scriptstyle I'} \quad \dfrac{}{C,C,C,W,[A]\vdash[A]}{\scriptstyle I'}}{C,C,C,C,C,C,W,W,[A],[A]\vdash[A]\otimes[A]}{\scriptstyle \otimes R}}{C,C,C,W,[A]\vdash[A]\otimes[A]}{\scriptstyle \text{Contract}}}{C,C,C,C,C,C,W,W,[\Gamma]\vdash[A]\otimes[A]}{\scriptstyle Cut}}{C,C,C,W,[\Gamma]\vdash[A]\otimes[A]}{\scriptstyle \text{Contract}}$$

Fig. 8. Cut on a contracted formula in IMLL2

$$\dfrac{\dfrac{\dfrac{\begin{matrix}\vdots\\ C,C,C,W,[\Gamma]\vdash[A]\end{matrix} \quad \dfrac{}{C\vdash C}{\scriptstyle I}}{C,C,C,C,W,[\Gamma]\vdash C\otimes[A]}{\scriptstyle \otimes R} \qquad \dfrac{\dfrac{\dfrac{\dfrac{\dfrac{\dfrac{}{C,C,C,W,[A]\vdash[A]}{\scriptstyle I'} \quad \dfrac{}{C,C,C,W,[A]\vdash[A]}{\scriptstyle I'}}{C,C,C,C,C,C,W,W,[A],[A]\vdash[A]\otimes[A]}{\scriptstyle \otimes R}}{C,C,C,W,[A],[A]\vdash[A]\otimes[A]}{\scriptstyle \text{Contract}}}{C\otimes[A],C,C,W,[A]\vdash[A]\otimes[A]}{\scriptstyle \otimes L}}{C\otimes[A],C\otimes[A],C,W\vdash[A]\otimes[A]}{\scriptstyle \otimes L}}{(C\otimes[A])\otimes(C\otimes[A]),C,W\vdash[A]\otimes[A]}{\scriptstyle \otimes L}}{(C\otimes[A])\multimap((C\otimes[A])\otimes(C\otimes[A])),C,C,C,C,C,W,W,\Gamma\vdash[A]\otimes[A]}{\scriptstyle \multimap L}}{\dfrac{C,C,C,C,C,C,W,W,[\Gamma]\vdash[A]\otimes[A]}{C,C,C,W,[\Gamma]\vdash[A]\otimes[A]}{\scriptstyle \text{Contract}}}{\scriptstyle \forall L}$$

Fig. 9. The result of standard IMLL2 cut-elimination on the proof in Figure 8

The point is that the result of IMLL2 cut elimination is not the image of the reduced LJ2 proof. The standard IMLL2 cut-elimination procedure reduces IMLL2 cuts very efficiently, but the resulting proofs have no direct correspondence to the reduced LJ2 proofs. It is possible to imagine a non-standard cut-elimination process for IMLL2 that would treat derived rules Contract and Weaken as units. However, it would be impossible to tell without direct reference to an LJ2 proof whether an IMLL2 rule application corresponds to a Contract step or the reduction of a similar-looking LJ2 formula.

4. REPRESENTING CLASSICAL LOGIC

In this section we represent second-order classical predicate logic LK12 in MLL12 and in MALL12. This yields the undecidability of both of these fragments of second-order linear logic and thus improves somewhat Amiot's earlier result that MLL12 (and MALL12) with function symbols are undecidable [2].

One uses essentially the same encoding given above for LJ2, but for the classical connectives and sequents with extension to include first-order quantifiers. The additional clauses are $[False]\triangleq\bot$, $[\neg H]\triangleq[H]^{\bot}$, $[\exists X.A]\triangleq\exists X.[A]$, $[\forall x.A]\triangleq\forall x.[A]$, and $[\exists x.A]\triangleq\exists x.[A]$.

Soundness and completeness of the representation are obtained as in the previous section.

Lemma 4.1 *If* $\Sigma \vdash \Delta$ *is provable in* LK12, *then* $C, C, C, W, [\Sigma] \vdash [\Delta]$ *is provable in* MLL12.

Proof. As before, C and W encode Contraction and Weakening on the left. Observe that the negation rules are derivable in MLL12, namely from $\Gamma, A \vdash \Delta$ one can infer $\Gamma \vdash A^{\perp}, \Delta$, and from $\Gamma \vdash A, \Delta$ one can infer $\Gamma, A^{\perp} \vdash \Delta$. Thus Contraction and Weakening on the right as well as other inference rules in LK12 become derived rules in MLL12. ∎

Lemma 4.2 *If* $C, C, C, W, [\Sigma] \vdash [\Delta]$ *is provable in* MALL12, *then* $\Sigma \vdash \Delta$ *is provable in* LK12.

Proof. As before, there exist C' and W' such that $[C'] = C$ and $[W'] = W$. Thus since every proof rule in MLL12 is a proof rule of LK12, and C' and W' are provable LK12 formulae, completeness follows. ∎

Theorem 4.3 MLL12 *and* MALL12 *with at least one binary predicate variable are undecidable.*

5. SUBSEQUENT WORK AND OPEN PROBLEMS

Our encoding of LJ2 into IMLL2 critically depends on the native 'intuitionistic' assumption in IMLL2. There is no obvious way to prevent classically valid but intutionistically invalid formulas from being provable under embeddings into MLL2 or MALL2 in this vein. Certainly MLL2 is not conservative over IMLL2 nor is MALL2 conservative over IMALL2. For example, our encoding of Peirce's law $C, C, C, W \vdash (\forall X.\forall Y.(((X \multimap Y) \multimap X) \multimap X))$ is provable in MLL2 but not IMLL2.

The undecidability of MALL2 is shown in subsequent work of Lafont [18] by using, instead of our formula C, a weaker formula

$$\psi \triangleq \forall X.(X \& 1) \multimap (X \otimes X).$$

The additive formula $X\&1$ may be seen as an approximation of the exponential formula $!X$ in the encoding of register machines in propositional linear logic that is described by Kanovich [15] and goes back to our earlier work [21]. Using the approximation $X\&1$ instead of $!X$ yields an encoding of register machines in MALL2. A particularly interesting aspect of Lafont's work [18] is that the faithfulness of this encoding of register machines in MALL2 is obtained as an application of the soundness theorem for phase semantics of linear logic [7, 9]. Lafont's work [18], however, uses additives in two crucial ways: in the formula $X\&1$ and in the encoding of the zero-test instruction in a register machine. Therefore this encoding does not address the problem of the undecidability of MLL2.

The undecidability of MLL2 is established in Lafont and Scedrov [19] by using a multiplicative approximation $\#X \triangleq X \otimes (X \multimap 1)$ and the formula

$$\rho \triangleq \forall X.\forall Y.(\#X \otimes \#Y) \multimap (X \otimes X \otimes \#Y \otimes Y)$$

instead of ψ. Furthermore, a new encoding of register machines is developed in [19] that utilizes second-order polymorphism rather than additives. This encoding is perhaps more basic conceptually than the encoding in quantifier-free linear logic that comes from [21]. The faithfulness of the encoding is again an application of phase models.

Similar methods are used in the subsequent work of Kopylov [17] to show that the undecidability is maintained if the structural rule of Weakening is added to the system. In contrast, in the quantifier-free case propositional linear logic is undecidable [21] but adding Weakening results in a decidable system, as shown in Kopylov [16].

The exact relationships among MLL2, MALL2, and full second-order propositional linear logic, as well as the relationships among their "intuitionistic" versions, are as yet unknown. For instance, it is not known whether full second-order propositional linear logic is conservative over MALL2 and whether MALL2 is conservative over MLL2. In the later case, an interesting example is provided by Yves Lafont. Consider the formula $\exists Z.(Z \multimap E) \otimes (Z \multimap D) \otimes (A \multimap Z) \otimes (B \multimap Z)$ where $A \multimap E$, $B \multimap E$, $A \multimap D$, and $B \multimap D$ are all provable in MLL2. This formula is readily provable in MALL2 by letting $Z = E \& D$. On the other hand, in order to provide a MLL2 proof it would be necessary to find an interpolant in MLL2, which is problematic even in the relatively simple case when E and D are distinct atoms G and G' respectively, $A = (\forall U.U \multimap G) \otimes (\forall U.U \multimap G')$, and $B = (\forall U.U \multimap 1) \otimes G \otimes G'$.

The decidability of the "explicit" version of IMLL2 analogous to Harper and Mitchell's explicit ML [12], where the second-order quantification only ranges over quantifier-free formulas, is also open. By the corresponding result for MLL, the decision problem is at least NP-hard but no upper bound is as yet known.

Some restricted fragments of MALL2 are known to be decidable: MALL2 is decidable in the case that for each essentially existential variable, it occurs with only positive or only negative polarity.

The decidability of MELL, the fragment of quantifier-free propositional logic that contains multiplicatives and exponentials (modalities) but not additives, remains an interesting open problem.

The main surprise of this study is that the structural rules of Contraction and Weakening may be simulated by second-order propositional quantifiers and the multiplicatives. This gives rise to direct representations of second-order classical and intuitionistic logic in second-order linear logic without modalities. Primary among these encodings is the embedding of LJ2 in IMLL2, demonstrating the undecidability of IMLL2. Other embeddings show the undecidability of MALL2, MLL12, and MALL12. These embeddings preserve proof structure including cuts, and can be used to analyze the uses of structural rules in LJ2 proofs. This may have applications in the control structure of second-order logic programs.

Patrick D. Lincoln, Natarajan Shankar
Computer Science Laboratory
SRI International

Andre Scedrov
Department of Mathematics
University of Pennsylvania, Philadelphia

ACKNOWLEDGEMENTS

Lincoln's work is supported under NSF Grant CCR-9224858. Scedrov's work is partially supported by NSF Grant CCR-94-00907, by ONR Grant N00014-92-J-1916, and by a Centennial Research Fellowship from the American Mathematical Society. Shankar's work is supported under NSF Grants CCR-9224858 and CCR-9300444.

REFERENCES

1. S. Abramsky. Computational interpretations of linear logic. *Theoretical Computer Science*, 111:3–57, 1993.
2. G. Amiot. *Unification et logique du second ordre.* PhD thesis, Université Paris 7 (Denis Diderot), 1994.
3. N. Benton, G. Bierman, V. de Paiva, and J.M.E. Hyland. Linear lambda-calculus and categorical models revisited. In *Computer Science Logic '92, Selected Papers*, pages 61–84. Springer LNCS, vol. 702, 1993.
4. J. Chirimar, C. Gunter, and J. Riecke. Reference counting as a computational interpretation of linear logic. Manuscript, 1993. Available by anonymous ftp from the machine research.att.com in the file dist/riecke/linear-logic-journal.ps.
5. D.M. Gabbay. *Semantical investigations in Heyting's intuitionistic logic.* Reidel, Dordrecht, 1981.
6. V. Gehlot and C.A. Gunter. Normal process representatives. In *Proc. 5-th IEEE Symp. on Logic in Computer Science, Philadelphia*, June 1990.
7. J.-Y. Girard. Linear logic. *Theoretical Computer Science*, 50:1–102, 1987.
8. J.-Y. Girard. Light linear logic. Manuscript, 1995. Available by anonymous ftp from lmd.univ-mrs.fr as /pub/girard/LLL.ps.Z.
9. J.-Y. Girard. Linear logic: its syntax and semantics. In J.-Y. Girard, Y. Lafont, and L. Regnier, editors, *Advances in linear logic.* London Mathematical Society Lecture Note Series, Cambridge University Press, 1995. Available by anonymous ftp from lmd.univ-mrs.fr as /pub/girard/Synsem.ps.Z.
10. J.-Y. Girard, Y. Lafont, and P. Taylor. *Proofs and Types.* Cambridge Tracts in Theoretical Computer Science, Cambridge University Press, 1989.
11. J.-Y. Girard, A. Scedrov, and P.J. Scott. Bounded linear logic: A modular approach to polynomial time computability. *Theoretical Computer Science*, 97:1–66, 1992.
12. R. Harper and J. C. Mitchell. On the type structure of Standard ML. *ACM Transactions on Programming Languages and Systems*, 15(2):211–252, April 1993.
13. J.S. Hodas and D. Miller. Logic programming in a fragment of intuitionistic linear logic. *Information and Computation*, 110:327–365, 1994. Draft available using anonymous ftp from host ftp.cis.upenn.edu and the file pub/papers/miller/ic92.dvi.Z.
14. M.I. Kanovich. The complexity of Horn fragments of linear logic. *Annals of Pure and Applied Logic*, 69:195–241, 1994.
15. M.I. Kanovich. The direct simulation of Minsky machine in linear logic. In J.-Y Girard, Y. Lafont, and L. Regnier, editors, *Advances in Linear Logic.* Cambridge University Press, London Mathematical Society Lecture Notes, 1995.
16. A.P. Kopylov. Decidability of linear affine logic. In *10-th Annual IEEE Symposium on Logic in Computer Science, San Diego, California*, June 1995.

17. A.P. Kopylov. The undecidability of second order multiplicative linear affine logic. Manuscript, 1995.
18. Y. Lafont. The undecidability of second order linear logic without exponentials. Journal of Symbolic Logic, to appear, 1996. Available by anonymous ftp from lmd.univ-mrs.fr as /pub/lafont/mall2.ps.Z.
19. Y. Lafont and A. Scedrov. The undecidability of second order multiplicative linear logic. Information and Computation, to appear, 1996. Available by anonymous ftp from ftp.cis.upenn.edu as pub/papers/scedrov/mrs.[dvi,ps].Z.
20. P. D. Lincoln and J. C. Mitchell. Operational aspects of linear lambda calculus. In *Proc. 7-th Annual IEEE Symposium on Logic in Computer Science, Santa Cruz, California*, pages 235–246. IEEE Computer Society Press, Los Alamitos, California, June 1992.
21. P. D. Lincoln, J. C. Mitchell, A. Scedrov, and N. Shankar. Decision problems for propositional linear logic. *Annals of Pure and Applied Logic*, 56:239–311, 1992. Extended abstract in: 31st Annual IEEE Symposium on Foundations of Computer Science, Volume 2, St. Louis, Missouri, October, 1990, pages 662-671.
22. P. D. Lincoln and A. Scedrov. First order linear logic without modalities is NEXPTIME-hard. *Theoretical Computer Science*, 135:139–154, 1994. Available using anonymous ftp from host ftp.cis.upenn.edu and the file pub/papers/scedrov/mall1.[dvi,ps].Z.
23. P. D. Lincoln and N. Shankar. Proof search in first-order linear logic and other cut-free sequent calculi. In *Proc. 9-th Annual IEEE Symposium on Logic in Computer Science, Paris*, pages 282–291, July 1994.
24. P. D. Lincoln and T. Winkler. Constant-Only Multiplicative Linear Logic is NP-Complete. *Theoretical Computer Science*, 135:155–169, 1994. Available using anonymous ftp from host ftp.csl.sri.com and the file pub/lincoln/comult-npc.dvi.Z.
25. P.D. Lincoln, A. Scedrov, and N. Shankar. Decision problems for second order linear logic. In *10-th Annual IEEE Symposium on Logic in Computer Science, San Diego, California*, pages 476–485, June 1995.
26. M.H. Loeb. Embedding first-order predicate logic in fragments of intuitionistic logic. *Journal of Symbolic Logic*, 41:705–718, 1976.
27. D. Miller. The π-calculus as a theory in linear logic: Preliminary results. In E. Lamma and P. Mello, editors, *Proceedings of the 1992 Workshop on Extensions to Logic Programming*, number 660 in Lecture Notes in Computer Science. Springer-Verlag, 1993. Available using anonymous ftp from host ftp.cis.upenn.edu and the file pub/papers/miller/pic.dvi.Z.
28. D. Miller. A multiple-conclusion meta-logic. In *Proc. 9-th Annual IEEE Symposium on Logic in Computer Science, Paris*, pages 272–281, July 1994.
29. A. Scedrov. A brief guide to linear logic. In G. Rozenberg and A. Salomaa, editors, *Current Trends in Theoretical Computer Science*, pages 377–394. World Scientific Publishing Co., 1993.
30. A. Scedrov. Linear logic and computation: A survey. In H. Schwichtenberg, editor, *Proof and Computation, Proceedings Marktoberdorf Summer School 1993*, pages 379–395. NATO Advanced Science Institutes, Series F, Springer-Verlag, Berlin, 1995. Available from the host ftp.cis.upenn.edu as the file pub/papers/scedrov/mdorf93.[dvi,ps].Z. See pub/papers/TableOfContents.
31. G. Takeuti. *Proof Theory.* North-Holland, 1987. 2nd edition.
32. A.S. Troelstra. *Lectures on Linear Logic.* CSLI Lecture Notes No. 29. Center for the Study of Language and Information, Stanford University, 1992.

STEVO TODORČEVIĆ

COMPARING THE CONTINUUM WITH THE FIRST TWO UNCOUNTABLE CARDINALS

The purpose of this note is to survey some technical aspects of the two hypotheses $\mathfrak{c} = \aleph_1$ and $\mathfrak{c} = \aleph_2$ together with their joint negation $\mathfrak{c} > \aleph_2$. The reader is probably already puzzled by the fact that we decide to stop at the second uncountable cardinal rather than at some other value of the aleph–function. Well, there are two reasons for this. The first one is that presently there is a considerable lack of knowledge about statements like $\mathfrak{c} = \aleph_3$, $\mathfrak{c} = \aleph_4$, and so on. The other reason is that the present knowledge seems to suggest that indeed there is some "affinity" between $\mathfrak{c}$ and $\aleph_2$. It also seems that even a meager understanding of this affinity can be quite useful as we shall demonstrate this at the end of the article.

1. THE CONTINUUM HYPOTHESIS

This is perhaps one of the most widely known hypotheses of Set Theory. The root of its popularity is in its simplicity both as a statement and the way it is most frequently used. The existence of the very short well-ordering of the reals enables one to easily diagonalize and get some striking (or "highly implausible" as Gödel [11] calls them) objects such as (see [24]):

(C_1) An uncountable set of reals which has countable intersection with every nowhere dense set of reals.

(P_3) An uncountable subset X of the Hilbert cube $[0,1]^{\mathbb{N}}$ such that every uncountable $Y \subseteq X$ is projected onto the interval $[0,1]$ by all but finitely many of the projections.

(P_{10}) An infinite-dimensional subset of the Hilbert cube which contains no uncountable finite-dimensional subset.

The statement P_{10}, originally due to Hurewicz [15], is not only a consequence of CH but it also implies CH since any metric space of size $< \mathfrak{c}$ is 0-dimensional. The statement P_3 is another equivalent of CH. Proving this statement from CH did cause a considerable difficulties (see [24; pp. 189–193]) and we conjecture that this is so because of a too strong preoccupation with getting a statement equivalent to CH rather than just its consequence (like C_1). In fact, P_3 is so made (perhaps artificially) that the implication $P_3 \to$CH is a triviality. Relaxing P_3 by replacing the interval $[0,1]$ by the set of all countable ordinals makes it a rather direct consequence of C_1 (see [25; §6]) which itself is a rather simple consequence of CH. One thing one notices while studying the known consequences of CH is that they all can be reformulated to be

M.L. Dalla Chiara et al. (eds.), Logic and Scientific Methods, 145–155.

Σ_1^2-statements, i.e. statements asserting the existence of a set of reals whose properties are describable using only quantification over the reals themselves. For example P_3 says that there is a mapping $p: \omega_1 \times \omega_1 \to \mathbb{R}$ such that for every countable $A \subseteq \omega_1$ and $f: A \to \mathbb{R}$ there is γ in ω_1 such that for all β in ω_1 above γ there exists α in A such that $p(\alpha, \beta) = f(\alpha)$. The Σ_1^2-form of the statement P_{10} says that there is a subset of the Hilbert cube which has countable intersection with every 0-dimensional G_δ-subset of the cube. The following remarkable result of Woodin [32] (proved using a certain large-cardinal hypothesis) not only gives the explanation of this phenomenon but it also points out how one should look at any future application of CH.

Theorem 1 *If a Σ_1^2-statement is true in one forcing extension of the universe then it is also true in any other forcing extension which satisfies CH.*

Thus, if there is a statement for which we would like to know whether it is implied by CH or not then the first thing to do is to try to restate it as a Σ_1^2-statement. If this is possible then the statement is almost certainly a consequence of CH for we shall never be able to construct a forcing extension which would show that this is not so.

2. CODING THE REALS

The increasing sophistication in the area of iterated forcing resulted in a profusion of models of Set Theory built for the purpose of proving the independence of statements like Souslin's Hypothesis, Borel's Conjecture, and so on. It turns out that many such proofs, especially the deeper ones involving other than the finite support iterations, could not produce models where the continuum would have value different from $\aleph_2$. For example, a typical such proof is Laver's consistency proof ([19]) of Borel's Conjecture. It is therefore natural to ask whether the statements in question do pose some restriction on the continuum such as $\mathfrak{c} = \aleph_2$ or not. This kind of question is especially interesting when the considered statements are some kind of maximality principles, such as, for example, various forcing axioms asserting that we can have a sufficiently generic filter for every member of a certain class of posets. (See the footnote 23 of [11] regarding this point.) The purpose of this and the next section is to explain a pair of such maximality principles which jointly imply that the continuum is equal to $\aleph_2$ and which are jointly true in a large class of models discussed above and especially in those made for the purpose of proving the consistency of certain forcing axioms which, in their complexity, lie just above Martin's Axiom (which itself does not pose any restriction on the continuum except its regularity).

Let $\mathbb{N}^{\mathbb{N}}$ denote the structure of all sequences of integers ordered by the ordering $<^*$ of eventual dominance. Define osc: $(\mathbb{N}^{\mathbb{N}})^2 \to$Card by

$$\operatorname{osc}(a, b) = \#\{n \in \mathbb{N} : a(n) \leq b(n) \text{ and } a(n+1) > b(n+1)\}.$$

This is a quite interesting object whose properties are closely tied to the notion of unboundedness in $\mathbb{N}^{\mathbb{N}}$ under $<^*$ (see [26], [27]). For example, it has the property that a subset of $\mathbb{N}^{\mathbb{N}}$ which is at the same time σ-directed and unbounded under $<^*$ realizes all possible finite oscillations. Its properties can be summarized as follows where

$$o : (\mathbb{N}^{\mathbb{N}})^2 \to \mathbb{N}$$

is obtained by composing osc with some natural (recursive) mapping from $\mathbb{N}$ into $\mathbb{N}$ (see [26], [27; §1]):

Theorem 2 *Suppose A is a σ-directed unbounded subset of $\mathbb{N}^{\mathbb{N}}$ and that X is an unbounded subset of $A^{[k]}$ for some integer $k \geq 1$. Let D be a countable dense subset of X. Then there is $b \in A$ such that for every integer $n \geq 1$, every $h: k \times n \to \mathbb{N}$, and every $x \in A^{[n]}$ above b there is $d \in D$ such that $o(d_i, x_j) = h(i,j)$ for all $i < k$ and $j < n$.*

(Here $A^{[k]} = \{x \in A^k : x_i \neq x_j \text{ for all } i \neq j < k\}$.) In simple words, this result tells that osc and o have a remarkably rich behavior on any A which is σ-directed and unbounded in $\mathbb{N}^{\mathbb{N}}$. A typical such A would be an unbounded $W \subseteq \mathbb{N}^{\mathbb{N}}$ which is well-ordered by $<^*$. So let us fix such a W. For x in W let x^{+n} be its nth successor in W. A subset $X \subseteq W$ is *σ-complete* if for every increasing sequence $\{x_n\}$ of elements of X the least upper bound of $\{x_n\}$ in W is also a member of X.

Definition 1 *A set $X \subseteq W$* codes *a "real" $r \in \mathbb{N}^{\mathbb{N}}$ if for every n there is a countable decomposition $X = \bigcup_{i=1}^{\infty} X_i^n$ such that*

$$o(x^{+n}, y^{+n}) = r(n)$$

for all $x <^* y$ in X_i^n and $i \in \mathbb{N}$.

Let PCR_W be the statement that every real is coded by a σ-complete subset of W with no last element.

Lemma 1 *PCR_W implies that there is a 1–1 mapping from $\mathbb{R}$ into W.*

Proof: For $r \in \mathbb{R}$ $(= \mathbb{N}^{\mathbb{N}})$ fix a σ-complete $X_r \subseteq W$ which codes r. Note that if $r \neq s$ then X_r and X_s have different supremums in $W \cup \{\infty\}$

The Principle of Coding the Reals, PCR, which would state that PCR_W is true for every unbounded and well-ordered $W \subseteq \mathbb{N}^{\mathbb{N}}$ is an interesting hypothesis. To see what it tells us about the continuum let $\mathfrak{b}$ be the minimal order-type of any such $W \subseteq \mathbb{N}^{\mathbb{N}}$. This is a well-known cardinal characteristic of the continuum whose importance has been recognized very early (see [6]). It is clearly a regular uncountable cardinal $\leq \mathfrak{c}$. From Lemma 1 we have the following

Proposition 1 *PCR implies that $\mathfrak{c} = \mathfrak{b} > \aleph_1$.*

Proof: Note that the equality $\mathfrak{b} = \aleph_1$ is impossible because in a well-ordered set of type ω_1 all σ-complete subsets with no least elements must be unbounded, so at most one real can be coded.

The statement PCR is true in any situation where we can do some internal forcing. In fact, Theorem 2 is saying exactly that the poset of finite approximations to a set which codes a given real is very well-behaved which can be stated in the following more precise way.

Lemma 2 *Suppose $C \subseteq W$ is σ-complete and unbounded and that its order-type is a regular uncountable cardinal θ. Then for every real r the poset of finite approximations to the sequence of decompositions $C = \bigcup_{i=1}^{\infty} C_i^n$ $(n \in \mathbb{N})$ as in Definition 1 is a θcc poset.*

Proof: Clearly we may assume C consists only of limit points of W. For $n \in \mathbb{N}$ let C^{+n} be the set of all c^{+n} for c in C. Let C^* be the union of C^{+n}'s. Conditions of the poset $\mathcal{P}$ are finite structures which tell us some finite information about the decomposition of C. In particular, for each $p \in \mathcal{P}$ the set S_p of all elements of C^* which p mentions is finite. We have to show that every $\mathcal{F} \subseteq \mathcal{P}$ of size θ contains two compatible conditions p and q. It is not hard to see that we may assume that the elements of $\mathcal{F}$ have disjoint supports, and moreover that they all are isomorphic finite structures. This in particular means that their supports are of the same size k and that we may identify them with the corresponding elements of $(C^*)^{[k]}$ which enumerate them increasingly according to the well-ordering on W. Let X be the subset of $W^{[k]}$ thus obtained and let D be a countable dense subset of it. Let $b \in C$ be the bound given by Theorem 2. The disjointness of supports and the assumption on the order-type of C imply that we can fix a $q \in \mathcal{F}$ such that $x = S_q$ is above b. Now, to find $d = S_p$ in D such that p and q are compatible, the oscillation map o has to satisfy the requirements of Definition 1 on the union of S_p and S_q, i.e. certain requirements of the form $o(u,v) = r(n)$ for $u \in S_p$ and $v \in S_q$. Given the fact that members of $\mathcal{F}$ are isomorphic structures it is easy to find $h\colon k \times k \to \mathbb{N}$ such that if $d = S_p$ from D and $x = S_q$ satisfy the conclusion of Theorem 2 for this h then the requirements are met.

Thus, if we start with a fixed W and a real r and if we go to a forcing extension of the universe obtained by simply collapsing the continuum to $\aleph_1$ by the natural σ-closed poset then all the properties of the oscillation maps osc and o (as well as of the set W) remain undisturbed. This indestructibility is one property of osc not shared by any other known two-place function and it is very crucial to us at this point. In the forcing extension, we can easily find $C \subseteq W$ as in Lemma 2 of order-type ω_1 so the corresponding poset of finite approximations is ccc. Thus if we are able to internally force with a σ-closed poset (in fact only the standard one which collapses $\mathfrak{c}$ to $\aleph_1$) followed by a ccc poset, then we can prove that PCR is true. It turns out that the forcing axiom for σ-closed$*$ccc posets is one of the weakest known and most frequently used extensions of MA. The reader is referred to [1], [2], [3], [5], [9], [23], [27] and [31] for more about this kind of axioms.

3. OPEN COLORING AXIOM

For a set of reals (or more generally for a topological space X) let OCA_X denote the following regularity property: For every open symmetric relation $K \subseteq X^{[2]}$ either

1. there is a countable decomposition $X = \bigcup_{n=1}^{\infty} X_n$ such that $X_n^{[2]} \cap K = \emptyset$ for all n, or
2. there is an uncountable $Y \subseteq X$ such that $Y^{[2]} \subseteq K$.

Note that OCA is preserved under continuous images and that if in (ii) we require Y to be a perfect set (homeomorphic to the Cantor set) we get a statement, call it OCA_X^*, which can be considered as a natural 2-dimensional analogue of the classical Perfect-Set Property (see [16; 3.1]). Note also that this stronger form OCA_X^* is true for some simple X such as the reals or the irrationals and therefore for all analytic sets of reals. It turns out that the requirement that K is open rather than, say, closed is crucial even when $X = \mathbb{R}$ (see [28]). Let OCA be the statement that OCA_X is true for every separable metric space X, or equivalently, for every set of of reals X. This has proved to be a very useful axiom for a wide range of problems (see [27]). Its most surprising consequence is the following fact discovered in [27] during the course of proving certain results about gaps in the structure $(\mathbb{N}^{\mathbb{N}}, <^*)$.

Proposition 2 *OCA implies that* $\mathfrak{b} = \aleph_2$.

A gap in $(\mathbb{N}^{\mathbb{N}}, <^*)$ is a pair (A, B) of subsets of $\mathbb{N}^{\mathbb{N}}$ such that $A \cup B$ is a chain under $<^*$ and $a <^* b$ for all $a \in A$ and $b \in B$, but there is no $x \in \mathbb{N}^{\mathbb{N}}$ such that $a <^* x <^* b$ for all $a \in A$ and $b \in B$. If α is the cofinality of $(A, <^*)$ (i.e., the minimal order-type of a subset of A cofinal in A) and if β is the cofinality of $(B, >^*)$ then we call (α, β^*) the *type of* (A, B). Gaps of type (ω_1, ω_1^*) are called *Hausdorff gaps* because Hausdorff was the first to show their existence ([13], [14]). Gaps of type (θ, ω^*) or (ω, θ^*) are called *Rothberger gaps*. Their existence was also first proved by Hausdorff [13], and later rediscovered by Rothberger [22]. For example, they showed that there is always an $(\mathfrak{b}, \omega^*)$-gap in $(\mathbb{N}^{\mathbb{N}}, <^*)$. They also showed that if there is a (θ, ω^*) gap then there is an unbounded θ-chain in $(\mathbb{N}^{\mathbb{N}}, <^*)$. The following Lemma about the influence of OCA on gaps first appeared in [27; §8].

Lemma 3 *If OCA is true then every gap* (A, B) *in* $(\mathbb{N}^{\mathbb{N}}, <^*)$ *is either Hausdorff or Rothberger.*

Proof: Let A^* be set of all elements obtained by taking finite changes of elements of A and let B^* be obtained from B in the same way. Let $A^* \otimes B^*$ be the set of all $(a, b) \in A^* \times B^*$ such that $a(n) \leq b(n)$ for all n. Let $X = A^* \otimes B^*$ considered as a subspace of the square of the Baire space $\mathbb{N}^{\mathbb{N}}$. Let $K \subseteq X^{[2]}$

be the set of all pairs $\langle\langle a,b\rangle,\langle a',b'\rangle\rangle$ from $X^{[2]}$ such that there is an $n\in\mathbb{N}$ for which $a(n)>b'(n)$ or $a'(n)>b(n)$. The alternatives of OCA give us the conclusion of Lemma 3.

The following Lemma also appears in [26] (as Lemma 3.10).

Lemma 4 *If $\mathfrak{b}>\aleph_2$ then there is a gap (A,B) in $(\mathbb{N}^{\mathbb{N}},<^*)$ which is neither Hausdorff nor Rothberger.*

Proof: Pick a subset A of $\mathbb{N}^{\mathbb{N}}$ which has order-type ω_2 under $<^*$ and extend it to a maximal $<^*$-totally ordered subset L of $\mathbb{N}^{\mathbb{N}}$. By the assumption of $\mathfrak{b}>\aleph_2$, the set A is not cofinal in L so let B be the upper half of the gap in L which A determines. Then (A,B) is also a gap in $(\mathbb{N}^{\mathbb{N}},<^*)$. Note that B does not have a minimal element for if $b=\min(B)$ then $b-1$ would be a member of $\mathbb{N}^{\mathbb{N}}$ which could be added to L contradicting its maximality. This phenomenon is specific for the structure $(\mathbb{N}^{\mathbb{N}},<^*)$ and it is crucial for us here. Our assumption $\mathfrak{b}>\aleph_2$ and the already quoted result of Hausdorff [13] and Rothberger [22] imply that the coinitiality of B must be uncountable. So (A,B) is neither a Hausdorff nor a Rothberger gap.

It should be noted that we had a considerable difficulty in getting the right form of Lemma 4 which may not be so visible from its relatively short proof. The difficulty was in getting rid of the influence of a very similar result, proved long time ago by a number of people including the author ([2; Theorem 4.4]), about the structure $(\mathcal{P}(\mathbb{N}),\subseteq^*)$ where the place of $\mathfrak{b}$ is played by the cardinal $\mathfrak{t}$, the minimal order-type of a nonextendable tower of subsets of $\mathbb{N}$. [A *nonextendable tower* is a set W of coinfinite subsets of $\mathbb{N}$ which is well-ordered by $\subseteq^*$ and which cannot be end-extended.] The older result, however, is quite useless in the present context because there is no oscillation theory of the structure $(\mathcal{P}(\mathbb{N}),\subseteq^*)$ where the corresponding form of Theorem 2 is true for a nonextendible tower. (See [30] for an explanation of this point.) To prove Proposition 2 we still owe to the reader an explanation of why OCA implies $\mathfrak{b}>\omega_1$. Well, assume $\mathfrak{b}=\omega_1$ and choose well-ordered unbounded $A\subseteq\mathbb{N}^{\mathbb{N}}$ of type ω_1. Let $K\subseteq A^{[2]}$ be the set of all pairs (a,b) such that $\operatorname{osc}(a,b)>0$. Then K is open and the fact that every unbounded $X\subseteq A$ realises oscillation 0 means that OCA is false for A and K. We finish this section with the remark that similarly to PCR considered in the last section, OCA is a consequence of the forcing axiom for the class of posets which can be written in the form σ-closed$*$ccc (se [27; §8]).

4. COVERING THE REALS WITH STRONG MEASURE ZERO SETS

The well-known unpublished manuscript of Gödel [12], written before the great sophistication in iterated forcing discussed above, contains interesting ideas which might be useful in explaining $\mathfrak{c}=\aleph_2$ in some of the models not covered by §§2 and 3. So the purpose of this section is to draw the reader's attention to some of these ideas. Gödel's approach was to cover the reals with a small family of "small sets" of reals. The "small sets" are sets of strong measure zero in the sense of E. Borel [4], i.e. the subsets of $\mathbb{R}$ which can be covered by a sequence $\{I_n\}$ of intervals of any prescribed sequence of lengths $\{\epsilon_n\}$. As Gödel has realized, these sets are in close relation with the structures $\mathbb{R}^{\mathbb{N}}$ or $\mathbb{N}^{\mathbb{N}}$ when ordered by some variation of the ordering of eventual dominance. In particular, the dominating number $\mathfrak{d}$, defined as the minimal cardinality of a subset D of $\mathbb{N}^{\mathbb{N}}$ which is cofinal in $(\mathbb{N}^{\mathbb{N}}, <^*)$, should play the important role. For example, note that every strong measure zero set is included in a $G_{\mathfrak{d}}$-set of strong measure zero, i.e. in a strong measure zero set which is equal to the intersection of a family of no more than $\mathfrak{d}$ open sets of reals. In fact, even larger powers seem to be relevant as the following fact shows. [Here SMZ stands for the ideal of strong measure zero sets of reals ordered by the inclusion while $\mathcal{Z}$ stands for the family of all subsets of $\mathbb{N}$ of asymptotic density 0 (i.e. sets $A \subseteq \mathbb{N}$ with the property that $\lim_{n\to\infty} \#(A \cap \{1, \ldots, n\})/n = 0$) ordered by the relation $\subseteq^*$ of inclusion modulo the ideal of finite sets, and where $\mathcal{Z}^{\mathfrak{d}}$ is the set of all functions from the initial ordinal $\mathfrak{d}$ into $\mathcal{Z}$ ordered coordinatewise.]

Proposition 3 *There is a monotonic map $\Phi\colon \mathcal{Z}^{\mathfrak{d}} \to SMZ$ whose range is cofinal in SMZ.*

Proof: Let $\{q_n\}$ be an enumeration of the rationals and fix a dominating $D \subseteq \mathbb{N}^{\mathbb{N}}$ of size $\mathfrak{d}$. Choose $F\colon \mathrm{SMZ} \to \mathcal{Z}^D$ so that

$$X \subseteq \bigcap_{m=1}^{\infty} \bigcup \{(q_n - 1/d(n), q_n + 1/d(n)) : n \in F(X)(d),\ n \geq m\}$$

for all $X \in \mathrm{SMZ}$ and $d \in D$. For $z \in \mathcal{Z}^D$, set $\Phi(z) = \bigcup \mathcal{F}_z$, where

$$\mathcal{F}_z = \{X \in \mathrm{SMZ} : F(X)(d) \subseteq^* z(d) \text{ for all } d \in D\}.$$

It is easily checked that Φ is as required.

The reader interested in more information about objects such as SMZ, $\mathcal{Z}$, $\mathbb{N}^{\mathbb{N}}$ is referred to [10] which gives an excellent synthesis of results of this sort. One of the crucial points of Gödel's approach rests on a proper understanding of Perfect-Set Properties of the following kind:

Problem 1 *When one can conclude that every $G_{\mathfrak{d}}$-set of reals of size $> \mathfrak{d}$ (or $> \mathfrak{d}^+$ if this makes a difference) contains a nonempty perfect subset?*

This is so because in any situation where $G_{\mathfrak{d}}$-sets of reals have this kind of the Perfect-Set Property, the size of the continuum reduces to the covering number of the ideal SMZ, i.e. to the minimal cardinality of a family $\mathcal{F} \subseteq \mathrm{SMZ}$ such that $\bigcup \mathcal{F} = \mathbb{R}$. (Note the relevance of Borel's Conjecture in this context.) One way of understanding this number might be through the study of powers like $\mathcal{Z}^{0}$, but more likely the real information will come from a deeper study of SMZ itself.

The idea of using strong measure zero sets to cover the reals in order to get some information about the size of the continuum can perhaps be transfered to other notions of smallness as far as no perfect set of reals satisfies it. A natural candidate would be the closely related notion of universal measure zero sets of reals. These are the sets of reals which get value zero under any nonatomic Borel measure on $\mathbb{R}$. They are not as small as the sets of strong measure zero since the analogue of Borel's Conjecture is false for this family, i.e. there is always an uncountable universal measure zero set of reals. Hence for this class of sets it might be easier to avoid the phenomenon encountered above where Borel's Conjecture trivializes any attempt to cover $\mathbb{R}$ with a small family of strong measure zero sets.

5. WHEN THE CONTINUUM IS BIGGER THAN $\aleph_2$

For an integer n a family $\mathcal{F}$ of sets is *n-disjoint* if it can be decomposed in no more than n disjoint subfamilies. It is clear that by compactness $\mathcal{F}$ is n-disjoint iff every finite subfamily of $\mathcal{F}$ is n-disjoint. In 1946 Richard Rado [21] was able to considerably improve this for certain families of sets $\mathcal{F}$. He proved that a family of intervals (or more generally convex sets) $\mathcal{F}$ of some fixed linearly ordered set is n-disjoint iff every subfamily of $\mathcal{F}$ of size at most $n+1$ is n-disjoint. This kind of compactness is a prominent theme in the field of Combinatorics. For example, the famous Dilworth chain decomposition theorem is also a result of this sort: a partially ordered set $\mathcal{P}$ can be covered by a family of $\leq n$ chains iff this is true for every $\mathcal{P}_0 \subseteq \mathcal{P}$ of size at most $n+1$. The pleasing behavior of families of convex subsets of linearly ordered sets has led Rado to conjecture the similar behavior at the next level:

(RC) A family of convex subsets of a linearly ordered set is σ-disjoint iff each of its subfamilies of size at most $\aleph_1$ is σ-disjoint.

(Here "σ-disjoint" means "the union of countably many disjoint subfamilies".) This turns out to be a very interesting compactness principle unique in the sense that it works at the level where any other known compactness principle fails badly. Its relevance to the question about the size of the continuum is revealed in the following fact proved in [29].

Theorem 3 *To every sequence $e_\alpha : \alpha \to |\alpha|$ $(\alpha < \omega_2)$ of 1–1 mappings we can associate a family $\mathcal{F}(\vec{e})$ of convex subsets of some linearly ordered set such that*

1. *$\mathcal{F}(\vec{e})$ can be written as the union of an increasing sequence $\mathcal{F}_\xi(\vec{e})$ $(\xi < \omega_2)$ of σ-disjoint subfamilies,*
2. *if $\mathcal{F}(\vec{e})$ itself is σ-disjoint then $\mathbb{R} \subseteq L[\vec{e}]$.*

(Here $L[\vec{e}]$ is the constructible subuniverse generated by the sequence $\vec{e}$.) Note that this, in particular, means that RC implies $\mathfrak{c} \leq \aleph_2$. (It is interesting that RC does not decide whether $\mathfrak{c} = \aleph_1$ or $\mathfrak{c} = \aleph_2$.) It follows that while the negation of CH, $\mathfrak{c} > \aleph_1$, is extremely poor in its consequences as far as "striking objects" are concerned this cannot be said about its apparently slight strengthening $\mathfrak{c} > \aleph_2$. The assumption $\mathfrak{c} > \aleph_2$ gives us a very striking object indeed, a family $\mathcal{F}$ of size $\mathfrak{c}$ of convex subsets of some linearly ordered set which is not σ-disjoint but every subfamily $\mathcal{F}_0$ of $\mathcal{F}$ of size at most $\aleph_1$ is σ-disjoint. We refer the reader to our paper [29] for further information about RC which we choose not to include here. Instead we chose to give an application of these ideas to an old problem of Erdös and Hajnal [7]:

Proposition 4 *For every decomposition of the family $\mathbb{N}^{\omega_2}$ of all integer-valued functions defined on ω_2 into countably many pieces one of the pieces must contain two almost disjoint functions.*

(Two functions $f, g \in \mathbb{N}^{\omega_2}$ are *almost disjoint*, written as $f \perp g$, if there is $\beta < \omega_2$ such that $f(\alpha) \neq g(\alpha)$ for all $\alpha \geq \beta$.) In other words, the graph $\mathcal{G}_{\omega_2} = (\mathbb{N}^{\omega_2}, \perp)$ is not countably chromatic which is the way Erdös and Hajnal have stated their problem. A close examination of the graph $\mathcal{G}_{\omega_2}$ reveals that this is an interesting object sensitive to many properties of the second uncountable cardinal. For example, note that $\mathrm{Chr}(\mathcal{G}_{\omega_2})$ is bounded by the size of any ultrapower of the form $\mathbb{N}^{\omega_2}/\mathcal{U}$ where $\mathcal{U}$ is a uniform ultrafilter on ω_2. Measuring the sizes of ultrapowers have been a reoccuring theme in Set Theory as well as in Model Theory. We mention the recent remarkable result of Foreman [8] which has produced a model of Set Theory (assuming the consistency of a certain large cardinal axiom) in which the minimal size of ultrapowers of the form $\mathbb{N}^{\omega_2}/\mathcal{U}$ for $\mathcal{U}$ a uniform ultrafilter on ω_2 is equal to $\aleph_1$. So, in particular in this model $\mathrm{Chr}(\mathcal{G}_{\omega_2})=\aleph_1$ which we now know to be the minimal possible valve. Another interesting property of $\mathcal{G}_{\omega_2}$ is that each of its subgraphs of size $\aleph_1$ is countably chromatic and that $\mathcal{G}_{\omega_2}$ contains a copy of every graph of size $\aleph_2$ which has this property. Thus, $\mathcal{G}_{\omega_2}$ is sensitive to any kind of incompactness of the chromatic number at this level. This fact was used by Baumgartner, Shelah, Komjath and others to show that in models of Set Theory in close proximity to the constructible universe the chromatic number of $\mathcal{G}_{\omega_2}$ can be rather large and in particular can have the maximal possible value $2^{\aleph_2}$. The sensitivity of $\mathcal{G}_{\omega_2}$ to almost any kind of incompactness has led us to discover the proof of Proposition 4 by connecting it with Theorem 3. The proof of Proposition 4 spits into two cases:

Case I: The power of the continuum is at most $\aleph_2$. This case is essentially covered already in the Erdös–Hajnal paper [7] who state their result assuming CH. That their argument also covers the case $\mathfrak{c} = \aleph_2$ was, according to Hajnal, first noticed by P. Komjath. Here is a Lemma whose hypothesis is true in both cases $\mathfrak{c} = \aleph_1$ and $\mathfrak{c} = \aleph_2$ and whose proof is essentially a repetition of the old Erdös–Hajnal argument.

Lemma 5 *If there is a sequence $g_\alpha : \omega_2 \to \mathbb{R}$ $(\alpha < \mathfrak{c}^+)$ of pairwise almost disjoint functions then Chr($\mathcal{G}_{\omega_2}$) is uncountable.*

Proof: For $\alpha < \beta < \mathfrak{c}^+$ define $f_{\alpha\beta} \in \mathbb{Z}^{\omega_2}$ by

$$f_{\alpha\beta}(\xi) = \Delta(f_\alpha(\xi), f_\beta(\xi)) \cdot \operatorname{sgn}(f_\alpha(\xi) - f_\beta(\xi))$$

where for two reals x and y, we let $\Delta(x, y)$ be the minimal n such that the nth rational number is in between x and y (for some fixed enumeration $\{q_i\}$ of $\mathbb{Q}$). Then $f_{\alpha\beta} \perp f_{\beta\gamma}$ whenever $\alpha < \beta < \gamma$, so we are done by $\mathfrak{c}^+ \to (3)^2_\omega$.

Case II: The power of the continuum is bigger than $\aleph_2$. Consider the family $\mathcal{F}(\vec{e})$ from Theorem 3 and for each $\xi < \omega_2$ fix a σ-disjoint decomposition $g_\xi: \mathcal{F}_\xi(\vec{e}) \to \mathbb{N}$. This enables us to define

$$\Phi: \mathcal{F}(\vec{e}) \to \mathbb{N}^{\omega_2}$$

by the following rules:

(a) $\Phi(I)(\xi) = 1$ for $I \notin \mathcal{F}_\xi(\vec{e})$,

(b) $\Phi(I)(\xi) = g_\xi(I)$ for $I \in \mathcal{F}_\xi(\vec{e})$.

Then $I \cap J \neq \emptyset$ implies $\Phi(I) \perp \Phi(J)$ for every $I, J \in \mathcal{F}(\vec{e})$, so the conclusion of Proposition 4 follows from the fact that under $\mathfrak{c} > \aleph_2$ the family $\mathcal{F}(\vec{e})$ is not σ-disjoint.

We refer the reader to a projected paper of A. Hajnal and the author which will contain information about the chromatic number of $\mathcal{G}_\theta = (\mathbb{N}^\theta, \perp)$ for some other values of θ.

Matematički Institut
Belgrade

Department of Mathematics
University of Toronto

REFERENCES

1. J. E. Baumgartner, Iterated Forcing, pp. 1–59 in [20]
2. J. E. Baumgartner, Applications of the Proper Forcing Axiom, Chapter 21 of [18]
3. M. Bekkali, Topics in Set Theory, Lecture Notes in Math., #1476 Springer–Verlag 1991
4. E. Borel, Sur la classification des ensembles de mesure nulle, Bull. Soc. Math. France 47 (1919), 97–125
5. K. J. Devlin, The Yorkshireman's guide to proper forcing, pp. 60–115 in [20]
6. E. K. van Douwen, The integers and topology, Chapter 3 of [18]
7. P. Erdös and A. Hajnal, On chromatic number of infinite graphs, in Proceedings of the Tihany Graph Theory Symposium (Publ. House of the Hungarian Acad. Sci, 1966). 83–98
8. M. Foreman, An $\aleph_1$-dense ideal on $\aleph_2$, preprint 1994
9. D. H. Fremlin, Consequences of Martin's Axiom, Cambridge Univ. Press 1984
10. D. H. Fremlin, The partially ordered sets of measure theory and Tukey's ordering, Noti di Matematica, Vol XI (1991), 177–214
11. K. Gödel, What is Cantor's Continuum Problem?, Amer. Math. Monthly (1947), 515–525, revised and expanded version in: P. Benacerraf and H. Putnam (eds), Philosophy of Mathematics, Selected Readings, Englewood Cliffs, N. J., Prentice Hall 1964, 258–273

12. K. Gödel. Some considerations leading to the probable conclusion that the true power of the Continuum is $\aleph_2$, unpublished manuscript, 1970
13. F. Hausdorff, Die Graduierung nach dem Endverlauf, Abhandlun. König. Sächsis. Gessellsch. Wissenschaften, Math.-Phys. Kl. 31 (1909), 296–334.
14. F. Hausdorff, Summen von $\aleph_1$-mengen, Fund. Math. 26 (1930), 241–255
15. W. Hurewicz, Une remarque sur l'hypothése du continu, Fund. Math. 19 (1932), 8–9
16. A. Kanamori, The Higher Infinite, Springer–Verlag, Berlin 1994
17. K. Kunen, (κ, λ^*)-gaps under MA, note of Aug. 1976
18. K. Kunen and J. E. Vaughan (eds), Handbook of Set–Theoretic Topology, North–Holland Amsterdam 1984
19. R. Laver, On the consistency of Borel's Conjecture, Acta Math., 137 (1976), 151–169
20. A. R. D. Mathias (ed), Surveys in Set Theory, Cambridge Univ. Press, 1983
21. R. Rado, Covering theorem for ordered sets, Proc. London Math. Soc. (2) 50 (1946), 504–535
22. F. Rothberger, Sur la familles indenombrables de suites de nombres naturels et les problemes concernant la propriete C, Proc. Cambridge Phil. Soc. 37 (1941), 109–126
23. S. Shelah, Proper Forcing, Lecture Notes in Math. #940, Springer–Verlag, 1982
24. W. Sierpiński, Hypothèse du Continu, Second Edition, Chelsea Publ. Co. New York 1956
25. S. Todorčević, Partitioning pairs of countable ordinals, Acta Math., 159 (1987), 261–294
26. S. Todorčević, Oscillations of real numbers, in: Logic Colloquium '86 (F. R. Drake and J. K. Truss, eds) Elsevier Sci. Publ. (1988), 325–331
27. S. Todorčević, Partition Problems in Topology, American Math. Soc., Providence 1989
28. S. Todorčević, Two examples of Borel partially ordered sets with the countable chain condition, Proc. Amer. Math. Soc. 112 (1991), 1125–1128
29. S. Todorčević, Conjectures of Rado and Chang and Cardinal Arithmetic, in: Finite and Infinite sets in Combinatorics (N. W. Sauer et al., eds), Kluwer Acad. Publ. (1993), 385–398
30. S. Todorčević, Oscillations of Sets of Integers, preprint 1994
31. B. Veličković, Forcing Axioms and Stationary sets, Adv. Math 94 (1992), 256–284
32. W. H. Woodin, Σ^2_1 absoluteness, notes of May 1985

3
Recursion Theory and Constructivism

KEVIN T. KELLY AND OLIVER SCHULTE

CHURCH'S THESIS AND HUME'S PROBLEM

ABSTRACT.
We argue that uncomputability and classical scepticism are both reflections of inductive underdetermination, so that Church's thesis and Hume's problem ought to receive equal emphasis in a balanced approach to the philosophy of induction. As an illustration of such an approach, we investigate how uncomputable the predictions of a hypothesis can be if the hypothesis is to be reliably investigated by a computable scientific method.

1. RELATIONS OF IDEAS AND MATTERS OF FACT

Following an ancient tradition, David Hume boldly divided the objects of inquiry into two kinds: *relations of ideas* and *matters of fact* (Hume, 1984). Relations of ideas embrace all mathematical and logical inquiry, whereas matters of fact are the principal concerns of empirical science and daily life. The view that mathematics concerns relations of ideas has two important consequences. First, mathematical questions can be answered independently of all empirical data and second, the ideas upon which such questions depend can be scanned all at once by the "mind's eye," resulting in certainty concerning their relations. After a brief discussion of this happy situation in mathematics, Hume turned to the apparently more problematic case of inquiry concerning matters of fact. Here, a general law covers a potentially unbounded stream of empirical data that can refute it at any time in the future, so there is no time by which certainty about the law can be achieved. In sharp contrast with the concepts exhaustively scanned by the mathematician's inner eye, the unbounded observations relevant to an empirical generalization always outrun the scope of the empirical scientist's outer eye. *Hume's problem* is both an observation and a challenge. The observation is that inductive[1] inferences can never be certain or demonstrative in the way that mathematical inferences were supposed to be and the challenge is to explain what is good about them, other than that we habitually draw them.

Church's thesis is the proposition that the intuitively algorithmically computable functions are all recursive, or equivalently, Turing computable. Turing showed that the thesis follows from locality conditions and finite bounds on the resources and perceptual abilities of a human following an algorithm.[2] The locality and perceptual boundedness assumptions are virtually identical to those invoked in classical arguments for inductive skepticism. For example:

[1]In this paper, *induction* will always refer to empirical inquiry rather than to mathematical induction.

[2]For an extended discussion of the relevance of Turing's argument, cf. (Sieg, 1994).

M.L. Dalla Chiara et al. (eds.), Logic and Scientific Methods, 159–177.

I assume then that the computation is carried out on one-dimensional paper, i.e. on a tape divided into squares. I shall also suppose that the number of smbols which may be printed is finite. If we were to allow an infinity of symbols, then there would be symbols differing to an arbitrarily small extent. ... The difference from our point of view between the single and compound symbols is that the compound symbols, if they are too lengthy, *cannot be observed at one glance.*

The behavior of the computer at any moment is determined by the symbols which he is observing, and his "state of mind" at that moment. We may suppose that *there is a bound B to the number of symbols or squares which the computer can observe at one moment.* If he wishes to observe more, he must use successive observations.[3]

This is very different from Hume's picture in which the mathematician's inner eye exhaustively scans two ideas and recognizes with certainty that the one is contained in the other. Indeed, Turing's computer is in a situation much more analogous to that of Hume's empirical scientist. Both the scientist and Turing's computer are limited by a finitely bounded perspective: the empirical scientist can see only a finite initial segment of an infinite data stream at any given time and the Turing machine can have seen only finitely many squares on an infinite tape by any given time. In a typical sort of computability argument, the Turing machine generates instances of some formal relation successively, waiting for a test condition to be met. It is natural to speak as though the machine is seeking to "verify" or "refute" some formal hypothesis in light of formally generated "data" that arise on its tape and that cannot be anticipated in advance due to the machine's bounded perspective.

Despite this intuitive analogy, there is still a strong tendency to treat induction and computation as disparate topics.[4] In the case of formal inquiry, one commonly seeks procedures guaranteed to arrive at a correct result on any input, and uncomputability is taken seriously. In the case of empirical inquiry, it is much more usual to speak of justification, probability, confirmation, and historical case studies. Computational considerations are either deemed irrelevant or are tacked onto the end of an ideal account, as when numerical computations are used to approximate ideal Bayesian updating.[5] We question this pervasive tendency. The purpose of this paper is to argue that both Hume's problem and Church's thesis are reflections of assumptions that give rise to inductive underdetermination, and to investigate some surprising methodological issues that arise when uncomputability and inductive scepticism are accorded equal status.

---

[3](Turing, 1936), our emphasis.

[4]For a recent example of this tendency, cf. (Levi, 1990).

[5]There are notable exceptions to this claim. On the formal side, there has been intererest in random algorithms for primality testing. On the empirical side, Hilary Putnam (1963; 1965) explicitly proposed adopting a computation theoretic perspective on induction, and his suggestion has been developed by computability theorists and linguists under the rubric of formal learning theory, cf. (Osherson *et al.*, 1986; Angluin and Smith, 1983). Formal learning theory is the inspiration for this paper.

2. CONVERGENCE AND CERTAINTY

One intuitive distinction between algorithms and inductive methods is that the former confer certainty whereas the latter do not. This certainty derives from two factors: (1) a logical guarantee that the algorithm will produce the right answer on each input in a specified class, and (2) the fact that the algorithm halts, thereby signalling to the user in an unambiguous way what its output is. Hume's problem might be expressed by saying that there is no procedure for inductive inference that has both properties, where inputs are taken to be the infinite data streams that may arise for all the scientist knows. The standard response to this difficulty is to exempt inductive inference from condition (1). For example, it may be required only that the method have a high personal probability of returning a correct conclusion, or that the overall outputs of the procedure be mutually coherent, explanatory, or simple. We will refer to this as the *standard approach* to inductive inference. Once the standard approach is adopted, the theory of inductive inference looks very different from the theory of computability: the former becomes preoccupied with probability distributions and their accompanying analysis, while the latter employs a distinctive kit of tools—diagonalization, dovetailing and reduction—characteristic of mathematical logic. The many differences between these alternative conceptual schemes in turn reinforce Hume's dichotomy between empirical and formal reasoning.

But an alternative response is to relax condition (2) without relaxing condition (1), so that an inductive method is guaranteed to *converge* to the right answer, but need not inform the user when it has done so. William James, for example, endorsed this alternative, which we will refer to as *logical reliabilism*[6]:

> We may talk of the *empiricist* and the *absolutist* way of believing the truth. The absolutists in this matter say that we not only can attain to knowing truth, but we can know when we have attained to knowing it; while the empiricists think that although we may attain it, we cannot infallibly know when. To *know* is one thing and to know for certain *that* we know is another.

Weakening the sense of convergence to an output so that there is no longer a determinate sign that the truth has been found (e.g., halting) can enlarge the range of empirical hypotheses about which we can be guaranteed to find the truth. But this strategy is not restricted to the empirical case. Even though algorithms confer certainty when we have them, Church's thesis informs us that most formal problems do not have algorithmic solutions, just as Hume's problem informs us that most empirical questions cannot be eventually decided by the data. Weakening the notion of convergence to an output in the formal case can likewise expand the scope of formal inquiry without relaxing the demand that the method succeed on every possible input.

In fact, various successively weaker notions of convergence to an answer have been entertained in both empirical and in computational contexts. Con-

---

[6](James, 1948, pp.95–96)

sider a system that produces a growing sequence of outputs through time. The system might be a Turing machine with an infinite output tape or it might be an inductive method producing test results in response to increasing data. The system converges to a value (say 1) *with certainty* just in case it provides a determinate sign that it has found the truth. Turing machines usually signal certainty by going into a designated halting state that the user can observe. An equally good convention would be to have some special symbol (e.g., '!', for 'Eureka!') such that the first output following the first occurrence of the special symbol is the output the system is certain of. Every subsequent output is dismissed as noise, as though the method has switched itself off after finishing its job, and every output before the first occurrence of '!' is uncertain. This works in both the Turing case and in the empirical case.

Alternatively, a system converges to an output n *in the limit* just in case after some finite period of vacillation, it eventually produces only n. Finally, a system capable of outputting rational numbers converges to an output n *gradually* just in case for each rational r, there is a time after which the conjectures of the system all remain within distance r of n.

3. VERIFICATION, REFUTATION, AND DECISION

Three concepts of convergence have been defined: certain, limiting, and gradual. We turn next to the important notion of *reliability*. Empirical and formal inquiry take on different, parallel forms depending on the application. For example, a Turing machine may be called upon to *assess* a given claim (e.g., 397 is the 20th prime) or to *generate* the value of a function (e.g. produce the 20th prime). Similarly, a scientific method may be required to test a given hypothesis or may be required to generate its own hypothesis from data. In the empirical case, generation corresponds, for example, to estimation or theory discovery.[7] In this paper, we will focus on assessment problems only. On the computational side, we assume the usual definitions of Turing machines and their inputs and outputs. On the empirical side, our model of hypothesis assessment is simplified and abstract. A *data stream* is a total function on the natural numbers, to be viewed as an infinite sequence of code numbers arriving discretely through successive moments of time. The code numbers may be thought of as representing discrete experimental outcomes. An *empirical hypothesis* will be identified with the set of all data streams of which it is true. A *hypothesis assessment method* is just a mapping from finite sequences of natural numbers (i.e., finite data sequences) to rationals in the unit interval, together with the certainty mark '!'. For the purposes of this paper it suffices to think of an assessment method as being dedicated to the investigation of a single hypothesis, so we needn't provide the hypothesis as an input.

[7] A discussion of such problems under the rubric of discovery problems may be found in (Kelly, 1995). Our other paper in this volume pertains to theory discovery.

The situation of a hypothesis assessment method is very similar to that of a Turing machine. Both the Turing machine and the inductive method face a range of possible inputs: numbers to be tested for a property in the case of the Turing machine and infinite data streams against which to check the truth of an empirical hypothesis in the case of the inductive method. In either case, there is a notion of *correct answer*: the correct answer for the Turing machine is 1 if the input number has the intended property and is 0 otherwise, whereas the correct answer for the inductive method is 1 if the hypothesis is true of the actual data stream and is 0 otherwise.

Given a criterion of convergence, there are three notions of reliable assessment that have been applied both in the theory of computability and in discussions of empirical method: verification, refutation, and decision. *Verification* requires that on every input the system converge (in the required sense) to 1 just in case 1 is the correct answer. Note that this requires that the system refuse to converge to 1 when the correct answer is 0. Similarly, *refutation* requires that on every input the system converge to 0 just in case the correct answer is 0. Finally, *decision* requires that the system converge to the correct answer whether it is 0 or 1. Combining notions of reliable assessment with notions of convergence yields nine different *standards of success*, such as refutation with certainty, verification in the limit, and gradual refutation.

On the empirical side, decidability with certainty seems to have been the standard of reliable success assumed in skeptical arguments from classical times up through Hume. Verification with certainty was briefly popular among the logical positivists, who insisted that cognitively significant theories be in principle verifiable. Karl Popper (1968) held that the scientific enterprise is characterized by its resolution to reject refuted theories with certainty rather than to protect them with conventionalistic stratagems. C.S. Peirce proposed an ideal of convergent success (1958) and Hans Reichenbach's vindication of the straight rule of induction (1949) was closely related to gradual verification.

On the computational side, numerical properties that are decidable with certainty are said to be *recursive*, those that are verifiable with certainty are said to be *recursively enumerable* or *r.e.*, and those that are refutable with certainty are said to be *co-r.e.* But the application of success standards to formal problems does not stop here. Hilary Putnam (1965) referred to properties formally decidable in the limit as *trial and error predicates*. E. Mark Gold (1965) referred to properties formally verifiable in the limit as *limiting r.e.* and to those that are refutable in the limit as *limiting co-r.e.* Finally, Peter Hàjek (1978) has examined formal problems that are gradually verifiable under the rubric of *experimental logics*. Putnam, Gold, and Hàjek clearly saw the analogy between their criteria of formal success and intuitions about convergence in empirical inquiry.[8]

[8] So did Putnam's student, Peter Kugel (1977).

There is a further analogy. In empirical applications, the scientist may start out with assumptions restricting the set of possible data streams he might encounter to some set $\mathcal{K}$. This has the effect of making the problem easier to solve, since there are fewer possibilities to contend with. Similarly, a formal decision problem can become easier if we restrict the range of natural numbers over which the machine is required to succeed, as when the halting problem is restricted to indices that either fail to halt on an input or that halt under a fixed complexity bound.[9]

4. THE HALTING PROBLEM AND INDUCTIVE GENERALIZATION

Sextus Empiricus recorded the following, classical argument for inductive skepticism.

> [The dogmatists] claim that the universal is established from the particulars by means of induction. If this is so, they will effect it by reviewing either all the particulars or only some of them. But if they review only some, their induction will be unreliable, since it is possible that some of the particulars omitted in the induction may contradict the universal. If, on the other hand, their review is to include all the particulars, theirs will be an impossible task .... (Sextus, 1985, p.105)

From the logical reliabilist point of view, Sextus' argument is actually a sketch for a proof of an elementary mathematical proposition. Suppose the hypothesis in question contains exactly one data stream ε. For example, the universal empirical hypothesis "You will always see 0" is a singleton containing the everywhere 0 data stream. Let α be an arbitrary empirical assessment method. A wily demon can adopt a strategy for presenting data to α in such a way that $\mathcal{H}$ is false but α becomes certain that $\mathcal{H}$ is true. The demon simply presents all 0s until such time as α declares 1 with certainty. Thereafter the demon presents all 1s. If α never says 1 with certainty, α fails on the everywhere 0 data stream. If α does, then α fails on a data stream that has all 0s prior to announcing certainty and all 1s thereafter. In fact, the same argument works for any non-open hypothesis (in the standard, infinite product topology on ω). A singleton is just one example of a non-open set.

In the formal domain, the most basic negative result is that the complement of the halting problem (the set of all natural numbers n such that the nth Turing machine does not halt on input n) is not r.e. (i.e., not formally verifiable with certainty). It certainly seems as though the difficulty is the same as in Sextus' skeptical argument: as soon as the would-be verification procedure becomes sure that a machine will never halt on its own index, the machine may halt at some later time (for all the verification procedure knows in light of its bounded perspective on the simulated computation). But despite the strong heuristic analogy, the usual proof of the fact in question bears little resemblance to Sextus' skeptical argument. Turing machines do not accept

[9] Cf. exercise 2-34 in (Rogers, 1987).

infinite inputs, so there is no place for a demon to feed misleading data to. Instead, we are presented with a static, Cantorian diagonal argument in which the complement of the halting problem is depicted as the counterdiagonal of an infinite table in which each characteristic function of an r.e. set is a row. This is elegant, but it gives the beginning student the impression that epistemological analogies are ultimately misleading.

They aren't. Let M_i be an arbitrary Turing machine (that we may think of as aspiring to verify the complement of the halting problem with certainty). Now we construct a "demonic machine" M_d dedicated to misleading M_i about whether it will halt. On a given input, it throws away the input and then passes along the result of simulating the aspiring positive test M_i on input d, where d is the demonic machine's own index. (That a program that feeds its own index to another program exists is an immediate consequence of the Kleene fixed point theorem.) Now we have a situation just like that in Sextus' argument: if the aspiring positive test ever declares certainty (by halting), the demonic program is total, so the aspirant is mistaken; and if the aspiring test never declares its certainty by halting, the demonic program halts on no input, so again the aspirant is mistaken. Just like the inductive demon, the demonic machine out-waits the would-be verification procedure.[10]

5. HIERARCHIES AND REDUCIBILITY

Impose the discrete topology on ω. Let $\mathcal{N}$ denote the set of all total functions on ω with the usual product topology induced by the discrete topology on ω. $\mathcal{N}$ is referred to as the Baire space. Let $\mathcal{R}$ be a relation involving n function arguments and m number arguments. $\mathcal{R}$ is open just in case it is open in the product space $\mathcal{N}^n \times \omega^m$. $\mathcal{R}$ is r.e. just in case there is a Turing machine such that for each input tuple in $\mathcal{N}^n \times \omega^m$, the machine eventually halts on the tuple just in case $\mathcal{R}$ holds of the tuple. In this definition, the functions are input to the machine on infinite read-only tapes, only finitely much of which can be read prior to halting.

The arithmetical complexity classes can now be defined as follows.[11] Let Σ^0_0 denote the collection of recursive relations. Then a Σ^0_{n+1} relation is a relation that can be expressed as $\exists y.\neg\mathcal{R}(\bar{\varepsilon}, \bar{x}, y)$, where $\mathcal{R} \in \Sigma^0_n$. A relation is Π^0_n just in case its negation is Σ^0_n. Finally, a relation is Δ^0_n just in case it is both Σ^0_n and Π^0_n. The Borel complexity classes of finite order are defined in exactly the same way, except that the bold-face Greek characters $\boldsymbol{\Sigma}^0_n$, $\boldsymbol{\Pi}^0_n$, and $\boldsymbol{\Delta}^0_n$ are

[10] Similar arguments can be given for a wide range of negative results. For demonic arguments against limiting computation cf. (Kelly, 1995). This inductive representation of computational intractability is common in learning-theoretic arguments (Case and Smith, 1983).

[11] Cf. (Rogers, 1987),(Moschavakis, 1980), or (Hinman, 1978).

used and the recursive relations are replaced with the clopen relations in the base case.[12]

The Borel hierarchy was studied at the turn of the century by analysts interested in whittling down the space of all set theoretic functions to some large class with nice analytical properties.[13] In the 1940's, S. Kleene developed the arithmetical hierarchy in order to study the new concept of computability that had evolved in response to Gödel's incompleteness theorems, apparently without knowledge of the earlier developments in analysis. J. W. Addison (1955) recognized the strong mathematical analogies between the two hierarchies: continunity is like computability, open sets are like r.e. sets, clopen sets are like recursive sets, the closure laws are similar, and the initial countable segment of the Borel hierarchy can be defined just like the arithmetical hierarchy except that the r.e. relations are replaced with open relations. Addison's (Σ, Π, Δ) notation for the hierarchies underscores these analogies.

The relevance of all this for our purposes is that the arithmetical hierarchy exactly characterizes the various senses of *formal* problem solvability defined above, and the corresponding cells in the Borel hierarchy exactly characterize the corresponding senses of *empirical* problem solvability.

Proposition 1 [14] *Table 1.1 matches each sense of problem solvability defined above with the complexity classification that characterizes the problems solvable in that sense, both for empirical problems (empirical hypotheses) and for formal problems (sets of numbers).*

In other words, the arithmetical hierarchy is a classification of formal problems according to the sense in which they are computably solvable and the Borel hierarchy is a parallel classification of hypothesis assessment problems according to the sense in which they are empirically solvable. That uncomputability and inductive skepticism should generate such similar mathematical structures provides a further analogy between Church's thesis and Hume's problem.

The concept of computable many-one reducibility is basic to the theory of computability. Generally speaking, a many-one reduction of a set of functions $\mathcal{A}$ to a set of functions $\mathcal{B}$ is an operator that takes elements of $\mathcal{A}$ to elements of $\mathcal{B}$ and nonelements of $\mathcal{A}$ to nonelements of $\mathcal{B}$. Here again there is an analogy. We may think of a reliable inductive assessment method as a continuous many-one reduction among sets of functions (i.e., of the hypothesis to the set of all conjecture streams that converge in the relevant sense[15]). The results summarized in the second column of proposition 1 therefore amount

[12] In the Baire space, every open set is a countable union of clopen sets. In spaces lacking this property (e.g., the real line), one must initiate the induction with the open sets, finally defining $\mathbf{\Sigma}^0_0$ to be $\mathbf{\Delta}^0_1$.

[13] For a summary of these issues, cf. (Moschavakis, 1980).

[14] The computational results summarized in this proposition are in (Putnam, 1965), (Gold, 1965) and (Hàjek, 1978). For an expanded presentation, cf. (Kelly, 1995).

[15] Assume an encoding of rationals in the unit interval by natural numbers.

Sense of Success	*Maximum Problem Complexity*	
	Empirical case	*Formal case*
decidable with certainty	$\mathbf{\Delta}^0_1$	Δ^0_1
verifiable with certainty	$\mathbf{\Sigma}^0_1$	Σ^0_1
refutable with certainty	$\mathbf{\Pi}^0_1$	Π^0_1
decidable in the limit	$\mathbf{\Delta}^0_2$	Δ^0_2
verifiable in the limit	$\mathbf{\Sigma}^0_2$	Σ^0_2
refutable in the limit	$\mathbf{\Pi}^0_2$	Π^0_2
decidable gradually	$\mathbf{\Delta}^0_2$	Δ^0_2
verifiable gradually	$\mathbf{\Pi}^0_3$	Π^0_3
refutable gradually	$\mathbf{\Sigma}^0_3$	Σ^0_3

TABLE 1.1. A characterization of the solvable empirical and formal problems

to continuous reducibility completeness theorems for the respective convergence criteria in their corresponding Borel complexity classes. An inductive demon continuously reduces the convergence criterion to the complement of the hypothesis. Since the convergence criterion is complete in the corresponding Borel complexity class, the existence of both a demon and a successful assessment method for a given hypothesis would imply collapse of the Borel hierarchy. So skeptical arguments fit naturally into the computability theorist's standard repertoire of moves.[16]

6. THE RICE-SHAPIRO THEOREM

The Rice-Shapiro theorem is a standard tool for proving that an index set (i.e., the set of all indices of some collection of partial computable functions) is not r.e. From the present point of view, however, it illustrates a strong connection between computability and ideal[17] empirical inquiry.

Assume a fixed, computable encoding of pairs of numbers into nonzero numbers. Let ϕ be a partial computable function. A data stream for ϕ is an enumeration of code numbers of ordered pairs in the graph of ϕ, possibly interspersed with 0s. (The 0s ensure that the everywhere undefined function will have at least one data stream, namely, the everywhere 0 data stream). Say

[16] Continuous reducibility is a standard topic in descriptive set theory; cf. (Moschavakis, 1980).

[17] By 'ideal' we mean subject to no computational limitations.

that the empirical hypothesis generated by an index set is the set of all data streams of functions with indices in the set. Then the Rice-Shapiro theorem[18] says:

Proposition 2 (Rice-Shapiro) *If an index set is formally verifiable with certainty, then the hypothesis generated by the index set is empirically verifiable with certainty given that the data stream is for a partial computable function.*[19]

Proposition 2 permits one to prove a purely computational result (that a set of numbers is not r.e.) by means of a purely empirical argument (a skeptical, demonic argument showing that a certain empirical hypothesis is not verifiable with certainty from empirical data).

7. COMPUTABILITY, EMPIRICAL INQUIRY AND COMPUTABLE EMPIRICAL INQUIRY

We began with a common view among inductive methodologists, according to which formal problems are relatively unproblematic and computational limitations are either ignored or are deferred until after some ideal norms governing inductive inference have been framed. In response, we argued that Hume's problem and Church's thesis share a common motivation, namely locality constraints on perception; whether on an empirical data stream or an internal data structure. It was also argued that virtually the same notions of convergence and reliability apply both in the formal and in the empirical domain, which gives rise to similar complexity hierarchies in the two subjects.

But there is a deeper sense in which computation and empirical inquiry are analogous. Say that a problem is *underdetermined* if there is no reliable method for solving it. *Inductive* underdetermination arises due to a finitely bounded perspective on an unbounded structure. It is immediate that classical sceptical arguments are about inductive underdetermination. What we have seen in our discussions of the Rice-Shapiro theorem and the halting problem, however, is that uncomputability arguments are as well.

All inductive methodologies are responses to inductive underdetermination. If inductive methodologists did not take inductive underdetermination seriously, then they would simply insist that everyone always believe the complete truth. But we have seen that uncomputability also stems from inductive underdetermination arising from similar locality and boundedness assumptions. It is therefore an oddly imbalanced position to ignore computability as an inessential accident in inductive methodology when one's primary focus hinges on the same underlying structure. Such a position resembles that of a physicist interested in pressure who disregards changes in temperature as annoying accidents when in fact the same underlying structure—the kinetic

[18](Cutland, 1986, Theorem 2.16, p.130)

[19]Cf. (Kelly, 1995, Corollary 6.8).

behavior of molecules—gives rise to systematic and essential relations between the two.

Given that uncomputability and inductive skepticism both stem from inductive underdetermination and that computability considerations should therefore not be ignored in the study of inductive methodology, there remains the question of how the two factors should be analyzed. The usual theory of computability requires procedures to eventually produce the right answer. The standard approach to induction is very different, exchanging guaranteed convergence to the truth for other considerations such as probability, confirmation, coherence, simplicity, or social pressure. These distinct conceptualizations of induction and computability do not fit together well. Probabilistic coherence requires logical omniscience, but logical ignorance may be unavoidable when logical relations are uncomputable. And even for effectively solvable problems, logical ignorance persists until the algorithm's output is obtained. Probabilists can say that such constraints are not to be taken literally, but this strategy must stop somewhere because probabilistic coherence is nothing without logical omniscience over some algebra.

If we adopt the logical reliabilist approach, however, then we obtain a uniform perspective on both the inductive and the formal aspects of computable inductive inquiry; a perspective in which the structures of computation and induction interleave without artificial obstacles, as the following proposition illustrates.

Proposition 3 *The characterization of ideal hypothesis assessment presented in proposition 1 holds in the case of hypothesis assessment by a computable method if we substitute arithmetical complexity classes for the corresponding Borel complexity classes.*

The proposition says that the arithmetical hierarchy incorporates both the purely inductive and the purely formal aspects of computable inductive inquiry and that it does so in a manner entirely parallel to the Borel characterization of ideal (uncomputable) inductive inquiry. Arithmetical complexity can be high because a hypothesis is inductively underdetermined by the empirical data, because it forces the scientist to perform impossible computations in order to check it against the data, or because of some complicated combination of the two factors. Moreover, a given amount of formal complexity may either add to the purely empirical complexity of the hypothesis or interleave with it so that the overall arithmetical complexity is not increased, depending on the particular case. Thus the analogy with pressure and temperature is apt: there is an underlying, fundamental structure (arithmetical complexity) against which the interactions between uncomputability and the problem of induction are essential dependencies rather than accidents to be dismissed. In the following sections we provide a systematic study of relations of this sort with some surprising consequences.

8. COMPUTABLE INQUIRY AND UNCOMPUTABLE PREDICTIONS

The traditional distinction between matters of fact and relations of ideas invites the inductive methodologist to separate formal questions from empirical ones. One of the most obvious formal problems facing an empirical scientist is to derive predictions from the hypothesis under test for comparison with the data. Recall that in our present setting, an empirical hypothesis is a set of data streams $\mathcal{H}$. We say that $\mathcal{H}$ predicts observation b at stage n given finite data sequence e just in case each data stream in $\mathcal{H}$ that extends e has b in position n. Then we write $Pred_{\mathcal{H}}(e, n, b)$. As a special case, say that a hypothesis is *empirically complete* just in case it unconditionally predicts a unique datum for each position in the data stream. An empirically complete hypothesis therefore contains a single data stream (namely, the one agreeing with all of the unconditional predictions). Finally, the range of a hypothesis $\mathcal{H}$ is the set of outcomes that may ever arise if $\mathcal{H}$ is true.

The following is perhaps the simplest and most obvious methodological advice imaginable: *To test a hypothesis, derive its predictions, compare them against the data, and reject the theory when a mismatch is found.*

The advice is correct when the hypothesis is empirically complete and there is a computable procedure for deriving the predictions of $\mathcal{H}$ (i.e., when $Pred_{\mathcal{H}}$ is recursively enumerable), for then the advice amounts to a computable method that refutes $\mathcal{H}$ with certainty.

But what happens when the predictions of the hypothesis under test are not computably derivable? Then it seems as though no computable method could refute the hypothesis with certainty. If the predictions are not computable, then every computable method for deriving them must either be unsound (it mistakenly derives the wrong prediction) or incomplete (it hangs in an infinite loop, never yielding the intended prediction). In the former case, the data might conspire to agree with the mistakenly derived prediction so that the hypothesis is false but the method never notices. In the latter case, the method never finds out what the theory predicts about some datum, and the truth of the theory may depend entirely on what this prediction is.

But remarkably enough, the argument just given is fallacious: there are empirically complete hypotheses that can be reliably assessed by computable methods even though their predictions are not even definable in arithmetic.[20]

Proposition 4 *There is a hypothesis $\mathcal{H}$ such that*

1. *$\mathcal{H}$ is empirically complete*
2. *$Pred_{\mathcal{H}}$ is not arithmetically definable*
3. *but $\mathcal{H}$ is computably refutable with certainty.*

[20] Complete proofs of the following propositions may be found in (Kelly and Schulte, 1995).

Due to the following proposition, the hypothesis that witnesses proposition 4 has an infinite range of predictions.

Proposition 5 *Suppose that $\mathcal{H}$ is empirically complete, computably verifiable in the limit, and of finite range. Then $Pred_{\mathcal{H}}$ is computably decidable with certainty.*

On the other hand:

Proposition 6 *There is a hypothesis $\mathcal{H}$ such that*

1. *$\mathcal{H}$ is empirically complete*
2. *the range of $\mathcal{H}$ is binary,*
3. *$Pred_{\mathcal{H}}$ is not arithmetically definable,*
4. *but $\mathcal{H}$ is computably refutable in the limit.*

No empirically complete hypothesis is open, so by proposition 1, no such hypothesis is verifiable with certainty. Thus we quickly arrive at a complete picture of how uncomputable the predictions of an empirically complete theory can be given that the truth of the theory can be reliably determined by computational means in a specified sense.

In light of the view that science should proceed by deriving predictions and checking them against empirical data, propositions 4 and 6 appear almost magical. Somehow or other, computable inquiry can proceed without being able to derive the predictions of the hypothesis, even though any one of these predictions might be wrong. But it underscores the analogy between computability and inductive methodology that these striking methodological facts are actually restatements of familiar results in the theory of computability. The key to seeing this is that the arithmetical complexity of $\{\varepsilon\}$ corresponds to the number of quantifier alternations in an implicit definition of ε in arithmetic and the arithmetical complexity of ε corresponds to the number of quantifier alternations in an explicit definition of ε in arithmetic. Let ν denote the characteristic function of the Gödel numbered truths of arithmetic. The fact that $\{\nu\} \in \Pi^0_2$ is just the implicit definability of arithmetical truth, which was already shown by Hilbert and Bernays. The fact that ν is not arithmetically definable is just Tarski's celebrated theorem on the undefinability of arithmetical truth. Proposition 6 follows immediately by proposition 3.

The proof of proposition 4 builds on the preceding argument (Hinman, 1978). Let $\langle \_, \_ \rangle$ denote an effective encoding of pairs of natural numbers as natural numbers and let $(\langle n, m\rangle)_1 = n$ and let $(\langle n, m\rangle)_2 = m$. Extend the coding to functions so that $(\varepsilon)_1(n) = (\varepsilon(n))_1$. Since $\{\nu\} \in \Pi^0_2$, there is a recursive relation $\mathcal{G}$ such that for each ε,

$$\varepsilon \in \{\nu\} \iff \forall x \exists y . \mathcal{G}(\varepsilon, x, y).$$

Then define

$$\delta(x) = \langle \nu(x), \mu y . \mathcal{G}(\nu, x, y)\rangle .$$

Hence, $\nu(x) = (\delta(x))_1$, so δ is not arithmetically definable. Finally, $\{\delta\}$ is definable as follows.

$$\varepsilon \in \{\delta\} \Longleftrightarrow \forall x.\mathcal{G}((\varepsilon)_1, x, (\varepsilon(x))_2) \text{ and } \forall x, y \text{ if } y < (\varepsilon(x))_2 \text{ then } \neg\mathcal{G}((\varepsilon)_1, x, y).$$

So $\{\delta\} \in \Pi^0_1$. Note that while the range of ν is binary, the minimization in the definition of δ makes the range of δ infinite.

Results relating implicit to explicit definability are standardly grouped together with the recursion theoretic basis theorems. A *basis* for a collection Γ of sets of functions (i.e., empirical hypotheses) is a collection of functions that shares an element with each nonempty set in Γ. Thus, proposition 6 says that the set of all arithmetically definable functions is not a basis for the set of Π^0_2 singletons of binary range, proposition 4 says that the set of all arithmetically definable functions is not a basis for the set of Π^0_1 singletons, and proposition 5 says that the set of all total computable functions is a basis for the Σ^0_2 singletons of finite range.

But the basis theorems are relevant to our topic only in the case of empirically complete hypotheses. When we move to empirically incomplete hypotheses, basis theorems say that each hypothesis of a given complexity is correct of a sufficiently simple data stream. But that is not our concern. We are interested in seeing how complex the predictions of a computably testable hypothesis can possibly be, which is something quite different. The empirical application therefore suggests a different generalization of the implicit definability issue than the one traditionally examined in the theory of computability.

In the empirically incomplete case, we can find hypotheses that are computably refutable with certainty whose predictions are even more complex than in the preceding examples. The *analytical* hierarchy is defined just like the arithmetical hierarchy, except that existential quantification over numbers is replaced with existential quantification over functions. The analytical hierarchy is indexed with superscript 1 to distinguish it from the arithmetical hierarchy. The relations in Δ^1_1 are said to be *hyperarithmetical.* A relation is *complete* in a complexity class just in case it is a member of the class and every member of the class is computably many-one reducible to it.

Proposition 7 *There is a hypothesis $\mathcal{H}$ such that*

1. *$Pred_{\mathcal{H}}$ is Π^1_1-complete*
2. *and $\mathcal{H}$ is computably refutable with certainty.*

In contrast, $\{\nu\}$ and $\{\delta\}$ are both Δ^1_1 and could not have been chosen outside of Δ^1_1. The construction of $\mathcal{H}$ is as follows. Let S be a Π^1_1-complete set of numbers. Let $\langle \_ \rangle$ effectively encode finite sequences of natural numbers into natural numbers. By a standard normal form theorem (Rogers, 1987, Ch.16,Cor.V), there is a recursive G such that for all x:

$$S(x) \Longleftrightarrow \forall \tau \exists n.G(\langle \tau|n \rangle, x)$$

Given sense in which $\mathcal{H}$ is empirically testable		Maximum complexity of $Pred_{\mathcal{H}}$			
		Empirically complete $\mathcal{H}$		General Case	
		Finite Range	Infinite Range	Finite Range	Infinite Range
decidable with certainty:	Δ^0_1	$\square$	$\square$	Δ^0_1	Π^0_1
verifiable with certainty:	Σ^0_1	$\square$	$\square$	Π^0_1	Π^0_1
refutable with certainty:	Π^0_1	Δ^0_1	Δ^1_1	Σ^0_1	Π^1_1
decidable in the limit:	Δ^0_2	Δ^0_1	Δ^1_1	Π^0_2	Π^1_1
verifiable in the limit:	Σ^0_2	Δ^0_1	Δ^1_1	Π^0_2	Π^1_1
refutable in the limit:	Π^0_2	Δ^1_1	Δ^1_1	Π^1_1	Π^1_1
	Σ^0_{n+3}	Δ^1_1	Δ^1_1	Π^1_1	Π^1_1

TABLE 1.2. The inductive and deductive complexity of empirical hypotheses.

where $\tau|n$ denotes the initial segment of τ of length n. Our hypothesis is then defined as follows:

$\varepsilon \in \mathcal{H} \iff$

1. $\varepsilon(2)$ encodes a unit sequence,
2. $\forall i \geq 2, \varepsilon(i+1)$ encodes an extension of length i of the sequence encoded by $\varepsilon(i)$,
3. and $\forall i \geq 2, \neg G(\varepsilon(i), \varepsilon(0))$.

By the form of the definition, $\mathcal{H} \in \Pi^0_1$. It turns out that each arithmetically definable hypothesis has Π^1_1 predictions. It remains to see that $Pred_{\mathcal{H}}$ reduces S. Suppose $x \in S$. Then $\forall \tau \exists n. G(\langle \tau|n \rangle, x)$. Hence, no data stream in $\mathcal{H}$ begins with x. So the unit data sequence (x) refutes $\mathcal{H}$ and $\mathcal{H}$ trivially predicts every possible outcome at position 1 given that (x) has been observed. On the other hand, suppose $x \notin S$. Then $\exists \tau \forall n. \neg G(\langle \tau|n \rangle, x)$. Hence, (x) is consistent with $\mathcal{H}$. Since $\mathcal{H}$ imposes no constraints on what happens at stage 1, any outcome may arise at stage 1 given only that (x) is observed. Therefore, $x \in S \iff Pred_{\mathcal{H}}((x), 1, 0)$.

Similar arguments lead to a complete table mapping out how intractable the predictions of a computably testable theory can be for each notion of inductive success (Table 1.2). The results differ sharply depending on whether or not the hypothesis is empirically complete and on whether or not the range of the hypothesis is finite. The $\square$ symbol indicates that the case in question is impossible.

9. IDEAL NORMS AND COMPUTATIONAL DISASTERS

So what, intuitively, is the magic that enables computable science to reliably test theories with uncomputable predictions? Say that a method that rejects a hypothesis with certainty as soon as it is refuted by the data is *consistent*. Consistency is recommended by almost all ideal methodologists, including hypothetico-deductivists, Bayesians and, more recently, belief revision theorists (Gärdenfors, 1988). But it turns out that:

Proposition 8 [21] *Suppose that $\mathcal{H}$ is assessable (in any of the senses defined above) by a* consistent *method α. Then we have the following consequences.*

1. *If α is in Σ^0_n, then $Pred_{\mathcal{H}} \in \Pi^0_{n+1}$.*
2. *If $\mathcal{H}$ is also empirically complete, then $Pred_{\mathcal{H}} \in \Sigma^0_n$.*
3. *If α is analytical, then $Pred_{\mathcal{H}}$ is of analytic complexity no greater than α's.*

It follows that hypotheses whose predictions are not arithmetically definable cannot be assessed in any of our senses by an arithmetically definable consistent method. Since the predictions of the hypothesis $\mathcal{H}$ exhibited in proposition 7 are not even hyperarithmetical, no consistent, hyperarithmetically definable method could even gradually refute or verify $\mathcal{H}$; a situation that contrasts sharply with the fact that $\mathcal{H}$ is refutable *with certainty* by a nonconsistent, *computable* method. Consistency imposes a severe restriction on the power of computable science.

Nonconsistent methods allow some time lag before noticing that the hypothesis under test has been refuted. This might suggest that the formal method deriving the predictions merely needs some extra time to do its work while further data are being read. But there is more to it than that, since nonarithmetical predictions cannot be derived by computable means no matter how much time delay is provided, or even in the limit. In each of our examples, the successful computable method succeeds by actually *using* future empirical data to detect the inconsistency of the hypothesis under test with earlier data. Let us return to the hypothesis $\mathcal{H}$ defined in the proof of proposition 7. To refute $\mathcal{H}$ with certainty, one begins by reading and saving the datum x in position 0. Then one ignores the datum in position 1 and checks whether the datum i in position 2 is the code number of a unit sequence and whether $\neg G(i, x)$. Thereafter, one checks whether the current datum j encodes a unit extension of the sequence encoded by the preceding datum and whether $\neg G(j, x)$. The important point is this: if the first datum encountered is in S, then $\mathcal{H}$ is already refuted by this datum, but our procedure won't notice that until some later datum is read and decoded. Moreover, it is clear that no hyperarithmetically definable formal method that reads only the first datum

[21] (Gaifman and Snir, 1990) contains a related result concerning the definability of Bayesian probability measures that maintain unit priors on aritmetical truths and nonextremal priors on contingent sentences in a first order language including arithmetic.

could ever determine, without looking at future data, whether this first datum already refutes $\mathcal{H}$. The same may be said of any arithmetically definable method that succeeds in any of our senses when faced with the hypotheses $\{\nu\}$ and $\{\delta\}$.

We have just seen that purely formal problems arising in the course of computable empirical inquiry may require future empirical data for their solution, so that the traditional distinction between the formal and inductive components of inquiry is blurred. This situation stands in sharp contrast with Hume's distinction between empirical matters of fact and a priori relations of ideas—the distinction with which this paper began.

10. CONCLUSION

The paper began with the traditional view that formal problems are to be bracketed or deferred in fundamental discussions on the logic of induction. In response, we have argued that Church's thesis is analogous to Hume's problem. Both are founded on material assumptions concerning bounded perception. Both arise out of similar reliabilist criteria of success. Both generate similar hierarchies of complexity. Both give rise to natural notions of reducibility and completeness. And most importantly, arguments for both uncomputability and inductive skepticism share a common structure; the structure of inductive underdetermination. Therefore it is unnatural for inductive methodology to abstract from computability considerations. Logical reliabilism provides a balanced perspective on uncomputability and inductive skepticism in the study of computable empirical inquiry.

As an illustration of this approach, we analyzed what seems to be the most elementary methodological recommendation conceivable: that to test a theory one should formally derive its predictions and then check them against the data. A striking result of this investigation is that for some hypotheses a computable inductive method must use future empirical data to determine the consistency of past data with the hypothesis, thereby further breaking down the traditional distinction between empirical and formal methodology. Another is that for empirically complete theories, inductive and deductive complexity are literally related as implicit and explicit definability in arithmetic. Still another is that inductive methodology suggests a natural generalization of implicit and explicit definability that differs from the standard basis theorems.

Department of Philosophy
Carnegie Mellon University
Pittsburgh PA

ACKNOWLEDGEMENTS

Oliver Schulte was supported by a grant from Siemens AG during the course of this research. We are indebted to our colleagues Clark Glymour and Wilfried Sieg for comments on a draft of this paper.

REFERENCES

Addison, J. (1955). "Analogies in the Borel, Lusin, and Kleene Hierarchies," *Bulletin of the American Mathematics Society* 61: 171-172.

Angluin, D. and Smith, C. "A survey of inductive inference: Theory and methods." *Computing Surveys* 15:237–289.

Case, J. and Smith, C. (1983)"Comparison of Identification Criteria for Machine Inductive Inference," *Theoretical Computer Science* 25: 193–220.

Cutland, N.J. (1986). *Computability: An introduction to recursive function theory.* Cambridge: Cambridge University Press.

Gaifman, H. and M. Snir (1982). "Probabilities Over Rich Languages, Testing and Randomness," *Journal of Symbolic Logic* 47: 495– 548.

Gärdenfors, P. (1988). *Knowledge In Flux: modeling the dynamics of epistemic states.* Cambridge: MIT Press.

Garber, D. (1983) "Old Evidence and Logical Omniscience in Bayesian Confirmation Theory," in *Testing Scientific Theories: Minnesota Studies in the Philosophy of Science*, vol. 10, ed. J. Earman. Minneapolis: University of Minnesota Press.

Gold, E.M. (1965). "Limiting Recursion," *Journal of Symbolic Logic* 30: 27–48.

Hàjek, P. (1978). "Experimental Logics and Π_3^0 Theories," *Journal of Symbolic Logic* 42: 515–522.

Hinman, P.G. (1978). *Recursion-Theoretic Hierarchies.* New York: Prentice-Hall.

Hume, D. (1984). *An Inquiry Concerning Human Understanding*, ed. C. Hendell. New York: Bobbs-Merrill.

James, W. (1948) "The Will To Believe," *Essays in Pragmatism.* ed. A. Castell. New York: Collier.

Kelly, K. (1995). *The Logic of Reliable Inquiry.* Oxford: Oxford University Press.

Kelly, K. and Schulte, O. (1995) "The Computable Testability of Theories Making Uncomputable Predictions,"*Erkenntnis.* 43:29–66.

Kugel, P. (1977). "Induction, Pure and Simple," *Information and Control* 33:276–336.

Levi, I. (1990). "Rationality Unbound," in *Acting and Reflecting,* ed. W. Sieg. Dordrecht: Kluwer.

Moschavakis, Y. (1980). *Descriptive Set Theory.* Amsterdam: North Holland.

Osherson, D., Stob, M. and Weinstein, S (1986). *Systems That Learn.* Cambridge, Mass: MIT Press.

Peirce, C.S. (1958). *Charles S. Peirce: Selected Writings*, ed. P. Wiener. New York: Dover.

Popper, K. (1968). *The Logic Of Scientific Discovery.* New York: Harper.

Putnam, H. (1963). " 'Degree of Confirmation' and Inductive Logic," In *The Philosophy of Rudolph Carnap*, ed. A. Schilpp. La Sall, Ill.: Open Court.

Putnam, H. (1965). "Trial and Error Predicates and a Solution to a Problem of Mostowski," *Journal of Symbolic Logic* 30: 49–57.

Reichenbach,H.(1949). *The Theory of Probability.* London: Cambridge University Press.

Rogers, H. (1987). *The Theory of Recursive Functions and Effective Computability.* Cambridge: MIT Press.

Sextus Empiricus (1985). *Selections from the Major Writings on Scepticism, Man and God,* ed. P. Hallie, trans. S. Etheridge. Indianapolis: Hackett.

Sieg, W.(1994). "Mechanical Procedures and Mathematical Experience," in *Mathematics and Mind*, ed. A. George. Oxford: Oxford University Press.

Turing, A. (1936). "On Computable Numbers, with an Application to the Entscheidungsproblem," *London Mathematics Society* 42: 230–265.

YIANNIS N. MOSCHOVAKIS

THE LOGIC OF FUNCTIONAL RECURSION[1]

1. INTRODUCTION

Recursive definitions of functions abound in mathematics and logic: in the classical theory of recursive functions on the set $\mathbb{N}$ of natural numbers—of course; in (so-called) abstract or generalized recursion theory; and, most significantly, in programming languages, where (especially since the work of Scott [9]) it has become clear that the main "programming construct" is *definition by recursion.* And so it is important to understand how properties of recursively defined functions—follow *logically* from their definitions. The paper could be called **the logic of programs**, but the title above is more specific, more general (in some ways), and also more honest, since the austere, mathematical formulation in which the problem is best understood is quite removed from actual programming practice. The formal proof systems and completeness results of this paper are related to "program verification" very much like predicate logic and its completeness are related to axiomatic set theory; they are certainly relevant, but not of much help in establishing specific, concrete results.

In its most general form, a recursive definition of a function is expressed by a *fixpoint equation* of the form

$$p(u) = f(u,p) \qquad (p : M \to W, f : M \times (M \to W) \to W), \tag{1}$$

in which the functional f provides a method for computing (or attempting to compute) each value $p(u)$ of $p : M \to W$, perhaps using ("calling") other values of p in the process. In applications, we deal mostly with *mutual recursive definitions* of the form

$$A = f_0(\vec{p}) \text{ where } \{p_1(\vec{u}_1) = f_1(\vec{u}_1,\vec{p}), \dots, p_n(\vec{u}_n) = f_1(\vec{u}_n,\vec{p})\}; \tag{2}$$

the idea here is that first we "solve" the system

$$p_1(\vec{u}_1) = f_1(\vec{u}_1,\vec{p}), \dots, p_n(\vec{u}_n) = f_n(\vec{u}_n,\vec{p}),$$

and then we apply the "head" functional f_0 to the "solutions" to get the required value. It is not hard to reduce such mutual recursions to the simpler (1), but their logic becomes more rather than less complex in the process and, in the end, it is best to take (2) as basic. So the language of functional recursion is, in effect, the language of the "where" construct, and it is formalized in the *Formal* (or *Functional*) *Language of Recursion* FLR introduced

[1]During the preparation of this paper the author was partially supported by an NSF Grant.

M.L. Dalla Chiara et al. (eds.), Logic and Scientific Methods, 179–207.

in [5]. It is possible (and useful) to think of FLR as a very abstract programming language, rich enough so that the more familiar, friendlier programming languages can be "faithfully" interpreted in it.

The basic *Monotone Fixpoint Theorem* insures the existence of a *least solution* for (1), when the range W of the "unknown" function p is a *complete poset*, and the functional f is *monotone*.[2] This (and its corollaries) suffice for the development of denotational semantics for deterministic programming languages. The situation is very different for languages with non-deterministic and "concurrent" constructs, whose interpretation requires the assignment of "canonical solutions" to (1), in situations where W is not a poset and so least solutions do not exist; it then becomes a serious problem how to justify the choice of one proposed method over another. Part of the motivation for this work is the hope that a better understanding of the laws of logic which govern deterministic recursion will help us understand, evaluate and compare the various approaches to "concurrency modeling."

The main contributions of this paper are: **first**, the introduction of a new class of least-fixed-point interpretations for FLR (and programming languages in general), a refinement of denotational semantics which is closer to our intuitive understanding of deterministic programs and mathematically simpler; **second**, the completeness and decidability of an equational proof system $\mathcal{E}_0$ for (absolute) identities between FLR expressions (program equivalence), with respect to the new interpretations; and, **third**, the completeness of a stronger system $\mathcal{E}$ for FLR-*sentences* which are valid consequences of a *first order theory*, suitably defined. The axiom systems $\mathcal{E}_0$ and $\mathcal{E}$ of Sections 7 and 10 are not complete (and could not be complete) for the general *consequence relation*, but they are strong enough so that many natural proofs of program correctness can (in principle) be formalized in them. They are also *sound* for the models of concurrent systems introduced in [6, 7, 8], and so the completeness results (especially) support the proposition that *the logic of non-deterministic and concurrent (recursive) programs is the same as that of deterministic programs.*

Acknowledgment. This paper generalizes and extends some of the basic results of the joint work [1], which deals with FLR_0, the *propositional* (or *equational*) part of FLR. I have omitted specific references to [1] (partly for reasons of space and also because the final form of [1] is not quite ready yet), and I have tried to make this paper self-contained, but, ideally, the two papers should be read together. I am grateful to my students Monica McArthur, Larry Moss and Glen Whitney and to Tonny Hurkens, for their many ideas and invaluable contributions to this project.

[2] A partially ordered set (poset) W is (directed) complete (a dcpo) if every directed $X \subseteq W$ has a least upper bound $\sup X$; $f : U \to V$ is monotone if $x \leq y \Longrightarrow f(u) \leq f(y)$; and (for complete U, V), $f : U \to V$ is *Scott-continuous* if for every non-empty, directed $X \subseteq U$, $f(\sup X) = \sup\{f(x) \mid x \in X\}$. There is no room in this paper to motivate or explain the basic ideas of poset theory and the *fixpoint theory of programs*, and so I will assume that the reader is familiar with them.

Term:	$A :\equiv$	$\mathsf{ff} \mid \mathsf{tt}$
		$\mid p(Z_1, \ldots, Z_n)$
		$\mid \mathsf{f}(Z_1, \ldots, Z_n, \pi_1, \ldots, \pi_m)$
		$\mid$ if A_0 then A_1 else A_2 fi
		$\mid A_0$ where $\{p_1(\vec{u}_1) = A_1, \ldots, p_n(\vec{u}_n) = A_n\}$
	$Z :\equiv$	$u \mid A$
ℓ-term:	$\pi :\equiv$	$p \mid \ell(u_1, \ldots, u_n)A$
Expression:	$E :\equiv$	$u \mid A \mid \pi$
Equation:	$\varepsilon :\equiv$	$u_1 = u_2 \mid u = A \mid A = u \mid A_1 = A_2 \mid \pi_1 = \pi_2$

TABLE 1. The syntax of FLR.

2. THE SYNTAX OF FLR

We fix formal, basic (individual) variables $\mathsf{u}_0, \mathsf{u}_1, \ldots$ and for each $n \in \mathbb{N}$, formal function variables $\mathsf{p}_0^n, \mathsf{p}_1^n, \ldots$ of arity n. An FLR **signature** is a pair $\boldsymbol{\tau} = (\{\mathsf{f}_i \mid i \in I\}, \text{kind})$ of a set of **functional symbols** $\{\mathsf{f}_i \mid i \in I\}$ and a mapping which assigns to each f_i a tuple $\text{kind}(\mathsf{f}_i) = \langle n, k_1, \ldots, k_m \rangle$. The general idea is that in the intended interpretations of FLR($\boldsymbol{\tau}$), each f_i will be interpreted by a functional $f_i(u_1, \ldots, u_n, p_1, \ldots, p_m)$, where each u_i ranges over the interpretation of the basic variables, and each p_j ranges over the interpretation of function variables of arity k_j. Functional symbols of kind $\langle n \rangle$ (with $m = 0$) will be interpreted by functions, including *nullary* functions when the kind is $\langle 0 \rangle$.[3]

The **expressions** of FLR($\boldsymbol{\tau}$) comprise the **variables**, the **terms**, and the **ℓ-terms**, which will be interpreted by functions and so have arities associated with them. They—and the (well formed) **equations**—are defined simultaneously by the inductive clauses in Table 1 and the following remarks, in which we also specify which variable occurrences are *bound*, the remaining ones being *free*. We use "metavariables" u, v for formal basic variables and p, q, r for function variables, sometime with subscripts and in vectors, $\vec{u} = u_1, \ldots, u_n$.

Notice that *basic variables are not terms, but function variables are ℓ-terms*, a mild eccentricity of the language which, however, fits in with the intended semantics.

(1) In $p(Z_1, \ldots, Z_n)$ and $\mathsf{f}(Z_1, \ldots, Z_n, \pi_1, \ldots, \pi_m)$, each Z_i is a basic variable or a term; p is a function variable of arity n; and f is a functional symbol of

[3]There are three inessential differences between this and the original presentation of FLR in [5]. First, only one sort of basic variables is allowed here, as is common in proof theory, and so the truth values cannot be assigned to basic variables, although they are allowed to be values of terms; second, the conditional is "typed" differently, i.e., its first argument must be a term (not a basic variable); and third, I write $\ell(u)$ rather than $\lambda(u)$, to avoid appealing inadvertently to classical properties of the λ-calculus which are not valid in all interpretations of FLR, some of which are strongly intensional. The two versions are equivalent in expressive power for the class of models considered in [5].

kind $\langle n, k_1, \dots, k_m\rangle$ with $k_j = \text{arity}(\pi_j)$. If f is a nullary symbol—a *constant*—we will write f rather than $\mathsf{f}()$.

(2) In the conditional expression, A_0, A_1 and A_2 are terms. We view this expression as an abbreviation of the more formal

$$\mathsf{cond}(\ell()A_0, \ell()A_1, \ell()A_2),$$

which makes it clear that the conditional will be interpreted as if it were a given functional symbol, of kind $\langle 0, 0, 0, 0\rangle$.

(3) In the recursive term A_0 where $\{p_1(\vec{u}_1) = A_1, \dots, p_n(\vec{u}_n) = A_n\}$, each A_i is a term, $p_1, \dots, p_n$ are distinct function variables, and each $\vec{u}_i$ is a (possibly empty) list of distinct basic variables of length arity(p_i). An occurrence of a variable x in the **head** A_0 or some **recursion part** A_i of this term is free, if it is a free occurrence of x in A_i, x is not some p_j, and also x is not one of the basic variables in the list $\vec{u}_i$. For an example, consider

$$\perp \equiv p() \text{ where } \{p() = p()\} \tag{3}$$

which plays a special role in the theory.

(4) In $\ell(\vec{u})A$, A is a term and $u_1, \dots, u_n$ are distinct, basic variables. This ℓ-term has arity n, and an occurrence of a variable u is free in it if it is free in A and not one of the u_is. We also count function variables among the ℓ-terms, basically identifying p with $\ell(\vec{u})p(\vec{u})$. We set $\ell(\vec{u})\ell(\vec{v})A \equiv_{\text{df}} \ell(\vec{u}, \vec{v})A$ when the lists $\vec{u}$ and $\vec{v}$ have no common variables, so we can meaningfully prefix ℓ-terms by the ℓ operator.

(5) In a well-formed ℓ-equation $\pi_1 = \pi_2$, the two sides must have the same arity.

For examples of expressions, see the list of axioms in Table 2.

2.1. Substitutions (replacements). By $E\{\vec{u} :\equiv \vec{Z}\}$ we denote the result of replacing in E every free occurrence of each u_i by the corresponding basic variable or term Z_i. If $E(\vec{u})$ stands for an expression in which the variables $\vec{u}$ may occur free, we will also write, as usual, $E(\vec{Z})$ for $E(\vec{u})\{\vec{u} :\equiv \vec{Z}\}$. The substitution $E\{p :\equiv \ell(\vec{u})A\}$ of ℓ-terms or function variables for function variables is defined by induction, using basic variable substitution in the one non-trivial clause, e.g.,

$$p(Z_1, Z_2)\{p :\equiv \ell(u_1, u_2)A\} \equiv A\{u_1 :\equiv Z_1', u_2 :\equiv Z_2'\},$$

where $Z_1' \equiv Z_1\{p :\equiv \ell(u_1, u_2)A\}$, and similarly with Z_2'. And with the abbreviated notation again: $E(\pi) \equiv E(p)\{p :\equiv \pi\}$, for an ℓ-term π.

3. STRICT SEMANTICS

3.1. Strict functionals. With each (non-empty) set M, we associate the *flat poset*

$$M^\dagger = M \cup \{\text{ff}, \text{tt}, \perp\}, \tag{4}$$

where the truth values ff, tt and $\perp$ (*bottom*, the "undefined") are fixed objects outside M; $\perp \leq x$ for every $x \in M^\dagger$; and all points other than $\perp$ are pairwise incomparable. An (extended) **partial function** on a set M is any function $p : M^n \to M^\dagger$, and a (partial, extended, monotone) **strict functional** on M of kind $\langle n, k_1, \ldots, k_m \rangle$ is any monotone mapping

$$f : M^n \times (M^{k_1} \to M^\dagger) \times \cdots \times (M^{k_m} \to M^\dagger) \to M^\dagger,$$

so that every n-ary partial function is a functional of kind $\langle n \rangle$. Perhaps the simplest functionals are the *calls* and the *strict conditional*,

$$\begin{aligned} \mathrm{call}_n(u_1, \ldots, u_n, p) &= p(u_1, \ldots, u_n), \\ \mathrm{cond}(p, q, r) &= \begin{cases} q() \text{ if } p() = \mathrm{tt}, \\ r() \text{ if } p() = \mathrm{ff}, \\ \bot \ \text{ otherwise}, \end{cases} \end{aligned} \tag{5}$$

but equally natural are the functionals which represent the operations of first order logic,

$$\begin{aligned} \mathrm{or}(p, q) &= \begin{cases} \mathrm{tt} \ \text{ if } p() = \mathrm{tt} \text{ or } q() = \mathrm{tt}, \\ \mathrm{ff} \ \text{ if } p() = q() = \mathrm{ff}, \\ \bot \ \text{otherwise}, \end{cases} \\ E_M(p) &= \begin{cases} \mathrm{tt} \ \text{ if } (\exists u \in M)[p(u) = \mathrm{tt}], \\ \mathrm{ff} \ \text{ if } (\forall u \in M)[p(u) = \mathrm{ff}], \\ \bot \ \text{otherwise}. \end{cases} \end{aligned}$$

All but the last of these are *continuous*, but E_M is *discontinuous* when M is infinite.

3.2. Functional structures. A (strict) *functional structure* of signature $\tau = (\{\mathsf{f}_i \mid i \in I\}, \mathrm{kind})$ is a tuple

$$\mathfrak{M} = (M, \{f_i \mid i \in I\}),$$

where each f_i is a functional on M with $\mathrm{kind}(f_i) = \mathrm{kind}(\mathsf{f}_i)$. By identifying each relation R_i on M with its characteristic function

$$\chi_{R_i}(\vec{u}) = \text{if } R(\vec{u}) \text{ then tt else ff},$$

we find among these the usual first-order structures of model theory

$$\mathfrak{M} = (M, R_1, \ldots, R_l, f_1, \ldots, f_m),$$

and, more specifically, the standard structure of arithmetic

$$\mathfrak{N} = (\mathbb{N}, 0, S, P, Z) \tag{6}$$

where $S(x) = x+1$, $P(x) = \text{if } (x = 0) \text{ then } 0 \text{ else } x - 1$, and $Z(x) \iff x = 0$. Structures with non-trivial functionals come up naturally as expansions of first order structures by functionals such as those above, e.g.,

$$(\mathfrak{M}, \mathrm{or}, E_M) = (M, \{f_i \mid i \in I\}, \mathrm{or}, E_M).$$

3.3. Strict denotations. A **valuation** (variable assignment) ω in a functional structure $\mathfrak{M}$ assigns some $\omega(\mathsf{u}_i) \in M$ to each basic variable, and some n-ary partial function $\omega(\mathsf{p}_i^n) : M^n \to M^\dagger$ to every function variable of arity n. We now associate with each expression E and each valuation ω a *denotation* $[\![E]\!]_s(\omega)$, so that $[\![E]\!]_s(\omega) = \omega(E)$ if E is a variable; $[\![A]\!]_s(\omega) \in M^\dagger$, if A is a term; and $[\![\pi]\!]_s(\omega)$ is an n-ary partial function, if π is an ℓ-term of arity n. We skip the construction, which is by induction on the definition of expressions and follows familiar lines. The salient points are:

(D1) ff and tt are interpreted by ff and tt.

(D2) Application is interpreted strictly,[4] i.e.,

$$[\![p(Z_1,\dots,Z_n)]\!]_s(\omega) = \begin{cases} \omega(p)([\![Z_1]\!]_s(\omega),\dots,[\![Z_n]\!]_s(\omega)) & \\ \qquad \text{if } [\![Z_1]\!]_s(\omega),\dots,[\![Z_n]\!]_s \in M, & \\ \bot \qquad \text{otherwise,} & \end{cases}$$

and similarly with the substitution of terms or variables in the given functionals.

(D3) Conditional terms are interpreted by the strict conditional (5).

(D4) Recursive terms are interpreted by the taking of mutual, least-fixed-points, and to justify this, we must interleave with the definition a proof that for each E, $[\![E]\!]_s(\omega)$ is *monotone in* ω. Notice that $[\![\bot]\!]_s = \bot$ for the closed term $\bot$ defined in (3).

(D5) $\ell(\vec{u})$ is interpreted by λ-abstraction.

3.4. Recursive definability. The partial functions definable on $\mathfrak{M}$ by closed ℓ-terms are the *recursive partial functions* of $\mathfrak{M}$, and in the case of $\mathfrak{N}$ they are precisely the classical, recursive partial functions on $\mathbb{N}$; on $(\mathfrak{N}, E_{\mathbb{N}})$ we get the *hyperartithmetical* functions, and, more generally, for each first-order $\mathfrak{M}$, the recursive partial functions of $(\mathfrak{M}, E_M)$ are precisely the $\mathfrak{M}$*-inductive* partial functions, with domains of definition the $\mathfrak{M}$-inductive relations on $\mathfrak{M}$ studied in [4]; and for every functional f on $\mathbb{N}$, the recursive, partial functions on $(\mathfrak{N}, f)$ are precisely those which are recursive in the sense of Kleene's type-2 recursion on $\mathbb{N}$. Kleene's higher-type recursion can also be represented (easily and naturally) as strict, FLR-definability on the appropriate structure.[5]

3.5. Strict consequence in FLR. Assuming that all equations listed are well-formed, we write

$$E_1 = F_1, \dots, E_m = F_m \models_s E = F$$

if for every functional structure $\mathfrak{M}$ and valuation ω in $\mathfrak{M}$,

$$[\![E_i]\!]_s(\omega) = [\![F_i]\!]_s(\omega) \text{ for } i = 1, \dots m \Longrightarrow [\![E]\!]_s(\omega) = [\![F]\!]_s(\omega);$$

this is the fundamental notion of **logical consequence** in the category of strict, monotone structures. We will not make a list of strict-valid consequences here since plenty will be put down later on, but notice that the β-**reduction** rule fails,

$$\ell(u)p(u,v) = \ell(u)q(v) \not\models_s p(A,v) = q(v), \tag{7}$$

by the trivial counterexample $p(u,v) = q(v) = \text{tt}$, $A \equiv \bot$.

[4]We might set instead $[\![p(Z)]\!]_s(\omega) = \text{ff}$ (or record an "error" in some other way) when $[\![Z]\!]_s(\omega)$ is ff or tt, but it is easier to handle such finer modelings of computation in the context of the more general $\jmath$-structures of Section 6.

[5]This is essentially the approach to recursion in higher types taken in [2].

4. CONTINUOUS SEMANTICS

4.1. CONTINUOUS POSET STRUCTURES. If W is a complete poset, then each function space

$$\mathrm{Cont}(W^n \to W) = \{p : W^n \to W \mid p \text{ is Scott-continuous}\}$$

is also a complete poset under the pointwise partial ordering, and so it makes sense to require that a functional

$$f : W^n \times \mathrm{Cont}(W^{k_1} \to W) \times \cdots \times \mathrm{Cont}(W^{k_m} \to W) \to W$$

is *continuous*. A *continuous poset structure* of signature $\boldsymbol{\tau} = (\{\mathsf{f}_i \mid i \in I\}, \text{kind})$ is a tuple

$$\mathfrak{W} = (W, \mathrm{ff}, \mathrm{tt}, \mathrm{cond}, \{f_i \mid i \in I\}), \tag{8}$$

where W is a complete poset; ff and tt are distinct, *discrete* points in W (maximal and comparable only with $\bot$); each f_i is a continuous functional on W with $\mathrm{kind}(f_i) = \mathrm{kind}(\mathsf{f}_i)$; and cond is a continuous functional of kind $\langle 0, 0, 0, 0\rangle$, such that

$$\left.\begin{aligned} \mathrm{cond}(\lambda()\mathrm{tt}, \lambda()\alpha, \lambda()\beta) &= \alpha, \\ \mathrm{cond}(\lambda()\mathrm{ff}, \lambda()\alpha, \lambda()\beta) &= \beta, \\ \mathrm{cond}(\lambda()\bot, \lambda()\alpha, \lambda()\beta) &= \bot. \end{aligned}\right\} \tag{9}$$

Perhaps the simplest example is the continuous structure

$$\mathfrak{M}^\dagger = (M^\dagger, \mathrm{ff}, \mathrm{tt}, \mathrm{cond}, R_1^\dagger, \ldots, R_l^\dagger, f_1^\dagger, \ldots, f_m^\dagger)$$

associated with each first order $\mathfrak{M}$, where cond is the strict conditional and each $f_i^\dagger$ is the strict extension of f_i to $M^\dagger$ (set $= \bot$ when one of its basic arguments is not in M).

4.2. STRICT VS. CONTINUOUS SEMANTICS. The continuous denotation function $[\![E]\!]_c(\omega)$ is defined exactly like the strict one, except that application is honest,

$$[\![p(Z_1, \ldots, Z_n)]\!]_c(\omega) = \omega(p)([\![Z_1]\!]_c(\omega), \ldots, [\![Z_n]\!]_c(\omega)), \tag{10}$$

and we must verify in each case of the inductive definition that for each E, $[\![E]\!]_c(\omega)$ is *continuous in* ω. Because of (10), and with the natural notion of **continuous consequence**, the β-reduction rule holds,

$$\ell(\vec{u})A = \ell(\vec{u})B \models_c A\{\vec{u} :\equiv \vec{C}\} = B\{\vec{u} :\equiv \vec{C}\}, \tag{11}$$

assuming, of course, that the substitutions are free. This is one, important difference between strict and continuous FLR-consequence. For a simpler one which involves recursion rather than application and no hypotheses, notice that

$$\models_s p(q()) \text{ where } \{p(u) = \mathrm{tt}, q() = q()\} = \bot \tag{12}$$

but

$$\models_c p(q()) \text{ where } \{p(u) = \mathrm{tt}, q() = q()\} = \mathrm{tt}. \tag{13}$$

From this we infer that neither of the equations in (12) and (13) should be provable in the logic of recursion, since we would want the theorems to be valid under both the strict and the continuous interpretations.

4.3. Domains. Suppose $\mathfrak{W}$ is a continuous structure as in (8) and there exists a homeomorphism

$$\lambda : \mathrm{Cont}(W \to W) \longrightarrow W \tag{14}$$

of W with its (continuous) function space, and consider the expansion

$$(\mathfrak{W}, \mathrm{id}, \mathrm{ap}, \lambda) = (W, \mathrm{ff}, \mathrm{tt}, \mathrm{cond}, \{f_i \mid i \in I\}, \mathrm{id}, \mathrm{ap}, \lambda),$$

where $\mathrm{id}(x) = x$ and $\mathrm{ap}(x, y) = \lambda^{-1}(x)(y)$. Structures of this type were first constructed by Scott [9] to model the (untyped, extensional) **λ-calculus** and its "applied" extensions, and we can see better the connection with these systems if we introduce the abbreviations

$$A \cdot B \equiv \mathrm{ap}(A, B), \ \ \lambda(u)A \equiv \lambda(\ell(u)A)$$

on the terms of $\mathrm{FLR}(\boldsymbol{\tau})$ for the relevant $\boldsymbol{\tau}$. The more complex posets of "domain theory" with "reflection" properties like (14) can also be represented in this way as continuous, FLR structures, with the various imbeddings and homeomorphisms among function spaces coded by functionals among the "givens" of the structure.

5. STREAMS

If a is a constant in the signature, then

$$\models_{s,c} p(\mathsf{a}) \ \mathsf{where}\ \{p(u) = p(\mathsf{a})\} = \bot \tag{15}$$

i.e., this equation is valid in both strict and continuous semantics, simply because the least-fixed-point of $p(u) = p(a)$ is the totally undefined function $\lambda(u)\bot$. On the other hand, it is quite easy to "program" the term $p(\mathsf{a})\ \mathsf{where}\ \{p(u) = p(\mathsf{a})\}$ in Pascal or C, and if a stands for some act, like "*ring the bell*", and we run the resulting program, we will then get an unending sequence of "rings" instead of a dull "hang"—at least until the stack overflows. We construct here some very simple models of FLR which account for the expected interpretation of (15), and also motivate the definition of the more general structures in the next section.

5.1. Stream structures. Fix a set A (of "acts") and a set M (of "values"), disjoint from A, with specified, distinct points $0, 1 \in M$—think of $M = \mathbb{N}$. A *convergent stream* is any finite (possibly empty) sequence $\alpha = \langle a_0, \dots, a_{n-1}, m\rangle$ with $a_0, \dots, a_{n-1} \in A$, and $m = [\alpha] \in M$, the *value returned* by α. A *divergent stream* is any finite or infinite sequence $\alpha = \langle a_0, a_1, \dots\rangle$ of acts, and it returns no value, $[\alpha] = \bot$. The set $\mathrm{Str} = \mathrm{Str}_{A,M}$ of all convergent and divergent streams is a complete poset under the "initial part" relation $\sqsubseteq$, and it carries the (continuous) operation of **sequential execution**,

$$\alpha\,;\beta = \begin{cases} \langle a_0, \dots, a_{n-1}, b_0, b_1, \dots\rangle & \text{if } \alpha = \langle a_0, \dots, a_{n-1}, m\rangle \\ & \qquad \text{is convergent,} \\ \alpha & \text{if } \alpha \text{ is a divergent.} \end{cases}$$

Suppose now $\{f_i \mid i \in I\}$ are continuous functionals on Str, let ff and tt be the (no-act) streams $\langle 0\rangle$ and $\langle 1\rangle$ respectively, and let

$$\mathfrak{S} = (M, \mathrm{Str}, \mathrm{ff}, \mathrm{tt}, \mathrm{cond}, \,;\,, \{\langle a, 1\rangle \mid a \in A\}, \{f_i \mid i \in I\}),$$

where each $\langle a, 1\rangle$ is is a constant (nullary) function and the (strict) conditional is defined by (5). For the semantics, we let the basic variables vary over M and the function variables over $(M \to \mathrm{Str})$, and we define the denotation function $[\![E]\!]_{str}(\omega)$ by simply copying all the clauses in the definition of the strict denotation $[\![E]\!]_s(\omega)$, except when $E \equiv p(\vec{Z})$ or $E \equiv f(\vec{Z}, \vec{\pi})$, when some of the Z_is are terms. For these cases, we first "execute" the acts of the stream arguments (from left to right) and then "pass" the returned values (if any) to the function, for example,

$$[\![\mathsf{p}(\mathsf{u}_1, A, \mathsf{u}_2, \mathsf{u}_3, B)]\!]_{str}(\omega) = \alpha\,;\beta\,; p(u_1, [\alpha], u_2, u_3, , [\beta]), \tag{16}$$

where $\alpha = [\![A]\!]_{str}(\omega)$, $\beta = [\![B]\!]_{str}(\omega)$, $p = \omega(\mathsf{p})$ and $u_i = \omega(\mathsf{u}_i)$. The definition is a bit messy in the general case, but very natural, and it assigns the correct, expected value to the term in (15), if a is the formal constant naming $\langle a, 1\rangle$.

5.2. Input dependent and interactive streams. The stream model is too simple to account for much of the behavior of programming languages with side-effects, e.g., it makes no room for state, input or interaction. One step up from it are the **input dependent streams**, which are functions $\alpha : S \to \mathrm{Str}_{A,M}$ from a given set of *states* S and so can account (partially) for input. We also assume that each act $a \in A$ induces a function $s \mapsto sa$ on the states, and then define sequential composition by:

$$(\alpha\,;\beta)(s) = \begin{cases} \alpha(s) & \text{if } \alpha(s) \text{ is divergent,} \\ \alpha(s)\,;\beta(sa_0a_1\cdots a_{n-1}) & \text{if } \alpha(s) = \langle a_0, \ldots, a_{n-1}, m\rangle. \end{cases}$$

Input dependent stream structures and the denotation function $[\![E]\!]_{ids}(\omega)$ are defined by a routine extension of the construction above.

Still richer in structure are the **interactive streams**, which are (suitably restricted) partial functions $\alpha : S^* \to (A \cup M \cup \{\bot\})$ from finite sequences of states to acts or values, and model the behavior of *reactive agents* in an interactive environment. (The agent α is reacting at each stage to a new state, produced by the execution of acts by other agents since the last time α acted.) We will not give any details of this construction here, since we do not need it, but it is natural and very simple.[6]

6. $\jmath$-SEMANTICS

In defining stream semantics, we emphasized the important principle "execute the acts first and then pass the returned value" but passed over without com-

[6]Input dependent streams are (essentially) the *procedures* and interactive streams the *behaviors* of [6], where these constructions were employed to build models of concurrency. A simpler language (with no value passing) was used in [6], but it is routine to extend the constructions given there to the present context.

ment the choice to evaluate the stream arguments *from left-to-right* in (16); but there is nothing sacrosanct about left-to-right evaluation, and it is clear that we would get a different denotation function if we chose to evaluate the streams from right-to-left, or (indeed) if we "interweaved" the evaluation of the stream arguments in some fixed way.[7] If the terms are interpreted in some poset more complex than Str, then more than "order of evaluation" comes into play: we have a real problem in assigning a value to $\mathsf{p}(A, B)$, when $[\![A]\!], [\![B]\!] \in W$, but the interpretation $p : M^2 \to W$ of p is defined only for arguments in M. The obvious way to handle this is to make **value passing** a primitive, one of the "givens" in the structure.

6.1. $\jmath$-STRUCTURES AND DENOTATIONS. A **monotone functional** on a set M to a complete poset W is any monotone mapping

$$f : M^n \times (M^{k_1} \to W) \times \cdots \times (M^{k_m} \to W) \to W,$$

and a $\jmath$-**structure** of signature $\boldsymbol{\tau} = (\{\mathsf{f}_i \mid i \in I\}, \text{kind})$ is a tuple

$$\mathfrak{M} = (M, W, \text{ff}, \text{tt}, \text{cond}, \{\jmath^n \mid n \in \mathbb{N}\}, \{f_i \mid i \in I\}), \tag{17}$$

where the following conditions hold:

($\jmath$1) M is a non-empty set and W is a complete poset, containing the discrete elments ff and tt. We make no special assumptions on the relationship between M and W—it may be that $M = W$, $M \subsetneq W$ or $M \cap W = \emptyset$.

($\jmath$2) cond is a monotone functional of kind $\langle 0, 0, 0, 0\rangle$ satisfying (9).

($\jmath$3) Each f_i is a monotone functional on M to W with $\text{kind}(f_i) = \text{kind}(\mathsf{f}_i)$.

($\jmath$4) Each $\jmath^n : (M^n \to W) \to \text{Mon}(W^n \to W)$ is a monotone mapping which assigns a monotone, n-ary function $\jmath^n(p) : W^n \to W$ to each n-ary function $p : M^n \to W$ and does not disturb the nullary functions, $\jmath^0(p) = p$. We will generally write

$$\jmath(p, \alpha_1, \ldots, \alpha_n) = \jmath^n(p)(\alpha_1, \ldots, \alpha_n)$$

for the values of these imbeddings, skipping the superscripts and placing the arguments from W on par with the function argument.

To interpret the FLR-expressions in a $\jmath$-structure $\mathfrak{M}$, we let the basic variables vary over M and the function variables over arbitrary functions $p : M^n \to W$. The denotation function $[\![E]\!]_\jmath(\omega)$ is defined exactly like the strict denotation in Section 3, except that the conditional is given by the structure and application uses the given imbeddings. The general definition is a bit messy and we will skip it, but the following adaptation of example (16) explains the idea:

$$[\![\mathsf{p}(\mathsf{u}_1, A, \mathsf{u}_2, \mathsf{u}_3, B)]\!]_\jmath(\omega) = \jmath^2(\lambda(ab)p(u_1, a, u_2, u_3, b), \alpha, \beta), \tag{18}$$

[7]The original manual for C [3] deliberately leaves the order of evaluation of expressions unspecified, and so delegates the definition of the precise semantics of the language to the specific compiler—unless we adopt a non-deterministic interpretation, which is by no means trivial because unrestricted recursion is allowed. (The concurrency model of [6] can account for this.) All the (simple) C-compilers I have used evaluate from right-to-left.

where $\alpha = [\![A]\!]_\jmath(\omega)$, $\beta = [\![B]\!]_\jmath(\omega)$, $p = \omega(\mathsf{p})$ and $u_i = \omega(\mathsf{u}_i)$. For another example, involving one of the givens,

$$[\![\mathsf{f}(A, \mathsf{u}, B, C, \pi_1, \pi_2)]\!]_\jmath = \jmath^3(\lambda(abc) f(a, u, b, c, \sigma_1, \sigma_2), \alpha, \beta, \gamma), \tag{19}$$

with the notation of (18) for the terms A, B, C and the variable u, and $\sigma_i = [\![\pi_i]\!]_\jmath(\omega)$ for the ℓ-terms π_1 and π_2.

6.2. Strict vs. continuous vs. $\jmath$-semantics. Each strict, functional structure has an obvious $\jmath$-associate

$$\mathfrak{M}_\jmath = (M, \{f_i \mid i \in I\})_\jmath = (M, M^\dagger, \mathrm{ff}, \mathrm{tt}, \mathrm{cond}, \jmath_s, \{f_i \mid i \in I\}),$$

where cond is the strict conditional and $\jmath_s$ is the strict extension map,

$$\jmath_s(p, \alpha_1, \ldots, \alpha_n) = \begin{cases} p(\alpha_1, \ldots, \alpha_n) & \text{if } \alpha_1, \ldots, \alpha_n \in M, \\ \bot & \text{otherwise,} \end{cases}$$

and it is also obvious that $[\![E]\!]_s(\omega) = [\![E]\!]_\jmath(\omega)$—the definitions of these two denotation functions are identical. The same is true for stream structures, which can be identified with $\jmath$-structures with the imbedding

$$\jmath_{str}(p, \alpha_1, \ldots, \alpha_n) = \alpha_1 ; \ldots ; \alpha_n ; p([\alpha_1], \ldots, [\alpha_n]).$$

The situation is just a bit more complex for continuous structures, because for them we restricted the function variables to vary over continuous functions, but the following is true:[8]

6.3. **Proposition.** *Suppose* $\mathfrak{W} = (W, \mathrm{ff}, \mathrm{tt}, \mathrm{cond}, \{f_i \mid i \in I\})$ *is a continuous structure in which the poset* W *has the Scott property, i.e., every two compatible points in* W *have a least upper bound, for each* $p : W^n \to W$ *let*

$$\jmath_c^n(p) = \sup\{q \mid q \in \mathrm{Cont}(W^n \to W), q \leq p\},$$

define $f_{i,c}$ *analogously, and let* $\mathfrak{W}_\jmath = (W, W, \mathrm{ff}, \mathrm{tt}, \mathrm{cond}, \jmath_c, \{f_{i,c} \mid i \in I\})$.

(1) *Each* $\jmath_c^n : (W^n \to W) \to \mathrm{Cont}(W^n \to W)$ *is defined and monotone.*

(2) *If* $[\![E]\!]_\jmath(\omega)$ *is the denotation function on* $\mathfrak{W}_\jmath$, *then for every expression* E *and every valuation* ω *which assigns continuous functions to the function variables,* $[\![E]\!]_c(\omega) = [\![E]\!]_\jmath(\omega)$.

We skip the (simple) proof of this proposition, but it is worth noting that *the imbeddings* $\jmath_c^n$ *are not continuous.* Perhaps this witnesses the known "inefficiency" of the continuous (call-by-name) interpretation of recursive definitions, but it also makes this unified approach to the known interpretations depend essentially on the full *Monotone Fixpoint Theorem*, rather than the much simpler special case for continuous functionals which is traditionally used in denotational semantics.

These simple observations imply that if $\Gamma \models_\jmath E = F$, then the consequence $\Gamma \models E = F$ is also valid under the strict, the continuous and the stream

[8] We could avoid the restriction of this proposition to posets with the Scott property by interpreting the n-ary function variables in a $\jmath$-structure over some (almost) arbitrary, complete subposet of $(M^n \to W)$, but it is not clear that this increased applicability of the theory is worth the additional complications it introduces.

interpretations of FLR. The converse is not true for any of these semantics. For example, easily,

$$\models_s p(a) \text{ where } \{p(u) = f(u,b)\} = q(b) \text{ where } \{q(v) = f(a,v)\}, \quad (20)$$

$$\models_{str} f(a,b) = p(a) \text{ where } \{p(u) = f(u,b)\}, \quad (21)$$

but (20) fails in stream semantics (and hence in $\jmath$-semantics also), and (21) fails in $\jmath$-semantics, using the right-to-left evaluation on streams. It is implicit in the proof of the Main Theorem 9.1 that (somewhat surprizingly), the input-dependent stream structures (with arbitrary, continuous givens) are sufficient to determine all $\jmath$-valid equations between FLR expressions.

7. THE SYSTEM $\mathcal{E}_0$

A **sequent** of the proof system $\mathcal{E}_0$ is a formal expression $\Gamma \Rightarrow \varepsilon$, where $\Gamma \equiv \varepsilon_1, \ldots, \varepsilon_n$ is a *set* of FLR($\boldsymbol{\tau}$) equations, the *hypotheses*, and $\varepsilon \equiv E = F$ is a single equation, the *conclusion*. The **provable sequents** (theorems) of $\mathcal{E}_0$ are defined inductively by the **axioms** and (closure) **rules** in Table 2 and by the comments and explanations below, and we write

$$\Gamma \vdash E = F \text{ if and only if } \Gamma \Rightarrow E = F \text{ is a theorem.}$$

In commenting on the axioms and rules, we will also discuss the more interesting points of the proof of the following result, whose details we will skip:

7.1. THE SOUNDNESS THEOREM. *If* $\Gamma \vdash E = F$, *then* $\Gamma \models_\jmath E = F$.

The *structural* axioms and rules (S1) – (S3) are standard, and so are the axioms for equality (FL1) – (FL4), where f^* in (FL3) is any of the functional symbols or the conditional. (FL5) is the *generality* or *abstraction rule*, and (FL6) is a very restricted form of **β-reduction** which requires some explanation.

7.2. IMMEDIACY AND LINEARITY. An ℓ-term $\ell(z_1, \ldots, z_n)A$ is **immediate** if $A \equiv p(\vec{u})$ where $\vec{u}$ is a sequence of basic variables; and **linear** if the variables $z_1, \ldots, z_n$ occur *exactly once* in A, and in the same order in which they occur in the prefix. (So $\ell()p(u,s,w,t,w)$, $\ell(s)p(u,s,w,t,w)$, $\ell(u,t)p(u,s,w,t,w)$ and every function variable p ($\equiv \ell(\vec{u})p(\vec{u})$) are immediate and linear, but $\ell(w)p(u,s,w,t,w)$ and $\ell(s,t)p(u,t,w,s)$ are not linear and $\ell(s)p(s,\mathsf{a})$ is linear but not immediate.)

It is quite simple to verify that (FL6) is valid in $\jmath$-semantics, but it cannot be strengthened by omitting either of the restrictions, because of (7) and:

$$\ell(u)p(u) = \ell(u)q(\mathsf{a},u) \not\models_{str} p(\mathsf{b}) = q(\mathsf{a},\mathsf{b}). \quad (22)$$

(Suppose a and b name the constants $\langle a,1\rangle$ and $\langle b,1\rangle$, and let $q(u,v) = \mathrm{tt} = \langle 1\rangle$; now $[\![q(\mathsf{a},u)]\!]_{str} = \langle a,1\rangle$ and we can validate the hypothesis by setting $p(u) = \langle a,1\rangle$; but then $[\![p(\mathsf{b})]\!]_{str} = \langle b,a,1\rangle$, while $[\![q(\mathsf{a},\mathsf{b}]\!]_{str} = \langle a,b,1\rangle$.)

The axioms (C1) – (C3) about the conditional are also natural (and $\jmath$-valid, by definition), but (FR1) – (FR7) about the recursion construct require some

(S1)	$\varepsilon \Rightarrow \varepsilon$
(S2)	From $\Gamma \Rightarrow \varepsilon$, infer $\Gamma, \varepsilon_1 \Rightarrow \varepsilon$
(S3)	From $\Gamma \Rightarrow \varepsilon_1$ and $\Gamma, \varepsilon_1 \Rightarrow \varepsilon$, infer $\Gamma \Rightarrow \varepsilon$
(FL1)	$\Rightarrow E = E$, $E = F \Rightarrow F = E$, $E = F, F = G \Rightarrow E = G$
(FL2)	$X_1 = Y_1, \ \ldots, \ X_n = Y_n \Rightarrow p(X_1, \ldots, X_n) = p(Y_1, \ldots, Y_n)$
(FL3)	$E_1 = F_1, \ \ldots, \ E_n = F_n \Rightarrow f^*(E_1, \ldots, E_n) = f^*(F_1, \ldots, F_n)$
(FL4)	$\Rightarrow p = \ell(\vec{u})p(\vec{u})$
(FL5)	From $\Gamma \Rightarrow A(\vec{z}) = B(\vec{z})$, infer $\Gamma \Rightarrow \ell(\vec{x})A(\vec{x}) = \ell(\vec{y})B(\vec{y})$ provided none of the variables in $\vec{z}$ occurs free in Γ
(FL6)	$\ell(\vec{x})A(\vec{x}) = \ell(\vec{y})B(\vec{y}) \Rightarrow A(\vec{Z}) = B(\vec{Z})$ provided *either* $\ell(\vec{x})A(\vec{x})$, $\ell(\vec{y})B(\vec{y})$ are immediate and linear *or* $\vec{Z}$ is a sequence of basic variables
(C1)	$\Rightarrow$ if tt then A else B fi $= A$
(C2)	$\Rightarrow$ if ff then A else B fi $= B$
(C3)	$\Rightarrow$ if $\bot$ then A else B fi $= \bot$
(FR1)	$\Rightarrow p(\vec{X}, A, \vec{Y}) = p(\vec{X}, r(), \vec{Y})$ where $\{r() = A\}$
(FR2)	$\Rightarrow f^*(\vec{E}, \ell(\vec{u})A, \vec{F}) = f^*(\vec{E}, r, \vec{F})$ where $\{r(\vec{u}) = A\}$
(FR3)	$\Rightarrow f(\vec{Z}, \vec{p}) = r(\vec{Z})$ where $\{r(\vec{u}) = f(\vec{u}, \vec{p})\}$
(FR4)	$\Rightarrow \big(B_0(\vec{q}, \vec{p})$ where $\{q_1(\vec{v}_1) = B_1(\vec{q}, \vec{p}), \ldots, q_m(\vec{v}_m) = B_m(\vec{q}, \vec{p})\}\big)$ where $\big\{p_1(\vec{u}_1) = A_1, \ldots, p_n(\vec{u}_n) = A_n\big\}$ $= B_0(\vec{r}, \vec{p})$ where $\big\{r_1(\vec{v}_1) = B_1(\vec{r}, \vec{p}), \ldots, r_m(\vec{v}_m) = B_m(\vec{r}, \vec{p}),$ $p_1(\vec{u}_1) = A_1, \ldots, p(\vec{u}_n) = A_n\big\}$
(FR5)	$\Rightarrow A_0(\vec{p})$ where $\big\{p_1(\vec{u}_1) = B_0(\vec{q}, \vec{p})$ where $\big\{q_1(\vec{v}_1) = B_1(\vec{q}, \vec{p}), \ldots, q_m(\vec{v}_m) = B_m(\vec{q}, \vec{p})\big\},$ $p_2(\vec{v}_2) = A_2(\vec{p}), \ldots, p_n(\vec{u}_n) = A_n(\vec{p})\big\}$ $= A_0(\vec{p})$ where $\big\{p_1(\vec{u}_1) = B_0(\vec{r}, \vec{p}), r_1(\vec{v}_1) = B_1(\vec{r}, \vec{p}), \ldots,$ $r_m(\vec{v}_m) = B_m(\vec{r}, \vec{p}), p_2(\vec{v}_2) = A_2(\vec{p}), \ldots, p_n(\vec{u}_n) = A_n(\vec{p})\big\}$ provided no variable in $\vec{u}_1$ is free in any B_i with $i > 0$
(FR6)	$\Rightarrow A_1$ where $\{p_1(\vec{u}_1) = A_1, \ldots, p_n(\vec{u}_n) = A_n\}$ $= p_1(\vec{u}_1)$ where $\{p_1(\vec{u}_1) = A_1, \ldots, p_n(\vec{u}_n) = A_n\}$
(FR7)	$\Rightarrow A$ where $\{\ \} = A$
(FC)	The Finite Closure Rule.
(RI)	The Recursion Inference Rule.

TABLE 2. The axiom system $\mathcal{E}_0$.

Note: All substitutions are assumed free, and the variables r, $\vec{r}$ on the right of (FR1) – (FR5) are fresh—i.e., they do not occur on the left of these equations.

attention, as they express the most basic equational properties of definition by functional recursion. In (FR2), f^* stands for any functional symbol or the conditional. Most interesting are (FR4) and (FR5), the **Bekič-Scott equations**, which reduce nested recursion to mutual recursion; and (FR6), the **fixpoint equation**, which asserts that the recursion construct indeed defines fixed points of the mutual recursion determined by its parts.[9]

The validity of these equations in all $\jmath$-structures can be established easily by standard arguments of fixpoint recursion.

This leaves the last two, basic rules of inference about recursion which require some explanation.

7.3. FINITE ITERATES. The least fixed point of a monotone functional $f(u,p)$ is obtained by iteration starting with $\perp$,

$$\overline{p}^0(u) = f(u, \lambda(u)\perp), \quad \overline{p}^\xi(u) = f(u, \sup_{\eta<\xi}\overline{p}^\eta),$$

and setting $\overline{p} = \overline{p}^\xi$, for the least ordinal ξ such that $\overline{p}^\xi = \sup_{\eta<\xi}\overline{p}^\eta$. The Finite Closure Rule (FC) asserts that if $\xi = k$ is a finite ordinal (a natural number), then $\overline{p} = \overline{p}^k$. It is a bit messy to express in FLR, partly because we must deal with mutual recursions, and (more substantially) because we do not have a general "replacement" rule, and so the finite iterates of a recursive term are also recursive terms.

For a given recursive term

$$A \equiv A_0(\vec{p}) \text{ where } \{p_1(\vec{u}_1) = A_1(\vec{u}_1, \vec{p}), \dots, p_n(\vec{u}_n) = A_n(\vec{u}_n, \vec{p})\},$$

set, by induction on k, simultaneously for $i = 0, \dots, n$

$$\begin{aligned} \alpha_i^0(\vec{u}_i, \vec{p}) &\equiv A_i(\vec{u}_i, \vec{r}) \text{ where } \{r_1(\vec{u}_1) = p(\vec{u}_1), \dots, r_n(\vec{u}_n) = p(\vec{u}_n)\} \\ \alpha_i^{k+1}(\vec{u}_i, \vec{p}) &\equiv A_i(\vec{u}_i, \vec{r}) \text{ where } \{r_1(\vec{u}_1) = \alpha_1^k(\vec{u}_1, \vec{p}), \\ &\qquad \dots, r_n(\vec{u}_n) = \alpha_n^k(\vec{u}_n, \vec{p})\}, \end{aligned}$$

where $\vec{r}$ are fresh variables. Now the (syntactic) kth iterate of A is

$$A^k \equiv \alpha_0^k(\vec{\perp}),$$

so that, for example,

$$A^0 \equiv A_0(\vec{r}) \text{ where } \{r_1(\vec{u}_1) = \perp, \dots, r_n(\vec{u}_n) = \perp\}.$$

It is easy to verify (directly from the definitions) that A^k denotes in $\jmath$-structures exactly the kth iterate of the recursion which determines $[\![A]\!]_\jmath$.

The Finite Closure Rule FC. For each recursive term A, with α^k and A^k defined as above,

$$\ell(\vec{u}_1)\alpha_1^2(\vec{u}_1, \vec{\perp}) = \ell(\vec{u}_1)\alpha_1^1(\vec{u}_1, \vec{\perp}), \dots, \ell(\vec{u}_n)\alpha_n^2(\vec{u}_n, \vec{\perp}) = \ell(\vec{u}_n)\alpha_n^1(\vec{u}_n, \vec{\perp}) \Rightarrow A = A^1.$$

This says only that if the recursion stabilizes in the second stage, then this

[9]The fresh variables $\vec{r}$ are needed in (FR4) and (FR5) to avoid conflicts when some q_j occurs free in some A_i; the simpler versions with $r_i \equiv q_i$ are theorems when no variable occurs both free and bound on the left, and we will tacitly assume this in applying the axioms in the sequel.

is the fixpoint, but the general fact about recursions which stop after k steps follows from it easily.

7.4. THE RECURSION INFERENCE RULE. Suppose we want to prove that *if $f(u,v,p)$ is a monotone functional and $f(u,v,p) = f(v,u,p)$, for all u, v, then*

$$p(u,v) \text{ where } \{p(u,v) = f(u,v,p)\}$$
$$= q(v,t,u) \text{ where } \{q(u,t,v) = f(u,v,q(\cdot,t,\cdot)), r(s) = g(s)\},$$

where $q(\cdot,t,\cdot)) = \lambda(u,v)q(u,t,v)$. First we notice that the occurrence or r is spurious—r never figures in the value of the expression on the right—and then that there is a plausible connection between p on the left and q on the right, namely that for all t,

$$\lambda(u,v)p(u,v) = \lambda(u,v)q(v,t,u). \tag{23}$$

To make the idea precise, we let $\overline{p}^{\xi}$ and $\overline{q}^{\xi}$ be the iterates of the recursions which determine the fixpoints $\overline{p}$ and $\overline{q}$, and (using the assumed commutativity of f) show by (trivial, transfinite) induction that for all u, v, ξ and t,

$$\overline{p}^{\xi}(u,v) = \overline{q}^{\xi}(v,t,u),$$

from which the desired equation follows immediately.

The Recursion Inference Rule formalizes this method of proof, and is viewed best as a classical *elimination rule* for the recursion construct. It is unfortunately a bit messy to make precise, and all we can do in this space is to put it down and mention that its soundness for $\jmath$-semantics is (quite easily) proved after the manner of the example. The "connections" used in it are generalizations of (23): finite systems of immediate ℓ-terms which relate some of the recursion variables of one recursion term with some of another, enough so that $A = B$ can be established in every $\jmath$-structure by (transfinite) induction, as in the example.

A (basic variable) **projection** is a map

$$\sigma(u_0, \ldots, u_{n-1}) = \langle u_{\pi(0)}, \ldots, u_{\pi(m-1)} \rangle \tag{24}$$

on variable tuples determined by a function $\pi : \{i \mid i < m\} \to \{j \mid j < n\}$, where (conventionally) $\sigma(\vec{v}) = \langle\ \rangle$ if $m = 0$. These are useful in describing variable substitutions: e.g., if $\sigma(u,v) = \langle u,v,u \rangle$ and $E(x,y,z)$ is an expression, then $E(\sigma(u,v)) \equiv E(u,v,u)$ and $E(\sigma(x,x)) \equiv E(x,x,x)$.

Suppose Γ is a set of equations and

$$\begin{aligned} A &\equiv A_0 \text{ where } \{p_1(\vec{u}_1) = A_1(\vec{u}_1), \ldots, p_m(\vec{u}_m) = A_m(\vec{u}_m)\} \\ B &\equiv B_0 \text{ where } \{q_1(\vec{v}_1) = B_1(\vec{v}_1), \ldots, q_n(\vec{v}_n) = B_n(\vec{v}_n)\}, \end{aligned} \tag{25}$$

in which the function variables $p_1, \ldots, p_m, q_1, \ldots, q_n$ are all distinct; and suppose the list $\vec{x} = x_1, \ldots, x_k$ includes all basic variables which are free in either A or B. A finite set of equations

$$\Sigma = \{\pi_1 = \rho_1, \ldots, \pi_c = \rho_c\}$$

between *immediate ℓ-terms* is a **preconnection** of A and B relative to Γ if:

(1) For each i, $\pi_i \equiv \ell(\vec{z})p_k(\sigma_i(\vec{z},\vec{x}))$ and $\rho_i \equiv \ell(\vec{z})q_l(\tau_i(\vec{z},\vec{x}))$, with suitable p_k, q_l, where the basic variables in $\vec{z}$ are fresh for A and B, and σ_i, τ_i are projections;

and its **claims** are the sequents

(2) $\Gamma, \Sigma \Rightarrow A_0 = B_0$, and $\Gamma, \Sigma \vdash A_k(\sigma(\vec{z},\vec{x})) = B_l(\tau(\vec{z},\vec{x}))$, for every equation $\ell(\vec{z})p_k(\sigma(\vec{z},\vec{x})) = \ell(\vec{z})q_l(\tau(\vec{z},\vec{x}))$ in Σ.

The Recursion Inference Rule RI. Suppose A and B are recursive terms as in (25), no p_i occurs free in any B_l, no q_j occurs free in any A_k, and none of the variables in $\vec{p}$, $\vec{q}$ are free in Γ: *from the claims in any preconnection between A and B relative to Γ, infer $\Gamma \Rightarrow A = B$.*

A **connection** is a preconnection whose claims are all provable, and, most often, we use **RI** in the following form:

RI. *If a connection exists between A and B relative to Γ, then $\Gamma \vdash A = B$.*

7.5. EXAMPLE. To see why the restriction on the variables $\vec{p}$ and $\vec{q}$ is needed, notice that if r does not occur in A, then $\vdash A = A\,\mathsf{where}\,\{r(\vec{u}) = B\}$, for any B; because $A = A\,\mathsf{where}\,\{\ \}$ is an axiom, and the empty set is a connection of $A\,\mathsf{where}\,\{r(\vec{u}) = B\}$ with $A\,\mathsf{where}\,\{\ \}$. (But it is not a connection if r occurs in A, which is why we cannot prove $r() = r()\,\mathsf{where}\,\{r() = r()\}$ (i.e., $r() = \bot$).

8. FORMAL PROOFS

After a few basic facts, we will prove here the key lemma we need for the Main Theorem in the next section.

8.1. **Lemma.** *With the syntactic conventions of Table 1 and assuming that all substitutions are free:*

*1. $\pi = \rho \vdash E(\pi) = E(\rho)$, *if π and ρ are linear and immediate ℓ-terms.*

*2. *The Part Replacement Property. If $\Gamma \vdash A_i = B_i$ for $i = 0, \dots, n$ and no p_i occurs free in Γ, then*

$$\Gamma \vdash A_0\,\mathsf{where}\,\{p_1(\vec{u}_1) = A_1, \dots, p_n(\vec{u}_n) = A_n\} \\ = B_0\,\mathsf{where}\,\{p_1(\vec{u}_1) = B_1, \dots, p_n(\vec{u}_n) = B_n\}.$$

*3. *The order in which the parts are listed in a recursive term is irrelevant.*

*4. *The Duplication Equation.*

$$A_0(q,r,\vec{p})\,\mathsf{where}\,\{q(\vec{u}_1) = B(q,\vec{p}), r(\vec{u}_1) = B(r,\vec{p}), \\ p_1(\vec{u}_2) = A_1(q,r,\vec{p}), \dots, p_n(\vec{u}_n) = A_n(q,r,\vec{p})\} \\ = A_0(q,q,\vec{p})\,\mathsf{where}\,\{q(\vec{u}_1) = B(q,\vec{p}), \\ p_1(\vec{u}_2) = A_1(q,q,\vec{p}), \dots, p_n(\vec{u}_n) = A_n(q,q,\vec{p})\}$$

*5. $\vec{X} = \vec{Y} \vdash E(\vec{X}) = E(\vec{Y})$, *if $\vec{X}$ and $\vec{Y}$ are basic variables or terms.*

*6. *If F is obtained from E by a renaming of bound variables, then $\vdash E = F$.*

*7. *The Bekič-Scott Lemma. With r fresh and $r(\vec{u}, \cdot) \equiv \ell(\vec{v})r(\vec{u},\vec{v})$,*

$$B_0(\vec{u},\vec{q}) \text{ where } \{q_1(\vec{v}_1) = B_1(\vec{q}), \dots, q_m(\vec{v}_m) = B_m(\vec{q})\}$$
$$= B_0(\vec{u},\vec{r}(\vec{u},\cdot)) \text{ where } \{r_1(\vec{u},\vec{v}_1) = B_1(\vec{r}(\vec{u},\cdot)),$$
$$\dots, r_m(\vec{v}_m) = B_m(\vec{r}(\vec{u},\cdot))\}.$$

*8. *The Extended Fixpoint Equation.*

$$p_1(\tau\vec{u}_1) \text{ where } \{p_1(\vec{u}_1) = A_1(\vec{u}_1)\dots\}$$
$$= A_1(\tau\vec{u}_1) \text{ where } \{p_1(\vec{u}_1) = A_1(\vec{u}_1)\dots\}.$$

*9. $\vdash A_0 \text{ where } \{p_1(\vec{u}_1) = p_2(\tau\vec{u}_1), p_2(\vec{u}_2) = A_2(\vec{u}_2), \dots\}$

$$= A_0 \text{ where } \{p_1(\vec{u}_1) = A_2(\tau\vec{u}_1), p_2(\vec{u}_2) = A_2(\vec{u}_2), \dots\}.$$

*10. *The Bottom Rule. If* $\Gamma \vdash A_i(\vec{u}_i,\vec{p}) \text{ where } \{\vec{p} = \vec{\bot}\} = \bot$, *for* $i = 1, \dots, n$ *and no variable in* $\vec{p}$ *or any* $\vec{u}_j$ *occurs free in* Γ, *then*

$$\Gamma \vdash A_0(\vec{p}) \text{ where } \{p_1(\vec{u}_1) = A_1, \dots, p_n(\vec{u}_n) = A_n\}$$
$$= A_0(\vec{p}) \text{ where } \{\vec{p} = \vec{\bot}\},$$

where $\vec{p} = \vec{\bot}$ *abbreviates the obvious.*

*11. $\vdash p(\vec{u}) \text{ where } \{p(\vec{u}) = \text{if } p(\tau\vec{u}) \text{ then } A \text{ else } B \text{ fi}\} = \bot$.

Proof. *1 is proved by induction on the length of the given expression $E(r)$, and all the cases are trivial, except perhaps for the following two.

$E(r) \equiv r(\vec{Z})$. If $\pi \equiv \ell(\vec{x})A(\vec{x})$ and $\rho \equiv \ell(\vec{y})B(\vec{y})$, then $E(\pi) \equiv A(\vec{X}^*)$ and $E(\rho) \equiv B(\vec{Y}^*)$, where $\vec{X}^* = \vec{Y}^*$ by the induction hypothesis; now $A(\vec{X}^*) = B(\vec{X}^*)$ by Axiom (FL6) (because $A(\vec{x})$ and $B(\vec{y})$ are linear and immediate), and $B(\vec{X}^*) = B(\vec{Y}^*)$ by (FL2).

$E(r) \equiv A_0(\vec{p},r) \text{ where } \{p_1(\vec{u}_1) = A_1(\vec{p},r), \dots, p_n(\vec{u}_n) = A_n(\vec{p},r)\}$. Now $E(\pi)$ and $E(\rho)$ are obtained by substituting π and ρ for r in each $A_i(\vec{p},r)$, the induction hypothesis gives us $A_i(\vec{p},\pi) = A_i(\vec{p},\rho)$, and we can get $E(\pi) = E(\rho)$ from *2—but we do not have *2 yet. Let instead

$$C \equiv A_0(\vec{q},\rho) \text{ where } \{q_1(\vec{u}_1) = A_1(\vec{q},\rho), \dots, q_n(\vec{u}_n) = A_n(\vec{q},\pi)\},$$

where the variables $\vec{q}$ are fresh. By the induction hypothesis,

$$\pi = \rho, \vec{p} = \vec{q} \vdash A_i(\vec{p},\pi) = A_i(\vec{q},\rho),$$
$$\vec{p} = \vec{q} \vdash A_i(\vec{p},\rho) = A_i(\vec{q},\rho),$$

for $i \le n$, and so $\{p_i = q_i \mid i = 1, \dots, n\}$ is a connection which establishes $A(\pi) = C$ and $A(\rho) = C$ from $\pi = \rho$.

*2, *3 and *4 are easily proved by the same method, of introducing fresh recursion variables and applying **RI** twice, with $\{q = r', r = r', \vec{p} = \vec{p}'\}$ the connection required for *4. *5 and *6 are proved by induction on E, using Part Replacement *2 when E is a recursive term, and for *7, we use *1 to verify that $\{\ell(\vec{z}_i)q_i(\vec{z}_i) = \ell(\vec{z}_i)r_i(\vec{u},\vec{z}_i) \mid i = 1, \dots, m\}$ is a connection which establishes the required identity. For *8, abstract on the fixpoint rule (FR6) by (FL5) and then use variable specification, (FL6).

For *9, first insure that no variable occurs both free and bound, and then (showing the variables and with only two parts, for simplicity) compute:

$$\begin{aligned} A_0(p_1,p_2)\ \mathsf{where}\ \{p_1(\vec{u}_1) = p_2(\tau\vec{u}_1), p_2(\vec{u}_2) = A_2(\vec{u}_2,p_1,p_2)\} \\ = A_0(p_1,p_2)\ \mathsf{where}\ \{p_1(\vec{u}_1) = r(\tau\vec{u}_1), \\ r(\vec{u}_2) = A_2(\vec{u}_2,p_1,r), p_2(\vec{u}_2) = A_2(\vec{u}_2,p_1,p_2)\} \\ \text{(by Duplication)} \\ = A_0(p_1,p_2)\ \mathsf{where}\ \{p_1(\vec{u}_1) = r(\tau\vec{u}_1)\ \mathsf{where}\ \{r(\vec{u}_2) = A_2(\vec{u}_2,p_1,r)\}, \\ p_2(\vec{u}_2) = A_2(\vec{u}_2,p_1,p_2)\},\ \text{(by (FR5))} \\ = A_0(p_1,p_2)\ \mathsf{where}\ \{p_1(\vec{u}_1) = A_2(\tau\vec{u}_1,p_1,r) \\ \mathsf{where}\ \{r(\vec{u}_2) = A_2(\vec{u}_2,p_1,r)\}, \\ p_2(\vec{u}_2) = A_2(\vec{u}_2,p_1,p_2)\}\},\ \text{(by *8)} \\ = A_0(p_1,p_2)\ \mathsf{where}\ \{p_1(\vec{u}_1) = A_2(\tau\vec{u}_1,p_1,r), \\ r(\vec{u}_2) = A_2(\vec{u}_2,p_1,r), p_2(\vec{u}_2) = A_2(\vec{u}_2,p_1,p_2)\} \\ \text{by (FR5)} \\ = A_0(p_1,p_2)\ \mathsf{where}\ \{p_1(\vec{u}_1) = A_2(\tau\vec{u}_1,p_1,p_2), p_2(\vec{u}_2) = A_2(\vec{u}_2,p_1,p_2)\} \\ \text{(by Duplication).} \end{aligned}$$

For *10, the hypothesis gives the antecedent for an application of the Finite Closure Rule **FC**, which then yields the required conclusion.

Finally, for *11, first compute:

$$\begin{aligned} \bot = \mathsf{if}\ \bot\ \mathsf{then}\ A\ \mathsf{else}\ B\ \mathsf{fi} \quad \text{(C3)} \\ = \mathsf{if}\ q()\ \mathsf{then}\ A\ \mathsf{else}\ B\ \mathsf{fi}\ \mathsf{where}\ \{q() = \bot\}\ \text{(FR2)} \\ = \mathsf{if}\ p(\tau\vec{u})\ \mathsf{then}\ A\ \mathsf{else}\ B\ \mathsf{fi}\ \mathsf{where}\ \{p(\vec{u}) = \bot\}, \end{aligned}$$

the last via the connection $\{\ell(\vec{u})q() = \ell(\vec{u})p(\vec{u})\}$. Now the Bottom Rule (BR) (with $C(p) \equiv p(\vec{u})$) gives

$$p(\vec{u})\ \mathsf{where}\ \{p(\vec{u}) = \mathsf{if}\ p(\tau\vec{u})\ \mathsf{then}\ A\ \mathsf{else}\ B\ \mathsf{fi}\} = p(\vec{u})\ \mathsf{where}\ \{p(\vec{u}) = \bot\},$$

and the right-hand-side is $= \bot$ by expanding $\bot$ as above and using another trivial connection. ⊣

We can use *7 to get a more general version of the Bekič-Scott Rule without the restriction on the variables $\vec{u}_1$—hence its name; notice that, despite appearances, $\vec{u}$ is not free in $B_1, B_2, \ldots$ on the right side of *7.[10]

8.2. **Lemma** (Chains of conditionals). *If for* $i = 1, \ldots, n$, $A_i \equiv p_{i+1}(\tau_i\vec{u}_i)$ *or* $A_i \equiv \mathsf{if}\ p_{i+1}(\tau_i\vec{u}_i)\ \mathsf{then}\ B_i\ \mathsf{else}\ C_i\ \mathsf{fi}$ (with $n+1 = 1$ and arbitrary projections τ_i), *then*

$$\vdash\ p_1(\vec{u}_1)\ \mathsf{where}\ \{p_1(\vec{u}_1) = A_1, \ldots, p_n(\vec{u}_n) = A_n\} = \bot.$$

Proof. The hypothesis (together with (C3)) sets up an application of the Bottom Rule *10, which then yields the required conclusion. ⊣

The significance of the next notion will not become clear until we use it in the next section.

8.3. Simplified terms. A recursive term (with no free function variables)

$$A \equiv A_0\ \mathsf{where}\ \{p_1(\vec{u}_1) = A_1, \ldots, p_m(\vec{u}_m) = A_m\} \tag{26}$$

is **simplified** if it satisfies (I) – (V) below, where, in (II) we also classify the recursion parts of these terms into *forms* and we assign to each of them a *character.*

---

[10]This more general Bekič-Scott Rule was taken as basic in [5].

(I) $A_0 \equiv p_1(\vec{x})$, with some basic variables $\vec{x}$, and these are the only free occurrences of variables of A.

(II) Each recursion part A_k is a **form** and has a **character**, as follows.

(SF1) *Fixed forms.* A_k is $\bot$, ff or tt, and it is its own character.

(SF2) *ȷ-forms.* $A_k \equiv p_i(Z_1, \ldots, Z_a)$, where each Z_j is either a basic variable or an immediate term; the character of A_k is the pair $\langle \mathsf{j}, s\rangle$, where s is the number of Z_js which are terms (rather than variables).

(SF3) *Functional forms.* $A_k \equiv \mathsf{f}_i(\vec{u}, \vec{\pi})$, where $\vec{u}$ is a list of basic variables and $\vec{\pi}$ is a list of immediate ℓ-terms; the character of A_k is $\langle \mathsf{f}, i\rangle$.

(SF4) *Conditional forms.* $A_k \equiv \mathsf{if}\ C_1\ \mathsf{then}\ C_2\ \mathsf{else}\ C_3\ \mathsf{fi}$, where C_1, C_2 and C_3 are immediate terms; the character of A_k is cond.

The character of a variable p_i is the same as the character of the form A_i which defines it.

(III) Every ȷ-form $A_k \equiv p_i(Z_1, \ldots, Z_a)$ is *strict*, i.e., its character is $\langle \mathsf{j}, n+1\rangle$ for some n (not every Z_j is a variable).

(IV) If $A_k \equiv \mathsf{if}\ p_i(\tau_k \vec{u}_k)\ \mathsf{then}\ C_2\ \mathsf{else}\ C_3\ \mathsf{fi}$ is a cond-form, then p_i does not have fixed character. (A_k is not *immediately resolvable.*)

(V) There is no *chain of conditionals*, i.e., no sequence of variables

$$q_1 \equiv p_{j_1}, \ldots, q_K \equiv p_{j_K},$$

so that for $i = 1, \ldots, K$, $A_{j_i} \equiv \mathsf{if}\ q_{i+1}(\tau \vec{u}_{j_i})\ \mathsf{then}\ B_i\ \mathsf{else}\ C_i\ \mathsf{fi}$ or $A_{j_i} \equiv q_{i+1}(\tau \vec{u}_{j_i})$ (with $K + 1 = 1$).

8.4. **Lemma.** *Every term without free function variables is provably equal to a simplified one.*

Proof. It is easy to see that by applying repeatedly the axioms (FR1) – (FR7), Lemma 8.1 (especially Part Replacement and the Bekič-Scott Lemma), we can transform every term without free function variables to a provably equal *almost simplified* term, i.e., one which satisfies (I) and (II) in 8.3. To insure that (III) – (V) also hold, we will use the following three facts, where A is almost simplified.

(T1) *Elimination of conditional chains.* If A has a conditional chain as in (V) and A^* is obtained from A by replacing each part $q_i(\vec{v}_i) = A_{j_i}$ by $q_i(\vec{v}_i) = \bot$, then $\vdash A = A^*$.

(T2) *Elimination of resolvable conditionals.* If A has a conditional part

$$p_k(\vec{u}_k) = \mathsf{if}\ p_i(\tau_k \vec{u}_k)\ \mathsf{then}\ C_2\ \mathsf{else}\ C_3\ \mathsf{fi},$$

where A_i is a fixed form, then $\vdash A = A^*$, where A^* is obtained by replacing the definition of p_k with its obvious value: $p_k(\vec{u}_k) = \bot$, if $A_k \equiv \bot$, $p_k(\vec{u}_k) = C_2$, if $A_k \equiv \mathsf{tt}$, and $p_k(\vec{u}_k) = C_3$, if $A_k \equiv \mathsf{ff}$.

(T3) *Reduction of non-strict ȷ-forms.* If A has parts $p_k(\vec{u}_k) = p_i(\tau_k \vec{u}_k)$, $p_i(\vec{u}_i) = A_i(\vec{u}_i)$ where $A_i(u_i)$ is not a ȷ-form, and if A^* is obtained from A by replacing the definition of p_k by $p_k(\vec{u}_k) = A_i(\tau_k \vec{u}_k)$, then $\vdash A = A^*$.

These are all easy to verify, using 8.2 for (T1), axioms (C1) – (C3) for (T2), and *9 for (T3). (To show (T1), for example, first insert K copies of the chain by the Duplication Equation *4, each defining the i'th term in the chain; and then isolate each of these by Bekič-Scott and replace it by $q_i(\vec{u}_i) = \perp$, using Lemma 8.2 and Part Replacement.)

The *rank* of an almost simplified term A is the triple $r = \langle o, c, b \rangle$, where o is the number of non-fixed forms in A; c is the number of cond-forms in A; and b is the number of non-strict $\jmath$-forms, those with character $\langle \mathsf{j}, 0 \rangle$. We order ranks *lexicographically*. It is clear that an application of (T1) lowers the rank, and so does an application of (T2)—although it may introduce new chains and increase b; and an application of (T3) also lowers the rank, *if A_i is not a* cond-*form*—although it may make some of the existing conditionals immediately resolvable. We now apply these transformations (with the restriction on (T3)), in any order, as long as it is possible, until we obtain in the end an almost simplified term A^*, provably equivalent to the given A, which satisfies (IV) and (V). It may still contain some non-strict $\jmath$-forms $p_k(\vec{u}_k) = p_i(\tau_k \vec{u}_k)$, with $p_i(\vec{u}_i) = A_i(\vec{u}_i)$ a cond-form, but now we can apply (T3) to eliminate them, clearly without violating (IV) or (V) in the result (although the rank will be increased). ⊣

9. COMPLETENESS OF $\mathcal{E}_0$ FOR EQUATIONS

In this section we will outline a proof of the following[11]

9.1. **Main Theorem.** (1) *If* $\models_\jmath E = F$, *then* $\vdash E = F$. (2) *An expression equation $E = F$ is $\jmath$-valid if and only if it is true in all continuous $\jmath$-structures.* (3) *The class of $\jmath$-valid expression equations is decidable.*

9.2. Reasonably free structures A $\jmath$-structure (17) is **reasonably free** if M has at least two members and the following conditions are satisfied.

(1) Every functional f_i of $\mathfrak{M}$ is injective and never takes on the values $\perp$, tt or ff; and for $i \neq j$, f_i and f_j have disjoint ranges.

(2) If $\beta = \text{cond}(p, q, r) = \text{cond}(p', q', r')$ and $p() \notin \{\perp, \text{tt}, \text{ff}\}$, then β is not in the range of any f_i; $\beta \notin \{\perp, \text{ff}, \text{tt}\}$; and $p = p'$, $q = q'$, $r = r'$.

(3) If p is n-ary with $n > 0$ and $\alpha_1, \dots, \alpha_n \in W$, then $\beta = \jmath(p, \alpha_1, \dots, \alpha_n) \notin \{\perp, \text{tt}, \text{ff}\}$; β is not in the range of any f_i; and $\beta \neq \text{cond}(p', q, r)$ if $p'() \notin \{\perp, \text{tt}, \text{ff}\}$. In addition, $\jmath$ is strongly injective on non-nullary arguments, i.e., if $\jmath(p, \alpha_1, \dots, \alpha_m) = \jmath(q, \beta_1, \dots, \beta_n)$ and $m, n > 0$, then $m = n$ and for $i = 1, \dots, m$, $\alpha_i = \beta_i$.

9.3. **Lemma.** *For each* FLR *signature $\boldsymbol{\tau}$, there exists a continuous, reasonably free $\boldsymbol{\tau}$-structure $\mathfrak{M}$.*

Proof. We let $M = \mathbb{N}$ and $W = (\mathbb{N} \to \text{Str}_{A,\mathbb{N}})$ be the set of functions from $\mathbb{N}$ to the set of streams which return integers and execute acts in the set

$$A = \{\mathsf{f}_i \mid i \in I\} \cup \{\mathsf{j}_{i+1} \mid i \in \mathbb{N}\} \cup \{\mathsf{ff}, \mathsf{tt}, \mathsf{cond}\}$$

[11]For the "propositional" part of FLR, this result is proved in [1], which also gives references to related, older work (about (2)).

of the formal symbols of FLR($\boldsymbol{\tau}$) and a symbol j_n for each $\jmath_n$ with $n > 0$. We let $\mathrm{tt} = \langle 1 \rangle$, $\mathrm{ff} = \langle 0 \rangle$, and we interpret each f_i by a functional which codes its name and all its arguments, e.g.,[12]

$$f_i(n,p,q)(s) = \langle \mathsf{f}_i, 1\rangle\,;\begin{cases} \langle n+2\rangle, & \text{if } (s)_0 = 0,\\ p()((s)_1), & \text{if } (s)_0 = 1,\\ q((s)_1,(s)_2)((s)_3), & \text{if } (s)_0 = 2,\\ \bot, & \text{otherwise.}\end{cases}$$

The imbedding is defined using the same idea, e.g.,

$$\jmath_2(p,\alpha,\beta)(s) = \langle \mathsf{j}_2, 1\rangle\,;\begin{cases} \alpha((s)_1), & \text{if } (s)_0 = 0,\\ \beta((s)_1), & \text{if } (s)_0 = 1,\\ p((s)_1,(s)_2)((s)_3), & \text{if } (s)_0 = 2,\\ \bot, & \text{otherwise.}\end{cases}$$

Ditto for the conditional, except that we give the right values when its first argument is $\bot$, tt or ff—in all other cases we start with **cond** and then we code all three arguments. ⊣

9.4. STRICT $\jmath$-FORMS. Suppose $A_k \equiv p_i(Z_1,\dots,Z_a)$ is a recursion part in a simplified term, so that by (III) of 8.3 at least one Z_i is not a variable. We call these *$\jmath$-forms* because their evaluation in $\jmath$-semantics involves the imbedding $\jmath$, and we will sometimes re-write them as

$$A_k \equiv \jmath(\ell(t_{b_1},\dots,t_{b_s})p_i(z_1,\dots,z_a), Z_{b_1},\dots,Z_{b_s}) \tag{27}$$

indicating how they are to be evaluated: here $Z_{b_1},\dots,Z_{b_s}$ are exactly those Z_j's which are terms (not variables), $t_{b_1},\dots,t_{b_s}$ is a sequence of fresh, distinct basic variables to refer to them, and

$$z_j \equiv \begin{cases} Z_j, \text{ if } Z_j \text{ is a variable,}\\ t_{b_j}, \text{ if } Z_j \text{ is a term.}\end{cases}$$

For example, we rewrite

$$p(s_1, r_2(s_1,s_3), s_4, r_3()) \equiv \jmath(\ell(t_2,t_4)p(s_1,t_2,s_4,t_4), r_2(s_1,s_3), r_3()).$$

If all the Z_i's are terms, then the $\jmath$-form is especially simple,

$$\begin{aligned} p(Z_1,Z_2,Z_3) &\equiv \jmath(\ell(t_1,t_2,t_3)p(t_1,t_2,t_3), Z_1,Z_2,Z_3)\\ &\equiv \jmath(p, Z_1,Z_2,Z_3).\end{aligned}$$

Notice that the ℓ-terms which appear in these $\jmath$-forms are linear; this is key both to the interpretation and to the proof theory of the system.

9.5. NOTATION AND TERMINOLOGY. If A is simplified as in (26) and $\mathfrak{M}$ is any structure, we will let $\overline{p}_1,\dots,\overline{p}_m$ be the least fixed points of the system which determines the value of A on $\mathfrak{M}$. Note that these functions do not depend on the interpretation of the free variables of A, which occur only in the head A_0.

[12]Here $(s)_i$ is is the i'th term of the sequence coded by s, relative to some fixed coding of integer-tuples by integers.

Suppose A and B are both simplified, $E = F$ is an equation in the function variables $p_1, \ldots, p_m, q_1, \ldots, q_n$ which occur in these terms, and $\mathfrak{M}$ is a structure. We will say that $E = F$ **is true** in $\mathfrak{M}$ if every valuation ω which assigns $\overline{p}_i$ to each p_i and $\overline{q}_j$ to each q_j satisfies $E = F$. In particular, a preconnection Σ between A and B is true in $\mathfrak{M}$ if every equation in Σ is true in $\mathfrak{M}$.

9.6. **Lemma.** *If A is simplified as in* (26), *$\mathfrak{M}$ is reasonably free and*

$$A_k \equiv \text{ if } p_i(\tau\vec{u}_k) \text{ then } C_2 \text{ else } C_3 \text{ fi} \tag{28}$$

is a cond*-form of A, then the functions $\overline{p}_i$ and $\overline{p}_k$ on M never take on values in the set $\{\bot, \text{ff}, \text{tt}\}$.*

Proof. Let us say that the cond-form A_k in (28) has *level* 0 if the character of p_i is not cond, and (inductively) *level* $r+1$ if A_i has level r. Now (V) implies that every cond-form has a level, and we can prove the Lemma by induction on the level.

If A_k has level 0, then A_i cannot be fixed by (IV), and so it is either a functional form or a $\jmath$-form; in the first case, $\overline{p}_i(\vec{u}_i)$ is a value of some functional f_j, and so outside $\{\bot, \text{ff}, \text{tt}\}$ by (1) of 9.2; in the second $\overline{p}_i(\vec{u}_i)$ is a value of $\jmath_n$ for some $n > 0$ since A_i is a strict $\jmath$-form by (III), and so again $\overline{p}_i(\vec{u}_i)$ is outside $\{\bot, \text{ff}, \text{tt}\}$ by (3) of 9.2; and, in either case, $\overline{p}_k$ does not take a value in $\{\bot, \text{ff}, \text{tt}\}$ by (2) of 9.2.

If A_k has level > 0, then $\overline{p}_i$ does not take on any of the forbidden values by the induction hypothesis, and so neither does $\overline{p}_k$ by (2) of 9.2 again. ⊣

The next, key Lemma deals only with (basically) **closed** preconnections, whose ℓ-terms have no free basic variables.

9.7. **Lemma.** *Suppose Σ is a closed preconnection between two simplified terms*

$$\begin{aligned} A &\equiv p_1(\vec{x}) \text{ where } \{p_1(\vec{u}_1) = A_1(\vec{u}_1), \ldots, p_m(\vec{u}_m) = A_m(\vec{u}_m)\} \\ B &\equiv q_1(\vec{y}) \text{ where } \{q_1(\vec{v}_1) = B_1(\vec{v}_1), \ldots, q_n(\vec{v}_n) = B_n(\vec{v}_n)\} \end{aligned} \tag{29}$$

which contains the identity

$$\ell(\vec{w})p_k(\sigma\vec{w}) = \ell(\vec{w})q_l(\tau\vec{w}), \tag{30}$$

and let $\mathfrak{M}$ be a reasonably free structure.

(1) *If either* (1a) *or* (1b) *are true, then Σ is not true in $\mathfrak{M}$.*

(1a) *A_k and B_l have different character.*

(1b) *$A_k \equiv \mathsf{f}_i(\sigma_k\vec{u}_k, \vec{\pi})$ and $B_l \equiv \mathsf{f}_i(\tau_l\vec{v}_l, \vec{\rho})$ have the same, functional character, but $\sigma_k\sigma\vec{w} \not\equiv \tau_l\tau\vec{w}$.*

(2) *If $A_k \equiv \mathsf{f}_i(\sigma_k\vec{u}_k, \pi_1(\vec{u}_k), \ldots, \pi_s(\vec{u}_k))$ and $B_l \equiv \mathsf{f}_i(\tau_l\vec{v}_l, \rho_1(\vec{v}_l), \ldots, \rho_s(\vec{v}_l))$ have the same functional character but* (1b) *does not apply, let*

$$\Sigma' = \Sigma \cup \{\ell(\vec{w})\pi'_j(\sigma\vec{w}) = \ell(\vec{w})\rho'_j(\tau\vec{w}) \mid j = 1, \ldots, s\},$$

where π'_j and ρ'_j are alphabetic variants of π_j and ρ_j with the same ℓ-prefix and fresh bound variables; then Σ is true in $\mathfrak{M}$ if and only if Σ' is true in $\mathfrak{M}$.

(3) *If* $A_k(\vec{u}_k) \equiv \text{if } C_1 \text{ then } C_2 \text{ else } C_3 \text{ fi}$ *and* $B_l(\vec{v}_l) \equiv \text{if } D_1 \text{ then } D_2 \text{ else } D_3 \text{ fi}$ *both have character* cond *and*

$$\Sigma' = \Sigma \cup \{\ell(\vec{w})\big(C_j\{\vec{u}_k :\equiv \sigma\vec{w}\}\big) = \ell(\vec{w})\big(D_j\{\vec{v}_l :\equiv \tau\vec{w}\}\big) \mid j = 1, 2, 3\},$$

then Σ *is true in* $\mathfrak{M}$ *if and only if* Σ' *is true in* $\mathfrak{M}$.

(4) *If* A_k *and* B_l *have the same* $\jmath$*-form* $\langle \mathsf{j}, s\rangle$, *so that after a renaming of the bound variables,*

$$A_k \equiv \jmath(\ell(t_{b_1}, \ldots, t_{b_s})p_i(x_1, \ldots, x_a), X_{b_1}, \ldots, X_{b_s}), \tag{31}$$

$$B_l \equiv \jmath(\ell(t_{b_1}, \ldots, t_{b_s})q_j(y_1, \ldots, y_a), Y_{b_1}, \ldots, Y_{b_s}), \tag{32}$$

let Σ' *be the extension of* Σ *by the equation*

$$\ell(\vec{w}, t_{b_1}, \ldots, t_{b_s})\big(p_i(x_1, \ldots, x_a)\{\vec{u}_k :\equiv \sigma\vec{w}\}\big) = \ell(\vec{w}, t_{b_1}, \ldots, t_{b_s})\big(q_j(y_1, \ldots, y'_a)\{\vec{v}_l :\equiv \tau\vec{w}\}\big)$$

and also all equations

$$\ell(\vec{w})\big(X_c\{\vec{u}_k :\equiv \sigma\vec{w}\}\big) = \ell(\vec{w})\big(Y_c\{\vec{v}_l :\equiv \tau\vec{w}\}\big) \quad (c = b_1, \ldots, b_s);$$

then Σ *is true in* $\mathfrak{M}$ *if and only if* Σ' *is true in* $\mathfrak{M}$.

Proof. (1) For (1a), notice that if p_k and q_l have different characters, then the functions $\overline{p}_k$ and $\overline{q}_l$ take values in disjoint sets, because $\mathfrak{M}$ is reasonably free (and using Lemma 9.6 if one of them has character cond). For (1b), the hypothesis of the Lemma implies that the identity

$$f_i(\sigma\vec{w}, \vec{\pi}) = f_i(\tau\vec{w}, \vec{\rho})$$

is true in $\mathfrak{M}$; but f_i is injective, and so this is not possible if the variable lists $\sigma\vec{w}$ and $\tau\vec{w}$ differ even in one place, because we could then assign distinct values to it (in m, which has at least two elements) and get back distinct values from f_i.

For parts (2) – (4), we need to show that if Σ is true, then so is Σ'.

(2) By the assumed truth of Σ and the fact that the mutual least fixed points satisfy their defining equations, we know that the identity

$$\mathsf{f}_i(\sigma\vec{w}, \pi'_1(\sigma\vec{w}), \ldots) = \mathsf{f}_i(\tau\vec{w}, \rho'_1(\tau\vec{w}), \ldots)$$

is true, and since f_i is injective, this means that for $j = 1, \ldots, s$, $\pi'_j(\sigma\vec{w}) = \rho'_j(\tau\vec{w})$ is true.

(3) By hypothesis, $A_k(\sigma\vec{w}) = B_l(\tau\vec{w})$ is true, and by Lemma 9.6, the values of $C_1\{\vec{u}_k :\equiv \sigma\vec{w}\}$ and $D_1\{\vec{v}_l :\equiv \tau\vec{w}\}$ are not in $\{\bot, \mathrm{ff}, \mathrm{tt}\}$; now (2) of 9.2 implies that

$$C_j\{\vec{u}_k :\equiv \sigma\vec{w}\} = D_j\{\vec{v}_l :\equiv \tau\vec{w}\}$$

is true, for $j = 1, 2, 3$, which by abstraction gives the required result.

(4) In the notation of 9.4, the hypothesis insures that the equation

$$\big(\jmath(\ell(t_{b_1}, \ldots, t_{b_s})p_i(x_1, \ldots, x_a), X_{b_1}, \ldots, X_{b_s})\big)\{\vec{u}_k :\equiv \sigma\vec{w}\} = \big(\jmath(\ell(t_{b_1}, \ldots, t_{b_s})q_j(y_1, \ldots, y_a), Y_{b_1}, \ldots, Y_{b_s})\big)\{\vec{u}_k :\equiv \sigma\vec{w}\}$$

is true in $\mathfrak{M}$, and from this, the required result follows easily by the injectivity of $\jmath$. ⊣

PROOF OF 9.1. If A and B are simplified terms with distinct recursion variables $\vec{p}$ and $\vec{q}$, and if $\vec{z}$ is a list of the variables which occur free in either A or B, then

$$A \equiv p_1(\sigma_h \vec{z}) \text{ where } \{p_1(\vec{u}_1) = A_1(\vec{u}_1), \ldots, p_m(\vec{u}_m) = A_m(\vec{u}_m)\}$$
$$B \equiv q_1(\tau_h \vec{z}) \text{ where } \{q_1(\vec{v}_1) = B_1(\vec{v}_1), \ldots, q_n(\vec{v}_n) = B_n(\vec{v}_n)\}$$

with suitable projections σ_h and τ_h, and the equation $A = B$ is true on a structure $\mathfrak{M}$ exactly when the closed preconnection

$$\Sigma_0 = \{\ell(\vec{z})p_1(\sigma_h \vec{z}) = \ell(\vec{z})q_1(\tau_h \vec{z})\} \tag{33}$$

is true in $\mathfrak{M}$. We will consider several preconnections between A and B, and we need the following, auxilliary notions.

(a) An equation $\ell(\vec{w})p(\sigma\vec{w}) = \ell(\vec{w})q(\tau\vec{w})$ between ℓ-terms is *tidy*, if every variable in the list $\vec{w}$ occurs either in $p(\sigma\vec{w})$ or in $q(\tau\vec{w})$, and a preconnection Σ is **tidy** if all its equations are tidy. It is clear that every equation is provably equivalent to a tidy one, and so the same is true of preconnections. The useful fact is that (up to alphabetic variance between terms) *there are only finitely many tidy preconnections between A and B*, with an upper bound on their number easily computable from the number and arities of the function variables $\vec{p}$ and $\vec{q}$.

(b) A preconnection Σ is **dead**, if some equation $\pi = \rho$ in it falls under Case (1) of Lemma 9.7.

(c) An equation $\ell(\vec{w})p_k(\sigma\vec{w}) = \ell(\vec{w})q_l(\tau\vec{w})$ in a preconnection Σ is **fulfilled**, if it does not fall under Case (1) of Lemma 9.7 and (briefly) $\Sigma' = \Sigma$, i.e., (the "tidyfication" of) every equation in the preconnection Σ' associated with Σ by the operative case (2) – (4) of Lemma 9.7 is an alphabetic variant of some equation in Σ. It is clear that *we can determine effectively whether* $\pi = \rho$ *is fulfilled in* Σ, since deleting empty abstractions and testing for alphabetic variance are effective operations.

We start now with Σ_0 in (33) and we apply Lemma 9.7 repeatedly to produce preconnections

$$\Sigma_0 \subsetneq \Sigma_1 \subsetneq \Sigma_2 \cdots \subsetneq \Sigma_e,$$

choosing a new unfulfilled identity at each stage, if one exists, and setting $\Sigma_{i+1} = \Sigma_i'$, as long as (1) of the Lemma does not apply to Σ_i; so, at each stage,

$$\Sigma_i \text{ is true } \iff \Sigma_{i+1} \text{ is true.} \tag{34}$$

Now, the procedure cannot continue indefinitely by (a) above, so that eventually we reach one of the following situations.

ALTERNATIVE 1. At Stage e, Σ_e is dead, and then by (34) Σ_0—and hence $A = B$—is not true, on every reasonably free structure.

ALTERNATIVE 2. Every equation in Σ_e is fulfilled, and then Σ_e is a connection between A and B.

The proof of this last fact is very simple now, but it is worth considering an example in Case (4) which involves the restricted form of β-reduction allowed in $\mathcal{E}_0$.

Suppose that $\ell(x,y)p_k(\sigma\langle x,y\rangle) = \ell(x,y)q_l(\tau\langle x,y\rangle)$ is in Σ_e, and that

$$\begin{aligned} A_k(\sigma_k\langle x,y\rangle) &\equiv p(p'(x,x), y, p''(y)) \\ &\equiv \jmath(\ell(t_1,t_2)p(t_1,y,t_2), p'(x,x), p''(y)), \\ B_l(\tau_l\langle x,y\rangle) &\equiv q(x,x,q'(y),y,q''(y,x)) \\ &\equiv \jmath(\ell(t_1,t_2)q(x,x,t_1,y,t_2), q'(y), q''(y,x)). \end{aligned}$$

We must show the equality of these two terms from Σ_e, which now contains by (4) the equations

$$\ell(x,y,t_1,t_2)p(t_1,y,t_2) = \ell(x,y,t_1,t_2)q(x,x,t_1,y,t_2), \tag{35}$$

$$\ell(x,y)p'(x,x) = \ell(x,y)q'(y) \tag{36}$$

$$\ell(x,y)p''(y) = \ell(x,y)q''(y,x). \tag{37}$$

From (35), by variable specification and abstraction, we get

$$\ell(t_1,t_2)p(t_1,y,t_2) = \ell(t_1,t_2)q(x,x,t_1,y,t_2),$$

which is a linear identity; and so by linear specification (FL6), we get

$$p(p'(x,x),y,p''(y)) = q(x,x,p'(x,x),y,p''(y)); \tag{38}$$

and then, from (36) and (37), using *5 of Lemma 8.1 we get

$$q(x,x,p'(x,x),y,p''(y)) = q(x,x,q'(y),y,q''(y,x))$$

which with (38) yields the required equation.

This completes the proof of the Main Theorem for simplified terms, and hence for arbitrary terms with no free function variables by Lemma 8.4. The result follows for arbitrary terms because $A(r) = B(r)$ is valid or provable exactly when $A(g) = B(g)$ is valid or provable, with g a fresh functional symbol, and then for equations between ℓ-terms, easily, by abstraction. ⊣

10. THE FULL SYSTEM $\mathcal{E}$

If we think of closed ℓ-terms as expressing programs, then the Main Theorem 9.1 says that the system $\mathcal{E}_0$ is complete for (denotational) *program equivalence.* But this is clearly not enough—we want to be able to prove more complex properties of programs than mere equivalence, and also to reason about programs from hypotheses. We describe here, briefly and without proofs, an extension of $\mathcal{E}_0$ which goes part-way towards achieving this.

10.1. FLR FORMULAS AND THEIR SEMANTICS. The FLR($\boldsymbol{\tau}$)-**formulas** are defined inductively by

$$\phi :\equiv \varepsilon \mid (\neg\phi) \mid (\phi_1 \,\&\, \phi_2) \mid (\phi_1 \vee \phi_2) \mid (\phi_1 \rightarrow \phi_2) \mid (\forall u)\phi \mid (\exists u)\phi \mid (\forall p)\phi \mid (\exists p)\phi,$$

where ε is an equation between FLR($\boldsymbol{\tau}$) expressions.

Formulas are interpreted on $\jmath$-structures of signature $\boldsymbol{\tau}$ in the obvious way. Notice that

$$(\forall p)[A(p) = B(p)] \not\models A(\ell(u,v)C(u,v)) = B(\ell(u,v)C(u,v)),$$

because in the stream structures of Section 5, the hypothesis is true and the conclusion fails when

$$A(p) \equiv p(a,b), B(p) = q(a) \text{ where } \{q(u) = p(u,b)\}, C(u,v) \equiv f(v,u).$$

However, obviously:

$$(\forall p)[A(p) = B(p)] \models A(p) \text{ where } \{p(\vec{u}) = C\} = B(p) \text{ where } \{p(\vec{u}) = C\},$$

even when p occurs in C, and we will need to include this among the axioms of $\mathcal{E}$, to replace the usual "instantantiation" property.

For each formula ϕ and terms $C_1, \ldots, C_n$, the formula

$$\phi \text{ where } \{p_1(\vec{u}_1) = C_1, \ldots, p_n(\vec{u}_n) = C_n\}$$

is constructed by "pushing" the recursion construct through the first-order operations (and $\ell(u)$) until it applies to FLR-terms, e.g.,

$$\big(A = B\big) \text{ where } \{p(\vec{u}) = C\} \equiv A \text{ where } \{p(\vec{u}) = C\} = B \text{ where } \{p(\vec{u}) = C\},$$
$$\big(\ell(\vec{v})A = \ell(\vec{v})B\big) \text{ where } \{p(\vec{u}) = C\} \equiv (\forall \vec{v})[\big(A = B\big) \text{ where } \{p(\vec{u}) = C\}],$$
$$\big(\neg\phi\big) \text{ where } \{p(\vec{u}) = C\} \equiv \neg\big(\phi \text{ where } \{p(\vec{u}) = C\}\big),$$

etc. It is clear that the denotation of ϕ where $\{p(\vec{u}) = C\}$ in a $\jmath$-structure can be computed by first determining the least-fixed-point $\overline{p}$ of $p(\vec{u}) = C$, and then evaluating ϕ for $p := \overline{p}$.

10.2. THE SYSTEN $\mathcal{E}$. **Sequents** (of formulas) are expressions of the form $\Gamma \Rightarrow \phi$, where $\Gamma \equiv \phi_1, \ldots, \phi_n$ is now a *set of formulas*, so they include the sequents of equations of $\mathcal{E}_0$, and (again), the **explicit sequents** are those in which where does not occur.

To define the axioms and rules of $\mathcal{E}$, we consider (classical) three-sorted structures of the form

$$(M, W, \{\mathcal{P}_n \mid n \in \mathbb{N}\}, \{\mathrm{ap}_n \mid n \in \mathbb{N}\}, \{\jmath^n \mid n \in \mathbb{N}\}, \{R_i \mid i \in I\}), \quad (39)$$

where M, W are arbitrary, non-empty sets; each $\mathcal{P}_n$ is a set of n-ary functions from M to W; each $\mathrm{ap}_n : M \times \mathcal{P}_n \to W$ stands for function application; the imbeddings $\jmath_n$ are operations of the same kind as in FLR-structures, but without the monotonicity hypothesis (which makes no sense now); and each R_i is a relation on arguments from M and functions on M to W, like the functionals on $\jmath$-structures, but, again, with no monotonicity hypotheses. Let $\boldsymbol{\tau}^*$ be the first-order signature for such structures which has infinitely many relation symbols for each kind, and consider the usual (first-order, three-sorted) language L^*. The terms of L^* are made up from function variables and the ap_n primitives, they can be identified with the simple terms of FLR (with no functional constants) and they are interpreted on structures of the form (39) using the imbeddings $\jmath_n$ to interpret composition, as in FLR. We now let T^* be the theory of structures of the form (39) which have enough functions to interpret their terms, i.e., they satisfy the obvious axioms which assert that the p's vary over functions and (the closure of) $(\exists p)(\forall \vec{u})[p(\vec{u}) = t(\vec{u})]$ for every term $t(\vec{u})$ (which is not a basic variable). The Completeness Theorem for (many-sorted) first-order logic yields a complete axiomatization of T^*.

The **provable formula sequents** of $\mathcal{E}$ are defined by combining the axioms of $\mathcal{E}_0$ with a complete axiomatization of T^*, as follows:[13]

($\mathcal{E}1$) The axioms and rules of $\mathcal{E}_0$ in Table 2 are part of $\mathcal{E}$, re-interpreted so that Γ stands for a finite set of formulas.

($\mathcal{E}2$) If $\Gamma \Rightarrow \phi$ is a valid sequent of T^*, and if $\Gamma' \Rightarrow \phi'$ is obtained from $\Gamma \Rightarrow \phi$ by replacing each prime formula $R(\vec{u}, \vec{p})$ throughout by an FLR-equation $E = F$ with the same free variables, then $\Gamma' \vdash \phi'$.

($\mathcal{E}3$) $(\forall p)\phi(p) \Rightarrow \phi\,\mathsf{where}\,\{p(\vec{u}) = C\}$ for each formula $\phi(p)$ and each term C.

It is clear that $\mathcal{E}$ is **sound** for $\jmath$-semantics, and (of course) it is not complete, but we can easily formalize in it proofs which combine basic properties of recursive definitions with first-order arguments whose soundness depends only on the fact that the p's stand for functions on the universe M to some set. For example:

10.3. LEMMA. *For all terms $A_1(\vec{u}_1), \ldots, A_n(\vec{u}_n)$ and each formula $\phi(\vec{p})$, if*

$$\Gamma, \text{ for } i = 1, \ldots, n, (\forall \vec{u}_i)[p_i(\vec{u}_i) = A_i(\vec{u}_i)], \vdash \phi(\vec{p})$$

and no p_i is free in Γ, then

$$\Gamma \vdash \phi(\vec{p})\,\mathsf{where}\,\{p_1(\vec{u}_1) = A_1(\vec{u}_1), \ldots, p_n(\vec{u}_n) = A_n(\vec{u}_n)\}.$$

10.4. EXAMPLE. Suppose we define addition by some standard recursion

$$m + n \equiv p(m, n)\,\mathsf{where}\,\{p(m, n) = A(m, n, p)\}$$

on the structure of arithmetic (6), and we want to show that it is commutative, $m + n = n + m$, from the usual axioms—including induction. By induction (on first-order formulas),

$$(\forall m, n)[p(m, n) = A(m, n, p)] \vdash p(m, n) = p(n, m),$$

from which, by the Lemma,

$$p(m, n)\,\mathsf{where}\,\{p(m, n) = A(m, n, p)\} = p(n, m)\,\mathsf{where}\,\{p(m, n) = A(m, n, p)\},$$

i.e., $m + n = n + m$.

Speaking a bit loosely, we can use the method of the Example to prove in $\mathcal{E}$ properties of recursive definitions which have unique fixed points, and that is what we usually expect from formal systems about recursion. The system is stronger than this, however, because it extends $\mathcal{E}_0$ which is already complete for (absolutely) valid FLR-equations, and so can verify many properties of recursions which do not have unique solutions.

[13] In a detailed exposition of this, one would naturally put down a specific set of axioms and rules, directly on FLR-formulas, and then prove the first-order completeness.

We end with an extension of the Main Theorem 9.1 which is still fairly weak (and none-too-elegant), but which suggests that there is more completeness here than meets the eye.

10.5. BASIC (FIRST-ORDER) THEORIES. A set of sentences T in FLR(τ) is a *basic* theory if:

(B1) Every function symbol f_i which occurs in a sentence of T has kind $\langle n \rangle$ (so that it stands for an n-ary function), and every θ in T is a classical, first order sentence—with no occurrences of $\mathbf{tt}$, ff, $\ell(\vec{u})$ or conditionals, no where, and no function variables.

(B2) If f_i occurs in some sentence of T, then T also contains the sentence

$$(\forall \vec{u})(\exists v)[\mathsf{f}_i(\vec{u}) = v],$$

so that f_i is interpreted in $\jmath$-structures by a function $f_i : M^n \to M$.

The *basic terms* of T are the explicit (first order) terms constructed from the function symbols which occur in T, so that they will always be interpreted in the basic universe M of every $\jmath$-structure. The *proper poset terms* are those with the opposite property, which will always take values in $W \setminus M$: they are defined by a new clause to Table 1, similar to that for terms except that only symbols f_i which do not occur in T are allowed while the Z_is can be arbitrary terms. For example, $f(g(u), v)$ is a basic term if f and g occur in T and a proper, poset term if f does not occur in T; and $f(p(u))$ where $\{p(u) = g(u, p(u))$ is not a basic term, and it is a proper, poset term only if neither f nor g occurs in T.

For example, consider the stream structure

$$\mathfrak{S} = (\mathbb{N}, \mathbb{N} \cup \mathrm{Str}, \mathrm{ff}, \mathrm{tt}, \mathrm{cond}, 0, 1, +, \cdot, Z, ;, \{\langle a, 1\rangle \mid a \in A\}, \{f_i \mid i \in I\}),$$

where $(\mathbb{N}, 0, 1, +, \cdot, Z)$ is classical arithmetic, and let T be Peano arithmetic in just these symbols. The basic formulas of T express first-order properties of the integers, while the proper, poset terms describe "computations" which "necessarily" involve the execution of acts.

10.6. **Basic Completeness Theorem.** *The system $\mathcal{E}$ is complete for $\jmath$-valid sequents $\Gamma \Rightarrow A = B$, where Γ is a basic theory and A, B are proper, poset terms of Γ.*

This is proved very much like the Main Theorem 9.1, (roughly) replacing variables by basic terms throughout.

Department of Mathematics
University of California
Los Angeles

REFERENCES

1. A. J. C. Hurkens, Monica McArthur, Yiannis N. Moschovakis, Lawrence Moss, and Glen T. Whitney. The logic of recursive equations. in preparation.

2. Alexander Kechris and Yiannis N. Moschovakis. Recursion in higher types. In Jon Barwise, editor, *Handbook of mathematical logic*, number 90 in Studies in Logic, pages 681–737. North Holland, 1977.
3. Brian. W. Kernighan and M. Ritchie, Dennis. *The C programming language.* Prentice-Hall, Englewood Cliffs, New Jersey, 1978.
4. Yiannis N. Moschovakis. *Elementary induction on abstract structures.* Studies in Logic, No. 77. North Holland, Amsterdam, 1974.
5. Yiannis N. Moschovakis. The formal language of recursion. *Journal of Symbolic Logic*, 54:1216–1252, 1989.
6. Yiannis N. Moschovakis. A model of concurrency with fair merge and full recursion. *Information and Computation*, 93:114–171, 1991.
7. Yiannis N. Moschovakis. Computable concurrent processes. *Theoretical Computer Science*, 139:243–273, 1995.
8. Yiannis N. Moschovakis and Glen T. Whitney. Powerdomains, powerstructures and fairness. In L. Pacholski and J. Tiuryn, editors, *Computer Science Logic*, number 933 in Lecture Notes in Computer Science, pages 382–396, Berlin, 1995. Springer-Verlag.
9. D. Scott and Strachey. Towards a mathematical semantics for computer languages. In J. Fox, editor, *Proceedings of the Symposium on computers and automata*, pages 19–46, New York, 1971. Polytechnic Institute of Brooklyn Press.

HELMUT SCHWICHTENBERG AND KARL STROETMANN

FROM HIGHER ORDER TERMS TO CIRCUITS[1]

1. INTRODUCTION

In his lecture at the congress, the first author gave a survey on some recent results relevant for computability theory in the context of partial continuous functionals (cf. (Scott, 1982; Ershov, 1977; Stoltenberg-Hansen *et al.*, 1994)):

- An abstract definition of *totality* due to Berger (cf. (Berger, 1990; Berger, 1993) and (Stoltenberg-Hansen *et al.*, 1994, Ch. 8.3)), and applications concerning density and effective density theorems.
- Bounded fixed points: one can have the flexibility of fixed point definitions *and* termination at the same time (cf. (Schwichtenberg and Wainer, 1995)).
- A notion of *strict* functionals as a tool to prove termination of higher order rewrite systems (cf. (van de Pol and Schwichtenberg, 1995)).

Since this work is published already, we do not give details here but rather concentrate on another "applied" aspect of computability theory in higher types (also mentioned in the lecture): its possible use for the simultaneous design (from given components) and formal verification of hardware.

The basic observation is that many hardware units can be viewed as stream transformers, converting some input (control or data streams) into an output stream. This is possible even for bidirectional circuits since, in most cases, these can be modeled as pairs of unidirectional circuits. Here a stream is simply a function from the natural numbers (used to model time, i.e. the system clock) into the booleans (for control streams) or into some kind of data.

We consider some simple schemata (explicit definition and a form of primitive recursion) to define computable functionals. The resulting terms can be thought of as stream transformers, i.e. as circuits. The form of our schemata then makes it possible to directly translate a term into a circuit. On the other hand, we now have the term as a compact formal representation of the circuit, in the context of a reasonable theory, i.e. computability theory for partial continuous functionals. This is particularly useful for purposes of formal verification.

Our approach offers a number of benefits.

[1]The first author is partially supported by the working group NADA (New Hardware Design Methods) of the EC, and the second author is partially supported by a grant from the German Federal Ministry of Education, Science, Research, and Technology under contract number 01IS519A (project KORSYS).

M.L. Dalla Chiara et al. (eds.), Logic and Scientific Methods, 209–220.

1. It opens the possibility to treat some questions on hardware synthesis and verification in a "mathematically civilized" setting. After all, it is an old experience in computability theory that it pays if one does not unnecessarily restrict (higher order) arguments. Moreover, it provides the proper mathematical framework to deal with undefined or error objects.
2. Both the design and the verification of a circuit can be done in a modular way. In particular, the design and verification can start from components that are only specified abstractly by higher order formulas (Gordon, 1986).
3. The approach sketched can deal with various types of data abstractions. For example, in the example presented we will abstract completely from the bit level representation of the data.

It is these benefits that distinguish our approach from others that are based on representing circuitry by finite automata.

2. TERMS

We consider *recursive definitions* of stream transformers C in the form

$$C(\vec{a}, t) = M$$

with $\vec{a}$ a list of stream variables and M a term with free variables among $\vec{a}, t$ which is built inductively by means of the clauses

$$C(\vec{a}, \mathrm{pred}^k(t)) \quad \text{with } k \geq 1,$$
$$a_i(\mathrm{pred}^k(t)) \quad \text{with } k \geq 0,$$
$$D(M_1, \ldots, M_n).$$

Here we have written $\mathrm{pred}(t)$ for $t-1$ (which will be convenient in section 3); hence $\mathrm{pred}^k(t)$ is undefined if $t < k$. D ranges over already defined constants, and clearly a_i denotes the i–th component of the list $\vec{a}$. As an important special case we have *explicit definitions*, where M is built using the last two clauses only; this special case will correspond to combinatorial circuits in section 3. Among the constants we always have **if–then–else–fi**.

As an example we pick the minmax unit proposed in (Claesen, 1990). We have taken its formulation from the IFIP WG 10.2 collection of circuit verification examples[2]; cf. (Kropf, 1995). It is described there as the first non–trivial example which has gained some popularity and reveals some problems arising in the area of digital signal processor verification. We quote from the WWW page:

"The minmax unit has an input signal *in* which consists of a sequence of integers in the range of −256 to +255. The minmax unit has three boolean control signals *clear*, *reset* and *enable.* The unit produces an output sequence *out* at the same rate as *in* in the following way.

[2]These examples can be found at the URL `http://i81fs1.ira.uka.de/benchmarks/`.

1. *Out* is zero if *clear* is true, independent of the other control signals.
2. If *clear* is false and *enable* is false then *out* equals the last value of *in* before *enable* became false.
3. If *clear* is false and *enable* is true and *reset* is true then *out* follows *in*.
4. If *reset* becomes false, then *out* equals, on each time point t, the mean value of the maximum and minimum value of *in* until that time point."

Recall that we model input streams as functions from time (here: discrete time modeled by the natural numbers) to data (here: natural numbers). For the reasons of both simplicity and generality we take arbitrary natural numbers, not just those with a fixed bit length. We assume that we have units (ALUs) computing the sum, maximum and minimum of two numbers. These functions are understood in the strict sense, i.e. an error in any of the arguments produces an error in the value. We also need a unit for the (strict) function `half` defined by

$$\mathtt{half}(2n) = \mathtt{half}(2n+1) = n.$$

The specification talks about the "last value of *in* before *enable* became false". In order to design a unit yielding that value, consider the following recursive definition $(*)$:

$$\begin{aligned}
\mathtt{last}(\text{in}, \text{enable}, 0) := \ & \textbf{if } \text{enable}(0) \\
& \quad \textbf{then } \text{in}(0) \\
& \quad \textbf{else } \text{undef}_{\mathtt{nat}} \\
& \textbf{fi}
\end{aligned}$$

$$\begin{aligned}
\mathtt{last}(\text{in}, \text{enable}, t+1) := \ & \textbf{if } \text{enable}(t+1) \\
& \quad \textbf{then } \text{in}(t+1) \\
& \quad \textbf{else } \mathtt{last}(\text{in}, \text{enable}, t) \\
& \textbf{fi}
\end{aligned}$$

Note that case distinction $0/t+1$ here is not really necessary; it also wrongly suggests that when describing a unit from these recursion equations we would need to perform a zero–test. We may write $(*)$ equivalently as $(**)$:

$$\begin{aligned}
\mathtt{last}(\text{in}, \text{enable}, t) := \ & \textbf{if } \text{enable}(t) \\
& \quad \textbf{then } \text{in}(t) \\
& \quad \textbf{else } \mathtt{last}(\text{in}, \text{enable}, t-1) \\
& \textbf{fi}
\end{aligned}$$

with the understanding that $0-1 = \text{undef}_{\mathtt{nat}}$. (Hence (i) the final argument t of *last* ranges over the extended natural numbers now, (ii) we have to require $t \neq \text{undef}_{\mathtt{nat}}$ in $(**)$ and (iii) we have to add $\mathtt{last}(\text{in}, \text{enable}, \text{undef}_{\mathtt{nat}}) = \text{undef}_{\mathtt{nat}}$).

Similarly we define the "the maximum and minimum value of *in* until that time point" (i.e. where *reset* becomes false) by

$$\begin{aligned}\texttt{max}(\text{in}, \text{reset}, t) := \ & \textbf{if } \text{reset}(t) \\ & \quad \textbf{then } \text{in}(t) \\ & \quad \textbf{else } \texttt{maximum}(\text{in}(t), \texttt{max}(\text{in}, \text{reset}, t-1)) \\ & \textbf{fi}\end{aligned}$$

$$\begin{aligned}\texttt{min}(\text{in}, \text{reset}, t) := \ & \textbf{if } \text{reset}(t) \\ & \quad \textbf{then } \text{in}(t) \\ & \quad \textbf{else } \texttt{minimum}(\text{in}(t), \texttt{min}(\text{in}, \text{reset}, t-1)) \\ & \textbf{fi}\end{aligned}$$

Now the minmax unit can be defined explicitly by

$$\begin{aligned}&\text{out}(\text{in}, \text{clear}, \text{reset}, \text{enable}, t) \ = \\ &\quad \textbf{if } \text{clear}(t) \\ &\qquad \textbf{then } 0 \\ &\qquad \textbf{else if } \text{enable}(t) \\ &\qquad\qquad \textbf{then } \texttt{half}\big(\texttt{max}(\text{in}, \text{reset}, t) + \texttt{min}(\text{in}, \text{reset}, t)\big) \\ &\qquad\qquad \textbf{else } \texttt{last}(\text{in}, \text{enable}, t) \\ &\qquad\quad \textbf{fi} \\ &\quad \textbf{fi}\end{aligned}$$

Note that among the constants for our example we have 0, +, `maximum`, `minimum`, and `half`.

3. TRANSLATION INTO CIRCUITS

We begin with recursive definitions, i.e., $C(\vec{a}, t) = M$ where M is a term with free variables among $\vec{a}, t$ built inductively by the three clauses above. We need a *register* unit

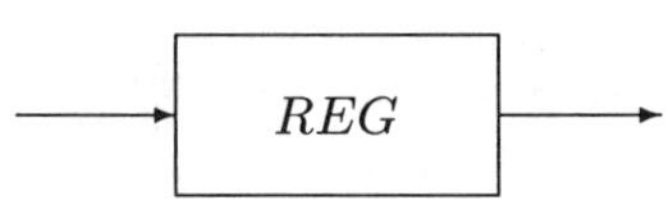

which at time $t+1$ gives out its input value at time t (and is undefined at time 0). For simplicity let us assume that $\vec{a} = a$. First for any such term $M[a, t]$ we inductively construct a circuit

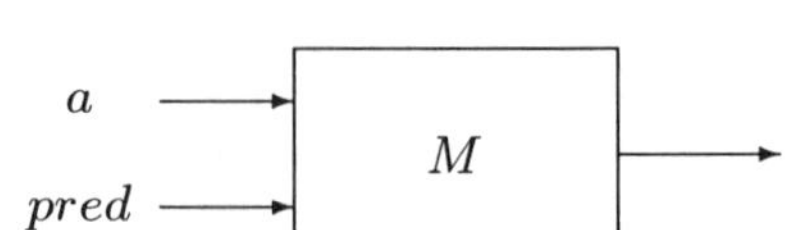

where both the a and the pred arrow may be missing. After this is done, we obtain the circuit for $C(a)$ by feedback:

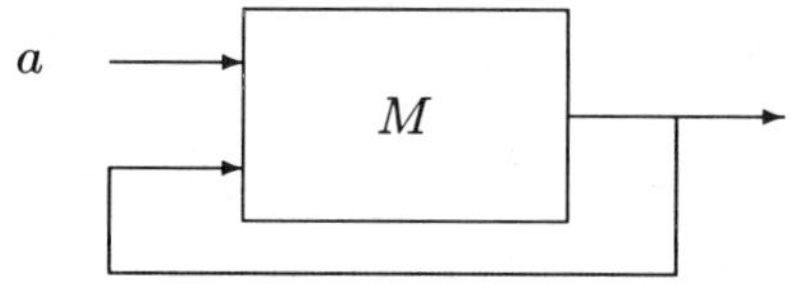

Case $C(a, \text{pred}^k(t))$ with $k \geq 1$. For $k = 1, 2, \ldots$ take

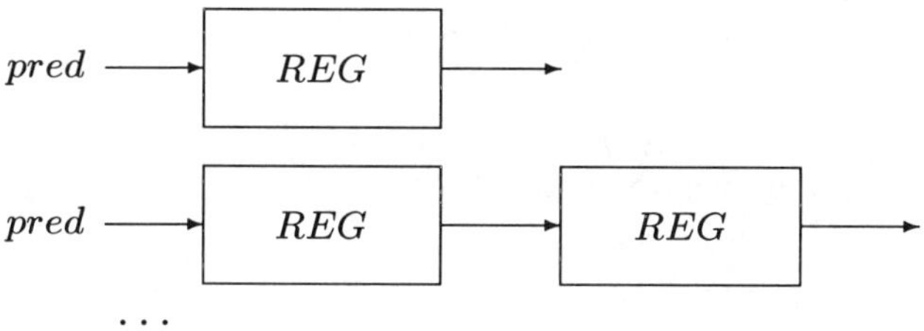

Case $a(\text{pred}^k(t))$ with $k \geq 0$. For $k = 0, 1, 2, \ldots$ take

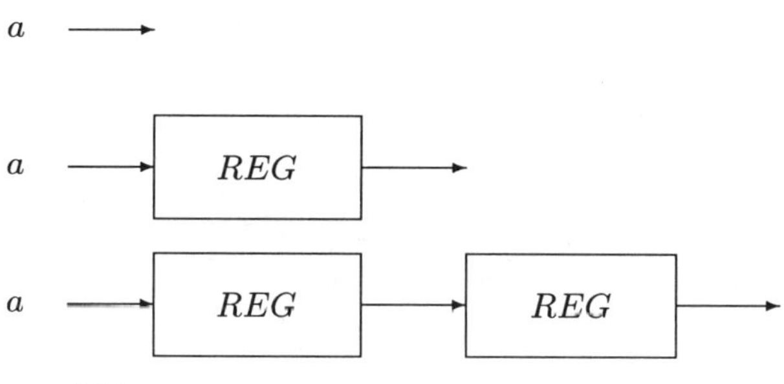

Case $D(M_1, \ldots, M_n)$.

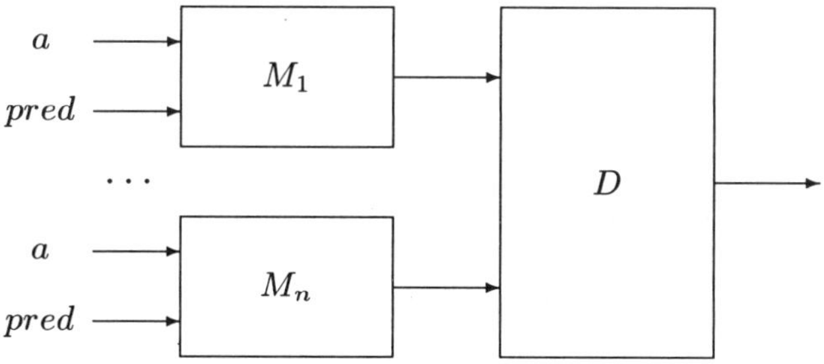

As an example, let us construct circuits for the recursive definitions of `last`, `max` and `min`. Here we need a *multiplexer* unit

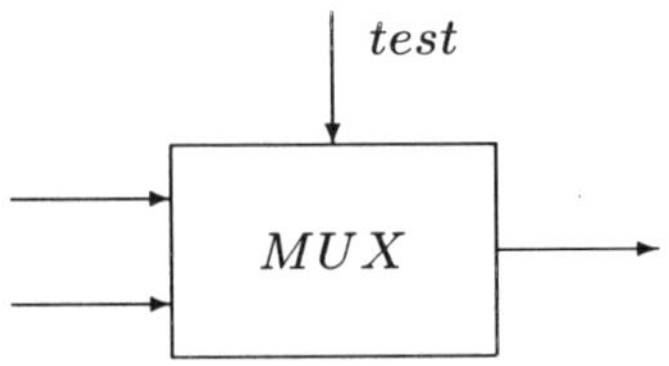

corresponding to **if–then–else–fi**. Recall the recursive definition of `last`:

$$\begin{aligned}\texttt{last}(\text{in}, \text{enable}, t) := \ & \textbf{if } \text{enable}(t)\\ & \quad \textbf{then } \text{in}(t)\\ & \quad \textbf{else } \texttt{last}(\text{in}, \text{enable}, \text{pred}(t))\\ & \textbf{fi}\end{aligned}$$

We now apply our inductive construction to the respective subterms of the right hand side.

$\texttt{last}(\text{in}, \text{enable}, \text{pred}(t))$:

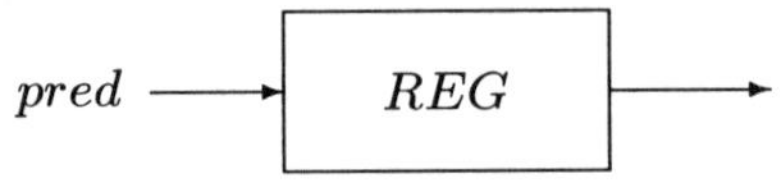

$\text{in}(t)$:

$\text{enable}(t)$:

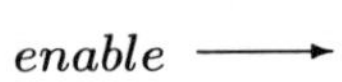

if $\text{enable}(t)$ **then** $\text{in}(t)$ **else** $\texttt{last}(\text{in}, \text{enable}, \text{pred}(t))$ **fi**:

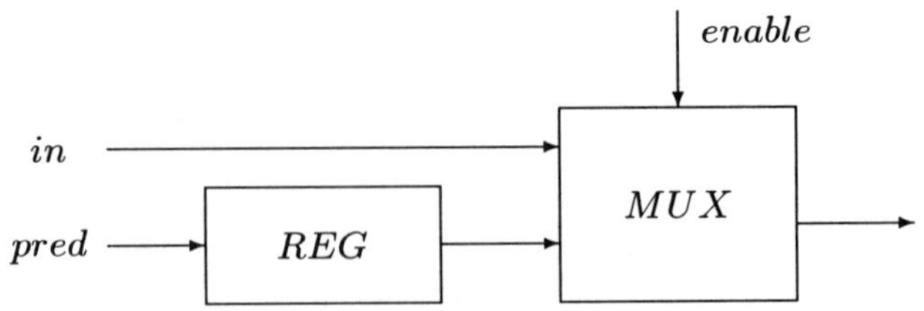

Finally our circuit for $\texttt{last}(\text{in}, \text{enable}, t)$ is built by feedback:

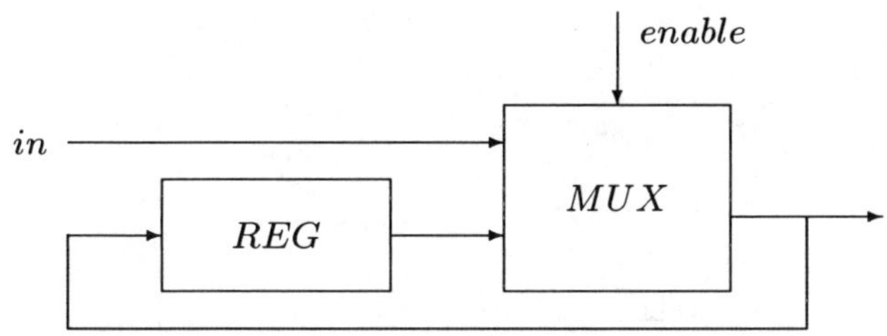

Similarly e.g. for

$$\begin{aligned}\texttt{max}(\text{in}, \text{reset}, t) := \ & \textbf{if } \text{reset}(t) \\ & \quad \textbf{then } \text{in}(t) \\ & \quad \textbf{else } \texttt{maximum}(\text{in}(t), \texttt{max}(\text{in}, \text{reset}, \text{pred}(t))) \\ & \textbf{fi}\end{aligned}$$

we obtain

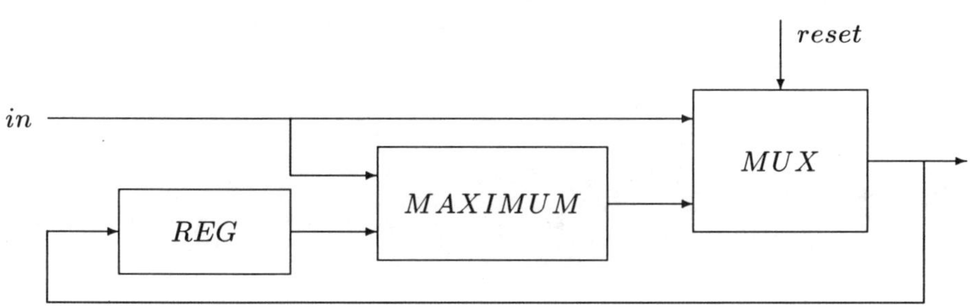

For explicit definitions $C(a, t) = M[a, t]$ things are even simpler. For any such M we inductively construct a circuit

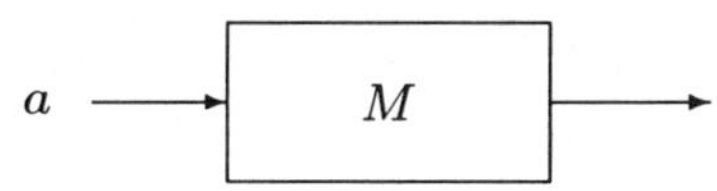

Case $a(\mathrm{t})$. Take $\mathtt{a}\longrightarrow$. *Case* $D(M_1,\ldots,M_n)$. Take

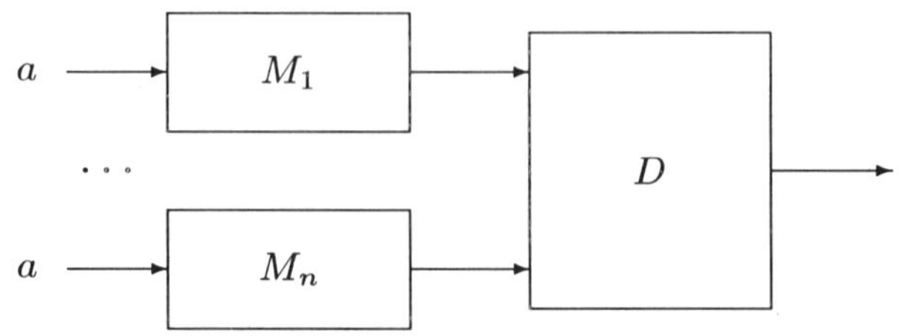

As an example, consider the explicit definition

$$\mathtt{mean}(\mathrm{in},\mathrm{reset},t) := \mathtt{half}(\mathtt{max}(\mathrm{in},\mathrm{reset},t) + \mathtt{min}(\mathrm{in},\mathrm{reset},t)).$$

We construct a circuit to be called MEAN from MAX, MIN, PLUS and HALF by

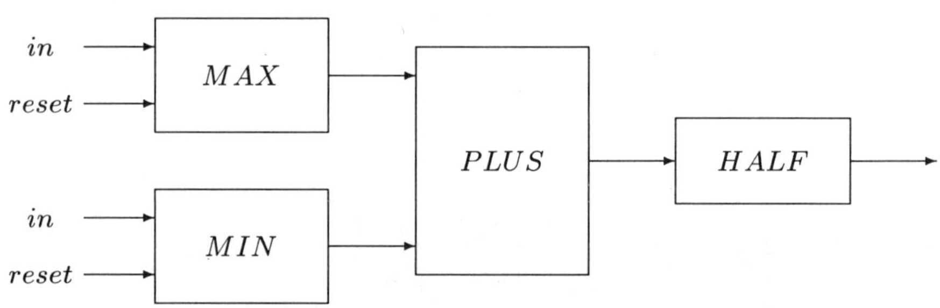

Now we can transform the explicit definition

$$
\begin{array}{l}
\mathrm{out}(\mathrm{in},\mathrm{clear},\mathrm{reset},\mathrm{enable},t) \;=\\
\quad \mathbf{if}\ \mathrm{clear}(t)\\
\qquad \mathbf{then}\ 0\\
\qquad \mathbf{else\ if}\ \mathrm{enable}(t)\\
\qquad\qquad\quad \mathbf{then}\ \mathtt{half}\big(\mathtt{max}(\mathrm{in},\mathrm{reset},t) + \mathtt{min}(\mathrm{in},\mathrm{reset},t)\big)\\
\qquad\qquad\quad \mathbf{else}\ \mathtt{last}(\mathrm{in},\mathrm{enable},t)\\
\qquad\qquad\ \mathbf{fi}\\
\quad \mathbf{fi}
\end{array}
$$

into a circuit:

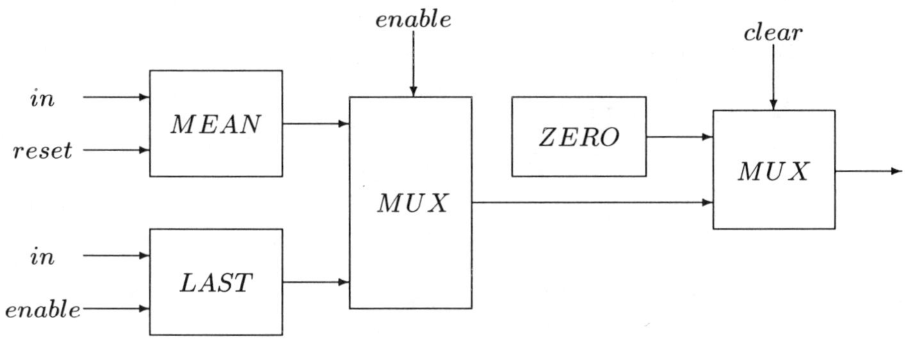

4. SPECIFICATION

For readability we leave out the stream arguments and e.g. write out(t) for

$$\text{out}(\text{in}, \text{clear}, \text{reset}, \text{enable}, t).$$

We also omit leading universal quantifiers. The required properties 1–4 of the informal description of the minmax unit in section 2 then translate into

$$\begin{array}{ll} & \text{clear}(t) = \texttt{true} \\ \to & \text{out}(t) = 0, \end{array} \tag{1}$$

$$\begin{array}{ll} & \text{clear}(t) = \texttt{false} \wedge \text{enable}(t) = \texttt{false} \\ \to & \text{out}(t) = \texttt{last}(t), \end{array} \tag{2}$$

$$\begin{array}{ll} & \text{clear}(t) = \texttt{false} \wedge \text{enable}(t) = \texttt{true} \wedge \text{reset}(t) = \texttt{true} \\ \to & \text{out}(t) = \text{in}(t), \end{array} \tag{3}$$

$$\begin{array}{ll} & \text{clear}(t) = \texttt{false} \wedge \text{enable}(t) = \texttt{true} \wedge \text{reset}(t) = \texttt{false} \\ \to & \text{out}(t) = \texttt{half}(\texttt{max}(t) + \texttt{min}(t)). \end{array} \tag{4}$$

Note that we need non–strict equality here.

The first thing to observe is that the assumption on reset(t) in (4) is not necessary, i.e. we can prove the following strengthened form (4*) of (4):

$$\begin{array}{ll} & \text{clear}(t) = \texttt{false} \wedge \text{enable}(t) = \texttt{true} \\ \to & \text{out}(t) = \texttt{half}(\texttt{max}(t) + \texttt{min}(t)). \end{array} \tag{4*}$$

However, this "strengthening" is slightly misleading, since we do need the assumption on reset(t) to prove that $\texttt{max}(t)$ and $\texttt{min}(t)$ have their expected properties, e.g. that $\texttt{max}(t)$ is the maximum value of *in* since *reset* became false. We split this up into two formulas: that $\texttt{max}(t)$ is an upper bound, and that it is the least upper bound. So we have to prove

$$
\begin{array}{ll}
 & \mathrm{reset}(t) = \mathtt{true} \\
\wedge & (\forall n\colon \mathtt{nat}.\ n \neq 0 \wedge n \leq l \to \mathrm{reset}(t+n) = \mathtt{false}) \\
\wedge & (\forall n\colon \mathtt{nat}.\ n \leq l \to \mathrm{in}(t+n)\downarrow) \\
\to & \forall n\colon \mathtt{nat}.\ n \leq l \to \mathrm{in}(t+n) \leq \mathtt{max}(t+l)
\end{array} \tag{5}
$$

$$
\begin{array}{ll}
 & \mathrm{reset}(t) = \mathtt{true} \\
\wedge & (\forall n\colon \mathtt{nat}.\ n \neq 0 \wedge n \leq l \to \mathrm{reset}(t+n) = \mathtt{false}) \\
\wedge & (\forall n\colon \mathtt{nat}.\ n \leq l \to \mathrm{in}(t+n) \leq k) \\
\to & \mathtt{max}(t+l) \leq k.
\end{array} \tag{6}
$$

Note that we have to require the definedness of $\mathrm{in}(t+n)$ in (5) (i.e. that $\mathrm{in}(t+n)$ is not $\mathrm{undef}_{\mathtt{nat}}$); otherwise (5) would not hold, since $\mathrm{undef}_{\mathtt{nat}} \leq \dots$ as well as $\dots \leq \mathrm{undef}_{\mathtt{nat}}$ are defined to be false.

Finally, we have to formulate the specification of $\mathtt{last}(t)$, i.e. that it is the "last value of *in* before *enable* became false":

$$
\begin{array}{ll}
 & \mathrm{enable}(t) = \mathtt{true} \\
\wedge & (\forall n\colon \mathtt{nat}.\ n \neq 0 \wedge\ n \leq l \to \mathrm{enable}(t+n) = \mathtt{false}) \\
\to & \mathtt{last}(t+l) = \mathrm{in}(t).
\end{array} \tag{7}
$$

5. FORMAL VERIFICATION

It is now more or less routine to prove that the minmax circuit constructed in section 3 meets the specification given in section 4. However, we have not confined ourselves with a *paper and pencil* proof but rather have checked the correctness of our proof with machine help. In the presence of error values for input streams this seems to be particularly advisable, since it is easy for a human to forget some cases.

In fact, we have done the formal verification twice: First with the interactive prover MINLOG (Schwichtenberg, 1993) developed by the first author, and then again with the help of the automatic theorem prover SEDUCT (Stroetmann, 1995) under development at Siemens.

The MINLOG system is designed to deal with terms denoting computable functionals over the partial continuous functionals; hence the proof went rather smoothly. However, MINLOG is an interactive prover with only limited automated support (except a mechanism to deal with equality logic which uses normalization of higher order terms (Berger and Schwichtenberg, 1991)). Therefore, it was a challenge to see to what extent a theorem prover with more automated components, but based on many sorted first order logic, could be used as well. It turned out that this was possible. We conclude with comments on some of the observations we made in the course of doing the proof.

The first problem is of course that the specification given in section 4 is written in *higher order logic* (Gordon, 1986), whilst SEDUCT is a theorem prover for many–sorted *first order logic*. We therefore have to translate this specification into first order logic. However, the only part of this specification that is not first order is the use of the variables *in*, *clear*, *reset*, and *enable* as higher order variables, e.g. in $out(t)$. Now a closer inspection reveals that this use of higher order variables is not essential: we can eliminate expressions of the form $out(t)$ by introducing a new function symbol @ with signature

$$@ : \texttt{n\_stream} \times \texttt{nat} \to \texttt{nat},$$

where n_stream is the type of *streams* of natural numbers. Of course, the intention is that for any stream of natural numbers f the value of $f @ t$ is the same as $f(t)$. Since second order variables occur only in this context we can eliminate them by replacing all expressions of the form $f(t)$ where f is a second order variable with the expression $f @ t$.

A point of general interest that we learned while carrying out the proof is the following. If the formulas to be proved contain quantified subformulas, then it seems to be a good idea to define predicates equivalent to these subformulas. For example, property (6) contains the universally quantified subformula

$$\forall n : \texttt{nat}.\ n \leq l \to in @ (t+n) \leq k.$$

To eliminate this subformula we have introduced the predicate bounded satisfying an appropriate lemma. This lemma could then be used to eliminate the above subformula from (6).

Helmut Schwichtenberg
Mathematisches Institut der Universität München

Karl Stroetmann
Siemens AG, Corporate Research and Development, München

ACKNOWLEDGEMENTS

The authors would like to acknowledge the fact that the paper has benefited from a number of comments made by Dr. Gerd Venzl.

REFERENCES

Ulrich Berger. *Totale Objekte und Mengen in der Bereichstheorie.* PhD thesis, Mathematisches Institut der Universität München, 1990.

Ulrich Berger. Total sets and objects in domain theory. *Annals of Pure and Applied Logic*, 60:91–117, 1993.

Ulrich Berger and Helmut Schwichtenberg. An inverse of the evaluation functional for typed λ–calculus. In Rao Vemuri, editor, *Proceedings of the Sixth Annual IEEE Symposium on Logic in Computer Science*, pages 203–211. IEEE Computer Society Press, Los Alamitos, 1991.

Luc Claesen, editor. *Formal VLSI Specification and Synthesis VLSI Design Methods I and II.* IFIP, sponsored by IMEC, North–Holland, Amsterdam, 1990.

Yuri L. Ershov. Model C of partial continuous functionals. In R. Gandy and M. Hyland, editors, *Logic Colloquium 1976*, pages 455–467. North–Holland, Amsterdam, 1977.

Mike Gordon. Why higher-order logic is a good formalism for specifying and verifying hardware. In G.J. Milne and P.A. Subrahmanyam, editors, *Formal Aspects of VLSI Design*, pages 153–177. Elsevier, Amsterdam, 1986.

Thomas Kropf. Benchmark–circuits for hardware–verification. In R. Kumar and T. Kropf, editors, *Theorem Provers in Circuit Design (TPCD '94)*, volume 901 of *Lecture Notes in Computer Science*, pages 1–12. Springer Verlag, Berlin, Heidelberg, New York, 1995.

Jaco van de Pol and Helmut Schwichtenberg. Strict functionals for termination proofs. In M. Dezani-Ciancaglini and G. Plotkin, editors, *Typed Lambda Calculi and Applications*, volume 902 of *Lecture Notes in Computer Science*, pages 350–364. Springer Verlag, Berlin, Heidelberg, New York, 1995.

Helmut Schwichtenberg. Proofs as programs. In P. Aczel, H. Simmons, and S.S. Wainer, editors, *Proof Theory. A selection of papers from the Leeds Proof Theory Programme 1990*, pages 81–113. Cambridge University Press, 1993.

Helmut Schwichtenberg and Stanley S. Wainer. Ordinal bounds for programs. In P. Clote and J. Remmel, editors, *Feasible Mathematics II*, pages 387–406. Birkhäuser, Boston, 1995.

Dana Scott. Domains for denotational semantics. In E. Nielsen and E.M. Schmidt, editors, *Automata, Languages and Programming*, volume 140 of *Lecture Notes in Computer Science*, pages 577–613. Springer Verlag, Berlin, Heidelberg, New York, 1982. A corrected and expanded version of a paper prepared for ICALP'82, Aarhus, Denmark.

Viggo Stoltenberg-Hansen, Edward Griffor, and Ingrid Lindström. *Mathematical Theory of Domains*. Cambridge tracts in Theoretical Computer Science. Cambridge University Press, 1994.

Karl Stroetmann. SEDUCT — a proof compiler for first order logic. In M. Broy and S. Jähnichen, editors, *KORSO: Methods, Languages and Tools for the Construction of Correct Software. Final Report*, volume 1009 of *Lecture Notes in Computer Science*, pages 299–316. Springer Verlag, Berlin, Heidelberg, New York, 1995.

ROBERT I. SOARE*

COMPUTABILITY AND ENUMERABILITY

ABSTRACT.
This paper contains a brief summary of the two parts of a lecture given at the international Congress for Logic, Methodology, and Philosophy of Science in Florence, August, 1995. More detailed versions will appear in [80] and [30], respectively.

In the first part we consider the informal concept of "computability" or "effective calculability" and two of the formalisms commonly used to define it, *"(Turing) computability"* and *"(general) recursiveness."* We consider their origin, exact technical definition, concepts, history, general English meanings, how they became fixed in their present roles, how they were first and are now used, their impact on nonspecialists, and how they should preferably be used.[1]

The second part of the ICLMPS lecture at Florence dealt with some recent results on the computably enumerable (c.e.) sets, especially their *definability* properties (as sets in the spirit of Cantor), their *automorphisms* (in the spirit of Felix Klein's *Erlanger Programm*), their *dynamic* properties, expressed in terms of how *quickly* elements enter them relative to elements entering other sets, and the *Martin Invariance Conjecture* on their Turing degrees, *i.e.*, their *information content* with respect to relative computability (Turing reducibility).

1. INTRODUCTION

All functions are on the nonnegative integers, $\omega = \{0, 1, 2, \ldots\}$, and all sets will be subsets of ω. Intuitively, a function is *"computable"* (also called *"effectively calculable"* or simply *"calculable"*) if it can be calculated by a finite mechanical procedure. Formally, (i) a function is *Turing computable* if it is definable by a Turing machine, as defined by Turing *1936* (see[78]); (ii) a set A is *computably enumerable (c.e.)* if A is $\emptyset$ or is the range of a Turing computable function; (iii) a function f is *recursive* if it is *general recursive*, as defined by Gödel *1934* (see also Kleene's variant [*1952*, p. 274]); (iv) a set A is *recursively enumerable (r.e.)* if A is $\emptyset$ or is the range of a general recursive function.[2]

For the moment we regard these terms strictly with their *intensional* meaning as above, and we do not *extensionally* identify them with each other or

*The author was partially supported by National Science Foundation Grant DMS 94-00825. Some of this material was given in the first part of a lecture at the 10th International Congress for Logic, Methodology and Philosophy of Science, Section 3: Recursion Theory and Constructivism, August 19–25, 1995.

[1]Helpful suggestions, comments, and criticisms on preliminary drafts of this part were made by: W. Calhoun, P. Cholak, B. Cooper, M. Davis, R. Downey, L. Fortnow, W. Gasarch, L. Harrington, C. G. Jockusch, A. H. Lachlan, S. Lempp, M. Lerman, J. Owings, J. Shoenfield, R. A. Shore, W. Sieg, T. Slaman, M. Stob, G. Tamburrini, and S. S. Wainer, among others.

[2]This terminology is the same as that introduced in the 1930's and used since then, except for the term "computably enumerable," recently introduced, because Turing and Gödel did not explicitly introduce a term for these corresponding *sets*, but just for the computable *functions*.

M.L. Dalla Chiara et al. (eds.), Logic and Scientific Methods, 221–237.

with other formal notions known to be mathematically equivalent (such as λ-definability, μ-recursiveness, or Post normal systems), nor with the *informal* notions of computable or effectively calculable under Church's Thesis or Turing's Thesis.

Several logicians have remarked that it is curious that the subject of computability theory (hereinafter called *"the subject"*) is called *"recursive function theory"* or *"recursion theory"* instead of the seemingly more natural, *"computability theory,"* as we shall call it here.

The purpose of the first part is to examine the meaning, origin, and history of the concepts "recursive" and "computable" with an eye toward reexamining how we use them in practice. The ultimate aim is to ask: *"What is the subject really about?"* For example, is it about computability, recursions, definability, or something else?

All background information can be found in Soare [78]. We cite references by their number in the order listed in our bibliography but also in the usual convention by author and year, *e.g.*, [Post, *1944*] or simply Post *1944*, with the year in italics. We omit from our bibliography some references which appear in Soare *1987*, van Heijenoort *1967*, Kleene *1952*, or or Davis *1965*.

2. A BRIEF HISTORY OF COMPUTABILITY

A *computation* is a process whereby we proceed from initially given objects, called *inputs*, according to a fixed set of rules, called a *program*, *procedure*, or *algorithm*, through a series of *steps* and arrive at the end of these steps with a final result, called the *output*. The concept of *computability* concerns those objects which may be specified in principle by computations, and includes relative computability (computability from an oracle).

The concept of *recursion* stems from the verb "recur," "to return to a place or status." The primary mathematical meaning of recursion has always been "definition by induction" (*i.e.*, by recursion), namely defining a function f at an argument x using its own previously defined values (say $f(y)$ for $y < x$), and also using "simpler" functions g (usually previously defined). The *concept of recursion* used here includes: (1) induction and the notion of *reflexive program call*, (including primitive recursion and also Kleene's Recursion Theorem); (2) the notion of a *fixed point* for some function, and the more general Platek style fixed points in higher types; (3) other phenomena related to (1) and (2) specified for certain situations and structures. However, the concept of recursion does *not* include the notion of "computable" or "algorithmic" as in the first paragraph.

Mathematicians have studied calculation and algorithms since the time of the Babylonians and the Greeks. In 1642 the French mathematician and scientist, Blaise Pascal, invented an adding machine which may be the first digital calculator. In 1671 the German mathematician and philosopher, Gottfried Wilhelm Leibniz, co-inventor with Newton of the calculus, invented a machine that performed multiplication. Around 1834 Babbage invented the idea of an "Analytic Engine," which could be programmed to perform long and tedious calculations, and formulated what Gandy [*1988*, p. 58] called "Babbage's Thesis," that "the whole of the development and operations of analysis are now capable of being executed by machinery."

These two trends of recursion and computability were brought together in the 1930's by Gödel, Church, Kleene, Turing, and others partly in response to questions raised earlier by Hilbert. Hilbert *1904* proposed proving the consistency of arithmetic by what became known by *1928* as his *finitist program*, which tended to reduce proofs to manipulation of finite strings of symbols which stimulated the development of mechanical processes to accomplish this.

Closely related was Hilbert's *Entscheidungsproblem*, the decision problem for first order logic, which emerged in the early 1920's in lectures by Hilbert and which was described in Hilbert and Ackermann *1928*. Hilbert characterized this as the fundamental problem of mathematical logic. Gödel *1931* proved his *first* incompleteness theorem which (stated in modern terms and with an improvement by Rosser) asserts roughly that any consistent extension T of elementary number theory is incomplete. By arithmetizing the proof Gödel obtained his *second* incompleteness theorem which asserts that such a T cannot prove its own consistency (see [45, §42]), which was a setback for Hilbert's program (see Gödel *1958*).

In his proof Gödel *1931* [23, p. 158] used the notion of a primitive recursive function (which he called "recursive" [*rekursiv Funktion*]) because these functions were easily representable in Gödel's formal system **P** for arithmetic, and were sufficient to enable him to "Gödel number" all the syntactic objects so that he could obtain self-reference and thereby incompleteness. Gödel realized, however, that the primitive recursive functions did not include *all* effectively calculable functions, and in *1934* he proposed a wider class of functions based on an earlier suggestion of Herbrand, and later called "general recursive" by Church and Kleene.

In 1930 Church had been studying a class of effectively calculable functions called *λ-definable functions*. Church's student, Kleene, showed by 1933 that a large class of number theoretic functions were λ-definable. On the strength of this evidence, Church proposed to Gödel in early 1934 that the notion of "effectively calculable" be identified with "λ-definable," a suggestion which Gödel rejected as "thoroughly unsatisfactory." Following this encounter with Gödel, Church changed formal definitions from "λ-definable" to "recursive," his abbreviation for Herbrand-Gödel general recursive, and Church presented on April 19, 1935, to the American Mathematical Society his famous proposition published in *1936* and known (since Kleene *1952*) as *Church's Thesis* which asserts that the effectively calculable functions should be identified with the recursive functions. On the basis of this Thesis, Church *1936* announced the unsolvability of Hilbert's *Entscheidungsproblem*.

Gödel, however, remained unconvinced of the validity of Church's Thesis through its publication *1936*. The flaw in Church's argument [*1936*, §7] for his thesis was this as Gandy [*1988*, p. 79] and especially Sieg [*1994*, pp. 80, 87] in their excellent analyses brought out. If the basic steps are stepwise recursive, then it follows easily by the Kleene Normal Form Theorem which Kleene had proved and communicated to Gödel before November, 1935 (see Kleene [*1987b*, p. 57]), that the entire process is recursive. The fatal weakness in Church's argument was the core assumption that the atomic steps were stepwise recursive, something he did not justify. It is exactly this stumbling block which Turing overcame by a totally new approach.

In the spring of 1935 a twenty-two year old student at Cambridge University, who had just given an independent proof of the Central Limit Theorem, heard the lectures of Professor M.H.A. Newman on Gödel's paper and the Hilbert *Entscheidungsproblem*. Turing worked on the problem for the remainder of 1935 and submitted to the incredulous Newman on April 15, 1936, his solution *1936*. Turing's monumental paper *1936* was distinguished because: (1) Turing analyzed an idealized *human* computing agent (a *"computor"*) which brought together the intuitive conceptions of a "function produced by a mechanical procedure" which had been evolving for more than two millenia from Euclid to Leibniz to Babbage and Hilbert; (2) Turing specified a remarkably simple formal device (*Turing machine*) and proved the equivalence of (1) and (2); (3) Turing proved the unsolvability of Hilbert's *Entscheidungsproblem* which established mathematicians had been studying intently for some time; (4) Turing proposed a *universal* Turing machine, one which carried within it the capacity to duplicate any other, an idea which was later to have great impact on the development of high speed digital computers and considerable theoretical importance. Gödel enthusiastically accepted Turing's Thesis and his analysis and always thereafter gave Turing credit for the definition of mechanical computability.

Turing proposed that intuitively computable functions be identified with those computable by Turing machine, a proposition which became known as "Turing's Thesis." However, both Church and Turing regarded their formalisms as *definitions* of the informal concept of effectively calculable, not as theses. They were not even called "theses" at all until Kleene [*1943*, p. 60] referred to Church's "definition" as "Thesis I," and then in *1952* Kleene referred to "Church's Thesis" and "Turing's Thesis." What is even more curious is that the phrase "Church's Thesis" came to denote also "Turing's Thesis" and perhaps others as well, thereby blurring all intensional distinctions. We propose now that Turing's Thesis be used as a *definition* of a computable function as Turing and Gödel suggested, and be separated from Church's Thesis. We argue for this position in [80].

3. DEVELOPMENT OF THE CONCEPTS AND TERMS

Both Turing and Gödel, the inventors of the two formal definitions and the two names, preferred the terminology "computable" for this class of functions, but the word "recursive" become preferred for it and for the subject. Turing and Gödel ceased to actively work in computability theory by the late 1930's and had little influence thereafter on its terminology.

The present terminology came from Church and Kleene. They had worked in the λ-definable functions until 1935 when they changed to recursive functions because it was more in the mathematical mainstream and had more audience appeal, as explained by Kleene *1987b*. They had both committed themselves to the new "recursive" terminology before they ever heard of Turing or his results. Furthermore, using "computable" in 1935 would not have increased audience appeal because a "computer" meant, even as late as 1946, a human being calculating with paper.

Later, Post *1941* and *1943* introduced a *second* and unrelated formalism called a *production* system and (in a restricted form) a *normal* system, which he explained again in *1944*. Post's (normal) canonical system is a *generational* system, rather than a *computational* system as in general recursive functions or Turing computable functions, and led Post to concentrate on *effectively enumerable sets* rather than computable functions. He proved that every effectively enumerable set is a normal set (one derived in his normal canonical system) and therefore formally equivalent to a recursively enumerable set.

In his famous and very influential paper *1944*, Post continued with the intuitive concepts of "effectively enumerable" and "generated set," which he explains again at some length. The *formalism* Post used was that of his own normal (production) system, *i.e.*, "normal set." He used the term "recursively enumerable set" (Church's term *1936*) as a name for *both* his *informal* "effectively enumerable set" and for his *formal* version, "normal set." However, the *concept* or formal *definition* of "recursive" does not enter Post's paper at all, only the *terms* "recursive" and "recursively enumerable." Post's use of the term "recursively enumerable" is one of several ambiguities in the subject (ambiguous from an *intensional* viewpoint).

In spite of this ambiguity, Post's entrance on the scene was fortunate for recursively enumerable sets and for the entire subject. Previously, the papers in the subject had been written in the very technical formalism of μ-recursive functions with little intuition. Recursively enumerable sets had attracted very little attention since their debut in 1936. Post's papers brought excitement, intuitive appeal, and an informal style of proof, much closer to ordinary mathematical proofs, and represented the real birth of the subject of recursively enumerable sets *1943* and *1944* and degrees of unsolvability *1948*. These papers of Post stimulated the entire subject for decades, but they simultaneously helped to fix the use of the terms "recursive" and "recursively enumerable" to acquire the additional meanings, "computable" and "computably enumerable."

In [80] we trace the further development of the concepts and terms "recursive" and "computable" in several branches of the subject including computably enumerable sets and recursion in higher types and Kleene's work.

4. CURRENT USAGE OF THE CONCEPTS AND TERMS

There is a current tendency in the subject to work in one formalism (usually that of Turing computable functions) but then to *name* the results using the terminology of *recursive* functions not *computable* functions. For example, Kleene wrote,

> "By *general recursive* functions (predicates) we mean ones whose values can be computed (decided) by ideal computing machines not limited in their space for storing information. A theory of such machines was given by Turing *1936* and in less detail by Post *1936*."
> —Kleene *1955b*

Consider from an *intensional viewpoint* the following review by Putnam [1995, p. 371] of Roger Penrose's new book *1994*

> "First Penrose provides the reader with a proof of a form of the Gödel Theorem due to Alan Turing, the father of the modern digital computer and the creator of the mathematical subject *recursion theory*, which analyzes what computers can and cannot in principle accomplish."

(Putnam's *own article* [71] is an excellent example of the modern use of computer related concepts and terms rather than recursive functions to describe computational processes. Putnam's review (like Penrose's book) is written *entirely* using the *Turing machine model*, speaking of "machines," "programs," "output," "lines of code," a "debugged" program, "Turing-machine action," and "programs which output theorems.") Turing certainly *never* used the term "recursion theory" or "recursive function theory" for the subject. Turing mentioned the term "recursive function" only very briefly in *1937b* and [*1939*, §2] to say that these functions were mathematically equivalent to his Turing computable functions, and then Turing *dismissed* general recursive functions with the phrase, "we will not be much concerned here with this particular definition." Turing certainly never used "recursive" to mean *"computable,"* and Turing did not refer to "recursive functions" again. Clearly from a *strictly intensional* viewpoint, the term *"recursion theory"* does not analyze anything about what *computers* can or cannot accomplish at all (contrary to Putnam's assertion); it deals with the properties of Herbrand-Gödel general recursive functions, the concepts of induction, recursion, reflexive program calls, and fixed points.

Gödel, who had invented *1934* the formal definition of general recursive function, abandoned it almost completely after seeing Turing machines *1936* and Turing's demonstration of Turing's Thesis. After 1936 Gödel rarely spoke of recursive functions, and never used the term "recursive" to mean "computable" or "decidable." Gödel often asserted later that Turing's was the correct definition of the notion of mechanical computability, and spoke often of the concept of computability *1946* [18, p. 84], *1951* [25, pp. 304–305], *193?* [25, p.168], *1968* [76, p. 88], *1936* [5, p. 82], *1964* [5, p. 71–72]. Both Turing and Gödel, even later in life, rejected "recursive" as a name for the subject and often for their results as explained in [80]. We also give a table of modern texts in the subject and how they usually define the concepts in terms of Turing computable functions (or a variant like register machines) and then proceed to use the *name* recursive for the final result. This leads the word "recursive" to have at least two meanings, one meaning computable (as understood by graduate students and reserchers) and the second meaning defined by induction (as understood by Hilbert and by undergraduates in computer programming courses). It is genuinely difficult to ascertain which meaning is intended in a given instance.

> The first rule of good taste in writing is to use words whose meaning will not be misunderstood; and if a reader does not know the meaning of the words, it is infinitely better that he should know he does not know it.
>
> —*Charles Sanders Peirce*, Ethics of Terminology, [64, p. 131]

We also explore [80] general English and scientific usage of the two terms.

5. ANALYSIS AND CONCLUSION ON TERMINOLOGY

Both of the concepts of recursion and computability have played a crucial role in the development of the subject and will continue to do so. The term and concept of "computable" is associated with the notion of computation, algorithm, and with the functions defined by (or sets enumerated by), and with relative Turing computability defined by Turing machines or register machines. The term and concept of "recursive" is associated with: definition by recursion (induction), general recursive functions in the sense in Herbrand-Gödel, fixed points as in the Kleene Recursion Theorem of Kleene's schema (S11) (see *1987c*), which Kleene believed included all possible recursions, and Kleene's μ-recursive functions.

Recommendation 1 The term "recursive" should no longer carry the additional meaning of "computable" or "decidable."

We already have better words to convey these meanings, namely "computable" and "decidable" themselves. Likewise, the term "computable" should not carry the extra meaning "recursive." Now we will be able to recognize clearly the concept of computability when it appears. Second, it reinforces the original meaning of "recursive" and induction as understood by Dedekind *1888*, Peano *1889* and *1891*, Hilbert *1904* and *1926*, Skolem *1923*, Gödel *1931* and *1934*, and Péter *1934* and *1951*, and most modern computer scientists, mathematicians, and physical scientists, and as expanded to fixed points, the recursion theorem, and to other kinds of recursion by Kleene, Platek, and others.

Recommendation 2 If functions are defined, or sets are enumerated, or relative computability is defined using Turing machines, register machines, or variants of these, then the name "computable" rather than "recursive" should be attached to the result.

Thus, the terms "recursive" and "computable" would reacquire their traditional and original meanings, and those understood by most outsiders. This would improve communication with many researchers outside the field. It would also give a new scientific precision to our discussions within the subject and would remove various ambiguities mentioned above.

Recommendation 3 We should distinguish between the intensional meaning of Church's Thesis (that all effectively calculable functions are general recursive) versus Turing's Thesis (that all intuitively computable functions are computable by a Turing machine). When we are writing a paper dealing with which classes of functions are Turing computable (*i.e.*, mechanistic), as in Gandy *1980* and in many other places, we should refer to *Turing's Thesis* (as in Sieg *1994* and Tamburrini *1995*) not to Church's Thesis.

Recommendation 4 We should call the subject *"Computability Theory* or simply *Computability* instead of *Recursive Function Theory* or *Recursion Theory.*

In 1935 when Church and Kleene adopted the terminology and formalism of "recursive functions" for the subject in preference to λ-definable functions, there was good reason to do so. Recursive functions lay in an established mathematical tradition and had more appeal to a general mathematical audience. Church was eager to establish mathematical logic as a field of mathematics and to avoid the inconsistencies discovered by Kleene and Rosser in Church's general λ-system. When Kleene and Church saw Turing's model in 1936, they had no desire to make another change so soon. Another conversion to the formalism of Turing would have had little increase in appeal at that time to a general audience because as late as 1946 the term "computer" meant a human being calculating with pencil and paper. However, in the modern world the term "computable" much more accurately describes the concept than "recursive."

5. COMPUTABLY ENUMERABLE SETS

We now summarize the second half of the lecture which will appear in full in [80]. Post *1943* introduced the formalism of a (normal) production system which was a *generational* system in contrast to the *computational* systems of Church, Gödel, Turing, and Kleene, and Post *1944* explored the properties of what he called *effectively enumerable sets* or *generated sets* rather than computable *functions*. He was particularly interested in the *information content* of a c.e. set A, namely what we also call its *degree*, usually under Turing reducibility ($A \leq_{\mathrm{T}} B$). Post's *problem* was whether there is a noncomputable c.e. set A which is *incomplete*, namely $\emptyset <_{\mathrm{T}} A <_{\mathrm{T}} K$, for K the complete set $\{W_e : e \in W_e\}$, where $\{W_e\}_{e \in \omega}$ is an effective listing of all c.e. sets.

To this end, Post considered certain "set-theoretic" properties of c.e. sets, *i.e.*, those definable as "sets", *e.g.*, using set inclusion, $\subset$. For example, Post defined a *simple set* A to be a coinfinite c.e. set which intersects every infinite c.e. set, a property which is definable in the language $L(\subset)$, because the predicate $Fin(A)$:"A is finite" is easily seen [78, Ch. X] to be $\subset$-definable (every subset of A is complemented). Let $\mathcal{E}$ denote the partial ordering of the c.e. sets under inclusion, $\mathcal{E} = (\{W_e\}_{e \in \omega}, \subset)$, and let $\mathcal{E}^*$ denote $\mathcal{E}$ modulo the class $\mathcal{F}$ of finite sets.

Among the properties originally considered by Post, simplicity and hh-simplicity are $\mathcal{E}$-definable, but h-simplicity is not (see [78]). (A set A is hh-simple if every superset is complemented modulo A by Lachlan *1968*.) Post was searching for a c.e. noncomputable A whose complement while infinite was so thin that it forced A to be incomplete. Post's Problem was solved by Friedberg and independently by Muchnik with the introduction of the priority method. Post's program remained open, however, and stimulated much research on the more general and fundamental question of studying the relationship between the algebraic structure of a c.e. set A and deg(A).

The set with the thinnest complement of all, a maximal set, was constructed by Friedberg *1958*, and Martin *1966b* proved that their degrees were exactly the high degrees, $\mathbf{H}_1$, those with jump $0'$. This led to a great deal of activity beginning in the 1960's in classifying the $\mathcal{E}$-definable properties of A and also led to results connecting these properties to the information content of A.

In §6 we describe a definable solution to Post's program, namely a nontrivial predicate $Q(A)$ definable in $L(\subset)$ over the c.e. sets which insures that A is incomplete. In contrast we describe an $L(\subset)$-definable property $T(A)$ which allows such a rapid flow of elements into A that A must be *complete* even though A may possess many other properties such as being promptly simple. We also present a related property $NL(A)$ which has a slower flow but fast enough to guarantee that A is not low, even though A may possess virtually all other related lowness properties (low$_2$ and others) and A may simultaneously be promptly simple. In §7 we describe a new method generating automorphisms of $\mathcal{E}$, which yields results that may be considered in opposition to the definability results and we study the possibility of coding a set into every nontrivial orbit. In §8 we consider an old conjecture of Martin about which classes of c.e. degrees are *invariant,* and we supply new evidence for the conjecture.

6. $\mathcal{E}$-DEFINABILITY AND INFORMATION CONTENT

We begin by producing an $\mathcal{E}$-definable property $Q(A)$ which guarantees that A is *incomplete* but noncomputable and thereby represents an $\mathcal{E}$-definable solution to Post's problem.

Definition 6.1 (i) A subset $A \subset C$ is a *major subset* of C ($A \subset_{\mathrm{m}} C$) if $C - A$ is infinite and $(\forall e)[\overline{C} \subseteq W_e \implies \overline{A} \subseteq^* W_e]$. (Note that if $A \subset_{\mathrm{m}} C$ then both A and C are noncomputable.)

(ii) $A \sqsubset B$ if there exists C such that $A \sqcup C = B$ (i.e., $A \cup C = B$ and $A \cap C = \emptyset$).

(iii) If $\{X_s\}_{s\in\omega}$ and $\{Y_s\}_{s\in\omega}$ are computable enumerations of c.e. sets X and Y define $X \setminus Y = \{z : (\exists s)[z \in X_s - Y_s]\}$, the elements enumerated in X before Y and $X \searrow Y = (X \setminus Y) \cap Y$, the elements enumerated first in X and later in Y.

Properties of c.e. sets stated in terms of X_s, $X \setminus Y$, or $X \searrow Y$ are called *dynamic* properties because they refer to the *order* of enumeration of elements in X and Y, but $\mathcal{E}$-definable properties are *static* because they refer only to the elements eventually in X or Y.

Theorem 6.2 (Harrington-Soare [27]) *There is a property $Q(A)$ which guarantees that $A <_{\mathrm{T}} K$ and which holds of some noncomputable set.*

Define the property $Q(A)$ by:

$$\begin{aligned} Q(A):\quad & (\exists C)_{A\subset_{\mathrm{m}}C}\ (\forall B \subseteq C)(\exists D \subseteq C)(\forall S)_{S\sqsubset C}\ [\\ & [B \cap (S - A) = D \cap (S - A)]\\ & \implies (\exists T)[\overline{C} \subset T\ \&\ A \cap (S \cap T) = B \cap (S \cap T)]]. \end{aligned}$$

(For intuition and details about $Q(A)$ and the other properties see [30].) Next we give an example of an $\mathcal{E}$-definable property which guarantees completeness but holds for noncreative sets, in particular for promptly simple sets, and which answers a conjecture on automorphisms.

Theorem 6.3 (Harrington-Soare [32]) *There is an $\mathcal{E}$-definable property T satisfied by a promptly simple set A such that for all W, $T(W)$ implies that $K \leq_{\mathrm{T}} W$.*

This also negatively answers a question raised by Cholak of whether for every promptly simple set A and high c.e. degree $\mathbf{d}$ there exists $B \in \mathbf{d}$ such that A is automorphic to B. Define $T(A)$ by:

$$T(A): \quad (\exists C \supseteq A)(\forall B \subset C)(\exists \text{ a computable set } R)_{(R-C \text{ not c.e.})}$$
$$[B \subset_s C \ . \Longrightarrow . \ A \supseteq (B \cap R) \ \& \ A \not\supseteq^* C \cap R].$$

The previous property $T(A)$ guaranteed such a rapid flow of elements from $C - B$ into A that A was complete. The next property $NL(A)$ is more subtle but guarantees a sufficiently large flow into A so that A is nonlow, but A can still be low$_2$ and hence incomplete. Define $NL(A)$ by:

$$NL(A): \ (\exists C \supseteq A)(\forall B_0, B_1)[$$
$$[B_0 \sqcup B_1 = C] \ \& \ [B_1 \subseteq A] \ . \Longrightarrow (\exists \text{ computable } R)[$$
$$[A \supseteq B_0 \cap R] \ \& \ [(B_1 \cup \overline{C}) \cap R \text{ is not c.e.}]]].$$

Theorem 6.4 *(i)* $(\forall W)[W$ *is low* $\Longrightarrow \neg NL(W)]$.
(ii) $(\exists A)[NL(A)$ & *A is promptly simple and low*$_2$ & $\overline{A}$ *is semi-low*$_{1.5}]$.

7. AUTOMORPHISMS OF $\mathcal{E}$

Automorphisms are useful for two reasons. First, if we are unable to exhibit an $\mathcal{E}$ definition of some property P, then we may be able to produce an automorphism Φ of $\mathcal{E}$ mapping some A with property P to some B with $\neg P$, thereby proving P is undefinable in $\mathcal{E}$. The second use is in the spirit of Klein's *Erlanger Programm,* which is to classify some mathematical object such as a geometry in terms of the properties left invariant under its automorphisms. The first application of the automorphism method was to classify orbits of maximal sets following Klein's program.

To answer a question of Martin and Lachlan, Soare *1974* produced a new method for constructing automorphisms of $\mathcal{E}$ and used it to prove that any two maximal sets are automorphic. The method begins by choosing an appropriate skeleton for the c.e. sets (*i.e.*, one member $U_{g(e)}$ from each class W_e^*) and then, by a fairly complicated construction, building an automorphism of $\mathcal{E}$ which is *effective* in the sense that there is a computable function $h(e)$ such that $\Phi(U_e) =^* W_{h(e)}$. An automorphism Φ is Δ_3^0 if there is a Δ_3^0 permutation h of ω such that $\Phi(W_e) = W_{h(e)}$ for all $e \in \omega$.

Recently, Harrington and Soare [28], and simultaneously Cholak [1] building on some conversations with Harrington, combined the essence of the effective automorphism method with the tree method of Lachlan *1975* to produce a powerful new method for constructing Δ_3^0-automorphisms, and used it to prove, for example, the following. Harrington and Soare [31] then applied the method further in coding of c.e. sets.

Theorem 7.1 (Harrington-Soare [28], Cholak [1]) *For every noncomputable c.e. set A there is a c.e. set B which is high (*i.e., *deg*$(B') = \mathbf{0}''$*) such that A is* Δ_3^0*-automorphic to B.*

Theorem 7.1 asserts that every nontrivial orbit contains a high set. This has some interesting corollaries for noninvariant classes as we shall see in the next section. Theorem 7.1 implies that $Q(A)$ which prevents A from being complete cannot be extended to cause A to be low or even nonhigh. It says if we are willing to extend the target set from the complete degree to the high degrees, then the automorphism builder (BLUE) wins. On the other hand, if we insist on mapping A to a complete set, what stronger hypotheses must we place on A? Cholak, Downey, and Stob [2] proved this result under the stronger hypothesis "A is a promptly simple *set*," and Harrington and Soare extended this with the much weaker hypothesis "of promptly simple degree". They then realized that the essence of the hypothesis is a much weaker promptness property still, which they named "almost promptly simple."

Theorem 7.2 (Harrington-Soare [28]) *If A is any c.e. set which is prompt (*i.e., *of promptly simple degree) or even if A is* almost prompt *then A is automorphic to a complete set.*

8. INVARIANCE

A property of c.e. sets or class $\mathcal{C} \subseteq \mathcal{E}$ is *invariant* if it is invariant under $\mathrm{Aut}(\mathcal{E})$, and *$\mathcal{E}$-definable* if there is a first order property in the language $L(\subset)$ which defines it over $\mathcal{E}$. A class $\mathbf{C}$ of c.e. degrees is *invariant* if it is the class of degrees of sets in some class $\mathcal{C} \subseteq \mathcal{E}$ which is invariant (e.g. if $\mathcal{C}$ is $\mathcal{E}$-definable). For $\mathbf{R}$ the c.e. degrees and $\mathbf{C} \subset \mathbf{R}$ define

$$\mathbf{H}_n = \{\mathbf{a} \in \mathbf{R} : \mathbf{a}^{(n)} = \mathbf{0}^{(n+1)}\},$$

$$\mathbf{L}_n = \{\mathbf{a} \in \mathbf{R} : \mathbf{a}^{(n)} = \mathbf{0}^{(n)}\},$$

$$\mathbf{L}_0 = \{\mathbf{0}\}, \ \mathbf{H}_0 = \{\mathbf{0}'\}, \ \text{ and } \overline{\mathbf{C}} = \mathbf{R} - \mathbf{C}.$$

The degrees in $\mathbf{H}_n$ ($\mathbf{L}_n$) are called *high$_n$* (*low$_n$*) and the high$_1$ (low$_1$) degrees are called *high* (*low*).

A set $A \in \mathcal{E}$ is *maximal* if A is maximal in the inclusion ordering, *i.e.*, $\neg(\exists W)[A \subset^* W \subset^* \omega]$, and a coinfinite A is *atomless* if A has no maximal superset. Martin *1966b* showed that the degrees of maximal sets are exactly $\mathbf{H}_1$, see [78, p. 217]. Lachlan *1968a* and Shoenfield *1976* showed that the degrees of atomless sets are exactly the nonlow$_2$ c.e. degrees $\overline{\mathbf{L}}_2$, see [78, p. 231]. Thus, $\mathbf{H}_1$ and $\overline{\mathbf{L}}_2$ are invariant. For the trivial jump classes corresponding to $n = 0$, $\mathbf{L}_0$ $\overline{\mathbf{L}}_0$, and $\mathbf{H}_0$ are invariant, while $\overline{\mathbf{H}}_0$ is noninvariant by Theorem 7.2. These three results and his work at the time on projective determinacy led Martin to make the following conjecture.

Conjecture 8.1 (Martin's Invariance Conjecture) *Among all the jump classes $\mathbf{H}_n$ and $\mathbf{L}_n$ for $n > 0$, and their complements $\overline{\mathbf{H}}_n$ and $\overline{\mathbf{L}}_n$, the invariant classes are exactly $\mathbf{H}_{2n-1}$ and $\overline{\mathbf{L}}_{2n}$.*

The following immediate corollary of the previous theorem confirms the Invariance Conjecture prediction for the *downward* closed jump classes for $n > 0$.

Corollary 8.2 *For all $n > 0$ the* downward *closed jump classes of c.e. degrees $\mathbf{L}_n$ and $\overline{\mathbf{H}}_n$ are noninvariant.*

For the *upward* closed classes $\mathbf{H}_n$ and $\overline{\mathbf{L}}_n$, $n > 0$, after the discovery of invariance of $\mathbf{H}_1$ and $\overline{\mathbf{L}}_2$, attention has been focused on $\overline{\mathbf{L}}_1$ because of the important role played by the low c.e. sets.

Theorem 8.3 (Harrington-Soare [33]) *(i) Every c.e. set A which is locally low is automorphic to a low set B.*

(ii) There is a c.e. set $D \in low_2\text{-}low_1$ such that every c.e. $A \leq_T D$ is locally low.

Corollary 8.4 *$\overline{\mathbf{L}}_1$ is not an invariant class of c.e. degrees.*

Proof. Suppose it is invariant. Choose D as in (ii). Now every $A \equiv_T D$ is locally low and hence by (i) A is automorphic to a low set B. ∎

Department of Mathematics
University of Chicago

REFERENCES

1. [Cholak, *1995*] , P. A. Cholak, Automorphisms of the Lattice of Recursively Enumerable Sets, *Memoirs of the Amer. Math. Soc.* **113**, 1995.
2. [Cholak-Downey-Stob, *1992*] P. Cholak, R. Downey, M. Stob, Automorphisms of the lattice of recursively enumerable sets: Promptly simple sets, *Trans. Amer. Math. Soc.* **332** (1992), 555–570.
3. [Church, *1936*] A. Church, An unsolvable problem of elementary number theory, *American J. of Math.*, **58** (1936), 345-363.
4. [Davis, *1958*] M. Davis, *Computability and Unsolvability,* Mc-Graw-Hill, New York, 1958; reprinted in 1982 by Dover Publications.
5. [Davis, *1965*] M. Davis, (ed.), *The Undecidable. Basic Papers on Undecidable Propositions, Unsolvable Problems, and Computable Functions*, Raven Press, Hewlett, New York, 1965.
6. [Davis, *1982*] M. Davis, Why Gödel did not have Church's Thesis, *Information and Control* **54** (1982), 3–24.
7. [Davis, *1988*] M. Davis, Mathematical logic and the origin of modern computers, In: Herken, *1988*, 149–174.
8. [Dedekind, *1888*] R. Dedekind, Was sind und was sollen die Zahlen? Braunschweig, (6th ed. 1930). Also in *Dedekind Gesammelte Mathematische Werke*, vol. III, pp. 335–391. Eng. tr. by Beman, The nature and meaning of numbers. loc. cit., pp. 31–105. (English trans. in Dedekind, *Essays on the Theory of Numbers*, Chicago, Open Court, 1901, 29–115.)
9. [Downey and Stob, *1992*] R. Downey and M. Stob, Automorphisms of the lattice of recursively enumerable sets : Orbits *Advances in Math.*, **92** (1992), 237-265.
10. [Feferman, *1988*] S. Feferman, Turing in the land of $O(z)$, In: Herken, *1988*, 113–147.
11. [Feferman, *1992*] S. Feferman, Turing's "Oracle": From absolute to relative computability–and back, In: *The Space of Mathematics*, J. Echeverria, et. al. eds., Walter de Gruyter, Berlin, 1992, pp. 314-348.

12. [Gandy, *1980*] R. Gandy, Church's thesis and principles for mechanisms, In: *The Kleene Symposium*, North-Holland, (1980), 123–148.
13. [Gandy, *1988*] R. Gandy, The confluence of ideas in 1936, In: Herken, *1988*, 55–111.
14. [Gödel, *1931*] K. Gödel, Über formal unentscheidbare sätze der Principia Mathematica und verwandter systeme. I, *Monatsch. Math. Phys.* **38** (1931) 173-178. (English trans. in Davis *1965*, 4–38, and in van Heijenoort, 1967, 592–616.
15. [Gödel, *1934*] K. Gödel, On undecidable propositions of formal mathematical systems, Notes by S. C. Kleene and J. B. Rosser on lectures at the Institute for Advanced Study, Princeton, New Jersey, 1934, 30 pp. (Reprinted in Davis *1965* [5, 39–74]).
16. [Gödel, *1936*] K. Gödel, On the length of proofs, In: Gödel *1986*, 397–399; reprinted in Davis *1965*, 82–83, with a *Remark* added in proof [of the original German publication].
17. [Gödel, *193?*] K. Gödel, Undecidable diophantine propositions, In: Gödel *1995*, 156–175.
18. [Gödel, *1946*] K. Gödel, Remarks before the Princeton bicentennial conference of problems in mathematics, 1946. Reprinted in: Davis *1965* [5], pp. 84–88.
19. [Gödel, *1951*] K. Gödel, Some basic theorems on the foundations of mathematics and their implications, In: Gödel *1995*, 304–323. (This was the Gibbs Lecture delivered by Gödel on December 26, 1951 to the Amer. Math. Soc.)
20. [Gödel, *1958*] K. Gödel, Über eine bisher noch nicht benütze Erweiterung des finiten Standpunktes, *Dialectica* **12** (1958), 280–287. (German and English trans. in Godel *1986*, 240–251, with introductory note by A. S. Troelstra, 217–241.)
21. [Gödel, *1964*] K. Gödel, Postscriptum to Gödel *1931*, written in 1946, printed in Davis *1965*, 71–73.
22. [Gödel, *1972*] K. Gödel, Some remarks on the undecidability results, (written in 1972); In: Gödel *1990*, 305–306.
23. [Gödel, *1986*] K. Gödel, *Collected works Volume I: Publications 1929–36*, S. Feferman et. al., editors, Oxford Univ. Press, Oxford, 1986.
24. [Gödel, *1990*] K. Gödel, S. Feferman et. al., editors, *Collected works Volume II: Publications 1938–1974*, Oxford Univ. Press, Oxford, 1990.
25. [Gödel, *1995*] K. Gödel, *Collected works Volume III: Unpublished essays and lectures*, S. Feferman et. al., editors, Oxford Univ. Press, Oxford, 1995.
26. [Herken, *1988*] R. Herken (ed.), *The Universal Turing Machine: A Half-Century Survey*, Oxford Univ. Press, 1988.
27. [Harrington and Soare, *1991*] L. Harrington and R. I. Soare, Post's Program and incomplete recursively enumerable sets, *Proc. Natl. Acad. of Sci. USA*, **88** (1991), 10242–10246.
28. [Harrington and Soare, *1996a*] L. Harrington and R. I. Soare, The Δ^0_3-Automorphism Method and Noninvariant Classes of Degrees, *Jour. Amer. Math. Soc.*, to appear.
29. [Harrington and Soare, *1996b*] L. Harrington and R. I. Soare, Dynamic properties of computably enumerable sets, In: *Computability, Enumerability, Unsolvability: Directions in Recursion Theory* (eds. S. B. Cooper, T. A. Slaman, S. S. Wainer), Proceedings of the Recursion Theory Conference, University of Leeds, July, 1994, London Math. Soc. Lecture Notes Series, Cambridge University Press, 1996.
30. [Harrington and Soare, *1996c*] L. Harrington and R. I. Soare, Definability, automorphisms, and dynamic properties of computably enumerable sets, *Bulletin of Symbolic Logic* **2** (1996), to appear.
31. L. Harrington and R. I. Soare, Codable sets and orbits of computably enumerable sets, J. Symbolic Logic, to appear.
32. L. Harrington and R. I. Soare, Definable properties of the computably enumerable sets, in preparation.
33. L. Harrington and R. I. Soare, Martin's Invariance Conjecture and low sets, in preparation.
34. [Hilbert, *1899*] D. Hilbert, *Grundlagen der Geometrie,* 7th ed., Tuebner-Verlag, Leipzig, Berlin, 1930.

35. [Hilbert, *1904*] D. Hilbert, Über die Grundlagen der Logik und der Arithmetik, In: *Verhandlungen des Dritten Internationalen Mathematiker-Kongresses in Heidelberg vom 8. bis 13. August 1904*, pp. 174–185, Teubner, Leipzig, 1905. Reprinted in van Heijenoort *1967*, 129–138.
36. [Hilbert, *1918*] D. Hilbert, Axiomatisches Denken, *Math. Ann.* **78** (1918), 405–415.
37. [Hilbert, *1926*] D. Hilbert, Über das Unendliche, *Mathematische Annalen* **95** (1926), 161–190. (English trans. in van Heijenoort *1967*, 367–392.)
38. [Hilbert and Ackermann, *1928*] D. Hilbert and W. Ackermann, *Grundzüge der theoretischen Logik*, Springer, Berlin, 1928 (English translation of 1938 edition, Chelsea, New York, 1950).
39. [Kleene, *1936*] S. C. Kleene, General recursive functions of natural numbers, *Math. Ann.* **112** (1936), 727–742.
40. [Kleene, *1936b*] S. C. Kleene, λ-definability and recursiveness, *Duke Math. J.* **2** (1936), 340–353.
41. [Kleene, *1936c*] S. C. Kleene, A note on recursive functions, *Bull. A.M.S.* **42** (1936), 544–546.
42. [Kleene, *1938*] S. C. Kleene, On notation for ordinal numbers, *J. Symbolic Logic*, **3** (1938), 150–155.
43. [Kleene, *1943*] S. C. Kleene, Recursive predicates and quantifiers, *Trans. A.M.S.* **53** (1943), 41–73.
44. [Kleene, *1944*] S. C. Kleene, On the forms of the predicates in the theory of constructive ordinals, *Amer. J. Math.* **66** (1944), 41–58.
45. [Kleene, *1952*] S. C. Kleene, *Introduction to Metamathematics*, Van Nostrand, New York (1952). Ninth reprint 1988, Walters-Noordhoff Publishing Co., Groningën and North-Holland, Amsterdam.
46. [Kleene, *1952b*] S. C. Kleene, Recursive functions and intuitionistic mathematics, In: Proceedings of the International Congress of Mathematicians, Cambridge, Mass., U.S.A., Aug. 30–Sept. 6, 1950, Amer. Math. Soc. (Providence, R.I.), 1952, Vol. 1, 679–685.
47. [Kleene, *1955*] S. C. Kleene, Arithmetical predicates and function quantifiers, *Trans. A.M.S.* **79** (1955), 312–340.
48. [Kleene, *1955b*] S. C. Kleene, On the forms of the predicates in the theory of constructive ordinals (second paper), *Amer J. Math.* **77** (1955), 405–428.
49. [Kleene, *1955c*] S. C. Kleene, Hierarchies of number-theoretical predicates, *Bull. A.M.S.* **61** (1955), 193–213.
50. [Kleene, *1959*] S. C. Kleene, Recursive functionals and quantifiers of finite type I, *Trans. A.M.S.* **91** (1959), 1–52.
51. [Kleene, *1962*] S. C. Kleene, Turing-machine computable functionals of finite types I, Logic, methodology, and philosophy of science: Proceedings of the 1960 international congress, Stanford University Press, 1962, 38–45.
52. [Kleene, *1962b*] S. C. Kleene, Turing-machine computable functionals of finite types II, *Proc. of the London Math. Soc.* **12** (1962), no. 3, 245–258.
53. [Kleene, *1963*] S. C. Kleene, Recursive functionals and quantifiers of finite type II, *Trans. A.M.S.* **108** (1963), 106–142.
54. [Kleene, *1981*] S. C. Kleene, Origins of recursive function theory, *Annals of the History of Computing*, **3** (1981), 52–67.
55. [Kleene, *1981b*] S. C. Kleene, The theory of recursive functions, approaching its centennial, Bull. A.M.S. (n.s.) **5**, (1981), 43–61.
56. [Kleene, *1981c*] S. C. Kleene, Algorithms in various contexts, *Proc. Sympos. Algorithms in Modern Mathematics and Computer Science* (dedicated to Al-Khowarizimi) (Urgench, Khorezm Region, Uzbek, SSSR, 1979), Springer-Verlag, Berlin, Heidelberg and New York, 1981.
57. [Kleene, *1987*] S. C. Kleene, Reflections on Church's Thesis, *Notre Dame Journal of Formal Logic,* **28** (1987), 490–498.

58. [Kleene, *1987b*] S. C. Kleene, Gödel's impression on students of logic in the 1930's, In: P. Weingartner and L. Schmetterer (eds.), *Gödel Remembered*, Bibliopolis, Naples, 1987, 49–64.
59. [Kleene, *1988*] S. C. Kleene, Turing's analysis of computability, and major applications of it, In: Herken *1988*, 17–54.
60. [Kleene-Post, *1954*] S. C. Kleene and E. L. Post, The upper semi-lattice of degrees of recursive unsolvability, *Ann. of Math.* **59** (1954), 379–407.
61. [Penrose, *1994*] R. Penrose, *Shadows of the Mind*, Oxford Univ. Press, Oxford, 1994.
62. [Peano, *1889*] G. Peano, *Arithmetices Principia, Nova Methodo Exposita*, Turin Bocca xvi+20 pp., 1889. English trans. In: van Heijenoort, 1967, 83–97.
63. [Normann, *1980*] D. Normann, Recursion on countable functionals, Lecture Notes in Mathematics, No. 811, Springer-Verlag, Heidelberg New York, 1980.
64. [Peirce, 1960] Charles S. Peirce, Book II. Speculative Grammar, In: *Volume II: Elements of Logic*, In: *Collected Papers of Charles Sanders Peirce* C. Hartshorne and P. Weiss (Eds.), The Belknap Press of Harvard Univ. Press, Cambridge, Mass. and London, England, 1960.
65. [Post, *1936*] E. L. Post, Finite combinatory processes–formulation I, *J. Symbolic Logic* **1** (1936) 103–105. Reprinted in Davis *1965*, 288–291.
66. [Post, *1941*] E. L. Post, Absolutely unsolvable problems and relatively undecidable propositions: Account of an anticipation. (Submitted for publication in 1941.) Printed in Davis *1965*, 340–433.
67. [Post, *1943*] E. L. Post, Formal reductions of the general combinatorial decision problem, *Amer. J. Math.* **65** (1943), 197–215.
68. [Post, *1944*] E. L. Post, Recursively enumerable sets of positive integers and their decision problems, *Bull. Amer. Math. Soc.* **50** (1944), 284–316. Reprinted in Davis *1965*, 304–337.
69. [Post, *1947*] E. L. Post, Recursive unsolvability of a problem of Thue, *J. Symbolic Logic* **12** (1947), 1–11. (Reprinted in Davis *1965*, 292–303.)
70. [Post, *1948*] E. L. Post, Degrees of recursive unsolvability: preliminary report (abstract), *Bull. Amer. Math. Soc.* **54** (1948), 641–642.
71. [Putnam, *1995*] H. Putnam, Review of Penrose *1994*, *Bull. A.M.S.* **32** (1995), 370–373.
72. [Sacks, *1990*] G. E. Sacks, *Higher Recursion Theory*, Springer-Verlag, Heidelberg New York, 1990.
73. [Shoenfield, *1967*] J. R. Shoenfield, *Mathematical Logic*, Addison-Wesley, Reading, Mass. (1967), 344 pp.
74. [Shoenfield, *1991*] J. R. Shoenfield, Recursion Theory, *Lecture Notes in Logic*, Springer-Verlag, Heidelberg New York, 1991.
75. [Shoenfield, *1995*] J. R. Shoenfield, The mathematical work of S. C. Kleene, *Bulletin of Symbolic Logic* **1** (1995), 8–43.
76. [Sieg, *1994*] W. Sieg, Mechanical procedures and mathematical experience, In: A. George (ed.), *Mathematics and Mind*, Oxford Univ. Press, 1994.
77. [Sieg and Byrnes, *1995*] W. Sieg and J. Byrnes, K-Graph machines: generalizing turing's machines and arguments, preprint, 1995.
78. [Soare, *1987*] R. I. Soare, Recursively Enumerable Sets and Degrees: A Study of Computable Functions and Computably Generated Sets, Springer-Verlag, Heidelberg, 1987.
79. [Soare, ta] R. I. Soare, The structure of computably enumerable objects as sets, In: *Handbook of Computability Theory*, North-Holland, Amsterdam, in preparation.
80. [Soare, ta] R. I. Soare, Computability and recursion, *Bulletin of Symbolic Logic*, to appear in 1996.
81. [Tamburrini, *1995*] G. Tamburrini, Mechanistic theories in cognitive science: The import of Turing's Thesis, Proceedings of the 10th International Congress of Logic, Methodology, and Philosophy of Science, August 19–25, 1995, Florence, Italy.

82. [Turing, *1936*] A. M. Turing, On computable numbers, with an application to the Entscheidungsproblem, *Proc. London Math. Soc.* ser. 2 **42** (Parts 3 and 4) (1936) 230–265; [Turing, *1937*] A correction, ibid. **43** (1937), 544–546.
83. [Turing, *1937b*] A. M. Turing, Computability and λ-definability, *J. Symbolic Logic*, **2** (1937), 153–163.
84. [Turing, *1939*] A. M. Turing, Systems of logic based on ordinals, *Proc. London Math. Soc.* **45** Part 3 (1939), 161–228; reprinted in Davis [1965], 154–222.
85. [Turing, *1948*] A. M. Turing, Intelligent machinery, In: *Machine Intelligence* **5**, 3–23. (Written in September, 1947 and submitted to the National Physical Laboratory in 1948.)
86. [Turing, *1949*] A. M. Turing, Text of a lecture by Turing on June 24, 1949, In: F. L. Morris and C. B. Jones, "An early program proof by Alan Turing", Annals of the History of Computing **6** (1984), 139-143.
87. [Turing, *1950*] A. M. Turing, Computing machinery and intelligence, *Mind* **59** (1950) 433–460.
88. [Turing, *1950b*] A. M. Turing, The word problem in semi-groups with cancellation, *Ann. of Math.* **52** (1950), 491–505.
89. [Turing, *1953*] A. M. Turing, Solvable and unsolvable problems, *Science News* **31** (1954), 7–23.
90. [Turing, *1986*] A. M. Turing, Lecture to the London Mathematical Society on 20 February 1947, In: B. E. Carpenter and R. W. Doran, eds., *A. M. Turing's ACE Report of 1946 and Other Papers*, Cambridge Univ. Press, 1986, 106–124.
91. [van Heijenoort, *1967*] J. van Heijenoort (ed.), *From Frege to Gödel, A Sourcebook in Mathematical Logic, 1879–1931,* Harvard Univ. Press, Cambridge, Mass., 1967.

GUGLIELMO TAMBURRINI*

MECHANISTIC THEORIES IN COGNITIVE SCIENCE: *THE IMPORT OF TURING'S THESIS*

1. INTRODUCTION

The current controversies about *mechanism* (or *computationalism*) in Cognitive Science (CS, for short) cannot be properly understood without having a relatively clear idea of what mechanism in CS is. Turing's thesis (TT, for short) contributes to clarifying the meaning of the expressions 'mechanistic theory' and 'mechanism'. This is the basic view, whose motivations are made explicit in section 2, of the import of TT for the philosophy of CS. In particular, TT enables one to set a necessary condition (hereinafter referred to as C_f) on theories in CS to count as mechanistic. This is a functional condition requiring that the laws accounting for the input/output behavior of the subjects in the domain of the theory be expressible in terms of Turing computable functions.

Various notions of theory in CS can be viewed as ramifications originating from C_f by the introduction of further constraints. The mechanistic theories satisfying additional constraints can be distinguished into different sorts of *restricted* mechanistic theories. These include Marr's competence theories, Newell and Simon's microtheories of problem solving and, with certain qualifications, connectionist theories. Also Newell's proposal that theories of cognitive tasks are to be constrained in ways depending on the functional architecture of the human cognitive system gives rise to a specialized notion of mechanistic theory. Condition C_f, and a related taxonomy of restricted mechanistic theories are discussed in section 3.

Various conceptual and empirical issues can be brought to bear on a discussion of the epistemological significance of C_f. In my view, a central issue concerns the "no homunculi" requirement (H) for theories in CS: A non-question-begging theory of intelligent behavior must not presuppose hidden homunculi for the execution of the tasks that the theory is supposed to account

*I am especially grateful to Wilfried Sieg for many invaluable suggestions and comments, and stimulating discussions —for over fifteen years now, and in many different places— on various issues bearing on the problems addressed here. I am glad that I have had the privilege to have discussed this paper with Robin Gandy, before his recent death. I wish to thank Aldo Aiello, Pantaleo Aloisio, Roberto Cordeschi, Marcello Frixione, Salvatore Guccione, Leen Spruit, Settimo Termini, and Giuseppe Trautteur for their helpful comments on an earlier draft. Materials from this paper were also presented in talks at Università di Salerno, Dipartimento di Filosofia, and Istituto di Cibernetica C.N.R., Reparto di Informatica. I thank each of these audiences.

M.L. Dalla Chiara et al. (eds.), Logic and Scientific Methods, 239–257.

for. A theory satisfying C_f meets (H) at least as far as input/output behaviors are concerned, since the mechanically calculable functions mentioned in C_f do not require intelligence for their execution. But according to a widespread view, discussed in section 4, also the converse is true: a theory in CS meets the no-homunculi requirement for explanations of intelligent behaviors only if it satisfies C_f. A basic claim of mechanism is construed as a thesis reflecting this view, and expressing a broad necessary condition on the form of theories in CS: adequate accounts of human cognitive tasks can only be given by theories satisfying C_f.

Challenging this claim requires showing that no adequate mechanistic theory for some cognitive task can be given, and that an alternative theoretical approach is viable. But the latter problem presupposes, in turn, showing that an adequate explanatory basis—meeting requirement (H), but not included in the class of functions and processes with which TT is concerned—is available. I examine in section 5 arguments by Gödel, Kreisel, and Penrose, that are precisely concerned with the problem of isolating cognitive tasks that cannot be accounted for in terms of mechanical procedures, or isolating alternative explanatory bases for CS. The latter arguments contribute to revealing the dogmatic character of the assertion that precise and genuinely explanatory theories of cognition can only be mechanistic, but do not make available any alternative conceptual framework to mechanistic theorizing.

2. TT AND CS: COORDINATES OF THE PROBLEM

1. *Any mechanically calculable function can be computed by a Turing machine.*

This is a fairly standard version of the claim that is now known as Turing's thesis (TT), which is equivalent to another claim, independently formulated in Church (1936), and now known as Church's thesis. Another familiar formulation of TT is the following:

2. *Any effectively calculable function can be computed by a Turing machine.*

From Turing's standpoint, there is only a verbal difference between these formulations. This can be seen from his retrospective account of what was achieved in Turing (1936):

> A function is said to be "effectively calculable" if its values can be found by some purely mechanical process. Although it is fairly easy to get an intuitive grasp of this idea it is nevertheless desirable to have some more definite, mathematically expressible definition... It was stated above that "a function is effectively calculable if its values can be found by some purely mechanical process". We may take this statement literally, understanding by a purely mechanical process one which could be carried out by a machine. It is possible to give a

mathematical description, in a certain normal form, of the structures of these machines.[1]

Clearly, Turing regarded the expression 'effectively calculable' as synonymous with 'mechanically calculable', and the class of functions computable by Turing machines as an adequate characterization of the mechanically calculable functions. This is a crucial qualification on the intended scope of Turing's work, in view of the fact that different senses of 'effectiveness' have emerged from conceptual distinctions drawn or properly sedimented only after his 1936 article was completed.

A well-known example of such distinctions, that I mention very briefly here and will turn to consider again in section 5, concerns the class of functions whose values can be determined by means of a finite *constructive* (more specifically, *intuitionistically acceptable*) procedure. Gödel suggested that Turing's analysis does not provide evidence for the inclusion of this class of number-theoretic functions in the class of total functions computable by a Turing machine, for the processes in terms of which these constructive functions and the mechanically calculable functions are respectively defined are not constrained in the same ways. In particular, Gödel pointed out that the execution of intuitionistic finite procedures may involve the use of mathematical terms on the basis of their meaning[2] whereas, according to Turing's analysis, a human agent executing a mechanical stepwise procedure is required only to possess the capability of recognizing and behaviorally reacting to the shape and spatial distribution of certain kinds of symbolic configurations.[3]

In view of these conceptual distinctions, and with the benefit of hindsight, I refer to (1) as the standard formulation of TT, sharply distinguishing between TT and its *variants*, which are concerned with classes of functions whose values can be determined by other sorts of allegedly effective procedures. One such variant is just Church's thesis for intuitionistic mathematics (CTI), which asserts that any total number-theoretic function whose values can be determined by an intuitionistically acceptable, finite procedure is general recursive. In addition to the notion involved in the formulation of CTI, other notions of "humanly" or "physically" effective processes have been introduced, and attempts to formulate appropriate variants of TT have been made.[4]

The significance of the distinction between TT and its variants in philosophical discussions of the mind-machine problem can hardly be overemphasized. Various senses of 'effectiveness' have been often conflated in expositions of

---

[1] Turing (1939), see Davis (1965, p. 160).

[2] See Gödel (1986, p. 370), and Gödel (1990, pp. 273–274).

[3] This assumption is made also in Gandy (1980) which provides new substantive evidence for TT. Gandy points out that some of Turing's restrictive conditions on human agents executing a mechanical procedure are not directly satisfied by some parallel computing devices. He replaces these conditions by weaker requirements, and shows that any device satisfying them can only determine the values of Turing computable functions.

[4] See, e.g., Kreisel (1972) and (1974), and the idea of non-uniform effectiveness in Kalmár (1959).

the meaning and scope of TT. And in turn, these extended interpretations of TT have misled some authors into deriving limitations to human mathematical powers solely from the incompleteness and undecidability theorems, whose general formulation makes an essential use of TT.[5] Gödel's observation that these results "do not establish any bounds for the powers of human reason"[6] can be viewed as a warning against these misinterpretations. Gödel, however, in contrast with the standpoint adopted here, claimed that the aims and the objective achievement of Turing's work do not coincide: Turing's work provides unquestionable evidence for TT, but he aimed (unsuccessfully) at establishing more than the equivalence of mechanically calculable and Turing computable functions, as he meant "to show that mental procedures cannot go beyond mechanical procedures."[7] A similar distinction between objective achievement and aim of Turing's work is drawn in Kreisel (1972, p. 318). This interpretive thesis conflicts with the passage in Turing (1939) quoted above, and is not supported, I believe, by other remarks of Turing[8]. However, I need not address this problem here, since it is immaterial to the main point I wish to make: both Gödel and Kreisel endorse the view that TT is what Turing's argument provides significant evidence for, and that TT and its variants are to be sharply distinguished.

The present interest of the distinction between TT and its variants is that they play different roles in discussions of mechanism in CS. On the one hand, insofar as it circumscribes the class of mechanically calculable functions, TT determines the horizon of the mathematical functions which can be used in mechanistic investigations of human cognition, and thus can contribute to clarifying the meaning of the very term 'mechanism', as the latter is used in CS. On the other hand, some variants of TT, which concern (functions definable by) classes of mental procedures whose extensional equivalence to mechanical ones is not obvious, seem to lend some form of support to mechanism in CS. CTI is a case in point: in view of Gödel's remarks on intuitionism and finitism mentioned above, the correctness of CTI entails that the execution of certain mental procedures, involving an understanding of the meaning of mathematical terms, does not enable one to go beyond the extensional mathematical powers of (Turing) machines. But a more careful assessment of the consequences for mechanism in CS of such variants of TT *presupposes* that the meaning of the term 'mechanism' be made relatively clear. It is my contention that TT can play a substantive role in such endeavour of conceptual clarification: since TT is a claim about a particular class of functions, it is used

[5]Classical examples of this line of argument can be found in a posthumously published article of Post (see Davis (1965, p. 418)), and in Kalmár (1959).

[6]Gödel (1986, pp. 369–371).

[7]Gödel (1990, p. 306).

[8]For an extensive discussion of the historical background, intended scope, and objective achievement of Turing's analysis, which is coherent with the standpoint adopted in this paper, see Sieg (1994). See also Sieg and Byrnes (1994, p. 7), for a discussion of Turing's notion of "state of mind" as the possible source of Gödel's misinterpretation.

here to formulate a *functional* condition on theories in CS; and a basic claim of mechanism is identified with the thesis that adequate theories of cognitive tasks must satisfy this condition. In the next section, I turn to examine the functional aspects of theories that are often regarded as paradigmatic examples of mechanistic theorizing in CS. The aim is that of arriving—by factoring out incompatible requirements and by appeal to TT—at a kernel condition on a theory in CS to count as mechanistic.

3. ON THE NOTION OF MECHANISTIC THEORY IN CS

(a) *Newell and Simon's theories of problem solving.* It is appropriate to begin this selective survey, in view of their pervasive influence on the development of AI and CS, from Allen Newell and Herbert Simon's pioneering investigations into human problem solving. These investigations, Newell and Simon argued, provide evidence that computer programs can be designed to simulate faithfully the problem solving behavior of subjects performing particular problem solving tasks. The empirical data for both designing and testing simulation programs were mainly provided by verbal reports of subjects asked to vocalize their conscious thinking activity when performing problem solving tasks.[9] Since problem solving behavior is not invariant over individuals and tasks, the range of these parameters is to be appropriately restricted to special classes of human subjects and problem solving tasks in order to achieve faithful simulations.

Each simulation program, Newell and Simon claimed, can be regarded as a *restricted empirical* theory (or *micro*theory) expressed as a computer program. It is a microtheory because its intended domain is limited to narrow classes of subjects and problem solving tasks. It makes empirical claims because the program can be used to derive predictions about the objects in its domain: which process they will execute next, which intermediate and final problem states, if any, will be reached. Expressibility as a mechanical procedure taking the form of a program in some computer language is a distinguishing feature of Newell and Simon's microtheories. Under a suitable coding of inputs and outputs, such a mechanical procedure can be viewed as defining a partial function over the natural numbers. And, according to TT, this is a Turing computable function. Thus, Newell and Simon's view entails the following functional condition on microtheories of problem solving: *a microtheory accounting for the input/output behavior of a limited class of subjects engaged in a particular problem solving task can be expressed under the form of a Turing computable function.*

(b) *Marr's theory of vision.* This functional condition on microtheories of problem solving is not applicable to Marr's theory of human vision, even

[9]For a recent discussion of verbal report analysis, see Ericsson and Simon (1993). A classic source is Newell and Simon (1972). For a comprehensive critical discussion of this methodology applied to the testing of simulation programs, see Cordeschi (1984).

though the latter claims that such a theory must provide an appropriate functional characterization of the task executed by the human visual system. Unlike Newell and Simon, Marr requires that this functional characterization be made abstracting from the ways in which human beings actually do (or, in view of their *limited* cognitive capacities, even fail to) accomplish particular instances of this task. An idealization from human performance similar to that involved in isolating the domain of *competence* theories in linguistics is, in Marr's view, of paramount importance for establishing a sound computational framework for the study of vision. Thus, the intended domain of a competence theory of vision *à la* Marr is formed by *idealized* human subjects—unhindered by memory limitations, distractions, and errors—whose capacities are only imperfectly approximated in actual human beings. According to Marr (1982), *such a theory must specify the functional transformations characterizing the i/o behavior of the idealized subjects correctly executing the task of vision, in addition to showing that a mechanical procedure is sufficient to carry out these mappings* (and thus, by TT, that these mappings are Turing computable).

Only when this functional characterization is provided, the theory can be further developed in order to address the problems of human performance. This requires undertaking, in Marr's terminology, an *algorithmic* (or, equivalently, *procedural*) level analysis of the task of vision, by asking how and within what limits a human being succeeds in carrying out, by means of appropriate representations and algorithms, the mechanically calculable correspondences isolated at the functional level.

Newell and Simon question the usefulness of a sharp distinction between competence and performance, and the related separation between a functional and an algorithmic level of analysis: in order to arrive at *predictive* theories of (problem solving) behavior, one has to take into account from the very beginning the overall limitations and individual variability in representation and information processing capacities. It is this different view of the significance of the competence/performance distinction—and more generally the issue of the appropriate idealizations that are needed in formulating theories of cognitive tasks—that chiefly sets apart Marr's and Newell and Simon's views.

(c) *another functional condition.* This divergence between two influential approaches to theory construction in CS may be regarded as a major obstacle towards the formulation of a general functional condition on mechanistic theories of cognitive tasks. However, I submit that when Marr's position is formulated with sufficient care, no evident conflict arises between his view and Newell and Simon's claim that a sharp competence/performance distinction is not useful in problem solving domains. A competence level analysis is, in Marr's view, most appropriate for "modularly organized" aspects of cognition: cognitive tasks involving the interaction of several cognitive modules may require a rather different theoretical treatment. He argued that vision, more than problem solving, can be regarded as a relatively "modular" component of human cognition: vision can deliver information about shape and position of objects in the visual field from retinical images, without interacting with the cognitive "modules" processing and storing information about the function of those objects. Marr cautioned that this sort of modularity may not be found in many "higher level" cognitive tasks: raising the problem of an abstract competence analysis for them may be premature or altogether impossible, as they may involve the interaction of several cognitive processes. In this connection, Marr (1977) introduced a distinction between type-1 and type-2 information processing theories. The latter concern non-modular tasks, and are not amenable to the sort of idealizations involved in setting up a type-1 competence theory: consideration of *how* the processes involved into a "type-2 domain" unfold and interact is the only way of arriving at a theory making substantive empirical predictions. He also pointed out that the type-1–type-2 distinction "is not a pure dichotomy, and there is a spectrum of possibilities between them," and this seems to suggest that competence theories are situated at one extreme of this spectrum.

The upshot of these considerations is that if one neither requires nor excludes that certain idealizations be made in characterizing the subjects in the domain of a theory, then a common core of Marr's and Newell and Simon's functional requirements can be isolated:

(C_f) The laws accounting for the input/output behavior of the class of subjects in the domain of a theory for a cognitive task t are expressible under the form of Turing computable functions.

I regard C_f as expressing a necessary condition on theories in CS to count as mechanistic. The ensuing discussion indicates that the boundary set by C_f accomodates, in addition to Marr's and Newell and Simon's theories, various other theoretical constructs in CS.

(d) *algorithmic level of analysis.* Pylyshyn (1984, pp. 87–88) maintains that i/o predictions are the weakest empirical claims that a theory of a cognitive task, unfettered by the methodological restrictions of behaviorism, has to make about the subjects in its domain: such a theory should also incorporate predictions about the mental states and processes that are put to work in achieving determinate i/o behaviors.

This sort of predictions is made by a microtheory of problem solving about the restricted classes of subjects in its domain. And clearly, predictions about representation structures and information processing methods is also what Marr expected from theories developed at the algorithmic level of analysis. One can thus distinguish, within the class determined by C_f, theories that merely account for i/o behaviors, such as Marr's competence theory of vision, and theories accounting for some of the cognitive processes and representation structures of the subjects in their domain by means of specific algorithmic processes that in principle enable one to compute the functions required by C_f. The former may be called *pure* mechanistic theories, and the latter *procedurally constrained* mechanistic theories.

(e) *architectural constraints.* Pylyshyn (1984) distinguishes also between two sources of procedural constraints to be embedded into theories of cognitive tasks. There are aspects of cognitive process that are task-dependent, insofar as they vary as a function of the subject's current goals and beliefs, and aspects that are not. The latter, he suggests, constitute the fixed repertoire of elementary cognitive processes and structures the human cognitive system makes available for the execution of any task whatsoever.

More recently, Newell (1990) has suggested that the methods made available over the last two decades by experimental psychology and neurophysiology make possible the formulation of empirical hypotheses on the fixed *architecture* of the human cognitive system, which enable one to isolate task-invariant procedural constraints to be embedded into each single-task (micro-)theory. Hypotheses on this fixed structure are testable, Newell claimed, primarily by means of experiments on immediate-response tasks (involving only about one second response time): just because there is time only to compose few elementary operations to attain the appropriate response behavior, these are the closest observatory on cognitive architecture.[10]

We have thus a further subclass of procedurally constrained, mechanistic theories: this is the class of *architecturally constrained* mechanistic theories, embodying empirical hypotheses about the fixed functional architecture of the human cognitive system. Clearly, one of the chief motivations for modeling cognitive function by means of brain-style processing units, so-called artificial "neurons", is that of arriving at appropriate procedural and architectural constraints. But do connectionist theories of cognitive tasks invariably satisfy C_f? It is this question that we now turn to consider.

(f) *connectionism.* Connectionist theories account for the behavior of the subjects in their domain in terms of the functions regulating the spreading of activation in networks of artificial neurons, and ultimately transforming the input vectors into output vectors. But are these functions necessarily Turing computable? In this regard, Smolensky (1988, p. 3) remarks:

> There is a reasonable chance that connectionist models will lead to the development of new somewhat-general-purpose self-programming, massively parallel analog computers, and a new theory of analog parallel computation: they may possibly even challenge the strong construal of Church's Thesis as the claim that the class of well-defined computations is exhausted by those of Turing machines.

Firstly, one has to notice that Smolensky's suggestion does not conflict with TT, and is neither supported nor excluded by Turing's analysis: TT is a claim about the class of functions calculable by *discrete* mechanical procedures, whereas analog computations may involve in essential ways *continuous* physical processes. And indeed, Smolensky is careful to point out that a prospective

[10]See Newell (1990), p. 238f, for a discussion of this testing methodology, and concrete examples of simulation programs of immediate response tasks embedded in SOAR.

theory of parallel analog computation based on an interpretation of artificial neurons as analog computers may only challenge a *strong construal* of Church's thesis (that is, in the terminology adopted here, some particular *variant* of TT or, equivalently, of Church's thesis, concerning analog computable functions).

Secondly, there are recent results (see, e.g., Siegelmann and Sontag (1994), Siegelmann (1995)) showing that a finite neural network, formed by elements connected by weights taking values in the real field and governed by a simple sigmoid activation function, is more powerful than any Turing machine. However, this mathematical result *per se* does not substantiate Smolensky's conjecture: it does become relevant if one shows that this network can provide a sensible theoretical model of a physical system, and, in turn, this physical system can be regarded as an analog computer under suitable stipulations associating observable physical magnitudes of the physical system to the input and output numbers of the network. If such an interpretation is provided, the network can be viewed as a theory of the analog computer in question. And, *according to this theory*,[11] the (i/o) behavior of this analog computer could not be simulated by any Turing machine. Unless a convincing physical interpretation of this sort is provided, these results cannot be viewed as a contribution to a theory of (analog) computability, no matter what their interest is from a mathematical point of view.

The much weaker point which is obviously supported by this kind of results is that the mathematical formalism of connectionism, unlike the classical neural network formalism *à la* McCulloch and Pitts, is sufficiently broad to provide a framework for a prospective theory of analog computability which goes beyond the horizon delimited by TT. This mere possibility, which cannot be excluded in the present state of our knowledge, indicates the difficulty of providing a straightforward answer to the question whether connectionist theories must *necessarily* fulfill C_f.

Let me now summarize the outcome of this selective survey. A mechanistic theory in CS must satisfy C_f. This condition neither requires nor excludes that particular mental states and processes be accounted for in terms of algorithmic structures and procedures postulated by the theory. Mechanistic theories meeting this more demanding requirement are the *procedurally constrained* mechanistic theories, while theories solely concerned with i/o behaviors are *pure* mechanistic theories. A subclass of the procedurally constrained theories is formed by the *architecturally constrained* mechanistic theories. Every connectionist theory formulated so far is included in the class of procedurally

[11] As Kreisel (1974) appropriately emphasizes in connection with the relationship between the notion of non-mechanistic physical theories and the behavior of the analog computers, if any, regimented by a theory of this sort. A different issue arises if one takes into account the measurement problem for analog computations. Physically realistic measurements have limited degrees of accuracy and precision. Accordingly, one can ask whether a Turing machine can simulate the i/o behavior of the analog computer to within some such degree.

constrained mechanistic theories, but the ultimate premises in terms of which connectionist explanations are framed do not ensure that a connectionist theory may be formulated which does not satisfy C_f.

4. A CENTRAL THESIS OF MECHANISM IN CS

So far, I have argued that TT is essential to the formulation of a necessary condition on a theory in CS to count as mechanistic, and I have explored a finer classification of mechanistic theories along the classical dimension of the distinction between input/output, generically procedural, and more specifically architectural predictions. This map may be enriched by considering other criteria of classification[12].

A wide variety of conceptual and empirical issues can be brought to bear on a discussion of the epistemological significance of C_f and the related classification of mechanistic theories. An issue that plays a significant role in the ensuing considerations is the *explanatory basis problem* in CS: What notions can be legitimately assumed as primitives in an explanatory theory of intelligence? A solution to this problem is constrained by the requirement (H) mentioned in the introduction: A theory of intelligent behavior must not presuppose hidden homunculi for the execution of the tasks that the theory is supposed to account for.[13]

The appeal of a theory satisfying C_f partly derives from the fact that the functions accounting for the i/o behaviors of the subjects in its domain are calculable by mechanical procedures, which do not require homunculi for their execution; moreover, the particular procedurally and architecturally constrained mechanistic theories mentioned in the previous section satisfy (H), as they purport to account for mental states and processes of subjects engaged in certain cognitive tasks in terms of specific algorithmic structures and procedures. But Dennett (1978, p. 83) has claimed that also the converse is true: any psychological theory meeting (H) *must* be mechanistic. Haugeland (1981, p. 2), partly motivated by the same problem of arriving at theories meeting (H), states: "...the guiding inspiration of cognitive science is that, at a suitable level of abstraction, a theory of "natural" intelligence should have the same basic form as the theories that explain sophisticated computer systems." And Boden (1988, p. 259) claims: "If a psychological science is possible at all, it must be capable of being expressed in computational terms. (If a scientific psychology is not possible, the computational approach is not the only loser: all attempts to develop a science of the mind would be doomed to failure.)"

I take these various claims as expressing a necessary condition on theories in CS: If it is possible to construct an adequate theory of some particular cognitive task or, more ambitiously, an overarching theory of human cognition,

[12]Massaro and Cowan (1993) provide an interesting taxonomy of information processing approaches to psychological theorizing along partly different dimensions.

[13]See, e.g., Crick (1979, p. 85), and Churchland (1986, pp. 406–407).

then this theory must be computational (or, in the terminology adopted here, *mechanistic*). Since C_f was introduced as a necessary condition on mechanistic theories in CS, the claims of Haugeland, Boden, and Dennett entail, in my view, the following schematic assertion about any cognitive task *t* whatsoever:

(M) If a theory T providing an adequate account of *t* can be given, then T satisfies C_f.

Thesis (M) falls short of asserting *unconditionally* that adequate mechanistic theories of every cognitive task can be given. But the cautious attitude reflected in the antecedent of this conditional statement is only fitting in view of the caveats of cognitive scientists on the prospects of isolating appropriate domains for psychological theorizing.[14] Notwithstanding this qualification, (M) still expresses a fairly strong claim on any adequate theory of a cognitive task: if any such theory is possible, then it satisfies C_f. I propose to identify a fundamental claim of mechanism in CS with thesis (M). Since CS is chiefly concerned with the problem of developing adequate theoretical accounts of (aspects of) human cognition, it is appropriate that this central claim be expressed as a thesis about theories. And the selective survey of the preceding section indicates that thesis (M) is sufficiently broad to be regarded as a reasonable candidate.

Let me now turn to consider the problem of providing a refutation of (M). This requires showing that (i) no adequate mechanistic theory for some task *t* can be given, and (ii) some non-mechanistic theory for *t* can be given. In order to illustrate some of the difficulties surrounding (ii), consider the following formal extension of C_f. Let A be an arbitrary set of natural numbers. One may require that the laws accounting for the input/output behavior of the subjects in the domain of a theory be expressible in terms of functions that are Turing computable in the set A. This functional condition—let us call it $C_{f\text{-}A}$—is more inclusive than C_f, as it allows for accounts of input/output behaviors in terms of functions that are possibly not Turing computable. In view of requirement (H), a theory satisfying $C_{f\text{-}A}$ for a non-decidable A contracts an explanatory debt, unless one shows that the values of the characteristic function of A can be determined without making appeal to unexplained intelligence. This may be done by exhibiting a process which does not require intelligent homunculi for its execution, and in principle enables one to determine the values of the characteristic function of A. Therefore, solving problem (ii) presupposes, in turn, showing that (iia) an explanatory basis—meeting requirement (H), but not included in the class of functions and processes with which TT is concerned—can be made available.

With this discussion of what is required for providing an outright refutation of thesis (M), one is in a better position to distinguish between different sorts of anti-mechanist claims and arguments. Firstly, there are arguments which do

[14]Paradigmatic examples of such caveats are in Fodor (1983).

not invest thesis (M) as a whole, no matter how fundamental are the problems they raise for some classes of restricted mechanistic theories. A pertinent example is Searle's (1992) criticism of the "rules and representations" paradigm in CS based on the formulation of his "connection principle": by Searle's own admission, his arguments do not concern connectionist approaches to theory construction which, with the only rather remote reservation discussed in the previous section, fall squarely within the mechanist paradigm.[15] Secondly, there are arguments challenging directly the functional restrictions on theories expressed by thesis (M). Among them, one can distinguish between weak and strong anti-mechanist arguments. The former support only claim (i)—that there are cognitive tasks t which cannot be adequately accounted for by any kind of mechanistic theory—, while the latter support, in addition to (i), claim (iia)—that an explanatory basis, meeting requirement (H), but not included in the class of processes with which TT is concerned, can be made available.[16]

Weak anti-mechanist arguments—examples of which can be found in Gödel's writings—do not directly conflict with thesis (M), because they are consistent with the *ignorabimus* claim that no adequate scientific theory, whether mechanistic or not, can be given for the tasks in question. (This *ignorabimus* view has been recently endorsed by, e.g., Putnam (1988).) However, weak anti-mechanist arguments jeopardize the significance of (M), as they purport to show that there are aspects of cognitive behavior that cannot be accounted for by a mechanistic theory. Strong anti-mechanist claims and arguments, supporting both (i) and (iia), have been discussed in several writings of Kreisel over the last thirty years, and have been more recently taken up again and argued for by Penrose.

5. CRITICAL REFLECTIONS ON WEAK AND STRONG ANTI-MECHANISM

The general formulation of the incompleteness theorems made possible by TT entails, Gödel claims in the posthumously published 1951 Gibbs lecture, a rigorous consequence for the mind-machine problem which can be expressed under the form of a disjunction (I refer to the first and second disjunct by D1 and D2, respectively):

> Either mathematics is incompletable in this sense, that its evident axioms can never be comprised in a finite rule, that is to say, the human mind (even within the realm of pure mathematics) infinitely

[15]Consider, for example, the remarks in Searle (1992, pp. 246–247). For a critical discussion of Searle's "connection principle", see the commentaries to Searle (1990).

[16]Obviously, this classification of anti-mechanist arguments is not meant to be exhaustive: one might be able to state other necessary (and sufficient) conditions on mechanistic theories, and mechanism as a whole might be challenged by criticisms of this more complete account of mechanistic theorizing in CS.

surpasses the powers of any finite machine, or else there exist absolutely unsolvable Diophantine problems of the type specified (where the case that both terms of the disjunction are true is not excluded, so there are, strictly speaking, three alternatives).[17]

There is a distinctly anti-mechanist thrust in D1. It might be viewed, at least *prima facie*, as expressing the claim that the human capability for generating evident mathematical axioms cannot be accounted for *in terms of a finite rule* (that is, with the terminology adopted in this paper, in terms of a function whose values can be determined by a mechanical procedure). Under this reading, establishing D1 would amount to a proof of (i) above, and thus would substantively contribute to undermine thesis (M): there is at least one cognitive task (the task of generating mathematical axioms) which cannot be accounted for by a mechanistic theory. But this conclusion is too hasty—for the "human mind" mentioned in D1 is a highly idealized notion. One of the elements involved in this idealization was brought out by Gödel himself, according to Hao Wang's reports of his conversations with Gödel:

> In my discussions with Gödel, I asked him some familiar questions from time to time. For example, does he mean by 'mind' that of a single person or the union of the minds of the human species (the 'collective mind')? In reply he said that he means an individual mind of an unlimited life span, and suggested the image of a person engaged in solving a whole set of problems. — He thinks that this is close to the real situation and remarks that people constantly introduce new axioms.[18]

But there is more to this idealized mind than the unlimited expanses of time (and space) familiar from the idealizations made in the theory of computability. This mind is to have the capability of generating *evident* axioms. The criterion for evidence is not elaborated on in detail by Gödel in the Gibbs lecture, even though he mentions the possibility of *perceiving* (with mathematical certitude) *that* certain axioms are correct. As is well known, in the 1964 supplement to "What is Cantor's continuum problem?", Gödel claimed that a particular form of perception, i.e., mathematical intuition[19], may enable one to set up new set-theoretic axioms leading, among other things, to "a decision of such problems as Cantor's continuum hypothesis". If the criterion for evidence is identified with a judgment based on a perception of the correctness of axioms, then Gödel's idealized mind must be capable of perceiving as correct a potentially infinite collection of axioms; otherwise, since any finite

[17] Gödel (1995, p. 310).

[18] Wang (1993, p. 121).

[19] On this notion of mathematical intuition, see Gödel (1990, p. 268); for discussion, see Parsons (1983, pp. 277–278) and (1995), and references therein. Another subsidiary, "inductive" criterion for the acceptability of an axiom is discussed by Gödel. See Gödel (1990, p. 261), and also Gödel (1990, p. 121).

set of sentences of a *formal* language is recursive (and I take it for granted that any finite set of mathematical sentences can be expressed in a formal language) then, contrary to D1, the evident axioms recognizable by the human mind would be comprised in a finite rule. Moreover, if one fixes a formal language such as the language of Zermelo-Fraenkel set theory (ZF) in which these axioms are to be stated, then the idealized mind must be capable of perceiving the correctness of sentences of arbitrary lenght (measured in terms of the number of primitive symbols of the language of ZF that are needed for their formulation). This follows from the fact that Gödel's idealized mind must be capable of generating a potentially infinite collection of axioms, and the language of ZF contains a fixed finite number of primitive symbols.

The latter conclusion about the lenght of axioms depends essentially on the hypothesis that a formal language, in which these axioms are to be formulated and containing a finite number of primitive symbols, is fixed once and for all. This hypothesis is not explicitly made by Gödel, even though he mentions the existence of "unexplored series of axioms of infinity" in set theory as showing "that ever more (and ever more *complicated*) axioms appear during the development of mathematics", and identifies the degree of complication of an axiom with the number of primitive symbols that are needed to formulate it.[20] An alternative to requiring, of the idealized mind mentioned in D1, the capability of recognizing the correctness of ever more complicated axioms is suggested by Gödel's remark that "*mind, in its use, is not static, but constantly developing*, i.e., that we understand abstract terms more and more precisely as we go on using them, and that more and more abstract terms enter the sphere of our understanding."[21] If one allows for this capability, which expresses yet another idealization from actual mathematical performance, the degree of complication of the new axioms—as measured in terms of the number of primitive symbols occurring in them—need not be unbounded, since this idealized mind might be capable of introducing ever new primitive symbols, and formulating the new axioms in an ever expanding language.

The upshot of these considerations is that, even just to understand what does it mean for D1 to hold, one has to interpret the "human mind" involved in its formulation as a highly idealized notion. Firstly, its workings require unbounded expanses of time and space. Secondly, it must be capable of expressing judgments, based on some form of mathematical intuition, on the correctness of a potentially infinite collection of (set-theoretic) axioms. If the axioms are to be formulated in a fixed formal language containing a finite number of primitive symbols, then this idealized mind must be capable of perceiving the correctness of sentences of arbitrary lenght. Alternatively, it must be capable of introducing and correctly using ever new primitive sym-

[20] Gödel (1990, p. 306). Emphasis mine.
[21] Gödel (1990, p. 306).

bols. Even though idealizations of this sort are clearly required for an understanding of D1, they are not sufficient, as the generative capacity of Gödel's "human mind" involves the difficult notion of a (non-sensual) perception *that* the axioms are correct, which may in turn involve a perception *of* the objects these axioms are about.[22]

Let us turn again to consider, after having made explicit some of the idealizations that are needed even just to understand the meaning of D1, what are the consequences of the supposition that D1 holds for thesis (M). According to D1, the generative capacity of Gödel's *idealized* mind cannot be accounted for in terms of the execution of a mechanical procedure, and thus cannot be accounted for by a mechanistic theory. But is this idealized entity relevant to an empirically adequate theory of mathematical reasoning? Two points emerging from the previous discussion of mechanistic theories bear on this issue.

First, I have pointed out that there is no consensus in the cognitive science community on the opportunity of introducing strong idealizations from actual performance in order to arrive at predictive theories of cognitive behaviors. The following critical remark of Feferman and Solovay (1990, p. 292) may be regarded as an expression of this sort of reservations about the usefulness of idealizations: "contrary to Gödel's view... there are simple "mathematical yes or no problems" which will probably never be settled by the human mind because they are beyond all remotely feasible computational power and provide absolutely no conceptual foothold. For example, we can ask the following question: Is it true that if w is the sequence of the first $2^{2^{2^{100}}}$ terms in the binary expression of $\pi - 3$, then the last term of w is 0?"

To this sort of criticism one may object that, e.g., Marr, and Chomsky before him, advocated the opportunity of abstracting from various sorts of malfunctioning, and from limitations of space and time on, say, memory, in characterizing the subjects in the domain of competence theories. However, and this is the second relevant point, the conceptual framework guiding Marr and Chomsky in the construction of competence theories has no counterpart in the scenario surrounding D1. Marr hypothesized that the visual system carries out the task of vision by applying a set of mechanical *procedures*, and suggested the opportunity of abstracting from the variety of limitations inherent in the human cognitive system that make humans less than perfect executors of these procedures. In Gödel's writings, however, there is no hint of any procedure, whether mechanical or non-mechanical, whose properties may suggest the appropriate idealizations that are needed for turning an actual mind into a perfect executor of such procedures.

[22]For a discussion of de dicto and de re mathematical intuition, see the references in footnote 19. In particular, for a critical view of de re intuition in the domain of higher set theory, see Parsons (1983, pp. 272–280).

Gödel was keenly aware of this problem in order to go beyond the unsatisfactory picture of a mind *unsystematically* producing ever new axioms, as is vividly shown by some of his posthumously published observations. He conjectured the existence of a *series* of undiscovered set-theoretic axioms of infinity,[23] and used this speculative claim as evidence against D2, namely, against the claim that there are mathematical problems absolutely undecidable for the human mind. But he also emphasized the difficulty of characterizing the "process" of forming stronger and stronger axioms of infinity in terms of a well-defined, non-mechanical procedure, and added the crucial qualification that the possibility of setting up a (non-mechanical) rule for generating axioms of infinity may emerge only at some future stage of the development of mathematics:

> This process, however, today is far from being sufficiently understood to form a well-defined procedure. It must be admitted that the construction of a well-defined procedure which could actually be carried out (and would yield a non-recursive function) would require a substantial advance in our understanding of the basic concepts of mathematics.[24]

Gödel's quest for non-mechanical rules characterizing human mathematical competence was not pursued exclusively within the horizon of higher set theory. He contended that other forms of mathematical reasoning may be governed by non-mechanical rules. As is well known, Gödel made this point on the basis of a distinction between constructive—more specifically, intuitionistic—mathematics and Hilbert's finitist mathematics. Referring to the view expressed in Bernays (1935) on the conceptual significance of the Gentzen-Gödel relative consistency proof for Peano arithmetic, Gödel stated that Hilbert's finitism is characterized by two constraints, a constructivistic and a more properly finitistic one, the latter excluding from the frame of finitist mathematics the abstract concepts allowed in intuitionistic mathematics.[25] Typically, the objects considered in finitist mathematics are finite and discrete configurations of symbols, whose meaning is totally irrelevant for finitist considerations. If one lifts this restriction—that Gödel called the "properly finitistic" constraint—one may still require that the objects considered be given by a construction, although they need not be finite configurations of spatio-temporal elements. Intuitionistic mathematics retains only the constructivistic element of Hilbert's finitism: the objects considered are numbers, proofs, propositions (rather than numerals, formal derivations, and well-formed formulas), and these are usually characterized as mental constructions. Thus, in particular, intuitionistic finite procedures defining number-theoretic functions are not restricted to methods for manipulating spatio-temporal configurations

[23] See, e.g., Gödel (1990, pp. 305–306).
[24] Gödel (1990, p. 306).
[25] See Gödel (1990, p. 274).

of elements, as they may "involve the use of abstract terms on the basis of their meaning"[26]. For this reason, the notions of constructive and mechanical procedure cannot be immediately identified.

The upshot of Gödel's considerations is characterizable by the following conditional statement: if the intuitionistic standpoint provides a correct picture of a part of human mathematical reasoning, then one cannot exclude that an adequate competence *theory* of this rule-governed behavior will be non-mechanistic.[27] Here, one of the difficulties surrounding Gödel's conjecture about the mind's capability of generating new set-theoretic axioms is apparently overcome: there are candidates for (non-mechanical) *procedures* in intuitionistic mathematics, which may guide the search for appropriate idealizations. But the difficulty of finding such idealizations, as Kreisel (1972, p. 317) observed, is still there.

Gödel's arguments, in view of their decidedly speculative character, do not provide evidence for claim (i) (stated at the end of section 4); at best, they play a sobering role in discussions of mechanism, by pointing to vast landscapes of *terra incognita* uncharted by (or resisting to) mechanistic analyses of human cognition, and against which the current plausibility of mechanism ought to be measured, too. And what about the strong anti-mechanist view expressed by the conjunction of claims (i) and (iia)? Let me examine this question from the viewpoint of a cognitive scientist committed to the ideal of arriving at precise and explanatorily adequate theories in CS. This cognitive scientist gets convinced (perhaps by new developments of Gödel's arguments) that the (idealized) human subjects executing a certain task t are capable of determining membership into a non-decidable set A. Thus, our scientist argues, an adequate theory for this task cannot be included within the boundary set by $\mathrm{C_f}$—it may be, e.g., a theory satisfying $\mathrm{C_{f\text{-}A}}$ for that A. And in view of condition (H), one has to show that the values of the characteristic function of A can be determined by a non-mechanical process which does not require intelligent homunculi for its execution. However, the search for a class of primitives in terms of which such non-mechanical process might be defined can presently rely only upon a rather long chain of possibilities and conjectures.

Kreisel (1974, p. 11) introduced a notion of mechanistic *physical* theory, asked whether current physical theories are mechanistic in this sense, and suggested the possibility that "some (of the unsolved) problems of current physical theories have nonmechanistic solutions". More recently, Penrose (1989, 1994) expressed a different view: present physical theories are most likely mechanistic, but non-mechanistic solutions to physical problems may still arise "in the no man's land between quantum and classical physics", as is suggested

[26]Gödel (1986, p. 370n).

[27]As is well known, there is no consensus on Gödel's view; rather different standpoints are expressed in, e.g., Kleene (1988, pp. 49–51), and Martin-Löf (1982, p. 156).

by an approach to quantum gravity reported in Geroch and Hartle (1986).[28] In either case, even if physical processes regimented by some non-mechanistic physical theory were discovered, our cognitive scientist would have to attack the further problems of showing that such processes play a role in the action of the human brain, and that they enable (idealized) human subjects to determine the values of a non-computable function. In view of these difficulties, the observations on non-mechanistic physical theories play only a very limited role in present discussions of mechanism: they reveal—as Kreisel (1966, p. 270) pointed out already thirty years ago—the dogmatic character of the assertion that precise and genuinely explanatory theories of human cognition can only be mechanistic. However, they provide no substantive reasons for abandoning or revising mechanistic approaches to theory construction in CS, which have crucially contributed to set high standards of precision and explanatory adequacy in this discipline.

Istituto di Cibernetica C.N.R.
Arco Felice, Italy

REFERENCES

Bernays, P. (1935), "Sur le platonisme dans les mathématiques", *L'Enseignement Mathématique* **34**, 52–69. English translation in P. Benacerraf and H. Putnam (eds.) *Philosophy of Mathematics*, 2nd edition, Cambridge U.P., Cambridge 1983.

Boden, M. A. (1988), *Computer Models of Mind*, Cambridge U.P., Cambridge.

Church, A. (1936), "An unsolvable problem of elementary number theory", *American Journal of Mathematics* **58**, 345–363.

Churchland, P.S. (1986), *Neurophilosophy*, The MIT Press, Cambridge, MA.

Cordeschi, R. (1984), "La teoria dell'elaborazione umana dell'informazione. Aspetti critici e problemi metodologici", in B. Continenza *et. al.*, *Evoluzione e modelli*, Editori Riuniti, Rome, 321–422.

Crick, F.H.C. (1979), "Thinking about the brain", *Scientific American* **241**, 181–188.

Davis, M. (ed.), (1965), *The Undecidable*, Raven Press, Hewlett, NY.

Dennett, D.C. (1978), *Brainstorms*, The MIT Press, Cambridge, MA.

Ericsson, K.A. and Simon, H.A. (1993), *Protocol Analysis. Verbal Reports as Data*, revised edition, The MIT Press, Cambridge, MA.

Feferman, S. (1995), "Penrose's Gödelian argument", *Psyche, An Interdisciplinary Journal of Research on Consciousness* **2**, http: //psyche.cs.monash.edu.au/ psyche/public/volume2-1/psyche-95-2-7-shadows-5-feferman.html.

Feferman S. and Solovay R.M. (1990), Introductory note to Gödel's "Another version of the undecidability theorem", in Gödel (1990), 287–292.

Fodor, J.A. (1983), *The Modular Mind*, The MIT Press, Cambridge, MA.

Gandy, R.O. (1980), "Church's thesis and principles for mechanisms", in J. Barwise *et al.* (eds.), *The Kleene Symposium*, North-Holland, Amsterdam, 123–148.

Geroch, R. and Hartle, J.B. (1986), "Computability and physical theories", *Foundations of Physics* **16**, 533–550.

Gödel, K. (1986), *Collected Works, vol. I, Publications 1929–1936*, S. Feferman *et al.* (eds.), Oxford U.P., Oxford.

[28] I am not concerned here with Penrose's argument for claim (i) based on Gödel's theorems. For diverging views on this argument see, e.g., Feferman (1995) and Guccione (1993).

Gödel, K. (1990), *Collected Works, vol. II, Publications 1938–1974*, S. Feferman *et al.* (eds.), Oxford U.P., Oxford.

Gödel, K. (1995), *Collected Works, vol. III, Unpublished Essays and Lectures*, S. Feferman *et al.* (eds.), Oxford U.P., Oxford.

Guccione, S. (1993), "Mind the truth: Penrose's new step in the Gödelian argument", *Behavioral and Brain Sciences* **16**, 612–613.

Haugeland, J. (ed.) (1981), *Mind Design*, The MIT Press, Cambridge, MA.

Herken, R. (ed.) (1988), *The Universal Turing Machine. A Half-Century Survey*, Oxford U.P., Oxford.

Kalmár, L. (1959), "An argument against the plausibility of Church's thesis" in A. Heyting (ed.) *Constructivity in Mathematics*, North-Holland, Amsterdam, 72–80.

Kleene, S.C., (1988), "Turing's analysis of computability, and major applications of it", in Herken (ed.), 17–54.

Kreisel, G. (1966), "Mathematical logic: What has it done for the philosophy of mathematics?", in E. Shoenman (ed.), *Bertrand Russell: Philosopher of the Century*, Allen & Unwin, London, 201–272.

Kreisel, G. (1972), "Which number-theoretic problems can be solved in recursive progressions on $\prod_1^1$-paths through O?", *J. of Symbolic Logic* **37**, 311–334.

Kreisel, G. (1974), "A notion of mechanistic theory", *Synthese* **29**, 11–26.

Marr, D. (1977), "Artificial Intelligence: a personal view", *Artificial Intelligence* **9**, 37–48.

Marr, D. (1982), *Vision*, Freeman, San Francisco, CA.

Martin-Löf, P. (1982), "Constructive mathematics and computer programming", in L.J. Cohen *et al.* (eds.), *Logic, Methodology, and Philosophy of Science VI*, North-Holland, Amsterdam, 153–175.

Massaro, D.W. and Cowan, N. (1993), "Information processing models: microscopes of the mind", *Annual Review of Psychology* **44**, 383–425.

Newell, A. (1990), *Unified Theories of Cognition*, Harvard U.P., Cambridge, MA.

Newell, A. and Simon, H.A. (1972), *Human Problem Solving*, Prentice Hall, Englewood Cliffs, NJ.

Parsons, Ch.D. (1983), *Mathematics in Philosophy*, Cornell U.P., Ithaca, NY.

Parsons, Ch.D. (1995), "Platonism and mathematical intuition in Kurt Gödel's thought", *The Bulletin of Symbolic Logic* **1**, 44–74.

Penrose, R. (1989), *The Emperor's New Mind*, Oxford U.P., Oxford.

Penrose, R. (1994), *Shadows of the Mind*, Oxford U.P., Oxford.

Putnam, H. (1988), *Representation and Reality*, The MIT Press, Cambridge, MA.

Pylyshyn, Z. (1984), *Computation and Cognition*, The MIT Press, Cambridge, MA.

Searle, J.R. (1990), "Consciousness, explanatory inversion, and cognitive science", *Behavioral and Brain Sciences* **13**, 585–642.

Searle, J.R. (1992), *The Rediscovery of the Mind*, The MIT Press, Cambridge, MA.

Sieg, W. (1994) "Mechanical procedures and mathematical experience", in A. George (ed.) *Mathematics and mind*, Oxford University Press, Oxford, 71–117.

Sieg, W. and Byrnes, J. (1994), "K-graph machines: generalizing Turing's machines and arguments", Report CMU-PHIL-55, Carnegie Mellon University, Pittsburgh, PA.

Siegelmann, H.T. (1995), "Computation beyond the Turing limit", *Science* **268**, 545–548.

Siegelmann, H.T. and Sontag, E.D. (1994), "Analog computation via neural networks", *Theoretical Computer Science* **131**, 331–360.

Smolensky, P. (1988), "On the proper treatment of connectionism", *Behavioral and Brain Sciences* **11**, 1–74.

Turing, A.M. (1936), "On computable numbers, with an application to the *Entscheidungsproblem*", *Proc. London Math. Soc., 2nd series* **42**, 230–265.

Turing, A.M. (1939), "Systems of logic based on ordinals", *Proc. London Math. Soc., 2nd series* **45**, 161–228.

Wang, Hao (1993), "On physicalism and algorithmism: can machines think?", *Philosophia Mathematica, 3rd series* **1**, 97–138.

5
Philosophical Logic

ETTORE CASARI

CONJOINING AND DISJOINING ON DIFFERENT LEVELS

A. INTRODUCTION

Both in ancient and in more recent times doubts have been raised about the legitimacy of arguments involving some kind of conjoining and disjoining. Up to this century, however, such doubts were generally not regarded as related to a possibly different notion of truth or as somehow suggesting a different perspective on the whole of the logical laws. Rather they were presented as an expression of uneasiness before particular arguments or kind of arguments which seemed to contradict *at least* our common linguistic usage.

Three basic properties concerning conjoining and disjoining are certainly those which in the proof-theoretic context are usually referred to as *contraction*, *permutation* (or *exchange*) and *weakening* (or *thinning* or *dilution*). In the algebraic-semantic context, current terms for the first two properties are respectively *idempotency* and *commutativity*; for the third one the term *boundedness* may be used. In particular we will say that *and* is *bounded* iff it satisfies the conditions: A *and* B implies A; A *and* B implies B. Dually, *or* is *bounded* iff A implies A *or* B and B implies A *or* B.

During the centuries, all these properties have been often rediscovered, reformulated and questioned. So, for instance, within the Stoic tradition weakening-boundedness was explicitly rejected, at least by some authors, as Sextus Empiricus [*Against the Mathematicians*, VIII, 431] testifies: ≪An argument becomes inconclusive by *redundancy* [κατὰ παρολκήν] when something external and superfluous is adjoined to the premises as in the following: "If it is day, it is light; but it is day; but virtue is beneficial; therefore it is light"≫. Also permutation-commutativity was set in doubt within that tradition, although nobody seems to have remarked this fact up to now. Actually, counterexamples arose in the context of the *changing arguments* [μεταπίπτοντες λόγοι] as it is clearly shown by this (slightly modified) paradoxical example, still by Sextus [*Outlines of Pyrrhonism*, II, 234], in which the 'third undemonstrated' schema $\dfrac{\alpha \quad \neg(\alpha \wedge \beta)}{\neg\beta}$ is invalidated by combining true premises and false conclusion:

$$\dfrac{\text{I made no assumption} \qquad \text{not both}\begin{pmatrix}\text{I made no assumption}\\ \text{and } 0 \neq 1\end{pmatrix}}{0 = 1}$$

M.L. Dalla Chiara et al. (eds.), Logic and Scientific Methods, 261–288.

It is indeed clear that changing their order, both premises become false and the paradox disappears. Even idempotency was well known to the Stoics, as the following text, still by Sextus [*Ag. Math.*, VIII, 230–3], clearly shows (remark, moreover, that the status of contraction rule in their system is a rather subtle question [see Frede 1974]): ≪From simple homogeneous syllogisms arises for instance the following argument: "If it is day, then if it is day it is light; but it is day; therefore, it is light" .... As we have two premises, first the conditional "If it is day, then if it is day it is light" which begins with the simple proposition "It is day" and concludes with the non simple conditional "If it is day, it is light", and second the antecedent in that conditional, *i.e.,* "It is day", so we may infer from them, by the first undemonstrable, the consequent of that conditional, i.e., "If it is day, it is light". We have this conditional as potentially inferred in the argument, although omitted in the presentation; by composing it with the minor premise of the presented argument, i.e., "It is day", we will also infer, again by the first undemonstrable, the "It is light", which was just the conclusion of the presented argument≫.

As more recent examples, just two passages by John Venn 1894. The first [p.60], on commutativity: ≪When however we say that this commutative law holds in common speech, we must remember that the idiosyncrasies of language insist upon exceptions here as elsewhere; and on the same ground, viz. the necessity of compressing a quantity of implication into our sentences, in addition to the direct assertions they contain. Thus when X implies Y we always use the form YX, rather than XY. For instance one would not direct a legacy to those old servants "who are destitute and alive"; but to those "who are alive and destitute"≫. The second [p.61], on idempotency ≪And that the same is true as regards the laws and usages of common speech must also be admitted. Or rather, it must be admitted under some exceptions; for there are signs of divergence towards the mathematical rule here, when we are dealing with comparative terms. Thus even in English, 'great great' does not mean quite the same as 'great' alone; and in some languages, as Italian for instance, adjectival repetition is really almost like mathematical multiplication, increasing or diminishing the effect according as the term is in itself an augmentative or diminutive. If this conventional rule prevailed universally in respect of all, or even most terms, the logician would have to give way to the grammarian here, with the result of having to abandon Symbolic Logic, as a formal science distinct from Mathematics≫.

In our century these questions underwent a radical change. Rejection of idempotency emerged in many-valued logics as a natural consequence of a systematical rethinking of the notion of truth-value. On the other hand, renewed attention to modal questions and chiefly the attempt to systematically account for questions of relevance in implications came to grips with boundedness. Remarkably enough, both in the many-valued and in the relevance tradition, attention was initially 'one-level' oriented, i.e., concentrated on *one* implication, *one* conjunction, *one* disjunction, disregarding the possibility of having *two* (even more) of them. And this despite the fact that, on one side, it is very easy to build up a proposition in which, for instance, 'and' occurs several times with evidently different meanings, i.e., expressing different kinds of conjunctive connection, and, on the other side, that it is not possible to interpret this phenomenon as a purely verbal question. In fact it is clearly a question of logical regularity, involved in every argument stating that such a proposition is equivalent or implies another one obtained from it by transformations, for instance by simplification or permutation of some of its conjunctive parts. Once the interest of simultaneous different levels was observed, this possibility was systematically exploited in the relevant tradition, where terminological distinction between *intensional* and *extensional* connectives was fixed. By the way, only scant attention seems to have been given to this possibility in the many-valued tradition; possibly because in the outstanding case of Łukasiewicz's logics, the 'extensional' connectives are explicitly definable from the 'intensional' ones. It is well known that in more recent years many other 'two levels' systems, differently motivated, have been developed. Also the terminology has been enriched. Besides the pair extensional-intensional, the pairs *multiplicative-additive*, *static-dynamic* and still others have been introduced; in some of our previous papers we used the pair *g(roup-theoretic)-l(attice-theoretic)*.

Notwithstanding its origins, what follows may be regarded as pertaining to this 'more-level logics' trend. In particular section B presents the propositional part of the 'two-level' logics we introduced to account for the logical mechanism of comparison; section C present the first order expansion of those logics; finally section D presents a 'three-level' expansion of propositional comparative logic. The material of the last two sections is entirely unpublished, whereas section B, which presents mainly results already published elsewhere, is intended to make the paper self-contained.

B. PROPOSITIONAL COMPARATIVE LOGIC

B1. *MOTIVATION*

Although today almost completely forgotten, comparative logic has a long and honorable history, which began with Aristotle who, especially in the *Topics,* discussed it at length and laid down the basis for its formal treatment. The ideas lying at the basis of our (propositional) comparative logic are the following.

[1] Propositions, besides admitting negation, are built up not only by means of the 'static' connectives $\wedge, \vee$ and $\rightarrow$ but also of 'dynamic' ones: $\oplus, \odot$ and $<$.

[2] A proposition has just one of a set of *truth-degrees* which, on their part, are partially ordered so that it may happen that the truth-degree of a proposition α *precedes* (in the inclusive sense) the degree of a proposition

β, and it may thus be said that α ***is at most so true as*** β or, equivalently, that β ***is at least so true as*** α, or still equivalently, α ***is at least so false as*** β as well as β ***is at most so false as*** α. By the way, this wording is old; Aristotle [*Metaphysics*, Γ, 5, 1008b, 33–36]: ≪We will not say that ... they are equally mistaken [ὁμοίως διέψευσται] ... he who thinks that four things are five and he who thinks that they are a thousand. But if they are not equally mistaken, it is clear that one is less [ἧττον] mistaken and thus he says more the truth [μᾶλλον ἀληθεύει]≫.

[3] Some truth-degrees are *positive*, in which case the proposition is *true*; some other are *negative* , in which case the proposition is *false*; however a truth-degree is not necessarily either positive or negative. In our early papers the name 'comparative logic' was reserved to the special case in which the ordering is total (and another peculiar restriction is satisfied); a case, to be sure, in which a straightforward reconstruction of relevant parts of the Aristotelian logic of comparison is possible. From now on we will use the name 'comparative logic' for the general case and refer to that special case by the name 'strong comparative logic'. In the present paper, however, we will usually not rest on the hypothesis that the truth-degrees are partially ordered, but will rather suppose that they are lattice-ordered.

[4] All negative degrees precede all positive degrees.

[5] Among the positive truth-degrees there is a least one which is the degree of the constant proposition *verum* ($\top$) and among the negative truth-degrees there is a greatest one which is the degree of the constant proposition *falsum* ($\bot$), so that being true is the same as having a degree which follows the degree of $\top$ and being false is the same as having a degree which precedes the degree of $\bot$.

[6] Negation is an involution on the degrees-order such that, in particular, the proposition $\neg\alpha$ is true iff α is false and it is false iff α is true.

[7] The truth-degrees of $\alpha \wedge \beta$ and $\alpha \vee \beta$ will be respectively the *inf* and the *sup* of the truth-degrees of their components, so that, in particular, $\alpha \wedge \beta$ is true iff both α and β are true and $\alpha \vee \beta$ is false iff both α and β are false, but $\alpha \wedge \beta$ may be false without any of its components being false and $\alpha \vee \beta$ may be true without any of its components being true; $\alpha \rightarrow \beta$ is just short for $\neg\alpha \vee \beta$.

[8] Truth-degrees may be added up and $\alpha \oplus \beta$ has as its truth-degree just the 'sum' of the truth-degrees of its components so that, in particular, it is true iff each one of α and β is at most so false as the other one is true, in other words iff α and β are such that taking them together we get in any case something true.

[9] $\alpha < \beta$ has the same degree as $\neg\alpha \oplus \beta$ so that, in particular, it is true iff the truth-degree of α precedes that of β, that is to say: iff α is at most so true as β iff ... (See [2]).

[10] $\alpha \odot \beta$ is short for $\neg(\alpha < \neg\beta)$ so that, in particular, it is false iff $\alpha < \neg\beta$ is true iff α is at most so true as β is false; *i.e.*, iff α and β are such that if we take both them, we get anyway something false.

B2. *PARTIALLY AND LATTICE-ORDERED PREGROUPS*

The preceding ideas can be given a mathematically quite satisfactory formulation.

B2.1 DEFINITION

[1] $\mathfrak{G} = \langle G, +, -, 0, \sqsubseteq \rangle$ is a *partially ordered pregroup* [*po-pregroup*] iff

[G1] $x + (y + z) = (x + y) + z$;

[G2] $- - x = x$;

[G3] $0 + 0 = 0$;

[G4] $x + -x = 0$;

[G5] $\langle G, \sqsubseteq \rangle$ is a poset;

[G6] $0 \sqsubseteq -x + y \leftrightarrow x \sqsubseteq y \leftrightarrow 0 \sqsubseteq y + -x$.

[2] $\mathfrak{G} = \langle G, +, -, 0, \sqsubseteq \rangle$ is an *Abelian po-pregroup* iff it is a *po*-pregroup satisfying

[G1′] $x + y = y + x$.

[3] $\mathfrak{G} = \langle G, +, -, 0, \sqsubseteq \rangle$ is an (Abelian) *lattice-ordered pregroup* [*l-pregroup*] iff it is an (Abelian) *po*-pregroup satisfying

[G5′] $\langle G, \sqsubseteq \rangle$ is a lattice.

B2.2 REMARK

An (Abelian) *po-(l)*-pregroup is an (Abelian) *po-(l)*-group iff $0 = -0$.

We collect in the following theorem some simple arithmetical consequences of the axioms.

B2.3 THEOREM

In every po-pregroup $\langle P, \sqsubseteq, +, -, 0 \rangle$:

[T1] $x \sqsubseteq y + z \leftrightarrow -z \sqsubseteq -x + y \leftrightarrow -y \sqsubseteq z + -x$, i.e., De Morgan's Theorem K;

[T2] $x + -0 = x = -0 + x$, i.e., $\langle P, +, -0 \rangle$ is a monoid;

[T3] $x \sqsubseteq y \leftrightarrow -y \sqsubseteq -x$, i.e., $\langle P, \sqsubseteq, - \rangle$ is an involutive poset;

[T4] $-(-x + y) \sqsubseteq z \leftrightarrow x \sqsubseteq y + z \leftrightarrow -(z + -x) \sqsubseteq y$, so both $-(-x + y) = \sqcap[z \mid x \sqsubseteq y + z]$ and $-(z + -x) = \sqcap[y \mid x \sqsubseteq y + z]$, i.e., $\langle P, \sqsubseteq, + \rangle$ is a coresiduated semigroup;

[T5] $x \sqsubseteq y \rightarrow (x + z \sqsubseteq y + z) \wedge (z + x \sqsubseteq z + y)$, i.e., $\langle P, \sqsubseteq, + \rangle$ is a *po*-semigroup;

[T6] $-(x + y) \sqsubseteq -y + -x$, $-(x + y) + 0 = -y + -x + 0$;

[T7] $-(-y + z) + -(-x + y) + (-x + z) = 0$.

B2.4 REMARK

A structure $\langle P, +, -, 0\rangle$ satisfying G1–G4, T2 and T7 is, by definition, a *pre-group* [Cf. Casari 89].

B2.5 THEOREM (Minari)

The following ten (eleven) equations define the class of (Abelian) l-pregroups as a variety in the signature $\langle +, -, 0, \sqcap\rangle$:

[E1] $x + (y + z) = (x + y) + z;$
([E1′] $x + y = y + x;$)
[E2] $- - x = x;$
[E3] $0 + 0 = 0;$
[E4] $x + -x = 0;$
[E5] $x + -0 = x;$
[E6] $x \sqcap (y \sqcap z) = (x \sqcap y) \sqcap z;$
[E7] $x \sqcap y = y \sqcap x;$
[E8] $x = x \sqcap -(-x \sqcap y);$
[E9] $x + (y \sqcap z) = (x + y) \sqcap (x + z);$
[E10] $x = x \sqcap (y + -(-x + y)).$

For a proof see Casari 91.

B3. *LANGUAGE AND SEMANTICS*

A *(propositional comparative) language* CL^0 is *BN*-defined by:
Atomic formulae $p \in A$
Formulae $\alpha \in FRM(A)$
$\alpha := p|\top|\bot|\neg\alpha|\alpha < \beta|\alpha \oplus \beta|\alpha \odot \beta|\alpha \wedge \beta|\alpha \vee \beta|\alpha \rightarrow \beta|$

B3.1 DEFINITION

If $\mathfrak{G} = \langle G, +, -, 0, \sqsubseteq\rangle$ is an Abelian l-pregroup, a *realization* of L in $\mathfrak{G}$ is a map $\rho : FRM \rightarrow G$ such that:

[1] $\rho(\top) = 0;$
[2] $\rho(\bot) = -0;$
[3] $\rho(p) \in G;$
[4] $\rho(\neg\alpha) = -\rho(\alpha);$
[5] $\rho(\alpha < \beta) = -\rho(\alpha) + \rho(\beta);$
[6] $\rho(\alpha \oplus \beta) = \rho(\alpha) + \rho(\beta);$
[7] $\rho(\alpha \odot \beta) = -(-\rho(\alpha) + -\rho(\beta));$
[8] $\rho(\alpha \wedge \beta) = \rho(\alpha) \sqcap \rho(\beta);$
[9] $\rho(\alpha \vee \beta) = \rho(\alpha) \sqcup \rho(\beta);$
[10] $\rho(\alpha \rightarrow \beta) = -\rho(\alpha) \sqcup \rho(\beta).$

B3.2 DEFINITION

[1] α is *true* in ρ $[\rho \models \alpha]$ iff $0 \sqsubseteq \rho(\alpha)$.
[2] ρ is a *model* of $M \subseteq FRM$ iff for all $\beta \in M$, $\rho \models \beta$.

[3] α is *valid* in $\mathfrak{G}$ [$\mathfrak{G} \models \alpha$] iff for all ρ in $\mathfrak{G}$, $\rho \models \alpha$.
[4] α is a *(logico-comparative) law* iff, for all l-pregroups $\mathfrak{G}$, $\mathfrak{G} \models \alpha$.
[5] α is a *(logico-comparative) consequence* of $M \subseteq FRM$ [$M \models \alpha$] iff, for every l-pregroup $\mathfrak{G}$ and realization ρ in $\mathfrak{G}$, if $\rho \models M$, then $\rho \models \alpha$.

B4. *SYNTAX*

The *Axioms of propositional comparative logic* are given by the following schemata (we write '$\alpha \lessgtr \beta$' for 'both $\alpha < \beta$ and $\beta < \alpha$')

[C1] $(\beta < \gamma) < ((\alpha < \beta) < (\alpha < \gamma))$;
[C2] $(\alpha < (\beta < \gamma)) < (\beta < (\alpha < \gamma))$;
[C3] $\mathrm{T} \lessgtr (\alpha < \alpha)$;
[C4] $(\alpha < \neg\beta) < (\beta < \neg\alpha)$;
[C5] $\neg\neg\alpha < \alpha$;
[C6] $\bot \lessgtr \neg\top$;
[C7] $\bot < \top$;
[C8] $(\bot < \top) < \top$;
[C9] $\alpha \oplus \beta \lessgtr \neg\alpha < \beta$;
[C10] $\alpha \odot \beta \lessgtr \neg(\alpha < \neg\beta)$;
[C11] $\alpha \wedge \beta < \alpha$;
[C12] $\alpha \wedge \beta < \beta$;
[C13] $(\alpha < \beta) \wedge (\alpha < \gamma) < (\alpha < \beta \wedge \gamma)$;
[C14] $\alpha \vee \beta \lessgtr \neg(\neg\alpha \wedge \neg\beta)$;
[C15] $\alpha \rightarrow \beta \lessgtr \neg(\alpha \wedge \neg\beta)$.

The *Rules* are *Modus Ponens* $\dfrac{\alpha \quad \alpha < \beta}{\alpha}$MP and *Adjunction* $\dfrac{\alpha \quad \beta}{\alpha \wedge \beta}$AR

B4.1 DEFINITION

[1] α *is (logico-comparatively) derivable from* $M \subseteq FRM(A)$ [$M \vdash \alpha$] iff there is a finite sequence of formulae each of which is either an axiom or an element of M or is inferred from preceding formulae by MP or RA.
[2] α is a *(logico-comparative) thesis* iff $\emptyset \vdash \alpha$.

B4.2 THEOREM (Strong Adequateness Theorem)

For all $M \subseteq FRM$ and $\alpha \in FRM$, $M \models \alpha$ iff $M \vdash \alpha$.

For a proof see Casari 89.

B5. *DUALPREGROUPS, EXPANDED PREGROUPS, ZEROIDS*

B5.1 DEFINITION

$\mathfrak{G} = \langle G, \cdot, -, \widehat{0}, \sqsubseteq\rangle$ is a *po-dualpregroup* iff

[D1] $x \cdot (y \cdot z) = (x \cdot y) \cdot z$;
[D2] $- - x = x$;
[D3] $\widehat{0} \cdot \widehat{0} = \widehat{0}$;

[D4] $x \cdot -x = \widehat{0}$;
[D5] $\langle G, \sqsubseteq \rangle$ is a poset;
[D6] $x \cdot -y \sqsubseteq \widehat{0} \leftrightarrow x \sqsubseteq y \leftrightarrow -y \cdot x \sqsubseteq \widehat{0}$

The notions of *Abelian po-dualpregroup* and of *(Abelian) l-dualpregroup* are defined in the obvious way.

B5.2 THEOREM
In a structure $\mathfrak{G} = \langle G, +, \cdot, -, 0, \widehat{0}, \sqsubseteq \rangle$ the following conditions are equivalent:

[1] $\langle G, +, -, 0, \sqsubseteq \rangle$ is a *po*-pregroup and moreover:
G7 $x \cdot y = -(-x + -y)$ and G8 $\widehat{0} = -0$.
[2] $\langle G, \cdot, -, \widehat{0}, \sqsubseteq \rangle$ is a *po*-dualpregroup and moreover:
D7 $x + y = -(-x \cdot -y)$ and D8 $0 = -\widehat{0}$.

B5.3 DEFINITION
A structure $\mathfrak{G} = \langle G, +, \cdot, -, 0, \widehat{0}, \sqsubseteq \rangle$ is an *expanded po-pregroup* iff it satisfies one of the conditions of the preceding theorem. The notions of *Abelian expanded po-pregroup* and *(Abelian) expanded l-pregroup* are defined in the obvious way.

B5.4 DEFINITION
An (Abelian) *po-(l-)*pregroup is an (Abelian) *po-(l-)zeroid* iff it satisfies $x + 0 = 0$. The notions of *(Abelian) po-dualzeroid*, *(Abelian) l-dualzeroid*, *(Abelian) expanded po-zeroid* and *(Abelian) expanded l-zeroid* are defined in the obvious way.

B5.5 DEFINITION (GRIŠIN)
A structure $\mathfrak{G} = \langle G, +, \cdot, -, 0, \widehat{0}, \sqsubseteq \rangle$ is an *L^0-algebra* iff

[L1] $x \cdot (y \cdot z) = (x \cdot y) \cdot z$;
[L2] $x \cdot y = y \cdot x$;
[L3] $x \cdot 0 = x$;
[L4] $x + (y + z) = (x + y) + z$;
[L5] $x + y = y + x$
[L6] $x + \widehat{0} = x$
[L7] $\langle G, \sqsubseteq \rangle$ is poset;
[L8] $x \sqsubseteq y \rightarrow x \cdot z \sqsubseteq y \cdot z$;
[L9] $x \sqsubseteq y \rightarrow x + z \sqsubseteq y + z$;
[L10] $x \cdot (y + z) \sqsubseteq (x \cdot y) + z$;
[L11] $-x \cdot x \sqsubseteq \widehat{0}$;
[L12] $0 \sqsubseteq -x + x$;
[L13] $\widehat{0} \sqsubseteq x$;
[L14] $x \sqsubseteq 0$.

An L^0-algebra is a *lattice-ordered L^0-algebra* (*l-L^0-algebra*) iff $\langle G, \sqsubseteq \rangle$ is a lattice.

B5.6 THEOREM

A structure $\mathfrak{G} = \langle G, +, \cdot, -, 0, \widehat{0}, \sqsubseteq \rangle$ is an Abelian *po*-(l-)zeroid iff it is an (l-)L^0-algebra.

B5.7 REMARKS

[1] Grišin writes '0' for '$\widehat{0}$' and '1' for '0'.

[2] Recalling Peirce 3.334 and 3.497, Grišin's axiom L10 deserves to be called *Peirce's law for* $\cdot$ *and* $+$. Recently other names such as *mixed associativity* or *weak distributivity* have been introduced (See Lambek 93, p.218).

[3] Peirce's law for $\sqcap$ and $\sqcup$, i.e., $x \sqcap (y \sqcup z) \sqsubseteq (x \sqcap y) \sqcup z$, is equivalent to lattice distributivity; indeed, if L is distributive, then $x \sqcap (y \sqcup z) = (x \sqcap y) \sqcup (x \sqcap z) \sqsubseteq (x \sqcap y) \sqcup z$; on the other side, if L satisfies Peirce's law, then from $x \sqcap (y \sqcup z) \sqsubseteq (x \sqcap y) \sqcup z$ we get $x \sqcap (x \sqcap (y \sqcup z)) \sqsubseteq x \sqcap ((x \sqcap y) \sqcup z)$ and so $x \sqcap (y \sqcup z) \sqsubseteq x \sqcap (z \sqcup (x \sqcap y))$; then, again by Peirce's law, $x \sqcap (y \sqcup z) \sqsubseteq (x \sqcap z) \sqcup (x \sqcap y))$.

B6. *SPECIALIZATIONS OF COMPARATIVE LOGIC*

We list here some important specializations of the notion of comparative logic together with the 'corresponding' varieties, i.e., the subvarieties of the variety of Abelian l-pregroups which supply them with a strongly adequate semantics. For each such pair of a logic and a variety some of the most representative characterizations are given. This is to be understood so that *any one* of the axioms (of the equations) when adjoined to the comparative logic (to the equations defining the Abelian l-pregroups) gives the indicated logic (variety). The inequalities in $\sqsubseteq$ are of course equivalent to equations; furthermore we refer in some cases to the expanded presentation of the variety. For more details see Casari 89 and 91.

B6.1 INCONSISTENT COMP. LOGIC

$[\bot,\ \alpha \oplus \top < \alpha,\ \alpha \oplus \beta < \alpha \odot \beta,\ (\bot < \alpha) < (\alpha \vee \beta)]$;
Abelian l-Groups $[0 = -0,\ x + 0 = x,\ x + y = -(-x + -y),\ x + 0 \sqsubseteq x \sqcup y]$.

B6.2 STRONGLY DISTRIBUTIVE COMP. LOGIC

$[(\alpha < \beta \vee \gamma) < (\alpha < \beta) \vee (\alpha < \gamma),\ (\alpha \wedge \beta < \gamma) < (\alpha < \gamma) \vee (\beta < \gamma),\ (\alpha \vee \beta < \alpha) \vee (\alpha \vee \beta < \beta),\ (\alpha < \alpha \wedge \beta) \vee (\beta < \alpha \wedge \beta)]$;
Abelian m-pregroups $[x + (y \sqcup z) = (x + y) \sqcup (x + z),\ (-(x \sqcup y) + x) \sqcup (-(x \sqcup y) + y) = 0]$.

B6.3 DISTRIBUTIVE COMP. LOGIC

$[\alpha \wedge (\beta \vee \gamma) < (\alpha \wedge \beta) \vee (\alpha \wedge \gamma),\ (\alpha \vee \beta) \wedge (\alpha \vee \gamma) < \alpha \vee (\beta \wedge \gamma),\ \alpha \wedge (\beta \vee \gamma) < (\alpha \wedge \beta) \vee \gamma,\ (\alpha \vee \beta) \wedge \gamma < \alpha \vee (\beta \wedge \gamma)]$;
Abelian d-pregroups $[x \sqcap (y \sqcup z) = (x \sqcap y) \sqcup (x \sqcap z),\ x \sqcup (y \sqcap z) = (x \sqcup y) \sqcap (x \sqcup z),\ x \sqcap (y \sqcup z) \sqsubseteq (x \sqcap y) \sqcup z,\ (x \sqcup y) \sqcap z \sqsubseteq x \sqcup (y \sqcap z)]$.

B6.4 Bounded Comp. Logic or Grišin's Contraction-free Logic
$[\alpha < (\beta < \alpha)$, $\alpha < (\alpha < \alpha)$, $\alpha < \top$, $\alpha < \alpha \wedge \top$, $\alpha < (\neg\alpha < \beta)$, $\alpha < (\neg\alpha < \alpha)$, $\bot < \alpha$, $\bot \vee \alpha < \alpha$, $\alpha < \alpha \oplus \beta$, $\alpha < \alpha \wedge (\alpha \oplus \beta)$, $\alpha < \alpha \oplus \alpha$, $\alpha \odot \beta < \alpha$, $\alpha \vee (\alpha \odot \beta) < \alpha$, $\alpha \odot \alpha < \alpha$, $\alpha \vee \beta \sqsubseteq \alpha \oplus \beta$, $\alpha \odot \beta \sqsubseteq \alpha \wedge \beta]$;
Abelian l-Zeroids or Grišin's l-L^0-Algebras $[x + 0 = 0$, $x \sqsubseteq 0$, $x \sqsubseteq x + y$, $x \sqsubseteq x + x$, $-0 \sqsubseteq x$, $x = x \sqcap (x + y)$, $x = x \sqcup (x \cdot y)$, $x \sqcup y < x + y]$.

B6.5 Łukasiewicz's ℵ-Logic
$[((\alpha < \beta) < \beta) < ((\beta < \alpha) < \alpha)$, $((\alpha < \beta) < \beta) \lessgtr (\alpha \vee \beta)]$;
Chang's MV-Algebras $[-(-x+y)+y = -(-y+x)+x$, $-(-x+y)+y = x \sqcup y]$.

B6.6 Classical Logic
$[(\neg\alpha < \alpha) < \alpha$, $(\alpha < (\alpha < \beta)) < (\alpha < \beta)$, $(\alpha \oplus \alpha) < \alpha$, $\alpha < (\alpha \odot \alpha)$, $\alpha \oplus (\beta \odot \gamma) < (\alpha \oplus \beta) \odot (\alpha \oplus \gamma)$, $(\alpha \oplus \beta) \odot (\alpha \oplus \gamma) < \alpha \oplus (\beta \odot \gamma)$, $(\alpha \oplus \beta) < (\alpha \vee \beta)$, $(\alpha \wedge \beta) < (\alpha \odot \beta)]$;
Boolean Algebras $[x + x = x$, $x + (x + y) = x + y$, $x + (y \cdot z) = (x + y) \cdot (x + z)$, $x + y = x \sqcup y$, $x \cdot y = x \sqcap y]$.

B6.7 TND Logic
$[\alpha \vee \neg\alpha]$;
Unperforated l-Pregroups $[0 \sqsubseteq x \sqcup -x]$.

B6.8 REMARK
The addition of one of the mentioned axiom-schemata may simplify the whole system. So, for instance, the *bounded comparative logic* is completely axiomatized by:

[Z1] $(\beta < \gamma) < ((\alpha < \beta) < (\alpha < \gamma))$;
[Z2] $(\alpha < (\beta < \gamma)) < (\beta < (\alpha < \gamma))$;
[Z3] $\alpha < (\beta < \alpha)$;
[Z4] $(\alpha < \alpha) < \top$;
[Z5] $(\neg\alpha < \neg\beta) < (\beta < \alpha)$;
[Z6] $\alpha \wedge \beta < \alpha$;
[Z7] $\alpha \wedge \beta < \beta$;
[Z8] $(\alpha < \beta) \wedge (\alpha < \gamma) < (\alpha < \beta \wedge \gamma)$;

$$\frac{\alpha \quad \alpha < \beta}{\beta}\text{MP};$$

[Z9] $\bot \lessgtr \neg\top$;
[Z10] $\alpha \oplus \beta \lessgtr \neg\alpha < \beta$;
[Z11] $\alpha \odot \beta \lessgtr \neg(\alpha < \neg\beta)$;
[Z12] $\alpha \vee \beta \lessgtr \neg(\neg\alpha \wedge \neg\beta)$;
[Z13] $\alpha \to \beta \lessgtr \neg(\alpha \wedge \neg\beta)$.

C. FIRST ORDER COMPARATIVE LOGIC

C1. *LANGUAGE*

A *first order comparative language* CL^1 is defined by:
Predicates $P^k \in PRD^k$ $\bot, \top \in PRD^0$
Variables $v \in VAR$
Formulae $\alpha \in FRM$
$\alpha := \bot | \top | P^k(v_1, \ldots, v_k) | \neg\alpha | \alpha < \beta | \alpha \oplus \beta | \alpha \otimes \beta | \alpha \wedge \beta | \alpha \vee \beta | \alpha \rightarrow \beta | \forall v\alpha | \exists v\alpha$

C2. *DERIVABILITY*

The axioms and rules of the *first order comparative logic* are given by the axiom-schemata and the rules of the propositional comparative logic plus the two schemata
[Q1] $\forall v\alpha < \alpha[v/w]$ and
[Q2] $\alpha[v/w] < \exists v\alpha$

and the two rules $\dfrac{\alpha < \beta, v \notin Fr(\alpha)}{\alpha < \forall v\alpha}\forall I$ and $\dfrac{\alpha < \beta, v \notin Fr(\beta)}{\exists v\alpha < \beta}\exists I$.

The notion of *comparative derivability* is thus defined in the obvious way and denoted by '$M \vdash \alpha$'.

C3. *SOME THESES*

The following easily provable theses throw light on the relationships between the quantifiers and the connectives. The obvious restrictions on v are omitted.

$$\begin{array}{cc}
\forall v(\alpha[v] \wedge \beta) \gtrless \forall v\alpha[v] \wedge \beta & \forall v\alpha[v] \odot \beta < \forall v(\alpha[v] \odot \beta) \\
\forall v\alpha[v] \vee \beta < \forall v(\alpha[v] \vee \beta) & \forall v(\alpha[v] \oplus \beta) \gtrless \forall v\alpha[v] \oplus \beta \\
\exists v(\alpha[v] \wedge \beta) < \exists v\alpha[v] \wedge \beta & \exists v(\alpha[v] \odot \beta) \gtrless \exists v\alpha[v] \odot \beta \\
\exists v(\alpha[v] \vee \beta) \gtrless \exists v\alpha[v] \vee \beta & \exists v(\alpha[v] \oplus \beta) < \exists v\alpha[v] \oplus \beta \\
\forall v\alpha \gtrless \neg\exists v\neg\alpha & \exists v\alpha \gtrless \neg\forall v\neg\alpha \\
\forall v(\alpha[v] < \beta) \gtrless (\exists v\alpha[v] < \beta) & \forall v(\alpha < \beta[v]) \gtrless (\alpha < \forall v\beta[v]) \\
\exists v(\alpha[v] < \beta) < (\forall v\alpha[v] < \beta) & \exists v(\alpha < \beta[v]) < (\alpha < \exists v\beta[v])
\end{array}$$

C4. *INFIMA AND SUPREMA IN po-PREGROUPS*

To set up semantics we need some strengthening of the notion of a *po-pregroup*. If $\mathfrak{G} = \langle G, +, -, 0, \sqsubseteq \rangle$ is a *po*-pregroup then, for $A \subseteq G$, let $L(A)$ and $U(A)$ be the sets of *lower* , respectively *upper bounds* of A and let $\sqcap A$ and $\sqcup A$ be the *Inf*, respectively the *Sup* of A.

C4.1 LEMMA
In every *po*-pregroup:

[1] If $\sqcap_{i\in I}x_i$ exists, then both $\sqcap_{i\in I}(x_i + x)$ and $\sqcap_{i\in I}(x + x_i)$ exist and $\sqcap_{i\in I}(x_i + x) = (\sqcap_{i\in I}x_i) + x$, $\sqcap_{i\in I}(x + x_i) = x + \sqcap_{i\in I}x_i$;
[2] If $\sqcap_{i\in I}x_i$ exists, then also $\sqcup_{i\in I} - x_i$ exists and $\sqcup_{i\in I} - x_i = - \sqcap_{i\in I} x_i$;
[3] If $\sqcup_{i\in I}x_i$ exists, then also $\sqcap_{i\in I} - x_i$ exists and $\sqcap_{i\in I} - x_i = - \sqcup_{i\in I} x_i$.

PROOF
[1] Suppose $\sqcap_{i\in I}x_i$ exists. By definition of $\sqcap_{i\in I}x_i$, it holds $\sqcap_{i\in I}x_i \sqsubseteq x_j$ for all $j \in I$; so, by T5, [i] $(\sqcap_{i\in I}x_i) + x \sqsubseteq x_j + x$ for all $j \in I$. On the other side, suppose $z \sqsubseteq x_j + x$ for all $j \in I$; then, by T4, $-(x + -z) \sqsubseteq x_j$ for all $j \in I$. Thus, by the definition of $\sqcap_{i\in I}x_i$, $-(x+-z) \sqsubseteq \sqcap_{i\in I}x_i$ and so, by T4, [ii] $z \sqsubseteq (\sqcap_{i\in I}x_i)+x$. From [i] and [ii] we conclude that $(\sqcap_{i\in I}x_i)+x$ is the infimum of the $x_j + x$ for all $j \in I$, i.e., $\sqcap_{i\in I}(x_i + x) = (\sqcap_{i\in I}x_i) + x$. Quite similarly we prove $\sqcap_{i\in I}(x + x_i) = x + \sqcap_{i\in I}x_i$. [2] and [3] hold in every involutive poset and so, by T3, in every *po*-pregroup.

C4.2 DEFINITION
A *po*-pregroup $\mathfrak{G}$ is *conditionally complete* [*c-c-pregroup*] iff for all $A \subseteq G$: $A \neq \emptyset \neq U(A)$ implies that $\sqcup A$ exists and $A \neq \emptyset \neq L(A)$ implies that $\sqcap A$ exists; it is *complete* iff $\sqcup A$ and $\sqcap A$ exist for all $A \subseteq G$.

Clearly, a complete *po*-pregroup is always an *l*-pregroup, whereas a *c-c*-pregroup is an *l*-pregroup iff it is directed, i.e., iff for all $x, y \in G$, $L(\{x,y\}) \neq \emptyset \neq U(\{x,y\})$. Moreover, in every *po*-pregroup the only possible top element is 0, i.e., $\forall x(x \sqsubseteq y) \rightarrow y = 0$. Indeed, from $\forall x(x \sqsubseteq y)$ it follows, on the one side, $0 \sqsubseteq y$, and, on the other side, $y + y \sqsubseteq y$; but from $y + y \sqsubseteq y$ it follows $y + y + -y \sqsubseteq y + -y$ and hence $y + 0 \sqsubseteq 0$; therefore, being always $y \sqsubseteq y + 0$, it is $y \sqsubseteq 0$ and so $y = 0$. Hence

C4.3 THEOREM
A *po*-pregroup is complete iff it is a *c-c-l*-zeroid.

C4.4 COROLLARY
Up to signatures, the notion of complete Abelian *po*-pregroup and the notion of complete Grišin's L^0-algebra coincide.

C4.5 DEFINITION
[1] A structure $\langle G, \cdot, \widehat{0}, \sqsubseteq\rangle$ is a *unital commutative quantale* iff

[Q1] $\langle G, \cdot, \widehat{0}\rangle$ is an Abelian monoid;

[Q2] $\langle G, \sqsubseteq\rangle$ is a complete poset;

[Q3] $x \cdot (\sqcup_{i\in I}x_i) = \sqcup_{i\in I}(x \cdot x_i)$.

[2] A structure $\langle G, \cdot, -, \widehat{0}, \sqsubseteq\rangle$ is a *Girard quantale* iff Q1-3 $\langle G, \cdot, \widehat{0}, \sqsubseteq\rangle$ is a unital commutative quantale; Q4 $- - x = x$; Q5 $x \sqsubseteq y \leftrightarrow x \cdot -y \sqsubseteq \widehat{0}$.

C4.6 THEOREM
A structure $\langle G, \cdot, -, \widehat{0}, \sqsubseteq\rangle$ is a complete Abelian *po*-dualpregroup iff it is a Girard quantale satisfying the additional condition Q6 $\widehat{0} \sqsubseteq x \cdot -x$.

C5. *BASIC SEMANTICAL NOTIONS*

C5.1 DEFINITION

[1] A *frame* is a pair $\langle V, \mathfrak{G} \rangle$, where V is a non empty set and $\mathfrak{G} = \langle G, +, -, 0, \sqsubseteq \rangle$ is an Abelian *c-c-l*-pregroup.
[2] A function $R : V^k \longrightarrow G$ is *bounded* iff its range is bounded.
[3] $BND(V, G)$ is the set of all bounded functions from V^k (for some k) to G.
[4] A *realization* of a CL^1 in a frame $\langle V, \mathfrak{G} \rangle$ is a map ρ from PRD into $BND(V, G)$ such that $\rho(P^k) : V^k \longrightarrow G$, $\rho(\top) = 0$ and $\rho(\bot) = -0$.
[5] An *assignment in* ρ is a map $\sigma : VAR \longrightarrow G$.

C5.2 LEMMA

If $\langle V, \mathfrak{G} \rangle$ is a frame, then $BND(V, G)$ is closed under the *elementary combinators* $\mathbb{Z}, \mathbb{T}, \mathbb{I}, \mathbb{V}, \mathbb{N}, \mathbb{K}, \mathbb{S}$ and $\mathbb{D}$ defined by:

$\mathbb{Z}R(a_1, \ldots, a_k) = R(a_2, \ldots, a_k, a_1)$;
$\mathbb{T}R(a_1, \ldots, a_k) = R(a_1, \ldots, a_{k-2}, a_k, a_{k-1})$;
$\mathbb{I}R(a_1, \ldots, a_k) = R(a_1, \ldots, a_k, a_k)$;
$\mathbb{V}R(a_1, \ldots, a_k) = R(a_1, \ldots, a_k, a_{k+1})$;
$\mathbb{N}R(a_1, \ldots, a_k) = -R(a_1, \ldots, a_k)$;
$\mathbb{K}(R, S)(a_1, \ldots, a_k, b_1, \ldots, b_i) = R(a_1, \ldots, a_k) \sqcap S(b_1, \ldots, b_i)$;
$\mathbb{S}(R, S)(a_1, \ldots, a_k, b_1, \ldots, b_i) = R(a_1, \ldots, a_k) + S(b_1, \ldots, b_i)$;
$\Delta S(a_1, \ldots, a_k) = \sqcup [S(a_1, \ldots, a_k, a_{k+1}) \mid a_{k+1} \in V]$.

C5.3 LEMMA

For any realization ρ in a frame $\langle V, \mathfrak{G} \rangle$ and any assignment σ in ρ, the *valuation* ρ^σ defined by the following conditions is a *total* map $\rho^\sigma : FOR \longrightarrow G$

$$
\begin{aligned}
\rho^\sigma(P(v_1, \ldots, v_k)) &= \rho(P)(\sigma(v_1), \ldots, \sigma(v_k)) \\
\rho^\sigma(\neg\alpha) &= -(\rho^\sigma(\alpha)) \\
\rho^\sigma(\alpha \oplus \beta) &= \rho^\sigma(\alpha) + \rho^\sigma(\beta) \\
\rho^\sigma(\alpha \odot \beta) &= \rho^\sigma(\alpha) \cdot \rho^\sigma(\beta) \\
\rho^\sigma(\alpha < \beta) &= -\rho^\sigma(\alpha) + \rho^\sigma(\beta) \\
\rho^\sigma(\alpha \wedge \beta) &= \rho^\sigma(\alpha) \sqcap \rho^\sigma(\beta) \\
\rho^\sigma(\alpha \vee \beta) &= \rho^\sigma(\alpha) \sqcup \rho^\sigma(\beta) \\
\rho^\sigma(\alpha \to \beta) &= -\rho^\sigma(\alpha) \sqcup \rho^\sigma(\beta) \\
\rho^\sigma(\exists v\alpha) &= \sqcup\{\rho^{\sigma(^v_a)}(\alpha) \mid a \in V\} \\
\rho^\sigma(\forall v\alpha) &= \sqcap\{\rho^{\sigma(^x_a)}(\alpha) \mid a \in V\}
\end{aligned}
$$

[where $\sigma(^v_a)$ is the assignment defined by $\sigma(^v_a)(w) = \begin{cases} \sigma(w) & \text{if } v \neq w \\ a & \text{otherwise} \end{cases}$].

This Lemma is an easy consequence of 5.2. and of the fact that, for any map ρ of PRD into the set of bounded functions from V into G, the class of functions elementarily definable (in the obvious sense) in ρ is the smallest class which contains $\rho(P^k)$ for all P^k and is closed under the elementary combinators. A proof of 5.2. can be recovered from Quine 1966.

C5.4 DEFINITION
For $\alpha \in FRM$ and $M \subseteq FRM$:

[1] ρ^σ *satisfies* α $(\rho^\sigma \models \alpha)$ iff $0 \sqsubseteq \rho^\sigma(\alpha)$;
[2] α *is true in* $\rho(\rho \models \alpha)$ iff for all σ in ρ, $\rho^\sigma \models \alpha$;
[3] ρ *is a model* of M $(\rho \models M)$ iff for all $\alpha \in M$, $\rho \models \alpha$;
[4] α is a comparative consequence of M (M$\models \alpha$) iff for every frame and realization ρ in it, if $\rho \models M$, then $\rho \models \alpha$.

C5.5 THEOREM (Soundness)
For all $\alpha \in M$ and $M \subseteq FRM$, if $M \vdash \alpha$ then $M \models \alpha$.

C6. *ARCHIMEDEAN DERIVABILITY*

The notion of comparative derivability is however too weak in order to completely capture the notion of comparative consequence. This depends upon the fact that *c-c-po*-pregroups enjoy a very strong closure property. Let us write nx for $\overbrace{x + \cdots + x}^{n \text{ times}}$); then

C6.1 DEFINITION
A *po*-pregroup is *integrally closed* iff $L(\{nx \mid n \geq 1\}) \neq \emptyset \to -0 \sqsubseteq x$. Clearly every *po*-zeroid is integrally closed. Generalizing a classical result by Kantorovitsch we get

C6.2 THEOREM
Every *c-c*-pregroup is integrally closed.

PROOF
Suppose that for some y, $y \sqsubseteq nx$ for all $n \geq 1$; therefore, by conditional completeness, $\sqcap_{n\geq 1} nx$ exists. By Theorem 3.5.1, $x + \sqcap_{n\geq 1} nx = \sqcap_{n\geq 1}(x + nx) = \sqcap_{n\geq 2} nx$ and also, clearly, $\sqcap_{n\geq 1} nx \sqsubseteq \sqcap_{n\geq 2} nx$; so $\sqcap_{n\geq 1} nx \sqsubseteq x + \sqcap_{n\geq 1} nx$. Therefore $\sqcap_{n\geq 1} nx + - \sqcap_{n\geq 1} nx \sqsubseteq x + \sqcap_{n\geq 1} nx + - \sqcap_{n\geq 1} nx$; so $0 \sqsubseteq x + 0$, i.e., $-0 \sqsubseteq x$.

C6.3 DEFINITION
M is an *Archimedean system* [ARCH(M)] iff it contains the first order comparative axioms and is closed under MP, AR, $\forall I$, $\exists I$ and the following

Archimedes' Rule:

$$\frac{\text{for some } \beta \begin{cases} \neg\alpha < \beta \in M \\ \neg\alpha < (\neg\alpha < \beta) \in M \\ \neg\alpha < (\neg\alpha < (\neg\alpha < \beta)) \in M \\ \vdots \end{cases}}{\bot < \alpha \in M}$$

or, equivalently

$$\frac{\text{for some } \beta \begin{cases} \alpha \oplus \beta \in M \\ \alpha \oplus \alpha \oplus \beta \in M \\ \alpha \oplus \alpha \oplus \alpha \oplus \beta \in M \\ \vdots \end{cases}}{\alpha \oplus T \in M}$$

C6.4 DEFINITION
$M \vdash_A \alpha$ iff $\alpha \in \cap\{N \mid M \subseteq N \wedge ARCH(N)\}$.

Remarking that $\bot < \alpha$ is simply one of the possible formulations of boundedness, we get

C6.5 COROLLARY
Every M closed under bounded first order derivability is Archimedean.

C7. *ADEQUATENESS*

The notion of Archimedean derivability turns out to be the right syntactical counterpart of first order comparative consequence. In fact it is easy to prove

C7.1 LEMMA
The set of formulae true in a ρ is an Archimedean system.

Hence

C7.2 THEOREM (SOUNDNESS OF ARCHIMEDEAN DERIVABILITY).
If $M \vdash_A \alpha$ then $M \models \alpha$.

But it also holds

C7.3 THEOREM (COMPLETENESS OF ARCHIMEDEAN DERIVABILITY).
If $M \cup \{\alpha\}$ is a set of sentences, then: if $M \models \alpha$ then $M \vdash_A \alpha$.

This is proved by showing that *for every Archimedean system* N there are an Abelian *c-c-l*-pregroup and a realization ρ in it s.t. for all sentences α, $\alpha \in N$ iff $\rho \models \alpha$. To this aim let us first remark that for each Archimedean N the relation $\alpha \approx \beta \in N$ is a congruence in the algebra of formulae; so we can define, on the quotient classes $|\alpha| := \{\beta \mid \alpha \approx \beta \in N\}$, the operations $+, -$ and 0 by $|\alpha| + |\beta| := |\alpha \oplus \beta|$, $-|\alpha| := |\neg\alpha|$ and $0 := |\top|$ and the relation $\sqsubseteq$ by $|\alpha| \sqsubseteq |\beta| := \alpha < \beta \in N$. It is easy to prove that the resulting structure —the *Lindenbaum algebra* $\mathfrak{L}(N)$ of N— is an Abelian pregroup which is a group iff

N is inconsistent (i.e., iff $\bot \in N$) and which is lattice-ordered by the relation $\sqsubseteq$, and precisely so that the *sup* and the *inf* of $\{|\alpha|, |\beta|\}$ are $|\alpha \vee \beta|$ and $|\alpha \wedge \beta|$, and moreover the sets $\{|\alpha[v/w]| \mid w \in VAR\}$ (defined up to renaming of bound variables) have both *Sup* and *Inf* and precisely $|\exists v\alpha|$ and $|\forall v\alpha|$. From the Archimedeicity of N it immediately follows that $\mathfrak{L}(N)$ is integrally closed. In order to construct, along familiar paths, the canonical model for N, we must now know that $\mathfrak{L}(N)$ can be embedded in a *c-c-l*-pregroup in such a way that all existing *Sup's* and *Inf's* are preserved. To this aim:

C7.4 DEFINITION

A *c-c-*(l)-pregroup $\mathfrak{G}^\circ = \langle G^\circ, \mathring{+}, \mathring{-}, 0^\circ, \mathring{\sqsubseteq}\rangle$ is a *completion* of the *po-(l)*-pregroup $\mathfrak{G} = \langle G, +, -, 0, \sqsubseteq\rangle$ iff there is a join-and-meet-dense embedding of $\mathfrak{G}$ into $\mathfrak{G}'$ [i.e., an embedding g of $\mathfrak{G}$ into $\mathfrak{G}'$ such that $\forall x \in G^\circ\ \exists A\ B \subseteq G(x = \sqcup g(A) = \sqcap g(B))$].

C7.5 THEOREM

Every integrally closed (Abelian) (directed) *po*-pregroup has one and only one (Abelian) (lattice-ordered) completion.

Since a join-and-meet-dense embedding (*jamd*-embedding, for short) clearly preserves all existing Suprema and Infima, this solves our problem. Let us remark that this theorem, which will be proved in the next section, generalizes on the one side a basic result of the theory of ordered groups which, according to Birkhoff 1948, has a very long history [Arnold (1929), Krull (1932), Kantorovitch (1935), Lorenzen (1939), Clifford (1940), Dieudonné (1941), Everett & Ulam (1945), Birkhoff (1948)]; on the other side, it generalizes Grišin's result about completability of Abelian *po*-zeroids (his L^0-algebras).

Now let $\mathcal{C}(\mathfrak{L}(N))$ be the completion of $\mathfrak{L}(N)$. Let $\rho(\top) = |\top|$, $\rho(\bot) = |\bot|$ and, for every predicate P^k, let $\rho(P)$ be the map VAR$\longrightarrow \mathcal{C}(\mathfrak{L}(N))$ defined by $\rho(P)(v_1, \ldots, v_k) = |P(v_1, \ldots, v_k)|$. Clearly $|\forall v_1 \ldots v_k P(v_1, \ldots, v_k)| \sqsubseteq |P(v_1, \ldots, v_k)| \sqsubseteq |\exists v_1 \cdots v_k P(v_1, \ldots, v_k)|$, so $\rho(P) \in BND(VAR, \mathcal{C}(\mathfrak{L}(N)))$. So ρ is a realization into the frame $\langle VAR, \mathcal{C}(\mathfrak{L}(N))\rangle$. It is now routine to show that for all α, $\rho^\iota(\alpha) = |\alpha|$ (where ι is the identity map on VAR) and hence, for every sentence α, $\alpha \in N$ iff $\rho \models \alpha$.

C7.6 COROLLARY

The notion of complete Abelian zeroid gives a strongly adequate semantics for bounded first order comparative logic.

C7.7 REMARK.

In spite of the fact that we have developed first order comparative logics only in the Abelian case, it is possible to do the same also in the non-Abelian case. This possibility has been worked out in detail by Minari in a still unpublished paper. We stress here that the completion theorem to be proved in the next section will be proved for the general case, and that this generality is not superfluous. Quite differently from the purely group-theoretical case, indeed, where Iwasawa (1944) has shown that a *c-c*-group is always Abelian, there is, as shown by Minari, a multitude of non-Abelian c-c-pregroups, even in the finite case.

C7.8 REMARK

Closure under Archimedes' rule has been required in order to give a smooth notion of model for which completeness could be proved. It is possible to avoid this somehow intriguing requirement by suitably modifying the notions of valuation and model along the way followed in Casari 87. In particular, we may admit, on one side, l- but not necessarily c-c-l-pregroups as ranges of valuations and, on the other side, the possibility for a valuation to be only a *partial* map of FRM into G. Truth of a formula in G and all related notions are then defined *with respect to those valuations which are total.* Completeness for these notions is thus easily recovered by the simple construction of the Lindenbaum algebra, because this algebra certainly contains all *Sup's* and *Inf's* which are needed in order to make the canonical valuation a total one. Summing up, we can say that the intuitive notion of first order comparative logical law turns out to be a rather complex one; it involves a strongly non-effective move which can be charged either to syntax or to semantics, but which seems to be unavoidable.

C8. *PROOF OF THE COMPLETION THEOREM*

We begin by recalling a basic result of the theory of posets (for a proof see e.g. Birkhoff 67, pp.126–7).

C8.1 THEOREM (DEDEKIND-MACNEILLE)

If $\langle G, \sqsubseteq\rangle$ is a poset and $\mathcal{C}(G)$ is the family of subsets of G defined by $\mathcal{C}(G) := \{A \subseteq G | A \neq \emptyset \neq U(A), A = L(U(A))\}$, then [1] $\langle \mathcal{C}(G), \subseteq\rangle$ is a c-c-poset, with $\sqcap\{A_i \mid i \in I\} = \cap\{A_i \mid i \in I\}$ and $\sqcup\{A_i \mid i \in I\} = L(U(\cup\{A_i \mid i \in I\}))$; [2] the map $h : G \longrightarrow \mathcal{C}(G)$ defined by $h(x) = (x]$ is a jamd-embedding of $\langle G, \sqsubseteq\rangle$ into $\langle \mathcal{C}(G), \subseteq\rangle$, i.e., an embedding such that for every $A \in \mathcal{C}(G)$ there are $B, C \subseteq G$ such that $A = \sqcup\{h(x) \mid x \in B\} - \sqcap\{h(x) \mid x \in C\}$). Moreover, if $\langle G, \sqsubseteq\rangle$ is directed, $\langle \mathcal{C}(G), \subseteq\rangle$ is a lattice.

C8.2 DEFINITION

Let $\mathfrak{G} = \langle G, +, -, 0, \sqsubseteq\rangle$ be a *po*-pregroup. On the power set $\mathcal{P}(G)$ of G we define the *complex operations* $+$ and $-$ by:

[1] $A + B := \{x + y \mid x \in A, y \in B\}$;
[2] $-A := \{-x \mid x \in A\}$.

It is easy to prove:

C8.3 LEMMA

[1] $A + (B + C) = (A + B) + C$;
[2] $- - A = A$;
[3] If $\mathfrak{G}$ is Abelian, $A + B = B + A$.

We write '$A \sqsubseteq B$','$A \sqsubseteq x$','$x \sqsubseteq A$' and '$x + A$' for '$\forall x(x \in A \to \forall y(y \in B \to x \sqsubseteq y))$', '$A \sqsubseteq \{x\}$', '$\{x\} \sqsubseteq A$' and '$\{x\} + A$' respectively. Moreover we write 'A°' for $L(U(A))$; remark that A° is a closure on $\mathcal{P}(G)$.

C8.4 LEMMA

[1] $L(A+B) = L(U(L(A)) + U(L(B)))$;
[2] $L(-A) = -U(A)$;
[3] $U(-A) = -L(A)$;
[4] $x \sqsubseteq (U(A) + -A) \to -(U(A) + -A) \sqsubseteq n(-x)$ for all $n \geq 1$.

PROOF

[1] It is both $A \subseteq U(L(A))$ and $B \subseteq U(L(B))$; so $(A+B) \subseteq U(L(A)) + U(L(B))$ and hence $L(U(L(A)) + U(L(B))) \subseteq L(A+B)$. On the other side suppose $x \in L(A+B)$; we have to show that if $y \in U(L(A))$ and $z \in U(L(B))$, then $x \sqsubseteq y+z$. From $x \in L(A+B)$ it follows that $x \sqsubseteq a+b$ for all $a \in A$ and all $b \in B$; thus, by T4, still for all $a \in A$ and $b \in B$, $-(b + -x) \sqsubseteq a$, i.e., $-(B + -x) \sqsubseteq A$. It follows that $-(B + -x) \subseteq L(A)$ and so, being $y \in U(L(A))$, it must be $-(B + -x) \sqsubseteq y$. Thus, for all $b \in B$, $-(b + -x) \sqsubseteq y$ and so, by two applications of T4, $-(-x + y) \sqsubseteq b$ for all $b \in B$, i.e., $-(-x+y) \sqsubseteq B$; but then $-(-x+y) \in L(B)$ and being $z \in U(L(B))$ it must be $-(-x+y) \sqsubseteq z$ and so by T4, $x \sqsubseteq y + z$.

[2] $x \in L(-A) \leftrightarrow x \sqsubseteq -A \leftrightarrow \forall y(y \in -A \to x \sqsubseteq y) \leftrightarrow \forall z(z \in A \to x \sqsubseteq -z) \leftrightarrow \forall z(z \in A \to z \sqsubseteq -x) \leftrightarrow A \sqsubseteq -x \leftrightarrow -x \in U(A) \leftrightarrow x \in -U(A)$.

[3] Dually.

[4] By induction on n. $[n = 1]$ From $x \sqsubseteq U(A) + -A$ it follows $-(U(A) + -A) \sqsubseteq -x = 1(-x)$. $[n = k+1]$ By inductive hypothesis $x \sqsubseteq U(A) + -A$ implies $-(U(A)+-A) \sqsubseteq k(-x)$; thus $-k(-x) \sqsubseteq U(A)+-A$ and so, for all $u \in U(A)$ and all $y \in A$, $-k(-x) \sqsubseteq u+-y$ and also, by T1, $y \sqsubseteq k(-x)+u$. Then, for all $y \in A$, $y \sqsubseteq k(-x) + U(A)$, i.e., $k(-x) + U(A) \subseteq U(A)$. Thus $k(-x) + U(A) + -A \subseteq U(A) + -A$. By hypothesis, $x \sqsubseteq (U(A) + -A)$; so a fortiori $x \sqsubseteq k(-x) + U(A) + -A$; but then, by T1, $-(U(A) + -A) \sqsubseteq -x + k(-x) = (k+1)(-x)$.

C8.5 DEFINITION

[1] $A \mathring{+} B := L(U(A) + U(B))$;
[2] $\mathring{-} A := L(-A)$;
[3] $0^\circ := \langle 0]$.

Being $L(U(L(A))) = L(A)$, it is for all $A, B \subseteq G : A \mathring{+} B \in \mathcal{C}(G)$, $\mathring{-}A \in \mathcal{C}(G)$ as well as $0^\circ \in \mathcal{C}(G)$; so $\mathcal{C}(G)$ is closed under these operations.

C8.6 MAIN LEMMA

If $\mathfrak{G} = \langle G, +, -, 0, \sqsubseteq \rangle$ is an (Abelian) integrally closed (directed) pregroup, then $\mathcal{C}(\mathfrak{G}) = \langle \mathcal{C}(G), \subseteq, \mathring{+}, \mathring{-}, 0^\circ \rangle$ is an (Abelian) *c-c-(l-)* pregroup

PROOF

By THEOREM 1, $\langle \mathcal{C}(G), \subseteq \rangle$ is a *c-c*-poset (lattice, if $\langle G, \sqsubseteq \rangle$ is directed).

[*ad* G1] $A \overset{\circ}{+} (B \overset{\circ}{+} C) = \{\text{by } 5.1\}\ L(U(A) + U(L(U(B) + U(C))))$ $= L(U(L(U(A))) + U(L(U(B) + U(C)))) = \{\text{by } 4.1\}\ L(U(A) + (U(B) + U(C))) = \{\text{by } 3.1\}\ L((U(A) + U(B)) + U(C)) = \{\text{by } 4.1\}\ L(U(L(U(A) + U(B))) + U(L(U(C)))) = L(U(L(U(A) + U(B))) + U(C)) = \{\text{by } 5.1\}$ $(A \overset{\circ}{+} B) \overset{\circ}{+} C$.

[*ad* G1′] Trivial.

[*ad* G2] It is $\overset{\circ}{-}\overset{\circ}{-}A = \{\text{by } 5.2\}\ L(-L(-A)) = \{\text{by } 4.2\}\ L(U(--A)) = L(U(A))$ $= \{\text{by } A \in \mathcal{C}(G)\}\ A$.

[*ad* G3] $0^\circ \overset{\circ}{+} 0^\circ = L(U(L(U\{0\}))) + U(L(U(\{0\})))) = L(U(\{0\}) + U(\{0\})) =$ $L([0\rangle + [0\rangle) = L([0\rangle) = 0^\circ$.

[*ad* G5] Trivial.

[*ad* G6] First of all remark that, for $A, B \in \mathcal{C}(G)$, $\overset{\circ}{-}A \overset{\circ}{+} B = L(-A + U(B))$. Indeed: $\overset{\circ}{-}A \overset{\circ}{+} B = \{\text{by } 5.1\text{–}2\}\ L(U(L(-A) + U(B)) = \{\text{by } 4.2\text{–}3\}\ L(-L(U(A)) + U(B)) = \{A \in \mathcal{C}(G)\}\ L(-A + U(B))$. So we must show $A \subseteq B \leftrightarrow \langle 0] \subseteq L(-A + U(B))$. [$\rightarrow$] Let be $A \subseteq B$. Suppose $z \in -A + U(B)$; then for some $-x \in A$ and $B \sqsubseteq y$, $z = x + y$. Since $A \subseteq B$, it is $-x \in B$; therefore $-x \sqsubseteq y$ and so $0 \sqsubseteq x + y$, i.e., $x + y \in [0\rangle$; so $-A + U(B) \subseteq [0\rangle$ and then $\langle 0] \subseteq L(-A + U(B))$. [$\leftarrow$] Suppose $\langle 0] \subseteq L(-A + U(B))$; then $0 \in L(-A + U(B))$, i.e., $0 \sqsubseteq (-A + U(B))$; thus for all $x \in A$ and $y \in U(B)$, $0 \sqsubseteq -x + y$ and so $x \sqsubseteq y$, i.e., $A \sqsubseteq U(B)$. So $U(B) \subseteq U(A)$ and therefore $L(U(A)) \subseteq L(U(B))$; being $A, B \in \mathcal{C}(G)$, it is $A \subseteq B$. Analogously we prove $A \subseteq B \leftrightarrow 0^\circ \subseteq B \overset{\circ}{+} \overset{\circ}{-}A$.

[*ad* G4] $\langle 0] \subseteq A \overset{\circ}{+} \overset{\circ}{-}A$ follows immediately from G6. Moreover, it is $A \overset{\circ}{+} \overset{\circ}{-}A =$ $\{by\ 5.1\text{–}2\}\ L(U(A) + U(L(-A)) = L(U(L(U(A))) + U(L(-A))) = \{by\ 4.1\}\ L(U(A) + -A)$; so we have to show $L(U(A) + -A) \subseteq \langle 0]$. Suppose $x \in L(U(A) + -A)$; then $x \sqsubseteq (U(A) + -A)$. By 4.4, $-(U(A) + -A) \sqsubseteq n(-x)$ for all $n \geq 1$ and, being $-(U(A) + -A) \neq \emptyset$, it is $L(\{n(-x) \mid n \geq 1\}) \neq \emptyset$. By hypothesis, $\mathfrak{G}$ is integrally closed; so $-0 \sqsubseteq -x$, i.e., $x \sqsubseteq 0$, i.e., $x \in \langle 0]$.

Remark that the hypothesis of integral closure is used only in the proof of $A \overset{\circ}{+} \overset{\circ}{-}A \subseteq 0^\circ$. It is also interesting to remark that in the expanded presentation of $\mathcal{C}(\mathfrak{G})$, it is $A \overset{\circ}{\cdot} B = \overset{\circ}{-}(\overset{\circ}{-}A \overset{\circ}{+} \overset{\circ}{-}B) = L(U(L(U(A)) \cdot L(U(B))))$, i.e., for $A, B \in \mathcal{C}(G)$, $A \overset{\circ}{\cdot} B = (A \cdot B)^\circ$. This explains way the usual proof of the completion theorem for *po*-groups (which are autodual!) works with the *definition* $A \overset{\circ}{+} B := (A + B)^\circ$. This identity is however not generally valid in *po*-pregroups.

C8.7 LEMMA
$\mathfrak{G}$ is jamd-embeddable in $\mathcal{C}(\mathfrak{G})$.

PROOF
By THEOREM 1 we already know that $h(x) = \langle x]$ is a *jamd*-embedding of the poset $\langle G, \sqsubseteq\rangle$ into the poset $\langle \mathcal{C}(G), \subseteq\rangle$; it remains to show that it also preserves the pregroup operations. This follows from the easily verifiable equations: $\langle x + y] = L([x\rangle + [y\rangle)$ and $-\langle x] = [-x\rangle$.

C8.8 REMARK

The completion is, up to isomorphisms, unique.
Indeed if $\mathfrak{G}' = \langle G', +', -', 0', \sqsubseteq' \rangle$ is a *c-c-po*-pregroup such that there is a jamd-embedding g of $\mathfrak{G} = \langle G, +, -, 0, \sqsubseteq \rangle$ into $\mathfrak{G}'$, then the map which associates to every element A of $\mathcal{C}(G)$ (which is a nonempty subset of G having in G an upper bound) the *Sup* of its image in G' under g, i.e., the map $f : \mathcal{C}(G) \longrightarrow G'$ defined by $f(A) = \sqcup'\{g(x) \mid x \in A\}$, is an isomorphism of $\mathfrak{G}^\circ$ onto $\mathfrak{G}'$.

C8.9

Whereas, for instance, the completion of a Boolean pregroup is always a Boolean pregroup, the completion of an *mv*-pregroup is not necessarily an *mv*-pregroup. For instance, the completion of the well known Chang's example of a linear but not simple *mv*-pregroup is not an *mv*-pregroup.

D. THREE LEVEL LOGIC

D1. *MOTIVATION*

The construction of a logic 'on three levels' originated from two different questions.

[1] To account in a reasonable and non trivial way for the formal aspects of *reasoning by analogy*, in particular to extend comparative logic in order to allow the discussion of arguments such as *If A is to B as C is to D, then A is to C as B is to D*, or of more intriguing ones. The endowing of the world of truth-degrees with an appropriate ratio seems to presuppose its endowing with an appropriate multiplication. A satisfactory theory of the inversion of such a multiplication is however still a *desideratum.*

[2] To develop a logic of compounded *action terms* in which besides the *static* level of composition by 'and' and 'or', characterized by idempotency and commutativity, there are two *dynamic* 'dimensions', one, let us freely say, of *concurrency*, characterized by giving up idempotency, though retaining commutativity and another one of *sequentiality*, on which both idempotency and commutativity are given up. Against Pratt's suggestion [see Pratt 91], however, the two dynamic dimensions should be, on their side, true 'levels', with the concurrent as the lower and the sequential as the higher one. One important marker of this difference should be the 'and'/'or' distributivity which (with the possible exception of the static one) does not act on the same level but only from the higher to the lower one. In particular the sequential 'or' distributes over the concurrent 'and' and, as far as possible, over the static 'and'; dually the sequential 'and' distributes over the concurrent 'or', and as far as possible over the static 'or'. The concurrent 'or' on its side distributes over the static 'and', and dually the concurrent 'and' distributes over the static 'or'. We are not completely satisfied with the results obtained up to

now; thus we will present here the basic ideas of the third level simply as an extension of comparative logic.

D2. *l- and f-PRERINGS*

We firstly introduce the structures to be used in evaluating the language.

D2.1 DEFINITION

[1] $\mathfrak{R} = \langle R, +, -, 0, \times, 1, \sqsubseteq\rangle$ is an *l-prering* iff

[R1] $\langle R, +, -, 0, \sqsubseteq\rangle$ is an Abelian l-pregroup;

[R2] $\langle R, \times, 1\rangle$ is a monoid;

[R3] $\begin{cases} x \times (y + z) = (x \times y) + (x \times z) \\ (y + z) \times x = (y \times x) + (z \times x) \end{cases}$;

[R4] $0 \times (x + 0) = 0 = (x + 0) \times 0$;

[R5] $x \times -(y + 0) = -((x \times y) + 0) = -(x + 0) \times y$;

[R6] $x \times -(y \times z) \sqsubseteq -(-(x \times -y) \times z)$;

[R7] $-0 \sqsubseteq x \to \begin{cases} (x \times y) \sqsubseteq x \times (y \sqcup z) \\ (y \times x) \sqsubseteq (y \sqcup z) \times x \end{cases}$;

[2] $\mathfrak{R} = \langle R, +, -, 0, \times, 1, \sqsubseteq\rangle$ is an *f-prering* iff it is an l-prering satisfying

[R7′] $-0 \sqsubseteq x \to \begin{cases} x \times (y \sqcup z) = (x \times y) \sqcup (x \times z) \\ (y \sqcup z) \times x = (y \times x) \sqcup (z \times x) \end{cases}$.

We collect in the following Theorem some simple arithmetical facts.

D2.2 THEOREM

In every l-prering:

[1] $0 \times 0 = 0$;
[2] $(0 \times x) + 0 = 0 = (x \times 0) + 0$;
[3] $x \times -0 = -0 = -0 \times x$;
[4] $-0 \sqsubseteq 0 \times x, x \times 0 \sqsubseteq 0$;
[5] $-0 \sqsubseteq x \wedge y \sqsubseteq z \to \begin{cases} (x \times y \sqsubseteq x \times z) \\ (y \times x \sqsubseteq z \times x) \end{cases}$;
[6] $-0 \sqsubseteq x \wedge -0 \sqsubseteq y \to -0 \sqsubseteq x \times y$;
[7] $0 \sqsubseteq x \to 0 \times x = 0 = x \times 0$;
[8] $x \sqsubseteq -0 \to 0 \times x = -0 = x \times 0$;
[9] $-0 \sqsubseteq x \sqsubseteq 1 \to -0 \sqsubseteq x \times x \sqsubseteq x \sqsubseteq 1$.

In every f-prering:

[10] $0 \times (x \sqcup -0) = 0 \times x$;
[11] $0 \times (1 \sqcup -0) = 0$.

PROOF

[1] By R4, $0 = 0 \times (-0 + 0) = 0 \times 0$.

[2] $(0 \times x) + 0 = (0 \times x) + (0 \times 0) = 0 \times (x + 0) = 0$.
[3] $x \times -0 = x \times -(0 + 0) = -((x \times 0) + 0) = -0$.
[4] from [2].
[5] From R7.
[6] From [5] and [3].
[7] If $0 \sqsubseteq x$, then $0 \times 0 \sqsubseteq 0 \times x$ and so, by [1] and [4], $0 \times x = 0$.
[8] As $-0 \sqsubseteq 0$, so $x \sqsubseteq -0$ implies $0 \times x \sqsubseteq 0 \times -0$ and then by [3] and [4], $0 \times x = -0$.
[9] If $-0 \sqsubseteq x$, so $-0 = -0 \times x \sqsubseteq x \times x$ and also, if $x \sqsubseteq 1$, $x \times x \sqsubseteq x \times 1 = x \sqsubseteq 1$.
[10] $0 \times (x \sqcup -0) = (0 \times x) \sqcup (0 \times -0) = (0 \times x) \sqcup -0 = 0 \times x$.
[11] By [10], $0 \times (1 \sqcup -0) = 0 \times 1 = 0$.

D2.3 THEOREM
The l-prerings (f-prerings) form a variety in the signature $\langle +, -, 0, \times, 1, \sqcap \rangle$.

PROOF
By THEOREM B2.5 and the easily provable fact that R7 and R7′ are respectively equivalent to:

$$\begin{cases} (x \sqcup -0) \times y \sqsubseteq (x \sqcup -0) \times (y \sqcup z) \\ y \times (x \sqcup -0) \sqsubseteq (y \sqcup z) \times (x \sqcup -0) \end{cases}$$

and

$$\begin{cases} (x \sqcup -0) \times (y \sqcup z) = ((x \sqcup -0) \times y) \sqcup ((x \sqcup -0) \times z) \\ (y \sqcup z) \times (x \sqcup -0) = (y \times (x \sqcup -0)) \sqcup (z \times (x \sqcup -0)) \end{cases}.$$

Clearly, if $0 = -0$ then an l-prering (f-prering) is simply an l-ring (f-ring, i.e., a subdirect product of totally ordered rings). As already seen, other equivalents of $0 = -0$ are, for instance, $x + 0 = x$ or $x + y = -(-x + -y)$. We now have other alternatives:

D2.4 LEMMA
In every l-prering $0 = -0$ iff $(-x) \times y = -(x \times y) = x \times (-y)$ iff $x \times y = (-x) \times (-y)$ iff $(-1) \times x = -x$ iff $x \times -(y \times z) = -(-(x \times -y) \times z)$.

PROOF
If $0 = -0$, then by R5, $x \times (-y) = -(x \times y) = (-x) \times y$. If $x \times (-y) = -(x \times y) = (-x) \times y$, then $x \times y = (--x) \times (--y) = -((--x) \times (-y)) = (-x) \times (-y)$. If $x \times y = (-x) \times (-y)$, then $0 \times 0 = -0 \times -0$ so, by 2.1 and 2.3, $0 = -0$. If $0 = -0$, then, by R5, $-1 \times x = -(1+0) \times x = -(1 \times x) + 0 = -x$; on the other side if $-1 \times x = -x$, then $-0 = -1 \times -0 = --0 = 0$. If $x \times -(y \times z) = -(-(x \times -y) \times z)$, then $-0 = -0 \times -(-0 \times -0) = -(-(-0 \times --0) \times -0) = 0$. On the other side, as already seen, if $-0 = 0$, then $x \times -(y \times z) = -(x \times (y \times z)) = -((x \times y) \times z) = -(-(x \times -y) \times z)$.

D3. *ZEROIDAL, DUAL AND EXPANDED PRERINGS*

The theory of zeroids can be transferred to the new situation. So

D3.1 DEFINITION
An l-(f)-prering $\mathfrak{R}$ is *zeroidal* iff it satisfies $x + 0 = 0$.

D3.2 LEMMA
$\mathfrak{R} = \langle R, +, -, 0, \times, 1, \sqcap \rangle$ is a zeroidal l-(f-)prering iff

[1] $\langle R, +, -, 0, \sqsubseteq \rangle$ is an Abelian l-zeroid;
[2] $\langle R, \times, 1 \rangle$ is a monoid;
[3] $\begin{cases} x \times (y + z) = (x \times y) + (x \times z) \\ (y + z) \times x = (y \times x) + (z \times x) \end{cases}$;
[4] $0 \times 0 = 0$;
[5] $x \times -0 = -0 = -0 \times x$;
[6] $x \times -(y \times z) \sqsubseteq -(-(x \times -y) \times z)$;
[7] $\begin{cases} (x \times y) \sqsubseteq x \times (y \sqcup z) \\ (y \times x) \sqsubseteq (y \sqcup z) \times x \end{cases}$;

respectively

[7′] $\begin{cases} x \times (y \sqcup z) = (x \times y) \sqcup (x \times z) \\ (y \sqcup z) \times x = (y \times x) \sqcup (z \times x) \end{cases}$.

D3.3 DEFINITION
$\mathfrak{R} = \langle R, \cdot, -, \hat{0}, *, \hat{1}, \sqsubseteq \rangle$ is an l-(f-)dualprering iff

[D1] $\langle R, \cdot, -, \hat{0}, \sqsubseteq \rangle$ is an Abelian l-dualpregroup;
[D2] $\langle R, *, \hat{1} \rangle$ is a monoid;
[D3] $\begin{cases} x * (y \cdot z) = (x * y) \cdot (x * z) \\ (y \cdot z) * x = (y * x) \cdot (z * x) \end{cases}$;
[D4] $\hat{0} * (x \cdot \hat{0}) = \hat{0} = (x \cdot \hat{0}) * \hat{0}$;
[D5] $x * -(y \cdot \hat{0}) = -((x * y) \cdot \hat{0}) = -(x \cdot \hat{0}) * y$;
[D6] $-(-(x * y) * z) \sqsubseteq x * -(-y * z)$;
[D7] $x \sqsubseteq -\hat{0} \to \begin{cases} x * (y \sqcap z) \sqsubseteq (x * y) \\ (y \sqcap z) * x \sqsubseteq (y * x) \end{cases}$;

respectively

[$D7'$] $x \sqsubseteq -\hat{0} \to \begin{cases} x * (y \sqcap z) = (x * y) \sqcap (x * z) \\ (y \sqcap z) * x = (y * x) \sqcap (z * x) \end{cases}$.

D3.4 DEFINITION
$\mathfrak{R} = \langle R, +, \cdot, -, 0, \hat{0}, \times, *, 1, \hat{1}, \sqsubseteq \rangle$ is an expanded l-prering iff

[1] $\langle R, +, -, 0, \times, 1, \sqsubseteq \rangle$ is an l-prering;
[2] $\langle R, \cdot, -, \hat{0}, *, \hat{1}, \sqsubseteq \rangle$ is an l-dualprering;
[3.1] $x \cdot y = -(-x + -y)$;
[3.2] $\hat{0} = -0$;
[4.1] $x * y = -(-x \times -y)$;
[4.2] $\hat{1} = -1$.

D3.5 REMARK
Instead of the schemata R6 and D6 we may assume *Peirce's Laws* for $\times$ and $*$, i.e., $x \times (y * z) \sqsubseteq (x \times y) * z$ [or, equivalently, $(x * y) \times z \sqsubseteq x * (y \times z)$].

D4. *EXAMPLES*

Let be $\mathfrak{Z} = \langle Z, +, -, 0, \times, 1, \leq \rangle$ the f-ring of the integers and $\mathfrak{R} = \langle R, \cup, -, V, \circ, \Delta, \subseteq \rangle$ a Boolean field of binary relations over a set V containing the identity Δ and closed under the relative product $\circ$. Then:

1. $\mathfrak{Z}$ is a (commutative) f-prering.
2. $\mathfrak{R}$ is a zeroidal f-prering in which – as a Boolean algebra – the 'concurrent' level collapses on the 'static' one (i.e., $x + y = x \sqcup y$).
3. The direct product $\mathfrak{Z} \otimes \mathfrak{R}$ is an f-prering both under the product and the lexicographical order.

In detail, the two models $\langle M, +, -, 0, \times, 1, \sqsubseteq \rangle$ and $\langle M, +, -, 0, \times, 1, \sqsubseteq' \rangle$ are defined by:

[1] $M := \{\langle a, r \rangle \mid a \in Z, r \in R\}$;
[2] $\langle a, r \rangle + \langle b, s \rangle := \langle a + b, r \cup s \rangle$;
[3] $0 := \langle 0, V \rangle$;
[4] $-\langle a, r \rangle := \langle -a, -r \rangle$;
[5] $\langle a, r \rangle \times \langle b, s \rangle := \langle a \times b, r \circ s \rangle$;
[6] $1 := \langle 1, \Delta \rangle$;
[7] $\langle a, r \rangle \sqsubseteq \langle b, s \rangle := a \leq b \wedge r \subseteq s$;
[7′] $\langle a, r \rangle \sqsubseteq' \langle b, s \rangle := a < b \vee (a = b \wedge r \subseteq s)$.

In particular it can be proved:

[8] $\langle a, r \rangle \sqcup \langle b, s \rangle = \langle max(a, b), r \cup s \rangle$;
[9] $\langle a, r \rangle \sqcap \langle b, s \rangle = \langle min(a, b), r \cap s \rangle$;

[8′] $$\langle a, r \rangle \sqcup' \langle b, s \rangle = \begin{cases} \langle b, s \rangle, & \text{if } a < b \\ \langle a, r \rangle, & \text{if } b < a \\ \langle a, r \cup s \rangle, & \text{otherwise} \end{cases}$$

[9′] $$\langle a, r \rangle \sqcap' \langle b, s \rangle = \begin{cases} \langle a, s \rangle, & \text{if } a < b \\ \langle b, r \rangle, & \text{if } b < a \\ \langle a, r \cap s \rangle, & \text{otherwise} \end{cases}$$

D4.1 REMARK

There are non trivial zeroidal l-prerings, i.e., such that the operations on the three levels are all distinguished. For instance (Minari), Ł$_3$ can be expanded into a (commutative) f-prering.

$+$	0	$\frac{1}{2}$	1
0	0	$\frac{1}{2}$	1
$\frac{1}{2}$	$\frac{1}{2}$	1	1
1	1	1	1

$\sqcup$	0	$\frac{1}{2}$	1
0	0	$\frac{1}{2}$	1
$\frac{1}{2}$	$\frac{1}{2}$	$\frac{1}{2}$	1
1	1	1	1

$*$	0	$\frac{1}{2}$	1
0	0	0	1
$\frac{1}{2}$	0	$\frac{1}{2}$	1
1	1	1	1

Remark: $x \ast y = (x \sqcup y) \cdot (x + y)$ and so $= -(-(-(-x + y) + y) + -(x + y))$

$\cdot$	0	$\frac{1}{2}$	1
0	0	0	0
$\frac{1}{2}$	0	0	$\frac{1}{2}$
1	0	$\frac{1}{2}$	1

$\sqcap$	0	$\frac{1}{2}$	1
0	0	0	0
$\frac{1}{2}$	0	$\frac{1}{2}$	$\frac{1}{2}$
1	0	$\frac{1}{2}$	1

$\times$	0	$\frac{1}{2}$	1
0	0	0	0
$\frac{1}{2}$	0	$\frac{1}{2}$	1
1	0	1	1

Remark: $x \times y = (x \sqcap y) + (x \cdot y)$.

D5. *LANGUAGE AND SEMANTICS*

A *three level propositional language* L^* is obtained from a propositional comparative language L^c by the adjunction of two new binary connectives $\boxdot$ and $\boxplus$, and two new propositional constants Δ and ∇. So:
$\alpha := p|\bot|\top|\Delta|\nabla|\neg\alpha|\alpha \wedge \beta|\alpha \vee \beta|\alpha \to \beta|\alpha \odot \beta|\alpha \oplus \beta|\alpha < \beta|\alpha \boxdot \beta|\alpha \boxplus \beta$.

D5.1 DEFINITION

[1] A *realization* of the language L^* in the f-prering $\mathfrak{R} = \langle R, +, -, 0, \times, 1, \sqsubseteq\rangle$ is a map ρ such that its restriction to the comparative sublanguage L^c is a comparative realization into the Abelian l-pregroup $\langle R, +, -, 0, \sqsubseteq\rangle$ and moreover $\rho(\Delta) = 1$; $\rho(\alpha \boxdot \beta) = \rho(\alpha) \times \rho(\beta)$; $\rho(\nabla) = -1$; $\rho(\alpha \boxplus \beta) = -(-\rho(\alpha) \times -\rho(\beta))$.
[2] $\rho \models \alpha$ iff $0 \sqsubseteq \rho(\alpha)$;
[3] $\rho \models M$ iff for all $\beta \in M$, $\rho \models \beta$;
[4] $\mathfrak{R} \models \alpha$ iff for all ρ in $\mathfrak{R}$ $\rho \models \alpha$;
[5] $M \models \alpha$ iff for all $\mathfrak{R}$ and ρ in it if $\rho \models M$, then $\rho \models \alpha$.

D6. *SYNTAX*

The *axioms* and *rules* of the *propositional three level logic* are given by the axiom-schemata and the rules of the propositional comparative logic plus the schemata

[A1] $\alpha \boxdot (\beta \boxdot \gamma) \gtrless (\alpha \boxdot \beta) \boxdot \gamma$

[A2] $\begin{cases} \alpha \boxdot \Delta \gtrless \alpha \\ \Delta \boxdot \alpha \gtrless \alpha \end{cases}$

[A3] $\begin{cases} \alpha \boxdot (\beta \oplus \gamma) \gtrless (\alpha \boxdot \beta) \oplus (\alpha \boxdot \gamma) \\ (\beta \oplus \gamma) \boxdot \alpha \gtrless (\beta \boxdot \alpha) \oplus (\gamma \boxdot \alpha) \end{cases}$

[A4] $T \boxdot T$

[A5] $\begin{cases} \bot < \alpha \boxdot \top \\ \bot < \top \boxdot \alpha \end{cases}$

[A6] $\begin{cases} \alpha \boxdot \top < \top \\ \top \boxdot \alpha < \top \end{cases}$

[A7] $\begin{cases} (\bot < \alpha \boxdot \top) < \top \\ (\bot < \top \boxdot \alpha) < \top \end{cases}$

[A8] $\begin{cases} (\alpha \boxdot \neg(\beta \oplus \top)) \between \neg(\alpha \boxdot \beta \oplus \top) \\ (\neg(\alpha \oplus \top) \boxdot \beta) \between \neg(\alpha \boxdot \beta \oplus \top) \end{cases}$

[A9] $\alpha \boxplus \beta \between \neg(\neg\alpha \boxdot \neg\beta)$

[A10] $\nabla \between \neg\Delta$

[A11] $\alpha \boxdot (\beta \boxplus \gamma) < (\alpha \boxdot \beta) \boxplus \gamma$

and the two rules

$$\frac{\bot < \alpha}{\alpha \boxdot (\beta \sqcup \gamma) \between (\alpha \boxdot \beta) \sqcup (\alpha \boxdot \gamma)} \quad \text{DR1}$$

$$\frac{\bot < \alpha}{(\beta \sqcup \gamma) \boxdot \alpha \between (\beta \boxdot \alpha) \sqcup (\gamma \boxdot \alpha)} \quad \text{DR2}$$

D6.1 REMARK

Instead of the rules we may assume one of the two pairs of schemata:

$$\begin{cases} (\alpha \vee \bot) \boxdot (\beta \vee \gamma) \between ((\alpha \vee \bot) \boxdot \beta) \vee ((\alpha \vee \bot) \boxdot \gamma) \\ (\beta \vee \gamma) \boxdot (\alpha \vee \bot) \between (\beta \boxdot (\alpha \vee \bot)) \sqcup (\gamma \boxdot (\alpha \vee \bot)) \end{cases}$$

$$\begin{cases} (\top \to \alpha) \boxdot (\beta \vee \gamma) \between ((\top \to \alpha) \boxdot \beta) \vee ((\top \to \alpha) \boxdot \gamma) \\ (\beta \vee \gamma) \boxdot (\top \to \alpha) \between (\beta \boxdot (\top \to \alpha)) \sqcup (\gamma \boxdot (\top \to \alpha)) \end{cases}$$

D6.2 THEOREM (STRONG ADEQUATENESS' THEOREM)

$M \vdash \alpha$ iff $M \models \alpha$.

The proof is routine.

D6.3 REMARK.

Our semantics refers to realizations in f-prerings; it is however clear that we could work as well with the more general case of l-prerings. Of course, the adequate axiomatization is then obtained by assuming

$$\frac{\bot < \alpha}{(\alpha \boxdot \beta) < \alpha \boxdot (\beta \sqcup \gamma)} \quad \text{DR}'1$$

and

$$\frac{\bot < \alpha}{(\beta \boxdot \alpha) < (\beta \sqcup \gamma) \boxdot \alpha} \quad \text{DR}'2$$

instead of DR1, DR2.

D7. *BOUNDED RINGOIDAL LOGIC*

The logic characterized by the zeroidal f-prerings is axiomatized by the schemata and the rule of the bounded comparative logic plus the schemata

[B1] $\alpha \boxdot (\beta \boxdot \gamma) \lessgtr (\alpha \boxdot \beta) \boxdot \gamma$;

[B2] $\begin{cases} \alpha \boxdot \Delta \lessgtr \alpha \\ \Delta \boxdot \alpha \lessgtr \alpha \end{cases}$;

[B3] $\begin{cases} \alpha \boxdot (\beta \oplus \gamma) \lessgtr (\alpha \boxdot \beta) \oplus (\alpha \boxdot \gamma) \\ (\beta \oplus \gamma) \boxdot \alpha \lessgtr (\beta \boxdot \alpha) \oplus (\gamma \boxdot \alpha) \end{cases}$;

[B4] $\top \boxdot \top$;

[B5] $\begin{cases} \neg(\alpha \boxdot \bot) \\ \neg(\bot \boxdot \alpha) \end{cases}$;

[B6] $\alpha \boxplus \beta \lessgtr \neg(\neg\alpha \boxdot \neg\beta)$;

[B7] $\nabla \lessgtr \neg\Delta$;

[B8] $\alpha \boxdot (\beta \boxplus \gamma) < (\alpha \boxdot \beta) \boxplus \gamma$

[B9] $\begin{cases} \alpha \boxdot (\beta \sqcup \gamma) \lessgtr (\alpha \boxdot \beta) \sqcup (\alpha \boxdot \gamma) \\ (\beta \sqcup \gamma) \boxdot \alpha \lessgtr (\beta \boxdot \alpha) \sqcup (\gamma \boxdot \alpha) \end{cases}$.

Università degli Studi di Firenze
Florence

ACKNOWLEDGMENTS

The author is greatly indebted to Pierluigi Minari for countless stimulating conversations on the topics covered by this paper.

REFERENCES

Birkhoff, G., *Lattice Theory*, AMS Colloquium Publ. 25, 2nd ed., 1948.

Birkhoff, G., *Lattice Theory*, AMS Colloquium Publ. 25, 3nd ed., 1967.

Casari, E., ≪Komparationstheorie und Mehrwertigkeit≫, *Tagungsberichte des math. Forschungsinstitutes Oberwolfach*, 1984 (1), p. 4.

Casari, E., ≪Comparative Logics≫, *Synthese* 73 (1987), pp. 421–449.

Casari, E., ≪Comparative Logics and Abelian l-groups ≫, *Logic Colloquium '88*, North-Holland, Amsterdam 1989, pp.161–190.

Casari, E., ≪Logics on Pregroups≫, *Atti del Congresso Soc. It. Log. Fil. d. Sc. 1990*, C.L.U.E.B, Bologna 1991, Vol.II, pp.39–58.

Frede, M., *Die stoische Logik*, Vandenhoeck & Ruprecht, Göttingen 1974

Grišin, V. N., ≪Predicate and set-theoretic Calculi based on Logic without Contractions≫, *Math. URSS Izvestija*, 18 (1982), No. 1, pp. 41–59.

Lambek, J., ≪From Categorial Grammar to Bilinear Logic ≫, in Schoeder-Heister & Došen 1993, pp. 207–37

Peirce, C. S., *Collected Papers*, Belknap Press of Harvard University Press, 1980, Vol. III.

Pratt,V., ≪Action Logic and Pure Induction≫, *LNCS* 478, Springer, Berlin 1990, pp. 97–120.

Quine, W. V., ≪Variables explained away≫, *Selected Logic Papers*, Random House, New York 1966, pp.227–235.

Schroeder-Heister, P. & Došen, K. (Eds.), *Substructural Logics*, Clarendon Press, Oxford 1993.

Venn, J., *Symbolic Logic*, 2nd ed. 1894, Chelsea, New York 1971.

KOSTA DOŠEN

LOGICAL CONSEQUENCE: A TURN IN STYLE

This talk summarizes some of the things that contemporary logic and, in particular, proof theory stemming from Gentzen have to say about the notion of consequence. It starts from very elementary facts, the understanding of which doesn't require any technical knowledge, to reach the more specialized areas of substructural logics and categorial proof theory. There, one may turn to a style of proof-theoretical investigation whose goal is not just the elimination of cut. Some tentative philosophical suggestions are drawn from this summary.

In connection with the notion of consequence logicians make two distinctions that other, less philosophically minded, mathematicians usually fail to make, or at least don't find so important. The first is the distinction between deduction and implication, the second the distinction between syntactical and semantical notions of consequence. We shall first briefly examine these two distinctions.

1. DEDUCTION AND IMPLICATION

The word 'consequence' belongs to a family of words in which one also finds 'deduction', 'derivation', 'inference', 'implication' and 'entailment'. The corresponding verbs are 'to deduce', 'to derive', 'to infer', 'to imply' and 'to entail', to which one should add 'to conclude' and 'to follow from', as well as the less formal 'to obtain from', 'to get from', 'to yield' and 'to give'. All these words have something to do with the activity of passing from sentences to sentences so that truth, or some related property like provability, is preserved. In case preservation of provability is envisaged, we presumably have to conform to some rules when we pass from sentences to sentences. We are then in the domain of syntax. In case preservation of truth is envisaged, we are in the domain of semantics.

The words 'deduction', 'derivation' and 'inference' may perhaps be distinguished by nuances — 'inference', for example, stressing an immediate step, and 'deduction' rather a chain of steps — but, otherwise, all three words are used in pretty much the same way. There are also no essential differences among the verbs 'to deduce', 'to derive', 'to infer', 'to conclude', 'to obtain from' and 'to get from'. So we may concentrate on one of these words, and let that be 'deduction'. A deduction is either the aforementioned activity or the result of this activity, which consists in having an array of sentences. Sentences coming later in the array are usually introduced with 'so', 'hence' or

M.L. Dalla Chiara et al. (eds.), Logic and Scientific Methods, 289–311.

'therefore', or are written beneath previously written sentences, sometimes with a line separating them. The sentences above are called 'premises' and the sentences below 'conclusions'; i.e., in a deduction, we pass from premises to a conclusion. There may be several premises from which we deduce a conclusion. Syntactically, a deduction is correct when it is done in conformity to some prescribed rules; semantically, a correct deduction, often called 'valid deduction', is one that preserves truth when passing from the premises to the conclusion. However, when one talks about 'deductions' in logic, syntactical aspects of the matter are usually more prominent.

In logic, the word 'implication' is reserved for the connective 'if __, then __' and sentences of the form 'if A, then B', usually written with symbols as '$A \rightarrow B$', or '$A \Rightarrow B$', or '$A \supset B$'. The verbs 'to imply', 'to follow from', 'to yield' and 'to give' are all tied to implication (of course, 'to follow from' is converse to the others). 'Entailment' and 'to entail' are used pretty much as 'implication' and 'to imply', except in those quarters of logic where they are reserved for a special, relevant, kind of implication. In the implication 'if A, then B', the sentence A is called the 'antecedent' and B the 'consequent'. (We also say that A is a sufficient condition for B, and B a necessary condition for A.) Conditionals, i.e. hypothetical sentences of the form 'if A, then B', need not all be implications in the logicians' sense: they may reflect relationships from the realm of the laws of nature or human behaviour, they may have to do with explanation, causation and dispositions. Among these other conditionals, a special place is occupied by counterfactual conditionals, whose if-clauses are known to be false. These matters may be germane to implication, but are nevertheless separate.

In classical logic, an implication is true when it is not the case that the antecedent is true and the consequent false. When an implication is true, the corresponding deduction, in which the antecedent of the implication is the premise and the consequent the conclusion, is a correct deduction (in a semantical sense; when the truth of the implication may be recognized by appealing only to the understanding of the language, the corresponding deduction is syntactically correct). Conversely, the correctness of a deduction should guarantee that the corresponding implication is true. When there are several premises in the deduction, we make a single sentence out of them with the help of the connective 'and', called 'conjunction', and we put this sentence as the antecedent of the corresponding implication. So we may take that premises in a deduction are joined by an implicit conjunction.

Although deduction and implication correspond to each other, as we have just indicated, they are separated by the essential difference that exists between, on the one hand, the activity of passing from sentences to sentences and, on the other, a connective, or sentences made with this connective. The purpose of Lewis Carroll's piece about the Tortoise and Achilles [1895] is presumably to stress this difference. The lesson is that there is no deduction without rules of deduction: these rules cannot be entirely replaced by implications (we need at least the rule that enables us to pass from an implication to the corresponding deduction, i.e. *modus ponens*).

The distinction between deduction and implication is not only philosophical or, perhaps, psychological. It has technical repercussions, too, as witnessed by Gentzen's mathematical achievements, which we shall examine below.

Close relatives of the words in our family are 'proof' and 'demonstration', which refer to a correct deduction in which the premises are true or, in more general terms, acceptable. A somewhat more distant relative is 'argument'. An argument is meant to convince somebody, and it may, but need not, be of a deductive sort.

The word 'consequence' refers either to the conclusion of a correct deduction or to the consequent of a true implication. Being a consequence is a relative property: in a correct deduction we say that the conclusion is a consequence *of* the premises, and in a true implication the consequent is a consequence *of* the antecedent. So a more basic notion than the property of being a consequence is the relation that obtains between the premises and the conclusion of a correct deduction, or between the antecedent and the consequent of a true implication. These relations are called 'consequence relations'.

2. SYNTACTICAL AND SEMANTICAL LOGICAL CONSEQUENCES

In logic, a consequence relation arising from a syntactically correct deduction is called a 'syntactical consequence relation'. If Γ is the collection of premises of a deduction and A the conclusion, one usually writes $\Gamma \vdash A$ to express that A is a syntactical consequence of Γ. However, when in logic one says 'consequence relation' *simpliciter*, one usually means 'semantical consequence relation' rather than 'syntactical consequence relation'. A semantical consequence relation corresponds to generalized implication explained by reference to truth values, as in classical logic. For a sentence A to be a semantical consequence of a collection of sentences Γ, which is usually written $\Gamma \models A$, we must have that if every sentence in Γ is true, then A is true.

If Γ consists of a single sentence, then a semantical consequence relation is the relation that exists between the antecedent and the consequent of a true implication. If Γ is a finite collection of sentences, then we may put all these sentences into a conjunction and have consequence corresponding to the implication whose antecedent is this conjunction and whose consequent is A. Short of introducing infinitary connectives (which indeed may be done in logic), we cannot reduce to implications consequence relations involving infinite collections Γ. We have such a consequence relation, for example, with Γ being made of the sentences '0 has a successor', '1 has a successor', '2 has a successor', etc. for each natural number; as a consequence of Γ we have the sentence 'Every natural number has a successor'. On a rather abstract level of logic, one may envisage a deduction corresponding to the consequence relation in this example (the rule justifying this deduction is called the *ω-rule*), but syntactical consequence relations, unlike this one, usually satisfy a property called 'compactness', which says that if A is a consequence of Γ, then A is a consequence of a finite collection of sentences included in Γ.

Besides studying matters that pertain to any notion of consequence, logic concentrates in particular upon consequence relations that hold in virtue of logical form. Logical form is obtained by keeping constant in sentences just certain words called 'logical constants' (they are also sometimes called 'logical words', or 'logical particles'). Everything else is replaced by schematic letters or variables. The resulting expressions are not called 'sentences' any more, but 'formulae'. Logical consequence relations are those that hold for these formulae, and hence for any of their instances. The study of the syntactical aspects of consequence belongs to proof theory, whereas semantical aspects are studied in model theory.

Consequence was studied in logic since ancient times, and a big corpus of medieval logic is devoted to the study of *consequentiae*. However, it is rather difficult to find in these old works a clear distinction between syntactical and semantical aspects of the matter, or between the activity of deduction and the connective of implication. As a matter of fact, these distinctions have not been quite clear even at the inception of modern logic. They don't appear in a form that would satisfy today's standards before the 1930s.

In [1936], Alfred Tarski gave a definition of the semantical notion of logical consequence (Bernhard Bolzano is usually credited of having anticipated this definition a hundred years earlier). The novelty of Tarski's definition is not the understanding of consequence in terms of a generalized implication — this is something everybody always had in mind. What he made precise is the notion of truth in a model, which replaces the intuitive notion of truth in the definition of semantical consequence given at the beginning of this section.

At roughly the same time, in a series of papers (see in particular [1930, 1930a]), Tarski studied logical consequence in an abstract axiomatic way, where purely syntactical aspects of the matter are prominent. In his syntactical studies, Tarski's basic notion was not a consequence relation between a collection of formulae and a formula, but a consequence operation **Cn**, which assigns to a set of formulae Γ the set of its consequences $\mathbf{Cn}(\Gamma)$. Properties satisfied by **Cn** are reminiscent of properties of a topological closure operation. From Tarski's axioms we can deduce:

$$\Gamma \subseteq \mathbf{Cn}(\Gamma);$$

$$\mathbf{Cn}(\mathbf{Cn}(\Gamma)) \subseteq \mathbf{Cn}(\Gamma);$$

$$\textit{if } \Gamma \subseteq \Delta, \textit{then } \mathbf{Cn}(\Gamma) \subseteq \mathbf{Cn}(\Delta);$$

$$\mathbf{Cn}(\Gamma) \subseteq \cup_{\Gamma_{\mathrm{fin}} \subseteq \Gamma} \mathbf{Cn}(\Gamma_{\mathrm{fin}}),$$

where Γ_{fin} is a finite set; the last inclusion expresses the compactness of consequence. Tarski's style of studying logical consequence via **Cn** has had a big influence in his native Poland (see [Wójcicki 1988]), but can hardly count as a universally accepted standard.

A more widely accepted style in the syntactical study of logical consequence was inaugurated by Gerhard Gentzen in his dissertation 'Investigations into Logical Deduction' [1935]. This work is the cornerstone of proof theory.

3. SEQUENTS AND STRUCTURAL PRINCIPLES

Gentzen's basic notion is a relation $\vdash$ between a sequence of formulae Γ and a formula A (Paul Hertz has studied this notion before Gentzen and has influenced him). He and a number of other authors write an arrow $\rightarrow$ instead of the turnstile $\vdash$. Expressions of the form $\Gamma \vdash A$ are called 'sequents', and in proof theory done in the style of Gentzen talk about sequents is more common than talk about consequence relations, though most logicians should be aware that asserting a sequent $\Gamma \vdash A$ amounts to asserting that the formula A is a consequence of the formulae in Γ. (The term 'sequent' could perhaps enter into the philosophical jargon to designate what remains in a deduction when it is disengaged from the activity of deducing, as the term 'proposition' is used for what remains in an assertion when it is disengaged from the activity of asserting.)

In $\Gamma \vdash A$, Gentzen takes Γ to be a sequence of formulae, but he assumes *structural* rules that permit him to transform $\Gamma \vdash A$ into $\Gamma' \vdash A$ where Γ' is obtained from Γ by permuting members of Γ or by omitting repetitions among these members. Following Gentzen, a logical principle concerning sequents is called 'structural' if and only if it can be formulated without mentioning any constant of the language to which belong the formulae we write on the left-hand side and right-hand side of the turnstile. (In the Polish logic literature, structural rules are those whose instances are closed under substitution; this notion should not be confused with Gentzen's notion of a structural rule.) The structural rule that permits permuting members of a sequence Γ is called 'permutation' (sometimes also 'interchange', or 'exchange') and the structural rule that permits omitting repetitions is called 'contraction'. Assuming permutation and contraction amounts to reducing sequences Γ in $\Gamma \vdash A$ to sets.

Gentzen's reason for working with sequences rather than sets is probably that he wanted sequents to be expressions of a language, i.e. strings of symbols. For the same reason, Gentzen's sequents $\Gamma \vdash A$ always have Γ finite. (Gentzen was a pupil of David Hilbert and wanted to reason about sequents in a finitistic way.) Because of the compactness property, which syntactical consequence relations usually have, this is not such an essential limitation. But, in general, it makes sense to extend Gentzen's approach to sequents $\Gamma \vdash A$ with Γ infinite.

Most logicians shun such infinitary sequents, as if they had finitistic scruples. However, one may expect to find such sequents useful in second-order or higher-order logic, in logic with infinitary connectives, in arithmetic with the ω-rule, or whenever finite axiomatizability fails. Of course, sequents are then something we talk *about*, rather than something we talk *in*.[1] But such are most object languages of logicians: they are studied at the metalevel, rather than actually used.

By taking sequents $\Gamma \vdash A$ where Γ is a set of formulae, perhaps infinite, and by taking that $\Gamma \vdash A$ is equivalent to $A \in \mathbf{Cn}(\Gamma)$, the principles concerning **Cn** mentioned in section 2 yield further structural principles for sequents:

(identity)	$\{C\} \vdash C$;
(cut)	*if* $\Gamma \vdash C$ *and* $\{C\} \cup \Delta \vdash B$, *then* $\Gamma \cup \Delta \vdash B$;
(thinning)	*if* $\Gamma \vdash B$, *then* $\Gamma \cup \Delta \vdash B$;
(compactness)	*if* $\Gamma \vdash B$, *then for some* $\Gamma_{\text{fin}} \subseteq \Gamma$, $\Gamma_{\text{fin}} \vdash B$.

These principles correspond to the four principles for **Cn**, but the correspondence is not exact; for example, to obtain $\Gamma \subseteq \mathbf{Cn}(\Gamma)$ from identity we need thinning (also called 'weakening'). One should note that in deductions corresponding to the implications of the last three principles, as well as in deductions arising from the structural rules of permutation and contraction, the premises and conclusions are themselves about deductions. For example, in cut, the premises $\Gamma \vdash C$ and $\{C\} \cup \Delta \vdash B$, and the conclusion $\Gamma \cup \Delta \vdash B$ are themselves about deductions. So we have here something we may conceive as a deduction of higher level, whose premises and conclusion are not sentences, but deductions.

If we have a noncompact consequence relation, we may envisage replacing cut by the following generalized cut

$$\textit{if } (\forall C \in \Theta)\Gamma \vdash C \textit{ and } \Theta \cup \Delta \vdash B, \textit{then } \Gamma \cup \Delta \vdash B.$$

If Θ is empty, $(\forall C \in \Theta)\Gamma \vdash C$ holds for every Γ, and we can derive thinning from our generalized cut. In fact, the structural principles of identity, generalized cut and thinning are all summed up by the equivalence

$$(\forall C \in \Theta)\, \Gamma \vdash C \textit{ if and only if } \forall \Delta \forall B(\textit{if } \Theta \cup \Delta \vdash B, \textit{then } \Gamma \cup \Delta \vdash B),$$

while identity and ordinary cut are summed up by the equivalence

$$\Gamma \vdash C \textit{ if and only if } \forall \Delta \forall B(\textit{if } \{C\} \cup \Delta \vdash B, \textit{then } \Gamma \cup \Delta \vdash B)$$

(other such equivalences, characterizing various sorts of consequence relations, are considered in [D. 1994]). At the bottom of these equivalences stands the fact that the reflexivity and transitivity of a binary relation R are converse to each other when these two properties are summed up by the equivalence

$$xRy \textit{ if and only if } \forall z(\textit{if } yRz, \textit{then } xRz).$$

All these equivalences show that the structural principle of identity may be conceived as converse to cut. (A related duality between identity and cut is explored in [Girard 1987, I.3] and [Hösli & Jäger 1994]; in this approach, identity is taken as saying that C as a premise yields C as a conclusion, whereas cut says that C as a conclusion yields C as a premise.)

[1] I owe this remark to Professor Dag Prawitz.

4. CUT ELIMINATION

Among structural principles, the cut rule has played by far the most important role in Gentzen's work and in the proof-theoretical tradition stemming from it. Cut is specific because, unlike all the other rules in Gentzen's formulation of logic, it has a formula (the formula C in the formulation of cut in the previous section), called the 'cut formula', which is in the premises but can disappear in the conclusion. Without cut we can establish the *subformula property*, which says that every provable sequent has a proof made of sequents in which occur only subformulae of the formulae occurring in our provable sequent. By constructing systems of sequents where the cut rule, or an analogue of it, is not primitive, but is nevertheless admissible (i.e. can be added to the system without increasing the stock of provable sequents), Gentzen produced a beautiful and extremely useful logical tool. This technique of *cut elimination* can serve to establish many important properties of systems, like decidability and completeness. Within the realm of Hilbert's program it serves for consistency proofs: Gentzen used it for his consistency proof of arithmetic. However, cut is a principle that should hold for most, if not all, reasonable notions of consequence: it expresses the transitivity of the consequence relation. To eliminate it does not mean to reject it. There is an advantage in not having it postulated as primitive, but it should nevertheless be admissible. And if we are not interested in exploiting cut elimination, or deal with a system where it cannot be accomplished, cut will figure among the postulates.

To obtain cut elimination in logical systems, Gentzen postulated sequent rules for logical constants in a symmetrical fashion. For each constant there are rules for introducing it on the left-hand side and on the right-hand side of the turnstile. If we read $\Gamma \vdash A$ as asserting that there is a *natural-deduction* proof of A with uncancelled hypotheses in Γ, rules for introducing logical constants on the left-hand side of the turnstile correspond to natural-deduction elimination rules, whereas rules for introducing them on the right-hand side correspond to natural-deduction introduction rules. A cut-free sequent proof gives instructions for building a natural-deduction proof in *normal form*, where, roughly speaking, eliminations are followed by introductions (see [Prawitz 1965]). If the structural principle of identity $\{C\} \vdash C$ is postulated only for atomic formulae C (so that for nonatomic formulae it has to be derived), the normal form will have only atomic formulae in between the eliminations and introductions (see [Prawitz 1971]). Actually, it is not necessary to eliminate all cuts for building natural-deduction proofs in such a normal form: it is enough if we eliminate cuts with nonatomic cut formulae. Besides this basic difference, there are others: cut elimination and conversion of natural-deduction proofs to normal form, i.e. *normalization*, are not exactly equivalent techniques. They are nevertheless closely related (see [Zucker 1974] and [Pottinger 1977]).

Advance in mathematics often consists in the invention of a good notation. Tarski and Gentzen often speak about the same matters concerning logical consequence, but Gentzen's notation was better chosen. It is doubtful whether one could have discovered cut elimination with Tarski's notation.

5. PLURAL CONSEQUENCE IN CLASSICAL LOGIC

For classical logic Gentzen used sequents of the form $\Gamma \vdash \Delta$, where on the right-hand side of the turnstile we don't have necessarily a single formula: both Γ and Δ are sequences of formulae. Let us call these sequents *plural* (they are sometimes called 'multiple-conclusion sequents'), whereas sequents where Δ has a single member or is empty are *singular* (these are sometimes called 'single-conclusion sequents'). The commas in the sequence Γ correspond to conjunction, whereas the commas in the sequence Δ to disjunction. There are structural rules for dealing with the right-hand side of sequents analogous to the structural rules mentioned above, which deal with the left-hand side. This permitted Gentzen to give an extremely elegant symmetrical axiomatization of classical logic, doing full justice to the underlying Boolean dualities.

He passes from this axiomatization of classical logic to an axiomatization of intuitionistic logic not by changing rules for logical constants, but just by restricting the language to singular sequents (in both axiomatizations cut can be eliminated). The restriction to singular sequents is tantamount to restricting the structural rule of thinning on the right-hand side so that we never have as a conclusion a sequent that is not singular. (There are, however, plural versions of intuitionistic sequent systems, where only rules for introducing implication, negation and the universal quantifier on the right-hand side fall under the singular-sequent restriction.)

Although the notion of plural consequence that underlies plural sequents is perfectly in tune with the spirit of classical logic, one very rarely hears about it outside proof theory done in the style of Gentzen. A semantical interpretation of plural sequents, which reads $\Gamma \vdash \Delta$ as saying that if all the formulae in Γ are true, then at least one of the formulae in Δ is true, was championed by Dana Scott in several papers from the early 1970s. In [1974] Scott remarked that a singular consequence relation, whose corresponding **Cn** operation satisfies Tarski's axioms, may be extended conservatively to different plural consequence relations, and he characterized the minimal and maximal ones among these. The minimal one $\Gamma \vdash_{min} \Delta$ obtains when for a member A of Δ we have $\Gamma \vdash A$ with the singular consequence, whereas the maximal one $\Gamma \vdash_{max} \Delta$ obtains when we have something that amounts to $\Gamma \vdash \vee\Delta$ with the singular consequence, $\vee\Delta$ being a disjunction (not necessarily finite) of the formulae in Δ. It can be shown that $\Gamma \vdash_{min} \Delta$ amounts to

$$(\exists A \in \Delta)\forall\Theta\forall B(\textit{if } \{A\} \cup \Theta \vdash B, \textit{then } \Gamma \cup \Theta \vdash B),$$

whereas $\Gamma \vdash_{max} \Delta$ amounts to

$$\forall\Theta\forall B(\textit{if } (\forall A \in \Delta)\{A\} \cup \Theta \vdash B, \textit{then } \Gamma \cup \Theta \vdash B),$$

which for Δ nonempty is equivalent to

$$\forall\Theta\forall B(\exists A \in \Delta)(\textit{if } \{A\} \cup \Theta \vdash B, \textit{then } \Gamma \cup \Theta \vdash B)$$

(cf. [D. 1994]). Gentzen's plural sequents for classical logic correspond to the maximal plural consequence relation of classical logic.

The semantical reading of plural sequents is much more natural than some proof-theoretical interpretations that have been proposed. In one of these proof-theoretical interpretations plural sequents are supposed to talk about natural-deduction proofs that are not trees: they branch disjunctively towards the conclusions, and not only conjunctively towards the premises, like ordinary natural-deduction trees (see [Shoesmith & Smiley 1978]). In another, more obvious, interpretation we have natural-deduction trees whose nodes are not made of single formulae but of sequences of formulae (see [Boričić 1985] and [Celluci 1992]). As suggested in [1956] by William Kneale (who was influenced both by [Gentzen 1935] and [Carnap 1943, § 32]), the philosophy behind these peculiar forms of deduction might be that in classical logic, instead of deducing single sentences, we should be deducing where is the field where truth lies. The problem is that it is doubtful whether this style of deduction could, or indeed should, be imposed on classical logicians, whereas nothing save habit seems to prevent them from accepting a plural notion of semantical consequence.

The lack of symmetry between premises and conclusions in ordinary deduction, which consists in having possibly several premises, but a single conclusion, may explain the lack of symmetry in intuitionistic logic, where conjunction is not dual to disjunction and the universal quantifier is not dual to the existential quantifier. Intuitionistic logic is truly the logic of ordinary, asymmetrical, deduction. This is demonstrated by the connection between singular sequents and natural deduction. The ordinary natural-deduction introduction and elimination rules for implication, which correspond to the Deduction Theorem and *modus ponens*, don't produce classical implication but intuitionistic implication. Classical logic, seen through the prism of plural sequents and the underlying symmetrical notion of plural consequence, seems to be rather about semantical matters involving the notion of truth.

The orientation of classical logic towards truth rather than deduction is manifested in the kind of formal system with which logicians are chiefly concerned. These systems, whose style is often associated with the name of Hilbert, concentrate on theoremhood rather than consequence. They have many axioms and a very few rules of inference (often only *modus ponens* and universal generalization). Syntactical consequence in these formal systems is only a derived notion, upon which one comes when one introduces proofs from hypotheses. In the model-theoretical tradition, which has been dominant in modern logic, one hardly feels the need for other formal systems save those in the style of Hilbert.

There is a gain in model theory in proving *strong completeness*, which is an equivalence between syntactical and semantical notions of consequence, over proving only ordinary completeness, which is the equivalence between theoremhood and validity, i.e. universal truth. With strong completeness, we can derive the model-theoretical Compactness Theorem from the compactness of the syntactical consequence relation of classical logic. However, once strong completeness for first-order classical logic has been proved, there don't seem to be many results in model theory that need to be stated in a strong form involving consequence relations, rather than an ordinary form involving theoremhood and validity. With compactness for our consequence relations and with the Deduction Theorem, the strong form of the result need not be more difficult to prove than the ordinary form: one can be inferred from the other. Consequence relations become interesting when completeness and compactness fail, as in second-order logic, or in the study of nonclassical logics, where one often finds that formal systems should axiomatize a particular notion of consequence rather than simply theoremhood.

6. SUBSTRUCTURAL CONSEQUENCE RELATIONS

Some nonclassical logics differ from classical logic, and from one another, not in what they assume about logical constants, but in what they assume about structural principles for their consequence relations. It is as if the structural part of logic were more fundamental: we change logic by changing this part. Logical constants are secondary: they are invariant, they play the same role in different structural contexts. These nonclassical logics have started being called 'substructural' in 1990. Individual substructural logics are much older (intuitionistic and relevant logic were first introduced in the 1920s). The term 'substructural' may convey that something has been subtracted in the structural part of logic, but it should also convey, according to the standard meaning of the word, that these logics are the product of interventions in the supporting structures of logic. (For references about substructural logics see [D. & Schroeder-Heister 1993].)

The most important substructural logic is intuitionistic logic, which, as was remarked in the previous section, may be obtained from classical logic by restricting thinning on the right-hand side. Rejecting thinning altogether produces a variant of relevant (or relevance) logic. The notions of consequence of these two logics are still based on *sets* of formulae Γ in $\Gamma \vdash A$. If, however, we reject the structural rule of contraction, as this is done in BCK logic, or both thinning and contraction, as in linear logic, the corresponding notions of consequence are based on *multisets* Γ in $\Gamma \vdash A$, a multiset being a set of occurrences of formulae, i.e. a set with possible repetitions. Finally, there is a logic derived from Joachim Lambek's calculus of syntactic categories, where permutation is rejected, too. In this logic *sequences* Γ in $\Gamma \vdash A$ are taken seriously. The weakest logic in this hierarchy is a nonassociative variant of

Lambek's calculus, where in $\Gamma \vdash A$ the collection Γ is not a sequence but a *term* made with a binary nonassociative operation joining formulae. This operation replaces the associative comma of sequences. In this logic we have rejected the structural rule of association, which is only implicit in Gentzen's sequents based on sequences.

These logics have various inspirations. Intuitionistic logic is a product of the constructivist side in the great debate on the foundations of mathematics. Relevant logic, though its earliest roots are mixed with the roots of intuitionistic logic, has a more philosophical inspiration: it grew out of attempts to formalize an implication devoid of the unintuitive features of material implication. The first reason for introducing the older contractionless logic, BCK logic, was to evade set-theoretical paradoxes while keeping unrestricted comprehension (systems of many-valued logic, where contraction is likewise restricted, have been used for the same purpose). The younger contractionless logic, linear logic, draws its inspiration and much of its appeal from quarters on the borderline of computer science. Finally, Lambek's calculus belongs primarily to mathematical linguistics, where it serves to build formal grammars called 'categorial grammars'. However, it has a natural algebraic inspiration, too, and it may be understood as a logic if we envisage deductions where we don't want to lose sight of the order of premises or of their exact arrangement on the top of deduction trees (in the case of Lambek's nonassociative calculus such trees are binary).

This information about the premises, which for most purposes we don't need, is lost by adding the structural rules of association and permutation. By adding contraction we lose information about how many times a premise was used. Finally, with thinning we introduce phantom premises that are not used at all, and we lose information about which premises were really used. It is presumably easier to work with less information, if we don't really need it. Substructural logics may be motivated by notions of deduction where we pay attention to information that we usually prefer to forget. So they are logics of deduction, and as such should be included in intuitionistic logic. However, there are also plural variants of substructural logics (indeed, in relevant, BCK and linear logic more attention has been paid to such variants).

In the production of substructural logics there has also been, no doubt, a certain amount of experimentation, of free play, with the infinitely many possibilities of Gentzen's apparatus. Interest in a particular logic, like linear logic, is then justified by its mathematical regularity rather than by applications. Like other mathematicians, logicians produce tools that may, but need not, find a user. In spite of what they claim when they go to the market, they are more concerned with the intrinsic than with the instrumental value of their products.

7. LOGICAL CONSTANTS

In classical logic and all substructural logics mentioned in the last section we can give the same postulates for the main logical constants. If we don't care about eliminating cut, the rules that stand behind the following equivalences will do for logics with association:

$$A, \Gamma \vdash \Delta, B, \Theta \textit{ if and only if } \Gamma \vdash \Delta, A \to B, \Theta;$$

$$(\Gamma \vdash \Delta, A, \Theta \textit{ and } \Gamma \vdash \Delta, B, \Theta) \textit{ if and only if } \Gamma \vdash \Delta, A \wedge B, \Theta;$$

$$(\Delta, A, \Theta \vdash \Gamma \textit{ and } \Delta, B, \Theta \vdash \Gamma) \textit{ if and only if } \Delta, A \vee B, \Theta \vdash \Gamma;$$

$$\Gamma \vdash \Delta, A, \Theta \textit{ if and only if } \Gamma \vdash \Delta, \forall x A, \Theta;$$

$$\Delta, A, \Theta \vdash \Gamma \textit{ if and only if } \Delta, \exists x A, \Theta \vdash \Gamma;$$

provided x does not occur free in the sequents on the right-hand sides of the last two equivalences. (We have something similar without association, but at the cost of some technical complications.) In the present context we need the structural rule of *substitution*, which enables us to pass from $\Gamma \vdash \Delta$ to $(\Gamma \vdash \Delta)^x_a$, obtained from $\Gamma \vdash \Delta$ by uniformly replacing the free variable x by the term a, with the usual provisos preventing the free variables of a to become bound after the substitution. Equality too can be completely characterized by an equivalence:

$$(\Delta, \Theta \vdash \Gamma)^x_a \textit{ if and only if } \Delta, x = a, \Theta \vdash \Gamma.$$

The connectives of conjunction and disjunction, $\wedge$ and $\vee$, behave like finite meet and join in a lattice, and the universal and existential quantifiers, $\forall$ and $\exists$, like infinite meet and join, the lattice ordering corresponding to $\vdash$ (however, in the absence of thinning or contraction, $\wedge$ will not be distributive over $\vee$). The propositional constants **T** and **F**, which behave like greatest and least element in a lattice, would satisfy the axioms $\Gamma \vdash \Delta, \mathbf{T}, \Theta$ and $\Delta, \mathbf{F}, \Theta \vdash \Gamma$. Lattice connectives are called 'additive' in linear logic and 'extensional' in relevant logic. In the absence of thinning or contraction, $\wedge$ will be distinguished from the product connective $\cdot$, which satisfies

$$\Delta, A, B, \Theta \vdash \Gamma \textit{ if and only if } \Delta, A \cdot B, \Theta \vdash \Gamma,$$

whereas in classical and intuitionistic logic these two connectives coincide. In plural variants of some substructural logics there may be a connective $+$ dual to $\cdot$, which replaces the comma on the right-hand side of the turnstile and is distinguished from $\vee$. Similarly, we may have a propositional constant **t**, distinguished from **T**, replacing the empty collection of formulae on the left-hand side, and **f**, distinguished from **F**, replacing the empty collection of formulae on the right-hand side of the turnstile. Algebraically, **t** and **f** behave like units for the multiplications $\cdot$ and $+$, respectively. We may define a negation $\neg A$ of an intuitionistic bent by $A \to \mathbf{f}$ or $A \to \mathbf{F}$, but the main proponents of relevant and linear logic prefer an involutive negation that satisfies De Morgan's laws, with respect to $\wedge$ and $\vee$, as well as with respect to $\cdot$ and $+$. The connectives

$\cdot$, $+$, **t** and **f**, together with implication $\rightarrow$, are called 'multiplicative' in linear logic and 'intensional' in relevant logic.

The essential algebraic fact concerning $\cdot$ and $\rightarrow$ is that we have

$$A \cdot C \vdash B \textit{ if and only if } C \vdash A \rightarrow B,$$

which means that algebraically implication behaves like a residual of the product. The product connective $\cdot$ stands for the comma on the left-hand side of the turnstile, and matches the way premises are joined, which in substructural logics is not necessarily an implicit conjunction corresponding to the connective $\wedge$. It is indistinguishable from such a conjunction in classical and intuitionistic logic. The implication connective $\rightarrow$ stands for the turnstile, and matches most directly the consequence relation. This explains why implication may be taken as the central connective of logic. The equivalences above suggest that all logical constants match some structural features of sequents, i.e. of the consequence relation. (Similar equivalences characterizing logical constants may be found in [Kneale & Kneale 1962, IX.4], [Bernays 1965], [Scott 1971, 1974] and [D. 1985, 1988]; a philosophy suggested by them is expounded in [D. 1989].)

8. CATEGORIAL PROOF THEORY

If we concentrate upon simple singular sequents of the form $A \vdash B$, where on both sides we have a single formula, then a sequent system may be conceived as a directed graph (i.e. directed multigraph with loops) whose vertices are formulae and where between a pair of vertices A and B there is an *arrow* if and only if $A \vdash B$ is provable in the system. Then, corresponding to different proofs of the sequent $A \vdash B$, we may distinguish between different arrows joining the same formulae A and B. We may introduce terms to code these proofs: atomic terms will correspond to axiomatic sequents and operations on terms to rules of inference. For example, the term $\mathbf{1}_A$ will code the axiomatic structural sequent of identity $A \vdash A$, which we write $\mathbf{1}_A : A \vdash A$, and for $f : A \vdash B$ and $g : B \vdash C$, by applying the binary partial operation of composition to g and f we obtain $gf : A \vdash C$. Composition corresponds to a simple form of the structural rule cut. Identity and cut are present in all the substructural logics of section 6.

This way we have come quite close to the mathematical notion of a category. It remains to replace formulae by arbitrary *objects*, to think about connectives as operations on objects, and to impose the categorial equations between arrows:

(1) For $f{:}A \vdash B$, $f\mathbf{1}_A = f$ and $\mathbf{1}_B f = f$.

(2) For $f{:}A \vdash B$, $g{:}B \vdash C$ and $h{:}C \vdash D$, $h(gf) = (hg)f$.

These equations have a clear proof-theoretical meaning. The equations of (1) correspond to the elimination of cuts with identity axioms, something we find in Gentzen's cut-elimination procedure. The equation (2) corresponds to the permutation of one cut with another cut; in $h(gf)$ we first apply cut with the cut formula B and then with the cut formula C, whereas in $(hg)f$ it is the other way round. Part of Gentzen's cut-elimination procedure consists in permuting cut with other rules, pushing it upwards where it will eventually disappear. Gentzen eschewed permuting cut with other cuts, because he starts, wisely, with a cut above which there are no other cuts. But he may as well have started with an arbitrary cut, which would require a permutation corresponding to (2). The exact moves in Gentzen's cut-elimination procedure are better described with the help of *multicategories.* In multicategories, instead of arrows, we have *multiarrows* of type $\Gamma \vdash A$, where, exactly as in Gentzen's sequents, Γ is a sequence of objects (see [Lambek 1989, 1993]).

In a category or multicategory for a system of sequents we don't have just a consequence *relation*, i.e. a *set of ordered pairs*, but rather an *indexed family of ordered pairs.*

The application of Gentzen's proof theory to category theory, and vice versa, was inaugurated by Lambek in the 1960s, and afterwards pursued mainly by him and his pupils. (Such an endeavour is suggested in [Lawvere 1969], and there is a Russian school of the same inspiration led by Mints since the 1970s; see references in [Lambek & Scott 1986] and [Szabo 1978]. One usually hears about 'categorical' proof theory and 'categorical' logic, rather than 'categorial', but 'categorical theory' is already reserved for a quite different notion in model theory.) Some important sorts of categories — cartesian, cartesian closed and bicartesian closed categories — correspond to fragments of intuitionistic logic. If these categories are formulated equationally, their equations between arrows are related to normalization of proofs in intuitionistic logic. Category theory enables us to get a unifying view of normalization through the concept of adjointness. This concept may explain the success of cut elimination and normalization. Adjoint functors are behind equivalences characterizing logical constants, like those of the preceding section.

The symmetric monoidal closed categories of Saunders Mac Lane [1971, VII], whose roots are in multilinear algebra, correspond to the multiplicative fragment of intuitionistic linear logic. Gentzen's methods were applied in the study of the coherence problem for these categories, i.e. the problem whether all diagrams of a particular sort commute (see [Kelly & Mac Lane 1971]). Quite natural sorts of categories correspond to other substructural logics, too, and clarify the properties of their consequence relations (see [D. & Petrić 1993]).

Natural-deduction proofs may also be coded by typed lambda terms in such a way that lambda conversion corresponds to normalization of proofs. This technique, called the 'formulae-as-types interpretation' or 'Curry-Howard correspondence', is closely related to coding of proofs in categories (see [Lambek & Scott 1986, part I], [Girard, Lafont & Taylor 1989] and [Stenlund 1972]). The connection between proof theory and combinatory logic is also manifested in the correspondence between the structural principles of identity, association, permutation, contraction and thinning, on the one hand, and the combinators I, B, C, W and K, on the other.

All these tools from category theory, the lambda calculus and combinatory logic are well adapted to intuitionistic logic and its substructural fragments, but not to classical logic. This may be explained by repeating something suggested in section 5: the consequence relations of the former logics are syntactical in inspiration, whereas the consequence relation of classical logic is semantical.

9. PROOF THEORY BEYOND CUT ELIMINATION

Many authors seem to think that cut elimination is not just a tool, but a goal — in fact, *the* goal of proof theory. This is presumably why they expend considerable effort to prove cut elimination even in cases where in the most natural sequent formulation cut is not eliminable, and where the system in question is already mastered with handier tools. Such is exactly the case with the modal logic S5, whose ordinary plural-sequent formulation has the modal rules

$$(\Box\vdash)\ \frac{A,\Gamma\vdash\Delta}{\Box A,\Gamma\vdash\Delta}\qquad\qquad(\vdash\Box)\ \frac{\Gamma\vdash\Delta,A}{\Gamma\vdash\Delta,\Box A}$$

provided that in $(\vdash\Box)$ every formula in Γ and every formula in Δ is prefixed with $\Box$. With these rules we have the proof

$$\dfrac{\dfrac{\dfrac{\Box A\vdash\Box A}{\vdash\Box A,\neg\Box A}}{\vdash\Box A,\Box\neg\Box A}\,(\vdash\Box)\qquad(\Box\vdash)\,\dfrac{A\vdash A}{\Box A\vdash A}}{\vdash A,\Box\neg\Box A}\,(\text{cut})$$

and the sequent proved cannot be proved without cut. However, other, very efficient, syntactical and semantical methods, in particular Kripke models, are available for the investigation of S5. It seems that with the help of these methods no important question concerning S5 has been left unanswered. In spite of that, a number of authors have pursued, and are still pursuing, a cut-free formulation of S5, producing sequent systems of a fairly complicated kind.

It is probably more interesting to learn that in the ordinary sequent formulation of S5 cut is not eliminable than to be given another complicated cut-free formulation of this logic. To find that cut is not eliminable in a natural sequent system is not a failure but a discovery. It is also quite interesting that, although in the ordinary sequent formulation of S5 cut cannot be eliminated, the subformula property can still be established (the point is that we can restrict all applications of cut to those where the cut formula is a subformula of the conclusion; see [Takano 1992]; cf. [Fitting 1983, chapter 5, # 8], [Smullyan 1968] and [Mints 1992, Theorem 2, p. 238]).

If the pursuit of cut elimination is not always justified on practical grounds, it can perhaps be justified on theoretical grounds (which are also sometimes aesthetical grounds). One may take that cut elimination and normalization are testimony of a 'harmony' between introduction and elimination rules: following Gentzen [1935, section 5.13], the premises of the former should equal in strength the conclusions of the latter (this harmony may be explicated by adjointness, as suggested in the previous section). If a formulation of a system lacks harmony, it would not characterize properly the constants of the system. Jean-Yves Girard has other suggestive things to say about the philosophy of cut elimination. For one, he says that cut-free proofs are for machines, whereas an intelligent being works with cut: he introduces lemmata in his proofs, and lemmata are cut formulae (cf. [1987, p. 95]). With cut, proofs become more manageable and shorter.

Granted that every 'good' system should have a cut-free formulation (and is hence 'good for machines'), we might wish to ascertain that some systems are not good because they cannot have such a formulation. If everything is good, goodness isn't worth much. However, it is more difficult to establish that a system cannot have a cut-free formulation than to find such a formulation if the system can have it. One can never be sure that one trick or another, perhaps extending the notion of sequent in some complicated way, will not produce the desired cut-free sequent system. There should be limits to what is permited as a sequent system if we want to be able to establish that cut is not always eliminable.

The question is whether there is anything else to do in proof theory besides eliminating cut, or proving related normalization theorems. Such has been the preponderance of this theme in proof theory that one may well doubt that anything else exists. Many authors take cut elimination to be such a bliss that after it is established there is hardly anything else to say (as in fairy tales there is nothing more to say after the young lady got married). Sometimes, but not always, a few corollaries may be drawn: the subformula property, decidability, interpolation... (We have primarily in mind here proof theory of logical systems in the tradition of Gentzen, and not proof theory of formalized fragments of mathematics in the tradition of Hilbert, which concentrates on consistency proofs.) Without pretending that they equal cut elimination in importance, let us mention a few other problems that have been or may be treated by proof-theorists.

There are, first, things with the flavour of universal algebra, like Tarski's results about **Cn** from the 1930s or Scott's results about plural consequences mentioned in section 5. (For results of a similar sort about definability and interpolation see [D. & Schroeder-Heister 1985, 1988].) These results don't depend on cut elimination.

Next, one may produce symmetrical sequent axiomatizations, like those with the equivalences characterizing logical constants in section 7, in which cut is not eliminable. Such axiomatizations should make clearer the articulation of logical systems, as the fact that the same equivalential principles for logical constants can be given for various substructural logics makes clearer the articulation of these logics. When tied to adjointness, this matter has also a categorical side.

It is especially in categorial proof theory that one may expect to find results that go beyond cut elimination and normalization. Composition, the simple form of cut found in categories, is not usually eliminable in categories corresponding to logical systems, but normalization is incorporated in the equations that hold between arrows. So results in categorial proof theory depend on something like cut elimination, but they may give more than what is usually inferred from cut elimination.

One such result is Lambek's *functional completeness* theorem for categories corresponding to fragments of intuitionistic logic (see [Lambek 1974] and [Lambek & Scott 1986, pp. 41-62]). This important theorem, whose applications one may study in [Lambek & Scott 1986], is a refinement of the deduction theorem. Roughly speaking, the deduction theorem says that the system is strong enough to express its own deductive metatheory: to deductions from hypotheses correspond arrows (i.e. sequents or implications) in the system. Functional completeness says that the deductive metatheory can be embedded in the system: there is a structure-preserving one-one map assigning arrows in the system to deductions from hypotheses (cf. [D. 1996]).

Functional completeness for categories corresponding to substructural logics and their modal extensions is treated in [D. & Petrić 1993], which follows [D. 1985, 1992]. In these papers, functional completeness is connected with systems of sequents of higher level. Such sequents, in which on the left-hand side and right-hand side of the turnstile one has sequents, were first introduced to analyze modal operators, by developing an idea of [Scott 1971]. They enable us to get a unique characterization of modal operators like the exponentials of linear logic, i.e. operators of the S4 or S5 type, which moreover may deliver the missing structural principles for formulae prefixed with a modal operator.

Functional completeness is a property of the same kind as what Haskell Brooks Curry called *combinatorial completeness* for systems of combinators; namely, the possibility of defining lambda abstraction in terms of combinators (see [Curry & Feys 1958, pp. 5, 186-194]). To prove versions of this property for categories corresponding to substructural logics (which is analogous to having restricted forms of lambda abstraction in systems of combinators where some of the combinators K, W, C or B may be missing), cartesian categories are presented in [D. & Petrić 1993] in a nonstandard fashion, in which the combinatory 'building blocks' hidden in the usual assumptions about projection arrows and the pairing operation on arrows are exhibited explicitly. (Such a nonstandard, combinatorially dissected, presentation of cartesian categories is used in [D. & Petrić 1994] for another purpose, which is mentioned below.)

It is shown in [D. & Petrić 1993] that assumptions made for cartesian and other categories corresponding to substructural logics are not only sufficient but also necessary for proving functional completeness. So, these categories may be characterized by functional completeness. One needs the same assumptions (in particular, Mac Lane's pentagonal and hexagonal commuting diagrams of natural associativity and commutativity; see [1971, VII.1, p. 158, and VII.7, p. 180]) to prove coherence for symmetric monoidal categories. Such is the case too if we consider coherence for cartesian and other categories corresponding to substructural logics. This leads to the conjecture that coherence and functional completeness are in general properties that entail each other.

As the deduction theorem is refined by functional completeness, so we may refine the *interpolation* theorem in categorial proof theory. If interpolation says that for every arrow $f: A \vdash B$ there are arrows $g: A \vdash C$ and $h: C \vdash B$, the formula C being in the intersection of the languages of A and B, in categories we may wish to have also that $f = hg$. This refinement is proved for bicartesian closed categories, i.e. for intuitionistic propositional logic, in [Čubrić 1994].

Categorial proof theory does not only suggest refinements of old results, like the deduction theorem and interpolation, but also new results, based on concepts that were not previously envisaged in proof theory. Such a concept, interesting both from a logical and philosophical point of view, is *isomorphism* between formulae. A formula A is isomorphic to a formula B if and only if there is a deduction f from A to B and a deduction f^{-1} from B to A such that f composed with f^{-1} reduces via normalization to the identity deduction $\mathbf{1}_A$ from A to A and f^{-1} composed with f reduces via normalization to the identity deduction $\mathbf{1}_B$ from B to B (these reductions are represented in categories by equations between arrows). That two formulae are isomorphic is equivalent to the assertion that the deductions involving one of them, either as premise or as conclusion, can be extended to deductions where this formula is replaced by the other, the deductions involving the first formula being in one-one correspondence with the deductions involving the second. Roughly speaking, whatever you can do in deductions with one of these formulae you can do as well with the other. Isomorphism is an equivalence relation stronger than the usual mutual implication. So, for example, $C \wedge D$ is isomorphic to $D \wedge C$ in intuitionistic logic, while $C \wedge C$ only implies and is implied by C, but is not isomorphic to it. (The problem is that the composition of the deduction from $C \wedge C$ to C, which is justified by either 'from $C \wedge D$ infer C' or 'from $D \wedge C$ infer C', with the deduction from C to $C \wedge C$ does not reduce via normalization to the identity deduction $\mathbf{1}_{C \wedge C}$ from $C \wedge C$ to $C \wedge C$.)

It seems reasonable to suppose that isomorphism analyzes propositional identity, i.e. identity of meaning for propositions: A and B stand for the same proposition, i.e., A means the same as B, if and only if A is isomorphic to B. This way we would base propositional identity upon identity of deductions, which is codified by equality between arrows.

Isomorphic formulae in the implication-conjunction fragment of intuitionistic logic, i.e. objects isomorphic in all cartesian closed categories, have been characterized in [Soloviev 1981] via the axiomatization of equations with exponentiation, multiplication and one, true for natural numbers. This fragment of arithmetic (upon which one comes in connection with Tarski's 'high-school algebra' problem) is finitely axiomatizable and decidable. (For related results about this and some other fragments of intuitionistic logic — in particular, the implication-conjunction fragment of second-order intuitionistic propositional logic — see the survey [Di Cosmo 1995] and references therein; [Soloviev 1993] and [D. & Petrić 1995] deal with the analogous product-implication fragment of linear logic, including the constant unit proposition **t**, i.e. with characterizing objects isomorphic in all symmetric monoidal closed categories.) The problem of characterizing isomorphic formulae in bigger fragments of intuitionistic first-order logic — in particular, in the whole intuitionistic propositional calculus, which corresponds to bicartesian closed categories — seems to be still open.

Another isomorphism problem is to characterize all arrows that are isomorphisms. This has been done in [D. & Petrić 1994] for the cartesian category that corresponds to the conjunctive fragment of intuitionistic logic (including the constant true proposition **T**, which in intuitionistic logic is indistinguishable from **t**, as conjunction $\wedge$ is indistinguishable from the product $\cdot$; this fragment of intuitionistic logic coincides with the corresponding singular-consequence fragment of classical logic). The isomorphisms of this category make exactly the symmetric monoidal category that corresponds to the product fragment of linear logic (including the constant unit proposition **t**). A procedure similar to cut elimination, in which we eliminate contraction and thinning, serves for deciding whether an arrow of the cartesian category is an isomorphism. All this may be taken as a justification of linear logic: it motivates the rejection of contraction and thinning by the wish to keep only structural principles that don't spoil isomorphism between premises (conjunction and product serve to join premises, while **T** and **t** replace the empty collection of premises).

By replacing talk about provable sequents with talk about arrows, categorial proof theory makes more tangible the things studied by proof theory since Gentzen's time. In a categorial setting we can express precisely and succinctly when two deductions are equal. This equality of deductions, which is a restatement of the reductions of cut elimination and normalization, happens to be related to important mathematical concepts, like adjointness. In categorial proof theory we are given more than a powerful technique like cut elimination: we are also led to find an explanation why this technique works. According to Descartes (*Regulae* XVI, 458), a mathematician differs from a computer by not being content with efficiency. The goal of mathematics is not power, but understanding and beauty — power is only a by-product.

10. CONSEQUENCE IN THE PHILOSOPHY OF LANGUAGE

Philosophy is as much concerned with understanding as mathematics. To reach this goal in its proper domain, twentieth-century philosophy has concentrated upon language, and has selected for particular attention two essential activities of language: referring and asserting. Gottlob Frege and Ludwig Wittgenstein found that when it comes to explaining how language functions, asserting is more basic than referring; i.e., sentences are more basic than nouns. A tangible reflection of that in logic is that the propositional calculus precedes the predicate calculus.

Philosophy has paid much less attention to the third essential activity of language — deducing, though the greater part of philosophy, a great part of mathematics and a considerable part of science, consist in this activity. It is hard to overestimate the influence of Frege and Wittgenstein in this matter. Michael Dummett, inspired by the natural-deduction style of introduction and elimination rules, has tried stressing the role of deduction in explaining key features of language, such as meaning (see [Dummett 1975]). However, a much more typical attitude in the philosophy of language of the twentieth century is exemplified by Willard Van Orman Quine's 'Two Dogmas' picture of language of [1951], which doesn't say much about deduction. Quine put logic in a body of sentences in the inner regions of language, far from the periphery of empirical sentences, rather than in the links of deduction, which bind sentences together in all regions of language and are not on the same level as these sentences. (In this respect Quine resembles Carroll's Tortoise.)

It seems that philosophy has not yet fully exploited ideas suggested by proof theory done in the footsteps of Gentzen. We have come upon a rather concrete suggestion of this kind in the previous section when we considered the analysis of propositional identity in terms of isomorphism. A less concrete suggestion may come from realizing that the part of logic that deals only with structural principles of deduction precedes both the propositional and predicate calculus. We might then be lead to speculate about giving deducing precedence over all other activities of language, or at least over asserting and referring. (Identity of deductions precedes identity of propositions in the analysis of the latter notion in terms of isomorphism; i.e., deducing precedes asserting.) This would be a precedence in the order of explaining language and not in the order of learning language (as the precedence of asserting over referring should not mean that we learn to assert before we learn to refer when we learn to speak). In such a philosophical conception, consequence relations would be more basic than sentences and nouns for explaining how language functions.

It seems indisputable that, in the order of explaining, a science that seeks laws accounting for phenomena has precedence over a taxonomical science (for example, physiology precedes anatomy, and theoretical linguistics precedes descriptive linguistics). This precedence is parallel to the precedence of asserting over referring. The precedence of logic over all sciences, championed by many philosophers, and in particular the precedence of logic over the rest of mathematics in foundational studies (which is not the same as the logicists' endeavour to reduce mathematics to logic) are parallel to the precedence of deducing over asserting, granted that, in contrast to other bodies of knowledge, logic draws its essence from deduction.

Mathematical Institute, Belgrade
and
IRIT, University of Toulouse III

REFERENCES

Bernays, P. [1965] Betrachtung zum Sequenzen-Kalkul, in A.-T. Tymieniecka ed., *Contributions to Logic and Methodology in Honor of J.M. Bocheński*, North-Holland, Amsterdam, pp. 1-44.

Boričić, B. [1985] On sequence-conclusion natural deduction, *Journal of Philosophical Logic* vol. 14, pp. 359-377.

Carnap, R. [1943] *Formalization of Logic*, Harvard University Press, Cambridge, Mass.

Carroll, L. [1895] What the tortoise said to Achilles, *Mind (N.S.)* vol. 4, pp. 278-280.

Ćelluci, C. [1992] Existential instantiation and normalization in sequent natural deduction, *Annals of Pure and Applied Logic* vol. 58, pp. 111-148.

Čubrić, D. [1994] Interpolation for bicartesian closed categories, *Archive for Mathematical Logic* vol. 33, pp. 291-319.

Curry, H.B., and Feys, R. [1958] *Combinatory Logic, Vol. I*, North-Holland, Amsterdam.

Di Cosmo, R. [1995] *Isomorphisms of Types: From λ-Calculus to Information Retrieval and Language Design*, Birkhäuser, Boston.

Došen, K. [1985] Sequent systems for modal logic, *The Journal of Symbolic Logic* vol. 50, pp. 149-168.

—— [1988] Sequent systems and groupoid models, I, *Studia Logica* vol. 47, pp. 353-385 (Addenda and corrigenda, *Studia Logica* vol. 49, 1990, p. 614).

—— [1989] Logical constants as punctuation marks, *Notre Dame Journal of Formal Logic* vol. 30, pp. 362-381 (slightly amended version in: D.M. Gabbay ed., *What is a Logical System?*, Oxford University Press, Oxford, 1994, pp. 273-296).

—— [1992] Modal logic as metalogic, *Journal of Logic, Language and Information* vol. 1, pp. 173-201.

—— [1994] On passing from singular to plural consequences, in: E. Orlowska ed., *Essays in Memory of Helena Rasiowa*, Kluwer, Dordrecht (to appear).

—— [1996] Deductive completeness, *The Bulletin of Symbolic Logic* (to appear).

Došen, K., and Petrić, Z. [1993] Modal functional completeness, in: H. Wansing ed., *Proof Theory of Modal Logic*, Kluwer, Dordrecht (to appear).

—— [1994] Cartesian isomorphisms are symmetric monoidal: A justification of linear logic (manuscript).

—— [1995] Isomorphic objects in symmetric monoidal closed categories, *Mathematical Structures in Computer Science* (to appear).

Došen, K., and Schroeder-Heister, P. [1985] Conservativeness and uniqueness, *Theoria* vol. 51, pp. 159-173 (Errata, *Theoria* vol. 52, 1986, p. 127).

—— [1988] Uniqueness, definability and interpolation, *The Journal of Symbolic Logic* vol. 53, pp. 554-570.

—— , eds [1993] *Substructural Logics*, Oxford University Press, Oxford.

Dummett, M.A.E. [1975] The philosophical basis of intuitionistic logic, in: H.E. Rose and J.C. Shepherdson eds, *Logic Colloquium '73*, North-Holland, Amsterdam, pp. 5-40.

Fitting, M. [1983] *Proof Methods for Modal and Intuitionistic Logics*, Reidel, Dordrecht.

Gentzen, G. [1935] Untersuchungen über das logische Schließen, *Mathematische Zeitschrift* vol. 39, pp. 176-210, 405-431 (English translation in *The Collected Papers of Gerhard Gentzen*, M.E. Szabo ed., North-Holland, Amsterdam, 1969).

Girard, J.-Y. [1987] *Proof Theory and Logical Complexity, Vol. I*, Bibliopolis, Naples.

Girard, J.-Y., Lafont, Y., and Taylor, P. [1989] *Proofs and Types*, Cambridge University Press, Cambridge.

Hösli, B., and Jäger, G. [1994] About some symmetries of negation, *The Journal of Symbolic Logic* vol. 59, pp. 473-485.

Kelly, G.M., and Mac Lane, S. [1971] Coherence in closed categories, *Journal of Pure and Applied Algebra* vol. 1, pp. 97-140.

Kneale, W. [1956] The province of logic, in: H.D. Lewis ed., *Contemporary British Philosophy*, Allen and Unwin, London, pp. 237-261.

Lambek, J. [1974] Functional completeness of cartesian categories, *Annals of Mathematical Logic* vol. 6, pp. 259-292.

—— [1989] Multicategories revisited, in: J.W. Gray and A. Scedrov eds, *Categories in Computer Science and Logic*, American Mathematical Society, Providence, pp. 217-239.

—— [1993] Logic without structural rules (Another look at cut elimination), in [Došen & Schroeder-Heister 1993], pp. 179-206.

Lambek, J., and Scott, P.J. [1986] *Introduction to Higher-Order Categorical Logic*, Cambridge University Press, Cambridge.

Lawvere, F.W. [1969] Adjointness in foundations, *Dialectica* vol. 23, pp. 281-296.

Mac Lane, S. [1971] *Categories for the Working Mathematician*, Springer, Berlin.

Mints, G. [1992] Lewis' systems and system T (1965-1973), in *Selected Papers in Proof Theory*, Bibliopolis, Naples, pp. 221-294 (English translation, with additions, of a paper published in Russian as an appendix to the Russian translation of R. Feys, *Modal Logic*, Nauka, Moscow, 1974, pp. 422-509).

Pottinger, G. [1977] Normalization as a homomorphic image of cut elimination, *Annals of Mathematical Logic* vol. 12, pp. 323-357.

Prawitz, D. [1965] *Natural Deduction: A Proof-Theoretical Study*, Almqvist & Wiksell, Stockholm.

—— [1971] Ideas and results in proof theory, in: J.E. Fenstad ed., *Proceedings of the Second Scandinavian Logic Symposium*, North-Holland, Amsterdam, pp. 235-307.

Quine, W.V.O. [1951] Two dogmas of empiricism, *Philosophical Review* vol. 60, pp. 20-43.

Scott, D.S. [1971] On engendering an illusion of understanding, *The Journal of Philosophy* vol. 68, pp. 787-807.

—— [1974] Completeness and axiomatizability in many-valued logic, in: L. Henkin et al. eds, *Proceedings of the Tarski Symposium*, American Mathematical Society, Providence, pp. 411-435.

Shoesmith, D.J., and Smiley, T.J. [1978] *Multiple-Conclusion Logic*, Cambridge University Press, Cambridge.

Smullyan, R.M. [1968] Analytic cut, *The Journal of Symbolic Logic* vol. 33, pp. 560-564.

Soloviev, S.V. [1981] The category of finite sets and cartesian closed categories (in Russian), *Zapiski nauchnykh seminarov LOMI* vol. 105, pp. 174-194 (English translation in *Journal of Soviet Mathematics* vol. 22, 1983, pp. 1387-1400).

—— [1993] A complete axiom system for isomorphism of types in closed categories, in: A. Voronkov ed., *Logic Programming and Automated Reasoning*, Lecture Notes in Artificial Intelligence no 698, Springer, Berlin, pp. 360-371.

Stenlund, S. [1972] *Combinators, Lambda Terms and Proof Theory*, Reidel, Dordrecht.

Szabo, M.E. [1978] *Algebra of Proofs*, North-Holland, Amsterdam.

Takano, M. [1992] Subformula property as a substitute for cut elimination in modal propositional logics, *Mathematica Japonica* vol. 37, pp. 1129-1145.

Tarski, A. [1930] Über einige fundamentale Begriffe der Metamathematik, *Comptes Rendus des Séances de la Société des Sciences et des Lettres de Varsovie* vol. 23, pp. 22-29 (English translation in [Tarski 1956]).

—— [1930a] Fundamentale Begriffe der Methodologie der deduktiven Wissenschaften, I, *Monatshefte für Mathematik und Physik* vol. 37, pp. 361-404 (English translation in [Tarski 1956]).

—— [1936] On the concept of logical consequence (in Polish), *PrzeglądFilozoficzny* vol. 39, pp. 58-68 (English translation in [Tarski 1956]).

—— [1956] *Logic, Semantics, Metamathematics: Papers from 1923 to 1938*, J.H. Woodger ed., Oxford University Press, Oxford (second edition, J. Corcoran ed., Hackett, Indianapolis, 1983).

Wójcicki, R. [1988] *Theory of Logical Calculi: Basic Theory of Consequence Operations*, Kluwer, Dordrecht.

Zucker, J. [1974] The correspondence between cut elimination and normalization, *Annals of Mathematical Logic* vol. 7, pp. 1-156.

SVEN OVE HANSSON AND DAVID MAKINSON

APPLYING NORMATIVE RULES WITH RESTRAINT

ABSTRACT.

Our purpose is to investigate the logic of applying normative rules. A commonsense distinction may be drawn between applications that are unrestrained, carried through in full even when they give rise to contradiction, and those that are restrained, carried through as fully as is compatible with a principled avoidance of contradictions or of other consequences specified as undesirable. Already, unrestrained application of a normative code is a little more subtle than mere closure under logical consequence. Restrained application in a situation is rather different, we suggest, from revision to accommodate the situation. On the pragmatic level, revision eliminates some normative rules from the code to make consistent place for new material, whereas restrained application leaves all rules intact in the code whilst ignoring some in specific situations. The paper proposes explicit formal definitions of both unrestrained and restrained application of normative rules. For the latter, the most delicate and interesting question is the level at which maximalization-and-choice is most appropriately carried out, in coordination with instantiation and closure under logical consequence.

1. A SIMPLE EXAMPLE

Consider the following simple example of the application of normative rules, which we shall often refer to in the paper.

Example 1: A doctor in a remote country district is working under a normative code which includes the following:

n_1: If a person is reported as having a heart attack, the doctor should visit him or her (henceforth briefly 'him') immediately,

n_2: If a person is reported as having a severed hand, the doctor should visit him immediately.

The inhabitants of the district are called $1, 1', 2, 3$, of whom 1 and $1'$ live next door to each other and so can be visited at almost the same time. On the other hand, they live far from 2 and 3, who also live far from each other, to such a degree that it is impossible to visit more than one of them immediately. One day the following situation is reported: 1 has a heart attack, $1'$ has a severed hand, 2 has a heart attack, and 3 has a severed hand. For simplicity, we suppose as constraints that these are the only rules, the only potential patients, and the only facts regarding the situation that are available, and that no means are accessible to supplement the information or the rules or call in additional help from outside. □

Already, intuitive commonsense tells us many things about the example that a formal analysis would be rash to ignore:

- Strictly speaking, the rules considered in themselves are not inconsistent. But they are in *conflict*, or as one might say, they are *potentially incon-*

M.L. Dalla Chiara et al. (eds.), Logic and Scientific Methods, 313–332.

sistent, in the sense that they can give inconsistent results when taken together with certain certain reports of real or possible situations. For instance, rules n_1 and n_2 conflict in the presence of the information that 1 has a heart attack and 3 has a severed hand.

- Indeed, even a single rule can conflict with itself. For instance, rule n_1 gives rise to inconsistency in the presence of the information that both 1 and 2 have heart attacks.
- Because of the above, an unrestrained application of the rules to the given situation will lead to inconsistent injunctions. With classical logic, that in turn requires the doctor to do everything and its opposite.
- In practice, however, there is little doubt in this example what a restrained or judicious application of the rules would be like, bearing in mind the constraints that were mentioned. The doctor should visit 1 and 1′ immediately, with second (unfulfillable) priority being to visit 2, and third priority to visit 3. Of course in other examples priorities may be less clear.

Our purpose in this paper is to articulate a formal analysis that respects such intuitions. It should be emphasized that we have no quarrel with classical logic. We do not propose to block the derivation of everything-and-its-opposite by replacing classical logic by some 'funny' truncated logic, be it paraconsistent, relevantist, intuitionistic, many-valued or other. In our view, such a path is a blind alley. Our approach, which we also believe to be far more in accord with practice, is to moderate (in a principled way with as little arbitrariness as possible) the process of application so as not to derive even a single contradiction from conflictual rules, whilst respecting classical logic. The analysis derives some of its key ideas from the AGM account of partial meet contraction and revision (see Alchourrón, Gärdenfors and Makinson 1985) and others from the work of Poole 1988 and Brewka 1989 on nonmonotonic explanations of data from within a predetermined range of possible hypotheses.

2. BACKGROUND FORMAL FRAMEWORK

We shall be working with a fragmented language, that is, one in which distinctions are made between different kinds of assertion. The basic ingredients that we shall need to work with are *descriptive assertions* (grouped in *situations*), *normative assertions, normative rules* (grouped in *normative codes*), the operation of *logical consequence*, and — of great importance — that of *instantiation*.

Any account of the application of normative rules that wishes to make contact with real-life practice must take into account the generality that is implicit in them. For this reason, our formal language will include individual variables, which are implicitly understood as universally quantified whenever they occur in an assertion or rule, and the language will also be closed under instantiating those variables to individual constants. However, for simplicity the language will not include explicit quantifications within assertions, whether universal or existential. All variables occurring in expressions will thus be free.

A *descriptive assertion* is one that can be formed by truth-functional connectives from *elementary descriptive assertions*, the latter being of the form $p(t_1, \ldots, t_n)$ where p is a predicate and the t_i are *terms*, that is, either individual constants (for which we shall use the numerals $1, 2, 3, \ldots$ in order to reduce notation) or individual variables $x, y, z, \ldots$.

On the other hand, *normative assertions* are truth-functional compounds of *elementary normative assertions*, the latter being of the form $O\alpha$ where α is a descriptive assertion and O is the deontic operator read as "it is obligatory that". Evidently, descriptive assertions are disjoint from normative ones (although of course they may be equivalent in the limiting cases of contradictions and tautologies).

We take a *normative rule* to be any expression that is either itself a normative assertion or is of the form $\alpha \rightarrow \varphi$ where φ is a normative assertion and α a descriptive one. Here, for simplicity, $\rightarrow$ is material implication. For brevity, we shall write normative rules as $[\alpha \rightarrow]\, \varphi$ where the square bracket indicates that its contents may be empty.

Assertions in general are truth-functional compounds of descriptive and/or normative assertions. Normative rules are thus a special kind of assertion.

A *normative code* is taken to be any set of normative rules. Usually, the elements of a normative code will contain no individual constants, only individual variables, but this will not be mandatory. Note that we are not assuming that a normative code is closed under any sort of consequence operation. Typically, it will be finite, and may be a singleton or empty. Normative codes in this respect (as well as the highly specific form of the normative assertions making them up) are quite unlike the belief sets or theories most commonly studied in investigations of belief change in the AGM tradition.

A *situation* is any set of descriptive assertions that contains no individual variables and is consistent under Cn. We do not assume that a situation is also closed under Cn (although in this case it would make only minor differences to presentation if such a requirement were imposed).

An *instantiation* is any operation that uniformly substitutes individual constants from a given domain for all individual variables in an expression. We thus understand instantiation in the sense of full instantiation, i.e. all variables are eliminated. For simplicity, partial instantiation is not considered in this paper. If n is a normative rule then we write $I(n)$ for the set of all instantiations of n (into the domain under consideration). For a set N of normative assertions, I(N) of course means $\cup\{I(n) : n \in N\}$. Note that an instantiation of any descriptive assertion (resp. normative assertion, normative rule, assertion) is of the same category.

We assume that we have available a consequence operation Cn to work on assertions, which includes classical consequence over the assertions, is compact, and satisfies disjunction in the premises, i.e.

$$\text{whenever } \gamma \in Cn(A \cup \{\alpha\}) \cap Cn(A \cup \{\beta\}) \text{ then } \gamma \in Cn(A \cup \{\alpha \vee \beta\}),$$

and satisfies the following deontic condition for descriptive assertions $\alpha_1, \ldots, \alpha_n$ and normative assertions $\varphi_1, \ldots, \varphi_m$:

$$\text{whenever } \bot \in Cn(\alpha_1, \ldots, \alpha_n, \varphi_1, \ldots, \varphi_m)$$
$$\text{then } \bot \in Cn(O\alpha_1, \ldots, O\alpha_n, \varphi_1, \ldots, \varphi_m).$$

This deontic principle is familiar in the case $m = 0$, less so when $m \geq 1$. Note that the parametric φ_i are *normative* assertions, so that all their predicates occur within the scope of the deontic operator; for descriptive assertions the condition would have neither plausibility nor standing in any recognized deontic logic. The reason for including the φ_i will be explained in section 4.1. The intersection of any non-empty family of consequence operations satisfying the above conditions clearly also satisfies them. When analyzing examples, we shall always take Cn to be the *least* operation satisfying the above conditions (and any others specific to the example).

3. UNRESTRAINED APPLICATION

Let N be a normative code and S a situation, and let D be a domain of individuals. By the *unrestrained application* $a(N, D, S)$ *of* N *to* D *in the light of* S, we mean the set $Cn(I(N) \cup S) \cap A_N$, where I is the set of all instantiations into D, and A_N is the set of all normative assertions without variables, all of whose individual constants are drawn from D. To simplify notation, we shall usually write $a(N, D, S)$ as $a(N, S)$, leaving the reference to the domain implicit. A possible weaker definition using detachment in place of Cn is discussed in Appendix I.

Example 1, in formal terms: In our medical example of section 1, N consists of two normative rules n_1 and n_2 where $n_1 = h(x) \rightarrow Ov(x)$ and $n_2 = s(x) \rightarrow Ov(x)$. The domain of individuals consists of persons 1, 1′, 2, 3, and Cn is understood to have $\neg v(j) \in Cn(v(i))$ whenever i, j are distinct elements of the domain and $\{i, j\} \neq \{1, 1'\}$. S consists of the descriptive assertions $h(1), s(1'), h(2), s(3)$. Clearly $I(N)$ consists of the normative rules $h(i) \rightarrow Ov(i)$ and $s(i) \rightarrow Ov(i)$ for all i in the domain. Finally, $a(N, S)$ consists of the normative assertions relating to $1, 1', 2, 3$ in $Cn(I(N) \cup S)$. It thus contains $Ov(i)$ for all i in the domain and so is inconsistent under Cn. □

Unrestrained application, so defined, shares some properties with logical consequence, and differs in others. On the negative side, it does not satisfy left or right inclusion, i.e. the inclusions $N \subseteq a(N, S)$ and $S \subseteq a(N, S)$ both fail — the latter because the elements of S are descriptive, not normative, assertions and the former because the elements of N may contain arrows (and variables) and so not be normative assertions (without variables). Also on the negative side, $a(N,S)$ is not closed under Cn except within the limits of A_N, that is, we have $Cn(a(N,S)) \cap A_N \subseteq a(N,S)$ but not $Cn(a(N,S)) \subseteq a(N,S)$. On the positive side — always assuming that the domain D is held constant — the monotony of I and of Cn clearly give us both:

(3) *Left monotony:* $a(N,S) \subseteq a(N',S)$ whenever $N \subseteq N'$.

(4) *Right monotony:* $a(N,S) \subseteq a(N,S')$ whenever $S \subseteq S'$.

Since no element of $a(N,S)$ contains any variables, clearly $I(a(N,S) = a(N,S)$. Using this, it is not difficult to check the following (*verification in appendix II*):

(5) *Left idempotence:* $a(N,S) = a(a(N,S),S)$.

From left monotony and left idempotence we clearly have left cumulativity $a(N,S) = a(N',S)$ whenever $N \subseteq N' \subseteq a(N,S)$ but this principle is of very little interest for the operation of application because it is almost vacuous — as noted above, $N \subseteq a(N,S)$ fails whenever there is an arrow or a variable in an element of N. Much more interesting for unrestrained application is the following variant (*verification in appendix II*):

(6) *Left stability:* $a(N,S) = a(N',S)$ whenever $N \subseteq N' \subseteq N \cup a(N,S)$.

Together, left stability, idempotence and monotony imply the following generalization of idempotence (*verification in appendix II*):

(7) *Strong left idempotence:*

$$a(N,S) = a(N',S) \text{ whenever } a(N,S) \subseteq N' \subseteq N \cup a(N,S).$$

From the above, we can say that the left projection operations $a_S(N)$ of $a(N,S)$, defined by the rule $a_S(N) = a(N,S)$, are like closure operations (alias Tarski consequence operations) in that they satisfy the following conditions:

Monotony: $a_S(N) \subseteq a_S(N')$ whenever $N \subseteq N'$

Idempotence: $a_S(N) = a_S(a_S(N))$

but are unlike them in that they do not satisfy inclusion. They do however satisfy the following conditions, which of course are of no interest for closure operations, where they collapse into cumulativity and idempotence respectively:

Stability: $a_S(N) = a_S(N')$ whenever $N \subseteq N' \subseteq N \cup a_S(N)$

Strong idempotence:

$$a_S(N) = a_S(N') \text{ whenever } a_S(N) \subseteq N' \subseteq N \cup a_S(N).$$

4. RESTRAINED APPLICATION: GENERAL CONCEPTUALIZATION

We come to the central concept of the paper - restrained application. We have already seen the urgent need for it when considering the medical example in section 1. In that example, unrestrained application of the normative rules in the given situation leads to inconsistent injunctions, whereas commonsense tells us that certain of the rules should be applied to certain of the instances whilst others are shelved.

4.1. *Two Kinds of Restraint*

Actually, restraint in an application can take either of two forms:

(i) restraint to avoid inconsistency $\perp$, or what is the same thing given the conditions on our background consequence operation Cn, to avoid the normative assertion $O\perp$;

(ii) restraint to avoid some normative assertion φ that is declared in advance to be undesirable.

Clearly, the former may be taken as a special case of the latter. But it is also possible to reduce the general case to the special one, by the trick of enlarging the background consequence operation Cn to ensure that the undesired normative assertion φ implies a contradiction, e.g. by introducing Cn' defined by $Cn'(X) = Cn(X \cup \{\neg\varphi\})$. It is trivial to check that Cn' is also a consequence operation, and that if Cn includes classical consequence (resp. is compact, satisfies disjunction in the premises, satisfies the deontic condition imposed on Cn) then the same is true for Cn'. The verification that Cn' satisfies the deontic condition makes essential use of the parameters $\varphi_1, \ldots, \varphi_m$; this is the reason why we have taken that condition in its less familiar strengthened form. For simplicity, in what follows, we shall focus on application that is restrained to avoid contradiction, calling it *consistent application*.

4.2. *Formal Desiderata*

Writing consistent application as $c(N, D, S)$ or more briefly as $c(N, S)$ with the domain D implicit and held constant, we begin by listing some of the formal properties that we regard as desirable in that they reflect the intended motivation, without treating them as shibboleths, nor as being exhaustive even on a formal level.

First, it should be a restriction of unrestrained application, that is:

(i) $c(N, S) \subseteq a(N, S)$ (restriction).

Furthermore, the outcome of the consistent application should be consistent:

(ii) $\perp \notin Cn(c(N, S))$ (consistency).

If $a(N,S)$ is already consistent, then it would be complaisant to forego any of its elements. Hence in this case $c(N,S)$ should be the same as $a(N,S)$:

(iii) If $\bot \notin Cn(a(N,S))$ then $c(N,S) = a(N,S)$ (non-complaisance).

The mode of describing a situation should not influence the outcome of applying a normative code to it:

(iv) If $Cn(S) = Cn(S')$ then $c(N,S) = c(N,S')$ (right equivalence).

Here it is understood that the domain D is held constant despite variation in the situation S.

Last, but not least, unwarranted dismissals of normative assertions should be avoided. In other words, $c(N,S)$ should not be unnecessarily small. Indeed, it should be maximal, in some appropriate respect that we shall seek to clarify.

4.3. *Pragmatic Comparison with Contraction and Revision*

The above desiderata are evidently infused by the same spirit as the AGM postulates for belief revision, and the question arises whether the operation of consistent application is really different. Differences between the formalizations will be discussed later in this paper, but for the present we draw attention to significant pragmatic differences in the intuitive concepts being formalized.

Application, restrained or not, is always *to a particular case requiring resolution*. On the other hand, contraction and revision — which in a normative context are often referred to as derogation and amendment — typically focus on the rules themselves, and are not directly concerned with individual cases. This basic difference manifests itself in several ways.

(a) When a rule is contracted out from a code, then it no longer has any place in the code. It cannot be used thereafter unless it is reinstated. On the other hand, when performing a consistent application, we merely shelve certain of the rules, not using them for the case in hand, but returning to the full set of rules as our starting point when we come to the next case.

(b) When restraining an application of a code, we take into account a much narrower descriptive situation than when derogating part of it. We may wish to derogate from a code if we realize that there is some possible situation in which it would give inconsistent or undesirable results. On the other hand, we restrain an application only when the code gives an inconsistent or undesirable result in the particular case under consideration. We don't care whether inconsistencies would arise in other real or imaginable cases.

(c) For these reasons, derogation and amendment are actions that are carried out by a legislature, but for which a judge usually has no authority. On the other hand, consistent application is a task of judges, but is not usually the concern of a legislature, whose business is to formulate general rules.

To be sure, these differences are pragmatic ones, but they also give rise to some purely formal differences, as we shall see in section 6.

Example 2: Consider a code governing inheritance in some legal system, and two cases that come before the courts, arising from disputes between relatives of the deceased persons, but quite separate from each other. The same code is operative in the two cases, but in each case we use only the facts known about the position of its parties. If the code is inconsistent in the context of the facts relating to the first case, the judge will be obliged to restrain its application. But the next day, called upon to examine the other case, the same judge will go back to the whole of the code and apply it to the facts relevant to that case, and if the code is consistent with them then the application can be unrestrained. If on the other hand there is inconsistency in the context of the facts of the second case, application must again be restrained. In general, the restraint may be quite different in the two cases, for the facts being quite different, the rules interacting with them to cause trouble may be so too. The judge will have no occasion to consider the consequences of the code along with all the facts known about the two cases taken together, for they are treated separately. Throughout the entire process the law remains the same. □

5. RESTRAINED APPLICATION: FORMAL DEFINITION

5.1. *Basic Questions about Maximalization and Choice*

Unrestrained application was formed with instantiation and logical consequence as principal component operations. Restrained application, as we shall define it, involves one further component, namely maximalizing within the limits of consistency and selecting from among the maximals — briefly *maximalization-and-choice.* There are two basic questions that arise immediately:

- *At what level should the maximalization take place?* Should we maximalize on the normative rules used, their instantiations, or the derived normative assertions? In other words, how should the three component operations be coordinated?
- Once maximalization is carried out, *how should we handle multiple maximals*, i.e. the existence of more than one maximal set of the kind envisaged?

5.2. *Dealing with Multiple Maximals*

The second question is of course familiar from an analogous one in the logic of belief change in the AGM tradition, and also various forms of nonmonotonic reasoning, and the main possible responses are already canvassed in those areas. If $\boldsymbol{M}$ is the collection of all maximal sets of a desired kind, then we use a selection function γ to choose a non-empty subset $\gamma(\boldsymbol{M})$ of $\boldsymbol{M}$, and we intersect the elements of $\gamma(\boldsymbol{M})$, putting $\cap\gamma(\boldsymbol{M})$ as the desired output. As a limiting case, γ may choose singletons, which amounts to choosing an element of $\boldsymbol{M}$ (this is known as *maxichoice*). As another limiting case, $\gamma(\boldsymbol{M})$ may be $\boldsymbol{M}$ itself, in which case the output is $\cap\boldsymbol{M}$ (known as *full meet* selection). If desired, it may be required that the selection function γ is determined by a relation over $\boldsymbol{M}$, on which constraints such as transitivity may be imposed.

In the case of belief change, the most reasonable response is usually taken to be partial meet selection if one is dealing directly with belief sets already closed under Cn, and maxichoice or partial meet selection if one is working on belief bases.

In the present situation, the set N of normative rules is not assumed to be closed under Cn or any other consequence operation (other than the identity operation), and so maxichoice selection appears attractive here too. However, on reflection, it appears that different kinds of normative code may require differing treatment. In our medical examples, the natural procedure does indeed seem to be the maxichoice one: the doctor seeks to apply as much of the normative code as may consistently be done, and if that implies making choices between patients in situations where there is no information to indicate which is the more urgent (nor time or opportunity to gather it), the choices must nevertheless be made. But, as Carlos Alchourrón has reminded the authors, in the context of a judge applying a legal code the picture may be quite different.

For example, the code may contain a rule that whenever persons are in certain circumstances they are obliged to perform a certain action. It may happen that more than one person is in those circumstances, but that it is impossible that more than one carry out the action. One maximal set would oblige one person to act, another maximal set would oblige another to act. Yet the judge may also be bound by a higher-order constraint: *never favour or prejudice one individual over another without sufficient reason*. In the absence of sufficient reason, neither of the two individuals may legitimately be singled out for the duty, and so maxichoice selection is inappropriate.

However, in such an example neither full nor, more generally, partial meet selection improves matters. For by the symmetry of the case, if one individual is relieved of the duty, so should be the other, in which case the desired act is not required of anybody.

It would appear that in such a case the only rational procedure is to pass into a new dimension of "creative application": either split the required action up into parts and oblige different individuals to perform different parts, or add to the rules available, or reinterpret their contents, or find additional facts, in order to render the code or the situation asymmetric, justifying selection of one individual over another. In this paper, we shall not consider such creative applications, which indeed are presumably resistant to formalization. We shall remain within the sphere of partial meet selection, working with its limiting case of maxichoice selection, whilst remembering that it may not be appropriate for certain cases.

As the formal apparatus for maxichoice selection is quite familiar since Alchourrón & Makinson 1982 and Alchourrón, Gärdenfors & Makinson 1985, we do not recount it here.

5.3. *When to Maximalize*

In contrast, the question *at what level to maximalize* does not arise in the AGM account of belief change, where instantiation has no part, although it does arise in the Poole/Brewka account of defeasible explanation where instantiation plays a central, if less analysed, role. For consistent application, at first sight there are three levels where one might think of performing maximalization-and-choice: at the level of (i) derived normative assertions, (ii) normative rules, and (iii) instantiations of normative rules. This is where the most interesting technical questions arise.

(i) *At the level of derived normative assertions*. Maximalization-and-choice at this level makes little sense, given our definition of $a(N, S)$. It would consist of choosing a maximal consistent subset of $a(N, S) = Cn(I(N) \cup S) \cap A_N$. But since the background consequence operation Cn includes classical logic, if $Cn(I(N) \cup S) \cap A_N$ is inconsistent, then it is just the set A_N (of all normative assertions without variables, all of whose constants are in D), and the maximal consistent subsets of A_N need have no relation at all with the particular code N and situation S that gave rise to the inconsistency.

(ii) *At the level of normative rules*. Maximalization-and-choice at this level amounts to forming a maximal subset N' of the given set N of rules such that $a(N', S) = Cn(I(N') \cup S) \cap A_N$ is consistent. This also faces difficulties. An obvious but easily remedied shortcoming is that if carried out alone, without later re-maximalization, it can be intuitively very incomplete.

Example 1a: Consider the medical example, but with domain of individuals $1, 2, 3, 4$ who are so located geographically that the doctor can visit any *two* of them immediately, but not more. The rules are as before: $h(x) \rightarrow Ov(x)$ and $s(x) \rightarrow Ov(x)$, but the situation consists of $h(1), s(2), s(3), s(4)$. Then $a(N, S)$ is inconsistent, and the only maximal subset $N' \subseteq N$ such that $a(N', S)$ is consistent is $\{n_1\}$. Now $Ov(1) \in Cn(I(n_1) \cup S) \cap A_N = a(n_1, S)$: the doctor should visit person 1. But intuition tells us that the output is seriously incomplete, as the doctor still has time to see some patient from among those with severed hands. Roughly speaking, after choosing a maximal set of rules, we should still apply other rules to those cases where we may consistently do so. □

If this were the only problem, we could get over it: after having carried out maximalization and choice at the level of rules, we could re-maximalize at a later stage. The procedure would then become lexicographical. However, there is also a separate problem with maximalization-and-choice on normative rules, less obvious and more recalcitrant: it can force intuitively incorrect results that no amount of re-maximalization at a later stage can rectify.

Example 1b: Again we consider a medical example with domain of individuals $1, 2, 3, 4$ who are so located geographically that the doctor can visit any two of them immediately, but not more. The rules are the same as in examples $1, 1a$, plus a third normative rule $n_3 = c(x) \to Ov(x)$, that if a person is reported as having a persistent cough then the doctor should visit him immediately. This time the situation is taken to be: $h(1), s(2), s(3), c(4)$. Now $Cn(I(\{n_1, n_2\}) \cup S)$ and $Cn(I(\{n_3, n_2\}) \cup S)$ are both inconsistent. The only maximal subsets N' of N such that $Cn(I(N') \cup S)$ is consistent are $N_1 = \{n_1, n_3\}$ and $N_2 = \{n_2\}$. Their outputs include $\{Ov(1), Ov(4)\}$ and $\{Ov(2), Ov(3)\}$ respectively. But given the relative gravity of the three medical conditions involved, the only acceptable courses of action for the doctor from the point of view of intuition are $\{Ov(1), Ov(2)\}$ or else $\{Ov(1), Ov(3)\}$ — each of which is inconsistent with each of the above outputs. In this example, maximalization-and-choice on rules thus gives us an output that is wrong and cannot be corrected by supplementary maximalization-and-choice at a later stage. □

(iii) *At the level of instantiations of normative rules*. At this level, maximalization-and-choice consists of choosing a maximal subset K of $I(N)$ such that $Cn(K \cup S) \cap A_N$ is consistent, and putting $c(N, S) = Cn(K \cup S) \cap A_N$. This indeed corresponds to the choice made, without discussion, by Poole 1988 in his treatement of the analogous question of explanation from hypotheses. The shortcoming here is that the operation appears to be rather unprincipled, leaving open too wide a range of possible outputs, in so far as it does not take into account the normative rules from which $I(N)$ was generated.

Example 1c: Consider the same individuals with the same geographic locations. The rules are the original two only: $h(x) \to Ov(x)$ and $s(x) \to Ov(x)$. The situation this time is $h(1), h(2), s(3), s(4)$. Then one maximal solution is to visit 1 and 2, who both have heart attacks, and another is to visit 3 and 4, both of whom have severed hands. We would expect the selection function γ to favour the former over the latter, given the relative urgencies of the medical conditions, but whichever is chosen, it is at least a principled solution. On the other hand, the maximal consistent solution of visiting, say, heart attack case 1 and severed hand case 3 appears to be unprincipled, following one rule for one individual and another rule for the other. Yet such a solution is available when we simply maximalize on subsets of $I(N)$. In the example one maximal subset K of $I(N)$ for which $Cn(I(N) \cup S) \cap A_N$ is consistent consists of $h(1) \to Ov(1), s(3) \to Ov(3)$, plus all the rule-instantiations whose antecedents are false in the situation S, and this puts both $Ov(1)$ and $Ov(3)$ in $a(N, S)$. □

However, the difficulty here is rather different from that arising at the level of normative rules. There, we saw that maximalization-and-choice can force outputs that are intuitively incorrect, whereas here it can *permit* outputs that are incorrect. To be sure, it does not leave open a completely unconstrained range of outputs, as was the case for maximalization-and-choice at the level of detached normative asertions, but still the range appears to be uncomfortably large, allowing for unprincipled as well as principled applications. A further constraint of some kind seems appropriate.

How, then, should maximalization-and-choice take place? Recall that any well-organized legal code contains internal priorities. For example, principles of constitutional law have a stronger standing than those of non-constitutional law, later laws have precedence over earlier ones of the same category, and so on. Following this guide, and in analogy with the refinement made by Brewka 1989 of Poole's account of explanation, we propose that maximalization-and-choice should continue to take place on instantiations of rules in N, but to avoid excessive arbitrariness it should be broken up into steps according to a ranking of the rules themselves according to their priority. Rigorously, in the finite case:

- Partition the set N of rules into mutually exclusive subsets and impose a linear ordering of priority $N_1, \ldots, N_k$ on them.
- Proceed lexicographically on the sets $I(N_1), \ldots, I(N_k)$, as follows: for each i with $0 \leq i \leq k-1$, put K_{i+1} to be an arbitrary maximal subset K of $I(N_{i+1})$ such that $K_1 \cup \ldots \cup K_i \cup K \cup S$ is consistent.
- Define $c(N, S) = Cn(K_1 \cup \ldots \cup K_k \cup S) \cap A_N$.

It is clear how to generalize this to the case that N is infinite. Strictly speaking, the construction defines a *family* of functions $c(N, S)$, one for each partition, linear ordering of its cells, and choice function fixing each K_{i+1}. Evidently, it covers maximalization-and-choice on $I(N)$ as a whole as a limiting case where $k = 1$. At the opposite extreme is the limiting case where the cells N_i are singletons.

It is easy to show (*verification in the appendix*) that each function $c(N, S)$ is well-defined and satisfies the formal desiderata (i) – (iv) listed in section 4.2. It should be emphasized, however, that satisfaction of the formal desiderata (i) – (iv) provides only partial reassurance as to the value and interest of the operation defined. There are a great many variant operations definable that satisfy them but give intuitively incorrect results in examples. If a construction is to be viable, as we hope that the one proposed may be, it must survive both kinds of test.

Evidently, the partitioning of N makes $c(N, S)$ highly dependent on the syntactic form of the norms in N. For example, if N contains a normative rule $\alpha_1 \vee \alpha_2 \rightarrow O(\gamma_1 \wedge \gamma_2)$, this may give a different value of $c(N, S)$ than if there were four separate norms $\alpha_i \rightarrow O\gamma_j$. This is so even though we are using a background logic Cn satisfying both supraclassicality and disjunction in the antecedent, and will continue to be so even if in addition we require $Cn(O(\gamma_1 \wedge \gamma_2)) = Cn(O\gamma_1 \wedge O\gamma_2)$. This is because breaking a compound norm of N into several components permits finer partitions than would otherwise be possible.

In the case that all norms in N are of the simple affirmative form $[\alpha \rightarrow] O\gamma$, this can be bypassed. Given the limited nature of our language, α, γ are purely truth-functional descriptive assertions, and we may thus avoid the variability of output by requiring that before $c(N, S)$ is calculated, each simple affirmative norm $[\alpha \rightarrow] O\gamma$ in N should be replaced by the set of norms $[\alpha_i \rightarrow] O\gamma_j$ $(i \leq m, j \leq n)$ where $\alpha_1 \vee \ldots \vee \alpha_m$ is the disjunctive normal form of α and $\gamma_1 \wedge \ldots \wedge \gamma_n$ is the conjunctive normal form of γ. However, when some of the norms in N are not of simple affirmative form, but more generally only of the form $[\alpha \rightarrow] \varphi$ for a normative assertion φ as defined in section 2, then logically equivalent codes may admit different partitions and so different values of $c(N, S)$ without any obvious recipe designating a "best" among them.

One might suggest that in moral codes, the partition should be rather more coarse than in legal codes. Indeed, one might argue that for moral codes a partition is quite inappropriate. For implicit in the lexicographic procedure using it, is a 'dominance' principle: compliance with any number of normative rules of lesser priority can never outweigh compliance with a single rule of higher priority. Thus, for instance, satisfaction of ten municipal ordinances cannot outweigh obeying a law passed by parliament. Whilst this is common enough in legal and administrative codes, it may find less frequent acceptance for moral ones, especially in so far as they are based on considerations of harm or utility allowing some kind of addition.

For such codes, we suggest the following alternative procedure for defining $c(N, S)$ — on the assumption that some recipe based on utility weights is available to give a linear ordering of the subsets of $I(N)$. For such subsets J, J', write $J < J'$ if it is preferred that J be the set of all non-violated rule instances in $I(N)$, than that J' should be so. Assume that this order satisfies the natural constraint that whenever J properly includes J' then $J < J'$. Then simply define $c(N, S)$ as $Cn(J_i \cup S) \cap A_N$, where J_i is the first set in the ordering that is consistent with S. Note that there is no need for iteration here. If J_i is as defined, then for all $k > i$ with $J_k \neq \emptyset$, if $J_i \cup J_k \cup S$ is also consistent then since $J_i \cup J_k$ is a proper superset of Ji we have by the constraint above that $J_i \cup J_k < J_i$, contrary to the choice of J_i as first in the list. This construction in effect transfers the main burden of work to whatever recipe the utilitarian offers to order the subsets of $I(N)$, and also bypasses the use of maximalization-and- choice, except in so far as it may appear in that recipe.

6. FORMAL COMPARISON WITH CONTRACTION AND REVISION

We have already noticed, in section 4.3, important pragmatic differences between the consistent application of a normative code, and revision. On the mathematical level it is interesting to compare our analysis of the former with the Alchourrón/Gärdenfors/Makinson 1985 (briefly, AGM) account of the latter. Evidently, they have a common inspiration, namely maximalization on a consistency condition followed by choice among the maximals, but there are also a number of major differences, of which a basic one is *fragmentation of the language*. The AGM approach works with an undifferentiated language of propositions. In the present account, the distinctions between descriptive assertions, normative assertions, and normative rules are all essential.

On the other hand, there are also formal accounts of revision, specifically intended for use as an operation of amendment of normative codes, that do make such differentiations. Such, in particular, is the classic account of Alchourrón and Bulygin 1971. Although they allow the elements of a normative code to be arbitrary assertions, their analysis makes vital use of distinctions similar to those above (see especially the discussion in their section IV.5). That account is thus closer to the present construction than is the AGM one. However, there remains an important difference, which indeed we believe must appear between *any* accounts of consistent application on the one hand compared to revision or amendment on the other, and this is the explicit and essential role of instantiation as a component of application.

Recall that revision is usually defined from contraction via the Levi identity $N * \alpha = Cn((N - \neg\alpha) \cup \{\alpha\})$. Evidently, as remarked by Hansson 1993, when S is a finite set of assertions we may likewise define $N * S = Cn((N - \neg \wedge S) \cup S)$. Likewise, we have defined restrained application by the rule $c(N, S) = Cn((N \hookleftarrow S) \cup S) \cap A_N$, where $N \hookleftarrow S$ is the operation that we used in section 5.3 to construct the set $K_1 \cup \ldots \cup K_k$ and which we shall call the *restrained instantiation of* N *in the light of* S. We may then note the following points of contrast:

- Rather trivially, when defining restrained application $c(N, S)$ from restrained instantiation $N \hookleftarrow S$, we do not use the whole of $Cn((N \hookleftarrow S) \cup S)$, but only its intersection with the set A_N (of all normative assertions without variables, all of whose constants are in the domain D). This gives rise to a formal difference between revision and consistent application: whereas for revision $S \subseteq N * S$, we do not have $S \subseteq c(N, S)$ — indeed we have $S \cap c(N, S) = \emptyset$.
- More significantly, there is also a difference between the underlying operations of contraction and restrained instantiation. For the former we apply maximalization-and-choice just once on the rules in N itself, whereas for the latter we order the rules in N but apply maximalization-and-choice k times to subsets $I(N_1), \ldots, I(N_k)$ of the set $I(N)$ of all instantiations of rules in N (into the domain under consideration). As a result, whereas $N - (\neg \wedge S) \subseteq N$, we have $N \hookleftarrow S \subseteq I(N)$.

In effect, we *are* using contraction, but it is buried as one of the steps within restrained application; it is not applied to N but to each of $I(N_{i+1})$ to produce $K_{i+1} = I(N_{i+1}) - \neg \wedge (K_1 \cup \ldots \cup K_i \cup S)$.

In brief, our definition of the application of a normative code N to a situation S is conceived in the general spirit of the AGM revision of N by S, but technically is quite different due to its essential use of a fragmented language. In this respect it rather closer to the Alchourrón/Bulygin account of amendment of N is the light of S, but nevertheless still differs from the latter as well as the former in the key role played by instantiation, with ensuing differences in the formal properties holding.

7. INTRODUCING AUXILIARY RULES FOR 'LEGAL PREDICATES'

The formal apparatus that we have worked with is too simple to handle certain kinds of example that arise in practice, particularly in legal and administrative contexts. Legal codes contain predicates such as "is a citizen", "is incapacitated", "is in good faith", "is liable to income tax", "is eligible for pension benefits", which are ostensibly descriptive, but whose content is partly or wholly determined by articles of the code itself, in a way that may be at variance with ordinary usage outside the code. Such predicates may occur on the right of one rule, and also on the left of another, in each case unmodalized by any deontic operator, thus permitting iterated detachment. They are sometimes referred to as 'legal predicates'. When such a predicates one occurs on the right, it is usually in the form of a simple assertion or denial, and the rule is often thought of as fixing its descriptive content, whilst the rules in which such a predicate occurs on the left fix its normative implications.

How should the formal framework of the preceding sections be adapted to take such auxiliary rules and 'legal predicates' into account? We sketch, tentatively, what seems to be the simplest approach, and which builds on the constructions above, rather than replace them.

The notion of a code needs to be widened. We define an *elaborated code* to be a pair (N, R) where N is a set of normative rules and R is a set of *auxiliary* rules. The latter are expressions of the form $[\alpha \rightarrow](\pm)p(t_1, \ldots, t_n)$, where α is a descriptive assertion and $p(t_1, \ldots, t_n)$ is an elementary descriptive assertion (these terms are defined in section 2). Note that there is an asymmetry here: the right hand side is always an elementary descriptive assertion or its negation, whilst the left hand side can be an arbitrary descriptive assertion. A situation S_0 none of whose predicates occurs on the right of any auxiliary rule in R is called *external for* R. It seems natural to require that all applications of (N, R) be performed in the light of an external situation.

Using these ingredients, we wish to define *the unrestrained application* of an elaborated code (N, R) to a situation S_0 external for R. Two definitions suggest themselves: one where the rules in R and N are processed together, and the other where the rules in R are processed before those in N. On the former option, we would simply put $a(N, R, S_0) = a(R \cup N, S_0)$ where the right hand side is defined as before. In other words, we put

$$a(N, R, S_0) = Cn(I(R \cup N) \cup S_0) \cap A_N \tag{1}$$

On the latter option we would put $a(N, R, S_0) = a(N, S)$ where $S = Cn(I(R) \cup S_0) \cap A_F$, taking A_F as the set of all descriptive assertions without variables, all of whose constants occur in D. Thus on this option we put:

$$a(N, R, S_0) = Cn(I(N) \cup (Cn(I(R) \cup S_0) \cap A_F)) \cap A_N \tag{2}$$

These two definitions are in fact equivalent, under the hypothesis that all individual constants occurring in S_0 are in D, i.e. that $S_0 \subseteq A_F$, which may naturally be regarded as the 'normal' case. For even without that hypothesis we have $RHS(2) \subseteq RHS(1)$, and with it we also have the converse (*verifications in appendix II*).

Nevertheless the two equivalent definitions of $a(N, R, S_0)$ suggest two quite different definitions of restrained application in the presence of auxiliary rules. On the philosophy of lumping together R and N and treating them in parallel, we would simply put $c(N, R, S_0) = c(R \cup N, S_0)$, whereas on the philosophy of differentiation and sequential processing, we would define $c(N, R, S_0)$ in two stages, the first of which irons out any inconsistencies arising from applying the rules in R, whilst the second eliminates any remaining consistencies provoked by N. Specifically, given R, N, S_0 the sequential approach would proceed as follows:

- Form a consistent application $S = c'(R, S_0)$ of the auxiliary rules to S_0. Here the operation c' is just like c in section 5.3 (with R in place of N) except that in the last step we intersect with the set A_F rather than with A_N.
- Put $c(N, R, S_0) = c(N, S)$ where S is as defined above and c is as defined in section 5.3.

It is easily checked, by adapting the verifications for $c(N, S)$ in the appendix, that both of these operations $c(N, R, S)$ also satisfies the formal desiderata (i) – (iv). We prefer the more refined one, that resolves inconsistencies in the auxiliary legal rules before tackling those induced by the normative rules.

8. SOME GENERAL METHODOLOGICAL REMARKS

Our work on applying normative rules has encountered two methodological phenomena that we believe to be endemic to investigations in the logic of norms.

First, the presence at each stage of the investigation, from the very definition of the languages to be used, of multiple options for procedure. Usually, some options appear more reasonable than others, but it is rare that one can declare confidently that one of them is "correct" or best whilst the others are "in error"" or inferior. A delicate middle path between dogmatism and complaisance needs to be steered.

Second, when weighing the merits of options we continually come up against the competing demands of theoretical simplicity and elegance, fidelity to informal human practice, and facility of computation. All three should be taken into account, but our personal prejudice is to give priority in the above order. Our strategy is to obtain a clear even if oversimplified first picture which is close to at least *some* significant aspects of informal human practice. From there, one may work "outwards", improving fidelity to other aspects of practice as far as appears profitable. But there is a limit, beyond which gain in fidelity is outweighed by the growth of notational overheads and loss of conceptual clarity. The formal theoretical structure should thus remain a rather abstract approximation to human practice, differing from it in certain respects whilst resembling it in others.

APPENDIX I: WEAKER NOTIONS OF APPLICATION

Our definition of unrestrained application $a(N, S)$ as $Cn(I(N) \cup S) \cap A_N$ makes strong use of the operation Cn of consequence. It is also possible to think of weaker forms of unrestrained application (and thus too of restrained application), defined using closures weaker than Cn.

The weakest such operation would appear to be mere detachment, giving a notion of weak (unrestrained) application defined by putting $a_d(N, S) = d(I(N), Cn(S))$ or eventually as $Cn(d(I(N), Cn(S)) \cap A_N$. Here d stands for detachment, i.e. $d(X, Y)$ is the set of all normative assertions which are either themselves in X or else occur as consequent of some conditional in X whose antecedent is in Y. Such an operation does not permit using the full force of classical logic as applied to $I(N) \cup Cn(S)$, nor such principles of deontic logic as may also be included in Cn.

Example 3: Suppose the background consequence operation is closed under replacement of mutually equivalent assertions and contains the deontic principle $Cn(O(\varphi \wedge \psi)) = Cn(O\varphi \wedge O\psi)$, both of which are familiar constraints. Let N consist of the following four normative rules:

$$p(x) \to Or(x) \qquad p(x) \to O(\neg r(x) \vee s(x))$$
$$q(x) \to Os(x) \qquad q(x) \to O(\neg s(x) \vee r(x)).$$

Let $S = \{p(1) \vee q(1)\}$. Then $a(N, S) = Cn(I(N) \cup S) \cap A_N$ contains $Or(1)$, $Os(1)$, and $O(r(1) \wedge s(1))$. However $d(I(N), Cn(S))$ is empty.

Given these limitations, we are inclined to refer disparagingly to the operation a_d as the "narrow-minded" or "bureaucratic" application of a normative code to a situation. As such, it may have some sociological interest. From a computational point of view, too, it permits rapid calculation. But from a logical point of view it is of secondary interest, and we do not study it further here.

APPENDIX II: VERIFICATIONS

Section 3: Verification of left idempotence of $a(N,S)$

We need to show that $a(N,S) = Cn(I(a(N,S)) \cup S) \cap A_N$. Suppose $\alpha \in LHS$. Since $a(N,S) = I(a(N,S))$ clearly $\alpha \in Cn(I(a(N,S)) \cup S)$. Also since $\alpha \in a(N,S)$ we have $\alpha \in A_N$ so $\alpha \in RHS$. For the converse, suppose $\alpha \in RHS \subseteq Cn(I(a(N,S)) \cup S) = Cn(a(N,S) \cup S) = Cn((Cn(I(N) \cup S) \cap A_N) \cup S) \subseteq Cn(I(N) \cup S)$. But $\alpha \in RHS$ also implies $\alpha \in A_N$, so $\alpha \in LHS$. □

Section 3: Verification of left stability of $a(N,S)$

Suppose $N \subseteq N' \subseteq N \cup a(N,S)$. We want to show $a(N,S) = a(N',S)$. We have the inclusion $a(N,S) \subseteq a(N',S)$ directly by left monotony, which also tells us that to show the converse it suffices to show that $a((N \cup a(N,S)),S) \subseteq a(N,S)$. Suppose $\alpha \in LHS$. Then $\alpha \in Cn(I(N \cup a(N,S)) \cup S) = Cn(I(N) \cup I(a(N,S)) \cup S) = Cn(I(N) \cup a(N,S) \cup S) = Cn(I(N) \cup (Cn(I(N) \cup S) \cap A_N) \cup S) \subseteq Cn(I(N) \cup Cn(I(N) \cup S) \cup S) = Cn(Cn(I(N) \cup S)) = Cn(I(N) \cup S)$. Also, since $\alpha \in LHS$ we have $\alpha \in A_N$. Thus $\alpha \in Cn(I(N) \cup S) \cap A_N = RHS$. □

Section 3: Verification of strong left idempotence of $a(N,S)$

Suppose $a(N,S) \subseteq N' \subseteq N \cup a(N,S)$. We want to show $a(N,S) = a(N',S)$. Now $a(N,S) = a(a(N,S),S) \subseteq a(N',S) \subseteq a(N \cup a(N,S),S) = a(N,S)$ where the first identity is by left idempotence, the two inclusions are by left monotony using the hypothesis, and the last identity is by left stability. □

Section 5.3: Verification that $c(N,S)$ *is well-defined and has properties (i) – (iv)*

To show that $c(N,S)$ is well-defined and also show property (ii) that $\bot \notin Cn(c(N,S))$, observe by induction on $i \leq$k that $K_1 \cup \ldots \cup K_i \cup S$ is consistent, recalling for $i = 1$ that S is assumed to be consistent (section 2).

To show property (i), that $c(N,S) \subseteq a(N,S)$, notice that each K_i is a subset of $I(N_i)$ so that $(K_1 \cup \ldots \cup K_k) \subseteq I(N)$ so $Cn(K_1 \cup \ldots \cup K_k \cup S) \subseteq Cn(I(N) \cup S)$ and thus $c(N,S) \subseteq a(N,S)$.

For property (iii), suppose $\bot \notin Cn(a(N,S)) = Cn(Cn(I(N) \cup S) \cap A_N)$. Then $\bot \notin Cn(I(N) \cup S)$ so by induction each $K_i = I(N_i)$ so $c(N,S) = Cn(I(N) \cup S) \cap A_N = a(N,S)$.

For property (iv), suppose $Cn(S) = Cn(S')$. Then the domain D being held constant, for each $i \leq k$, $I(N_i) = I'(N_i)$, and so clearly each $K_i = K_i'$ and thus $c(N, S) = c(N, S')$. □

Section 7: Verification that
$Cn(I(N) \cup (Cn(I(R) \cup S_0) \cap A_F)) \cap A_N \subseteq Cn(I(R \cup N) \cup S_0) \cap A_N$

Clearly, it suffices to show that the inclusion holds between the parts before intersection with A_N, so it suffices to show both $I(N) \subseteq Cn(I(R \cup N) \cup S_0)$ and $Cn(I(R) \cup S_0) \cap A_F \subseteq Cn(I(R \cup N) \cup S_0)$. But the former is immediate given that $I(N) \subseteq I(R \cup N)$, and the latter is also immediate given that $I(R) \subseteq I(R \cup N)$. □

Section 7: Verification that if $S_0 \subseteq A_F$ then
$Cn(I(R \cup N) \cup S_0) \cap A_N \subseteq Cn(I(N) \cup (Cn(I(R) \cup S_0) \cap A_F)) \cap A_N$

Clearly, it suffices to show that the inclusion holds between the parts before intersection with A_N, so it suffices to show that $I(R \cup N) \cup S_0 = I(N) \cup I(R) \cup S_0 \subseteq Cn(I(N) \cup Cn((I(R) \cup S_0) \cap A_F))$. But for $I(N)$ this is immediate, whilst for $I(R) \cup S_0$ it follows directly from $I(R) \subseteq A_F$ and the hypothesis $S_0 \subseteq A_F$. □

David Makinson
Ville d'Avray, France
and
Sven Ove Hansson
Uppsala University

ACKNOWLEDGEMENTS

Thanks are due to Carlos Alchourrón and Michael Freund for their penetrating and sometimes devastating critical remarks on various drafts. Carlos died on 13 January 1996, and this paper is dedicated to him. The text has also benefitted from discussions at LMPS-95.

REFERENCES

Alchourrón, Carlos E. and Eugenio Bulygin. 1971. *Normative Systems*. New York, Wien: Springer Verlag.

Alchourrón, Carlos E., Peter Gärdenfors and David Makinson. 1985. On the logic of theory change: partial meet contraction and revision functions. *The Journal of Symbolic Logic* 50: 510–530.

Brewka, Gerhard. 1989. Preferred subtheories: an extended logical framework for default reasoning. *Proceedings IJCAI-1991*: 1043–1048

Gärdenfors, Peter and David Makinson. 1988. Revisions of knowledge systems and epistemic entrenchment. In M. Vardi ed., *Proceedings of the Second Conference on Theoretical Aspects of Reasoning about Knowledge*. Los Altos: Morgan Kaufmann.

Hansson, Sven Ove. 1993. Reversing the Levi identity. *Journal of Philosophical Logic* 20: 637–669.

Poole, David. 1988. A Logical framework for default reasoning. *Artificial Intelligence* 36: 27–47.

10
Foundations of Logic, Mathematics and Computer Science

CHARLES PARSONS

WHAT CAN WE DO "IN PRINCIPLE"?*

In philosophy it is *can* in particular that we seem so often to uncover, just when we had thought some problem settled, grinning residually up at us like the frog at the bottom of the beer mug.

J.L. Austin, "Ifs and cans"[1]

I

It is characteristic of some quite important discourses in mathematics, having to do with computability and constructivity, to appeal to what appear to be possibilities of human thought and action, where, however, these possibilities go beyond what is "practically" possible. In particular, they do not rest on actual human abilities. Such possibilities are what I have in mind in asking what we can "in principle" do. Such possibilities are also invoked about non-human agents, in particular computers. I am primarily concerned, however, with real agents, human beings like you and me and computers like the ones we have on our desks, or at most supercomputers. But talk about the "in principle" capabilities of real agents is roughly equivalent to talk about the capabilities of ideal agents.

Usually we answer questions about what we can do by using information about what we and other human beings have done, what we have previously done, and general knowledge about human behavior. I know that I can teach a course on Kant's *Critique of Pure Reason* because I have done it several times and have no reason to believe I have lost the ability. On the other hand I can't teach one on quantum field theory. This may be something I could have done, if I had made different decisions about my career years ago. There are some things that I cannot and could not have done, that we say human beings are capable of, because they have been done. An example would be to write a body of music in the classical style, comparable to that composed by Mozart. Another example, of which unlike the Mozart example it can't be argued that

*Earlier versions of this paper were presented as the Skolem Lecture, University of Oslo, and to philosophy colloquia at Stanford and UCLA. I have received valuable comments from a number of hearers; I am painfully aware of not having done justice to all of them. Special thanks are owed to Joseph Almog, Tyler Burge, John Carriero, Solomon Feferman, Dagfinn Føllesdal, David Kaplan, Per Martin-Löf, and Grigori Mints. The earlier versions of the paper were written while I was a Fellow of the Center for Advanced Study in the Behavioral Sciences, with the support of the Andrew W. Mellon Foundation. I wish to thank both institutions.

[1] *Philosophical Papers* (Oxford: Clarendon Press, 1961), p.231.

M.L. Dalla Chiara et al. (eds.), Logic and Scientific Methods, 335–354.

no one in our own time could have done it, would be to have a major league baseball career in which I broke Ty Cobb's record for total number of hits. I have no reason to believe I had any talent at all as a composer and good reason to think that I lacked talent for baseball and sports generally. But we all know that others have such talents. Some things, however, though they look like extensions of the feats of great athletes or artists, no one can do. No one can jump over the Eiffel Tower or compose in five minutes a symphony comparable to Beethoven's Ninth or one of the symphonies of Mahler.

Now philosophers do quite often talk of things like this as *possible*. Though no one actually has the ability to jump over the Eiffel Tower, that someone jumps over the Eiffel Tower is often said to be "logically" or "metaphysically" possible.[2] And that is a possible meaning of the statement that I can "in principle" jump over the Eiffel Tower. In the usage of some writers, such possibilities may contradict even basic laws of physics. It might be argued that that is true of the possibility that a human being jumps over the Eiffel Tower, but it might be hard to argue this conclusively because the concept of a human being and of jumping are not concepts of fundamental physics, and if we specify precise physical limits on the size, muscular structure, and energy turnover of a human being, it might still be argued that these could have been different, or could become different by an evolutionary process that preserves the identity of the human species. However, it is not difficult to come up with other examples where there is a conflict with fundamental physics, such as travel between galaxies at speeds faster than light.

One thing that could be meant by the claim that "in principle" I can jump over the Eiffel Tower, or it is in principle possible that I should do so, is that this is logically or metaphysically possible, in whatever sense is intended by the philosophers who use these terms. Does this help? *Is* it logically or metaphysically possible that I should jump over the Eiffel Tower, or that anyone should? Is it possible in the relevant sense that I should compute the value of 257^{729} and write the value out in decimal notation? It might seem clear that the answers are affirmative, because each seems conceivable. It seems that the first can be imagined; I wouldn't even rule out the possibility that it might be filmed: that is, by tricks of the film-maker's art one might produce a short film in which someone who looks for all the world like me is seen to jump the height of the Eiffel Tower, pass over the building, come down, and land safely. Imagining the second poses a problem about time and complexity: there are so many steps to the computation that imagining them all takes

[2]In the course of explaining the notion of necessity that concerns him, Alvin Plantinga remarks of "leaping tall buildings in a single bound or ... travelling faster than a speeding bullet", "These things are impossible for us; but not in the broadly logical sense" (*The Nature of Necessity* (Oxford: Clarendon Press, 1974), p.2). In talking of "jumping", I am assuming that the jumper relies on his own power and does not use such aids as balloons. It's not clear that all talk of jumping satisfies this constraint; consider (where the problem is the downward journey rather than reaching a height) parachute jumping.

an amount of time of an order of magnitude close to the time required for the computation itself. But we don't usually demand for imagining something that all its details be imagined, and in a looser sense we can certainly imagine my doing this computation. Moreover, we certainly tell stories, such as fairy tales and certain myths and legends, in which much stranger things happen.[3] The latter fact tells us very little; there is no reason to think that what is recounted in stories has to be possible in any sense stronger than escaping blatant contradiction.

II

What is the relevance of this to mathematics? Consider a very everyday notion in mathematics and logic, that of a decision procedure, or, more generally, a computing procedure. Many of us would characterize a decision procedure as a mechanical procedure by which, given any one of a class of problems, *we can* produce a yes or no answer by applying the procedure to that problem. Consider standard decision procedures for validity in classical propositional logic: a formula is valid just in case it is a tautology, that is comes out true under every assignment of the values True and False to its sentence letters, and the procedure consists essentially in going through these possible assignments, of which there are 2^n if the formula contains n sentence letters, and for each one computing the truth-value of the formula, going on if it is true and stopping (with the answer of nonvalidity) if it is false.[4] If one has gone through all the assignments and got True every time, then at the end one has the answer that the formula is valid. Writers of textbooks seem to take for granted that given *any* formula, *we* can, by applying the procedure, determine whether the formula is valid.

Consider another simple example, from a standard work by an eminent logician:

> Let $f(x)$ and $g(x)$ be polynomials with given integral coefficients. Is $f(x)$ a factor of $g(x)$? We can divide $g(x)$ by $f(x)$. This division is *performable step by step* by a preassigned method. It will terminate *in a finite number of steps* (how many will depend on how we count the steps, but this could be made precise so that the number will depend on the degree of the equation and the size of the coefficients). *We* shall then have the remainder *before us*. This remainder we can

[3]The time demands of computing 257^{729} should not have daunted the Elves of *The Lord of the Rings*, although it is hard to imagine one of them devoting himself to such an undramatic task. The space demands might have posed a problem.

[4]I ignore possible shortcuts, but the fact that the problem is NP-complete shows that there are serious limits of what they can accomplish.

> recognize to be either 0 or different from 0. If it is 0, the answer to the question is "yes". If it is different from 0, the answer is "no".[5]

It appears that when Kleene says "performable step by step", he means performable by human beings of normal computational ability; indeed, usually part of the process of coming to understand such a procedure is working out examples. Of course for real human beings to perform them, the steps cannot be too complex. And yet the intent is clearly that the procedure can be applied to an *arbitrary* pair of polynomials, even of very high degree with very large coefficients, numbers so large that even recognizing identities and differences among them can be a very long process.

Kleene then makes a decisive move, characteristic of discussions of such procedures, in saying that in a given application the procedure will terminate in a finite number of steps. To establish that a decision problem is solvable, it has almost always been regarded as sufficient that the number of steps of a procedure offered should be finite. If that is the only restriction, then the number can be very large, even very large in relation to the size of the input (measured in some way). But that already cuts us loose from the actual abilities of real human beings. For the number may be so large that even if the person applying the procedure is as fast as the fastest digital computers, he will be dead long before that number of steps has been carried out (as will be true in the present case if the degree and coefficients of the polynomials are very large).

Let me give another example, also from Kleene. He is explaining L.E.J. Brouwer's claim that the law of excluded middle holds for propositions involving only finite domains but that we have no reason to hold it once infinite domains are involved. It is the less controversial, positive claim that concerns me:

> For definiteness, let us specify P to be a property such that, for any given member of D, *we can* determine whether that member has the property P or does not.
> Now suppose D is a finite set. Then we could *examine every member* of D in turn, and thus either find a member having the property P, or verify that all members have the property not-P. There might be practical difficulties, e.g. when D is a very large set having say a million members, or even for a small D when the determination whether or not a given member has the property P may be tedious. But the possibility of completing the search exists *in principle*. It is this possibility which for Brouwer makes the law of the excluded middle a valid principle for reasoning with finite sets D and properties P of the kind specified.[6]

[5] S.C. Kleene, *Introduction to Metamathematics* (New York: Van Nostrand, 1952), p.136, emphases mine.

[6] Ibid., p.47, emphases again mine. In formalized intuitionistic mathematics, it is possible to prove $A \vee \neg A$ for quantifier-free A, and then derive it for any A containing only bounded

In this case too, the finiteness of D is regarded as sufficient for its being the case that "we" can carry out the procedure for deciding whether some element of D has P. When it is recognized that there are "practical difficulties" if D is very large or the procedure for deciding P lengthy to execute, it is said that it is still possible "in principle" to carry out the search through D for an instance of P.

Apparently what the statement of logic textbook writers that we can always decide whether a formula is a tautology rests on is the fact that we know the number of assignments is finite (in fact, in this case we have the simple bound 2^n), and it would also be possible to say in straightforward terms how many steps it would take to compute the truth-value for a given assignment. And it is taken not to matter how many steps that is.

Through the work of a number of logicians in the 1930's, it came to be accepted that a function on natural numbers is computable if and only if it is recursive. As is well known, the class of recursive functions can be characterized in a number of ways. Generally one first defines a *computing procedure* of a certain canonical form (it is not claimed that every computing procedure is of this form, at most that every computing procedure can be simulated by one of this form). Laying down a convention about how numbers are to be represented as inputs and outputs to the procedure, we can think of any such procedure as computing a *partial* function f on the natural numbers: $f(n) = m$ if, given the number n as input, the procedure eventually terminates and yields the number m as output.[7] The partial functions thus computed are the partial recursive functions. Then the general recursive functions are the functions obtained if we require that the procedure always terminates and yields a number output.

As you all know, what is called Church's Thesis is the principle that the computable total functions of natural numbers are exactly the general recursive functions. Much of the argument used to buttress this principle, both by the founders of recursion theory and by those who came to the matter later, was concerned to show that all computable functions are recursive, usually by arguing that any computing procedure can be simulated by one of the particular canonical form under consideration.[8] I will consider later the most

quantifiers. The proof makes essential use of primitive recursion, but not in any very strong form. It suggests that it may be possible to look at the grounds for the truth of this restricted law of excluded middle in a different way.

[7]For simplicity I will assume that the procedure is deterministic and can thus yield at most one number as output. Functions of more than one argument can be accommodated by conventions about the representation of n-tuples of numbers.

It can of course happen that the procedure does not yield a number value for *any* number argument; that is f is undefined for all numbers. Moreover, for a given argument we may not be able to determine whether or not the procedure terminates. That, for a given canonical form, it is not effectively decidable whether a procedure terminates is one of the basic results of the theory of computability.

[8]In fact, sometimes it is only this direction that is called Church's Thesis, for example by Kleene, op. cit., p.300. The equivalence fits better with Church's original formulation.

important such argument, that offered by Turing. First I want to consider briefly the other direction of Church's Thesis: that all recursive functions are computable.

At first glance this requires us only to recognize that the procedures of the canonical form in a given definition are indeed *computing* procedures. In the most intuitive cases, such as Turing machines, this seems quite evident. In fact there is an element of mathematical idealization even at that stage, but that fact is not what concerns me now. I want to point to another well-known consideration appealed to in arguing that all recursive functions are computable. The principle "Finiteness is sufficient" is used in a very strong way. To say that the procedure terminates for a given argument is to say that there is a finite sequence of steps of the procedure, ending in termination, i.e. a situation where the computation cannot continue.[9] But nothing is said about the length of the sequence, or how we know it terminates. A function is general recursive if there is a procedure of the canonical form such that, for every number n as input, *there is* a terminating sequence of steps with a number output.[10]

It is well known that even in the more restricted domain of primitive recursive functions, there are functions that grow very fast, so that for arguments of modest size the values will be so large that it is beyond our capacities, and even those of supercomputers, to write down the values in a canonical notation such as the decimal. In the 1950's the Dutch mathematician D. van Dantzig questioned whether $10^{10^{10}}$ is a finite number. It was a problem like this that he had in mind: A number is finite if it is generated by application of the successor operation S beginning with 0 or 1, but we can't grasp the sequence of steps that leads to $10^{10^{10}}$, or even physically construct a representation by a numeral of the form S^n0.[11] However we understand the construction of a

He states that he is proposing a "definition of effective calculability" as either λ-definability or general recursiveness ("An unsolvable problem of elementary number theory", *American Journal of Mathematics* 58 (1936), 345–363, p.90 of the reprint in Martin Davis (ed.), *The Undecidable* (Hewlett, N.Y.: Raven Press, 1965); cf. also p.100 and "An unsolvable problem of elementary number theory: Preliminary report" (abstract), *Bulletin of the American Mathematical Society* 41 (1935), 332–333). On the other hand Church's most developed argument, in §7 of the paper, is for the claim that all effectively calculable functions are recursive.

[9]In the Turing machine case, the machine is in a "terminal state"; in the case of the formalism of recursive functions, we have reached an equation of the form $f(\mathbf{n}) = \mathbf{m}$, where f is the principal function letter.

[10]To be sure, this formulation fits best with an embedding of the notion of general recursiveness in classical mathematics. A constructivist would demand a *constructive proof* that the computation terminates in general. Cf. Kleene, op. cit., p.319.

[11]Van Dantzig writes, "Modern physics implies an upper limit, by far surpassed by $10^{10^{10}}$, for numbers which actually can be constructed in this way." ("Is $10^{10^{10}}$ a finite number?" *Dialectica* 9 (1955–56), 273–277, p.273). This point was later developed further by Robin Gandy. Although van Dantzig's point of view is akin to what is now called strict finitism, he does not propose to banish exponentiation from mathematics; rather he argues for a kind of relativity of the notion of natural number and the irreducibility of formal expressions

representation of this number, whether in a mentalistic or physicalistic way, such a construction is not a genuine possibility. Van Dantzig seems at the outset to dismiss metaphysical possibilities that might contradict the fundamentals of physics and apparently also the idea that the capacities of the mind that are relevant in this connection might be defined in such a way as not to be limited by the capacities of the mortal human organism embedded in the physical world. I will call such a conception of mental capacity *transcendental*.[12] One might, however, look at the matter from the other direction and hold that since it is evident that exponentiation is computable, the idealized capacities that van Dantzig dismisses should in fact be accepted. In my view, mathematical thinking has in fact made idealized conceptions of our capacities seem persuasive and natural. Whether they should be taken quite at face value is another matter.[13]

III

The question raised for us is whether, in taking it for granted that arbitrary primitive recursive functions are computable, and going on to attribute computability to general recursive functions no matter how long their computing procedures need to run to give answers, and no matter how it is established that they always terminate, mathematicians involve themselves in such metaphysics, i.e. either metaphysical possibilities surpassing the physical, or transcendental capacities of the mind that equally surpass the actual capacities of the embodied mind.

The mathematician might reply that discussions and theories of computability are just mathematics, and that the idea that computing procedures are executable by *us* to *find out* the answers to problems is not to be taken literally, at least in the context of general statements. In fact the principle "Finiteness is sufficient" was built into the formulation of decision problems very early. Consider two classical Hilbertian formulations, first of his Tenth

---

involving functors introduced by primitive recursion to canonical $0-S$ numerals, or even to expressions introduced earlier in a chain of primitive recursions. These ideas are developed in more recent strict finitist writing.

[12]Such a view is usually attributed to Brouwer, not unreasonably in spite of the strongly naturalistic tendencies in his philosophy. Cf. the remarks below on Heyting.

[13]John Carriero asked the reasonable question whether some notion of human capacity in principle already underlies the conception of mathematical construction in Euclid. Certainly lines, for example, are thought of in traditional geometry as indefinitely extendible, and we naturally think of the constructions postulated by Euclid as indefinitely iterable. Euclid's language in stating his postulates seems to me quite reserved; for example postulate 2 is stated "To produce a finite straight line continuously in a straight line". (T.L. Heath, *The Thirteen Books of Euclid's Elements*, 2d ed. (Cambridge University Press, 1926), vol. 1, p.154.) Although verbs of action are used, nothing is said about what this implies. The question is not addressed in Heath's commentary on Euclid or a couple of others I have consulted.

Problem and then of the decision problem for first-order logic, what was called the *Entscheidungsproblem* (appropriately translated "*the* decision problem"):

> Given a diophantine equation with arbitrary unknowns and rational integer coefficients: one is to give a procedure, according to which by means of a finite number of operations it can be decided whether the equation is solvable in rational integers.[14]
> The *Entscheidungsproblem* is solved if one knows a procedure which will permit one to decide, using a finite number of operations, on the validity, respectively the satisfiability of a given [first-order] logical expression.[15]

The emphasis on finiteness would suggest that the talk of permitting *one* to *decide* is metaphorical. J.B. Rosser could write in 1939:

> "Effective method" is here used in the rather special sense of a method each step of which is precisely predetermined and which is certain to produce the answer in a finite number of steps.[16]

Still, von Neumann could write at almost the same time as Hilbert and Ackermann that if a positive solution to the *Entscheidungsproblem* were given, "mathematics, as we now understand it, would cease to exist" and be replaced by "an absolutely mechanical prescription, with whose help everyone could decide of any given statement whether or not it can be proved."[17] Talk of "determining" or "deciding" the answer to a mathematical problem by applying a procedure of the kind described in mathematical literature on decision problems may have been metaphorical, but the metaphor was and is a very powerful one. This no doubt results in large part from the fact that there is an easy transition from computation procedures that are devised in a setting in which they will actually be executed (and without the aid of high technology) to general mathematical results about procedures of that kind, to a more abstract conception of computing procedure implicit in the mathematical results, to a situation where computing procedures and computations in general are themselves mathematical objects about which there is a mathematical theory. The concept of recursiveness or computability arose in the course of the development of this theory, in particular when questions were posed like the Tenth Problem and the *Entscheidungsproblem*, and the

[14]David Hilbert, "Mathematische Probleme", *Nachrichten von der Königlichen Gesellschaft der Wissenschaften zu Göttingen*, 1900, 253–297, Problem 10, my translation.

[15]David Hilbert and Wilhelm Ackermann, *Grundzüge der theoretischen Logik* (Berlin: Springer, 1928), p.73, translation from Robin Gandy, "The confluence of ideas in 1936," in Rolf Herken (ed.), *The Universal Turing Machine: A Half-Century Survey* (Oxford University Press, 1988), pp.55–111, at p.62.

[16]"An informal exposition of proofs of Gödel's theorem and Church's theorem," *The Journal of Symbolic Logic* 4 (1939), 53–60, p.225 of the reprint in *The Undecidable*.

[17]Gandy, op. cit., pp.66–7, quoted from J. von Neumann, "Zur Hilbertschen Beweistheorie," *Mathematische Zeitschrift* 26 (1927), 1–46; I have modified Gandy's translation.

question arose how one could possibly prove that a problem of that kind is *not* solvable.

IV

Before we take the story within mathematics further, let us turn to philosophy. I have already mentioned the fact that philosophers make statements about what is "metaphysically" or "broadly logically" possible, which sometimes allow as possible what would be ruled out by scientific laws. What I am directly interested in is the philosophical use made of the idea that recursive predicates are *in principle* decidable (and that the same might be true of an analogous class of nonmathematical predicates), and recursive functions in principle computable.

An example worth exploring, because it uses ideas from the foundations of mathematics, is the kind of theory of meaning that, according to Michael Dummett, would underlie an anti-realist stance in metaphysics. Here it is clear that decidability and computability by human beings are at stake. It makes little difference whether we allow them to have technological aids, i.e. computers, because the increase in power they give, vast as it is, is dwarfed by the speed with which some mathematical functions and the length of any computation of their values will increase.

Dummett finds an exemplar of such a theory, for the special case of (presumably regimented) mathematical language, in intuitionism. It has the important feature of explaining the meaning of a statement not in terms of the conditions under which it is true but in terms of what would be a proof of it. It was in those terms that Heyting originally gave explanations of the logical connectives in intuitionism. He and Kolmogorov, who did parallel work, were explicating what they took to be implicit in Brouwer; therefore the interpretation is now called the "BHK interpretation". The working out of this interpretation has led to different theories (corresponding to different semantic theories) and some real foundational difficulties. But for us it suffices to observe that the basic ideas (at least together with some mathematical assumptions that the intuitionists took to be unexceptionable) already involve us in regarding as well-defined and in some sense "effective" the same functions where we have already observed that determining their values surpasses our practical capacities.

This should be clear from the intuitionist explanation of the quantifiers, which I will suppose to range over natural numbers. A proof of $\forall x A$ is a function (or procedure) that, given a number n, yields a proof of $A\ x/\mathbf{n}$ ($\mathbf{n}$ a canonical numeral for n). A proof of $\exists x A$ is a pair $\langle n, p\rangle$ such that n is a number and p is a proof of $A\ x/\mathbf{n}$. Consider the proposition that m^n is defined for all m and n; i.e. for all m and n there is a k such that $k = m^n$. The latter predicate might be defined in a way that does not presuppose that m^n is defined, but then a proof that it is will, on this reading, yield a function that for any m and n yields a value of k such that $k = m^n$. That is just the exponential function. If, as intuitionists did, we accept the mathematical arguments to the effect that faster-growing functions are defined, then canonical proofs of them will have to be such that the functions are recoverable from them.

The "proofs" referred to in these definitions are not what a mathematical lecturer presents, or a mathematical author writes, nor are they formalized versions of such informal mathematical proofs, in the languages characteristically used to formalize intuitionistic theories. They are usually referred to as canonical proofs. Roughly, they are representations of the information that proofs are supposed to provide. A statement of the form "p proves A" has implications that it would require elaborate computational unpacking of p to obtain. The complexity of this computation is not any more limited than computations as they arise in the theory of computable functions in mathematics.

Clearly this results from the fact that intuitionism does not curb its mathematics in an extreme way, toward what is called "strict finitism"; it accepts primitive recursion and even higher recursions that go far beyond primitive recursion. In modern formulations it appears in effect as an axiom of intuitionist mathematics that a function f from a given domain $\mathbf{A}$ into itself can be iterated. Taking $\mathbf{A}$ as the natural numbers $\mathbf{N}$, and taking successor and composition as given, this will more or less immediately yield the primitive recursive functions. Looking at the matter another way, taking intuitionistic logic as given and formulating first-order arithmetic in that framework, it is well known that the same functions are provably general recursive as in the classical case.

The problem for intuitionism is whether, if "proofs" are as the BHK interpretation applied to actual intuitionist mathematics requires, they can have the epistemic significance they are supposed to have. In traditional intuitionism, as exemplified by Brouwer and Heyting, this problem emerges as the question how what has the structure that intuitionistic mathematics requires can still consist of "mental" constructions. In one of his general expositions, Heyting remarks that the existence of a number "signifies nothing but the possibility of actually producing a number with the requisite property."[18] Heyting insists that mathematical construction is a mental activity that makes no presuppositions about the existence outside our thoughts of its objects.[19] Then this "producing" must be done in thought, and it follows that the possibilities of mental construction must extend as far as the functions admitted in intuitionist mathematics call for. This formulation would commit Heyting to the transcendental view of mental capacity and is thus not so presuppositionless as it appears.

---

[18]A. Heyting, "The intuitionist foundations of mathematics," in Paul Benacerraf and Hilary Putnam (eds.), *Philosophy of Mathematics: Selected Readings* (2d ed., Cambridge University Press, 1983), p.54. (Translation from *Erkenntnis* 2 (1931).)

[19]E.g. ibid., p.53.

In Dummett's discussions of intuitionism and of the issues concerning realism in general, effectiveness is frequently invoked, in a way that pretty clearly requires that the procedures contained in intuitionistic proofs can be carried out, so that it has to be sufficient that they should be "in principle" capable of being carried out. The idea underlying the first (and in his view more promising) argument for intuitionism that he sketches in "The philosophical basis of intuitionistic logic" is that there cannot be anything in the meaning of mathematical statements that is not reflected in the use that is made of them, and this is by its nature something that can be manifested. Austin's "can", or some functional equivalent, turns up at crucial points:

> Thus the knowledge in which, on the platonistic view, a grasp of the meaning of a mathematical statement consists must, in general, be implicit knowledge, knowledge which does not reside in the capacity to state that which is known. But, at least on the thesis that use exhaustively determines meaning, and perhaps on any view whatever, the ascription of implicit knowledge to someone is meaningful only if he is *capable*, in suitable circumstances, of fully manifesting that knowledge. ... When the sentence is one which we have a method for effectively deciding, there is again no problem: a grasp of the condition under which the sentence is true may be said to be manifested by a mastery of the decision procedure, for the individual may, by that means, get himself into a position *in which he can recognise* that the condition for the truth of the sentence obtains or does not obtain ....[20]

Dummett goes on to suggest that it is in the mastery of the practice of computation and proof that our grasp of the meanings of mathematical statements consists, and that the notion of truth as "the central notion of the theory of meaning for mathematical statements" should be replaced by the notion of proof. In his further working out of this suggestion in other writings, the notion of proof as it is used in the BHK interpretation and work on the foundations of intuitionistic logic based on it figures prominently. Dummett only rarely invokes the possibility in principle of completing computations and deciding statements, and he does not give a very direct defense of the use of this notion that he does make. But if the result of the argument is to leave actual intuitionistic mathematics as justified, it seems that such possibilities are needed to cash in the capability of recognizing that A or that $\neg A$, where A is a statement like "$10^{10^{10}} + 1$ is prime".[21]

---

[20] "The philosophical basis of intuitionistic logic" (1973), in *Truth and Other Enigmas* (London: Duckworth, 1978), p.224, emphases mine.

[21] Cf. the opening remarks in "Wang's paradox" (1975), in *Truth and Other Enigmas*, p.248.

Commentators on Dummett, in particular Crispin Wright and Alexander George,[22] have claimed that Dummett needs to rebut strict finitist arguments against intuitionism, and Dummett's own argument in "Wang's paradox" that strict finitism is incoherent because it uses concepts that are subject to the Sorites paradox was pretty clearly intended as part of a defense on this point. In default of a direct defense of in principle possibility, Dummett is vulnerable to the version of the criticism offered by George, who argues that the intuitionist does not have an adequate reply to the strict finitist claim that Dummett's proposal for a theory of meaning would not meet his own standard and that this tells against the standard, that is against Dummett's reconstruction of the intuitionist case against classical mathematics.

V

I now want to comment on two possible approaches to explicating the notion of in principle possibility in the contexts that have concerned us. The first is that the "in principle" possibility of carrying out a computing procedure is something close to the philosopher's logical or metaphysical possibility. This is, I think, the purport of a proposal offered by Wright. He considers it inadequate, although partly for reasons rather specific to the context in which he advances it.

Wright puts forward the following principle P:

> An undecided statement S is decidable in principle just in case an appropriately large but finite extension of our capacities would confer on us the ability to verify it or falsify it in practice.[23]

One critical observation made by Wright about this suggestion amounts to pointing out that it has the questionable features of an appeal to metaphysical possibility.[24] There is no reason why the conception of P should be limited to verification, and to illustrate the style of argument it licenses, I will choose an example of a different kind. I can ("in practice") jump over a fence or similar obstacle up to two feet in height; that is after all not much of a feat. Here we have a reference to a capacity that involves a particular number, namely 2. It would, surely, be a finite extension of my capacity to allow replacement of this number by another, larger number. It follows by (the generalized) P that I can in principle jump over an obstacle up to 2000 feet in height. It would

[22]Crispin Wright, "Strict finitism," *Synthese* 51 (1982), 203–282, cited according to the reprint in *Realism, Meaning, and Truth*, 2d ed. (Oxford: Blackwell, 1993); Alexander George, "The conveyability of intuitionism: An essay in mathematical cognition," *Journal of Philosophical Logic* 17 (1988), 133–156.

[23]"Strict finitism," p.113. P should generalize readily to other activities than verifying or falsifying statements; in the text I understand it in this generalized way.

[24]In spite of these reservations, Wright curiously appeals to P in another essay in the same volume where in principle capabilities are at issue; see *Realism, Meaning, and Truth*, 2d ed., p.326.

be reasonable to conclude that I can in principle jump over the Eiffel Tower. Already with what from a mathematical point of view is a rather modest use of the "finite extension" condition, we let in something as an "in principle" possibility whose compatibility with the fundamentals of biology is not easy to see: what changes in my body would give rise to this 1000-fold increase in my capacity to jump high and land safely?[25] Even if an animal that can do this is a biological possibility, is it possible that such an animal could be me, or at least could have been me? Moreover, if we replace 2000 by a sufficiently large number, then fundamental physical facts would rule out my jumping to that height, so that again we have the task of making sense of a kind of possibility that has no anchor in the laws that actually govern the world, so far as we know them, and yet is still supposed to concern an activity as mundane as jumping. Deciding statements might be thought better off, but this is only because the objectivity of truth gives some handle on the idea that such deciding need not be an activity of human beings, so that once again we have the transcendental view of mental capacity.[26]

Wright pursues at some length problems that are raised by the use in P of the notion of a *finite* extension of our capacities, which he rightly says will not be common ground between the strict finitist and the classical or intuitionistic mathematician. One can see this problem from a usefully different angle by noting that the idea of a "finite extension of our capacities" is naturally interpreted by taking, as we did, a description of something we can do that contains a reference to a number and replacing it by reference to a larger number or perhaps by a variable. Such a procedure opens the way to mathematical argument with these notions, as indeed is already used when one uses induction to generate sorites paradoxes.

This observation gives rise to another suggestion, which in some of my earlier writing on mathematical intuition I implicitly embraced for questions of what can be intuited.[27] This is that the possibility in question is mathematical possibility. In practice, talk of mathematical possibility in this context amounts to allowing extensions of possibilities arrived at on other grounds by generalizing numerical references and by using mathematical reasonings. The reasoning used in the sorites paradox would be an example, if we generate it for a notion like "accessible", where we say that a number is accessible if it is possible to generate a canonical numeral for it. Evidently 0 is accessible; some will find it evident that if n is accessible, so is Sn, even if "accessible" is understood in terms of practical possibility. Then the problem arises because we can describe numbers that are not accessible on that interpretation.

---

[25] We are talking about *jumping* over the Eiffel Tower; the fact that a bird might be able to *fly* over it is presumably irrelevant. Cf. note 2.

[26] I should make clear that I do not claim to be refuting this view. For present purposes it should suffice to observe that it would not be acceptable to either Dummett or Wright.

[27] For example "Mathematical intuition," *Proceedings of the Aristotelian Society* N.S. 80 (1979–80), 145–168, pp.158–159.

The interpretation in terms of mathematical possibility simply embraces the conclusion that every number is accessible.

In what sense is this mathematical possibility of generating a canonical numeral for any number a genuine possibility? A first answer might be that as a mathematical object, the numeral exists. That is insufficient because it does not give any sense to the claim that it is possible to *generate* the numeral. In this case we have a proposal for remedying the insufficiency, because we have a simple mathematical model for the generation of numerals in time. In the case of my earlier concerns with intuition, this way of blunting the force of our question does not seem to be available. Do we really understand what it means to say that intuition of a canconical numeral for $10^{10^{10}}$ is mathematically possible? I will not directly ask the question where the matter stands for the case of computability and decidability. It is more helpful to look at analyses of computability, in particular that of Turing.

VI

In recent years some quite interesting work has been done on the history of the theory of computability, particularly the developments leading to the formulation of the notions of general recursiveness, λ-definability, and Turing computability and their acceptance as mathematically precise notions of effective computability. Something that this work has not emphasized, probably because it was taken for granted, is that the notions of algorithm and decision procedure that the actors in this history (say Hilbert, Bernays, Herbrand, Gödel, Church, Kleene, Post, and Turing) were working with were intended to be mathematical ones. The question whether there is a decision procedure for problems of a certain class was understood to be a mathematical problem. Existing results such as the early results on decidability of fragments of first-order logic were mathematical theorems. Even then, when the analysis of computability was not yet available, the "procedures" had to be described in mathematical terms, and it would violate the constraint if any substantive assumptions about, say, human capacities, had to be used to establish the results. The principle that finiteness is sufficient could be motivated by that consideration, since any further constraint would be arbitrary unless some appeal is made to physical or psychological limitations. It is noteworthy that the matter looks somewhat different today, after the development of the theory of computational complexity. Any positive solution to a decision problem still leaves a number of further problems open, whether the decision problem is in P, NP, or any one of a number of complexity classes. These concepts may be motivated by questions of physical possibility or some other consideration of feasibility, but in themselves they are perfectly mathematical notions.

Let us now turn to Turing. Turing gives, in §9 of his classic paper,[28] a very interesting argument for the claim that all (real) numbers "which could naturally be regarded as computable" are computable according to his definition, in other words computable by a Turing machine. We would put the matter in terms of number-theoretic functions, where the considerations are equivalent. For the problem of showing that certain functions are *not* computable, it will not matter if the definition is too generous provided that it includes everything that could legitimately be regarded as computable. That was the condition, in particular, for using the analysis to give negative solutions to decision problems, as Church and Turing were the first to do. It means that we won't find as explicit an argument in Turing for the other claim in what I called Church's thesis, that all functions computable by his definition are "naturally regarded as computable".

Turing asks, "What are the possible processes which can be carried out in computing a [real] number?"[29] He then considers the situation of a human carrying out a computation, by a procedure that is supposed to be mechanical. Following Robin Gandy and Wilfried Sieg, we may call such a person a computor. Computation works by writing symbols on paper, which Turing says "we may suppose ... divided into squares like a child's arithmetic book." The two-dimensional character is according to Turing not essential, so that he assumes the computation to be carried out on an effectively one-dimensional paper, a tape divided into squares.

Turing derives from certain finiteness assumptions the conclusion that the activity of the computor can be simulated by a Turing machine. The computor is supposed to be deterministic; what he does at any point is determined by his "state of mind" and the symbols that he is observing. It is assumed that

(1.1) There is a fixed finite number of symbolic configurations a computor can immediately recognize.
(1.2) There is a fixed finite number of states of mind that need to be taken into account.[30]

Concerning the operations of the computor, the assumptions are

(2.1) Only elements of observed symbolic configurations can be changed.
(2.2) The distribution of observed squares can be changed, but each of the new observed squares must be within a bounded distance L of an immediately previously observed square.

---

[28] "On computable numbers, with an application to the Entscheidungsproblem," *Proceedings of the London Mathematical Society* (2) 42 (1936–7), 230–265, cited according to the reprint in *The Undecidable*.

[29] Ibid., p.135.

[30] This formulation, as well as (2.1) and (2.2), are from Wilfried Sieg, "Mechanical procedures and mathematical experience," in Alexander George (ed.), *Mathematics and Mind* (Oxford University Pres, 1994), p.93. I am much indebted to this article as well as the one by Gandy cited in note 15.

Turing is credited by Gandy and Sieg with proving a *theorem*, which Sieg states as that a number-theoretic function that can be computed by a deterministic computor satisfying these conditions can be computed by a Turing machine.[31] In Turing's discussion, some sort of transition is made from remarks and asumptions about a certain human capability to a mathematical conclusion.[32] What is "the decisive move in the conjuring trick"? The question for us is whether there *is* a conjuring trick of the kind that has concerned us, in other words whether Turing has slipped in extravagant assumptions about what human beings might do.

The first point to observe is that Turing makes (and argues for) assumptions of *finiteness*; he does not attempt to go further and, perhaps on the basis of an empirical theory in cognitive psychology, to say what the actual bounds might be. A second point is that once we have the assumptions, the role of the more pictorial aspects of the analysis, and even of concepts that involve the mind like "observe" and "immediately recognize", effectively disappears. The conception is of a computor whose action is determined by what he observes and his state of mind, but since that is what is used the epistemic force of the word "observe" reduces, in effect, to a certain functional relation. Furthermore, since there are only finitely many states of mind, and their role is again exhausted by this relation, there is no need to inquire further into the question what these "states of mind" are.[33]

What these remarks amount to is the rather banal observation that Turing offers a *mathematical model* of the computor and argues that what would legitimately be called "mechanical" computation would be subsumed under this model. The process of computation is then treated in terms of the model. One feature of the latter is, of course, that the "tape" is unbounded: the computor can begin the computation with as long an expression on the tape as one pleases, and there is no restriction (other than finiteness) on the amount of tape that can be used in the course of a computation. This is an even more essential respect in which the model is "idealized" than the fact that the numbers instantiated by the finiteness assumptions can be arbitrary. Without the unboundedness of the initial expression on the tape, of course, we could not make non-trivial sense of the question what number-theoretic functions the computor can compute, since a bound on the length of the tape, in the presence of the bound on the number of symbols, would imply a finite bound on the number of distinct inputs the computor could operate on.

[31]In crediting Turing with a theorem at this point, Gandy anticipated Sieg, but Sieg's statement, given here, is more precise.

[32]Gandy and Sieg rightly emphasize that Turing in §9 was not analyzing computation by a *machine*, but "mechanical" computation by a human. It is another task, undertaken by Gandy in another paper, to analyze the notion of computation by a mechanical device.

[33]Such an inquiry might be relevant to the *justification* of the assumptions; cf. Gödel's questioning of (1.2) (*Collected Works*, volume II, *Publications 1938–1974*, Solomon Feferman et al, eds. (Oxford University Press, 1990), p.306), and Sieg's (in my view correct) defense of Turing, op. cit., p.98.

It is, I think, in terms of this model that Turing would consider the question whether all Turing computable functions are "really" computable. Turing machines are actually more constrained than the assumptions call for; for example the machine "observes" only a single square and can move only one step to the left or right at a time. The principle "finiteness is sufficient" is built into the model, since what it means to compute a value of a function is to go through a *finite sequence* of operations beginning with the argument and yielding the value. We can say, if we wish, that for the computor to compute the value is "mathematically possible", but what this really means is that in our mathematical model of the computor, he "computes the value", that is, there is a sequence of steps that, according to the terms of the model, is a computation of the value.

It is important to note how different Turing's approach to the concept of computability is from a typical philosopher's approach to the question of in principle capability, such as the one explored by Wright and discussed above. Turing makes highly idealizing assumptions, of which one might well ask whether they represent genuine possibilities. Such a question really misses the point of the assumptions, which is to get a mathematical structure that subsumes the realistic possibilities and that one can go on to work with, in his case show that the activity of his computor can be simulated by a Turing machine. I think we can acquit the theory of computability of the sort of dubious metaphysical commitment about which we were worried earlier on. But it does mean that we should not take mathematician's statements about decision procedures enabling us in principle to find out answers too literally. They are made against a background where (already long before Turing) algorithms were mathematical objects, and the question of the existence of an algorithm was a mathematical question about the existence of a certain kind of general solution to a problem. This does not mean that Turing's model has no application to the real world, any more than it is the case that the ideal models of classical mechanics or economic theory have no application to the real world. Nor does it mean that we cannot seek more constrained models of computation, in which conditions are laid down that yield more restricted notions of computability that not all recursive functions, and not even all primitive recursive functions, satisfy. However, these notions still have elements of mathematical idealization, such as the lack of a bound on the length of the input and parameters that can take arbitrary values, such as the degree of the polynomial in the concept of polynomial-time computability.[34]

[34]Cf. Christos H. Papadimitriou, *Computational Complexity* (Reading, Mass.: Addison-Wesley, 1993), pp.6–7.

The remarks in the text do not, I think, contradict Gödel's claim that the notion of computability has an absolute character, in the context of classical mathematics. (See in particular "Remarks before the Princeton Bicentennial Conference on Problems in Mathematics," in *Collected Works*, II, 150–153.) Turing computability was meant to answer to certain mathematical demands. Once one admits primitive recursive functions as com-

VII

The conclusion that the notion of computability is a sensible notion both for pure and for applied mathematics does not yet tell us anything about how to understand appeals to possibility in principle in *philosophy*. I will not be able to say very much about this question. A conclusion that is strongly suggested by our investigation is that it is far from being a unitary notion. We cannot expect a single answer to the question whether it is in principle possible for *me* to decide whether the polynomial $g(x)$ is a factor of $f(x)$. The affirmative answer given by the existence of a decision procedure appeals to a particular mathematical model, which captures some fundamental features of the process of computation but idealizes them in such a way that one abstracts from even very fundamental limitations. On the other hand other features of the model are motivated by the fact that it is to model mechanical computation and not, in particular, determining the answer by more "creative" mathematical thought.[35] If one builds in some of the limitations, as in complexity theory or in attempts to consider the fundamental physical limitations that a human computor or a computing machine would be faced with, one will obtain a different answer. Although these considerations are not a direct argument that there isn't a well-defined notion of metaphysical or broadly logical possibility that answers to the use made of it in modal realist literature, it is at least a reason to be skeptical.

Is our discussion of any help to Dummett, for example in making the meaning-theoretic argument for intuitionism persuasive? Dummett seems to me to face a problem that many years ago I claimed was a problem for Kant's conception of the possibility of experience.[36] Which comes first, the question what is possible for us or the question what mathematics we are to accept? If, in order to accept exponentiation as well-defined, we need to convince ourselves that we always *can* compute the value for any argument, then how does one answer the strict finitist objection that in fact we can't? I shall not argue here that Dummett is unable to answer this objection, but only that the sug-

putable (or, indeed, much less) and applies the principle "Finiteness is sufficient", then one is forced to the conclusion that all general recursive functions are computable. And the latter principle was certainly very well motivated from a mathematical point of view. More constrained notions question either "Finiteness is sufficient" or some accepted classical assumption (such as that exponentiation is well-defined). The motivation is either philosophical or for some application, such as to implementation of algorithms.

Mints, to whom I am indebted for questioning me on this point, remarks that although there is a unique notion of computable function, there is not a unique notion of *efficient* computation or computability.

[35] Cf. Gödel's remark that the undecidability results "do not establish any bounds for the powers of human reason, but rather for the potentialities of pure formalism in mathematics." (Postscriptum (1964) to "On undecidable propositions of formal mathematical systems," *Collected Works*, volume I, *Publications 1929–1936*, Solomon Feferman et al, eds., p.370.)

[36] "Infinity and Kant's conception of the 'possibility of experience'," first published 1964, in *Mathematics in Philosophy: Selected Essays* (Ithaca: Cornell University Press, 1983).

gestion I have made as to how to understand computability is principle does not provide him with an answer.[37]

Dummett is committed to the view that if we accept exponentiation as a well-defined function, then for any argument it is possible in principle to compute the value. The proposal is to understand this as meaning that it is computable in the sense of Turing's model, that is that it is computable by a human computor satsifying Turing's finiteness conditions. But now this is a mathematical statement, roughly to the effect that for a suitable program and input there is a sequence of steps leading to an output, such that if the input represents the pair m, n, the output represents m^n. Whether we can get out this result, however, depends on what we put in. The sequence of steps will be of an order of magnitude m^n; if the theory in which we undertake to prove that the computation always terminates does not already have exponentiation or something that would enable one to prove that exponentiation is well-defined, it will not be able to prove that a computation of the exponential function will always terminate. In the usual first-order arithmetic, in which the only given functions are successor, addition, and multiplication, we are of course able to define exponentiation as a relation and prove it is a total function. This is done by means of induction on a Σ_1^0 predicate. Proof-theoretically, admitting such induction is just equivalent to admitting primitive recursion, and in the BHK interpretation of quantification over numbers they come to the same thing. But it is possible to describe computations in weaker theories in which exponentiation cannot be proved to be total, such as certain forms of bounded arithmetic.[38] Turing's model of computability does not tell us whether it is to be applied in the context of a mathematics admitting primitive recursion or an equivalent resource or in the context of a weaker theory. Thus it does not give a prior insight into what is computable that will enable us to accept exponentiation on the ground that it is computable.

It should be admitted that this point concerns only Dummett's difference with strict finitism. A classical opponent will accept the mathematics needed to show it possible in principle according to the present proposal to compute the values of the functions the intuitionist admits to be total, at least for the standard intuitionistic theories. However, the strict finitist objection can also be put in the form of the question why the possibility of finding the value should not be assessed in terms of a more constrained model, since after all its relevance comes from the demand that the meaning a speaker gives to a

[37] I do not deal at all with one possible objection to this proposed use of Turing: it is a matter of doubt and controversy whether the effective procedures called for by the intuitionistic understanding of the mathematical notion of function must be mechanical and thus whether Church's thesis is an acceptable principle for intuitionistic mathematics. This question is clearly not relevant to the issues surrounding strict finitism.

[38] See Petr Hájek and Pavel Pudlák, *Metamathematics of First-Order Arithmetic* (Springer-Verlag, 1993), ch. V, especially theorem 4.16, p.332.

mathematical statement should be manifested in use. I need not pursue this point, since it has already been pressed by Wright and George.

The present paper has argued that reflection on Turing's analysis of computability should defuse the concern that the theory of computability might be committed to metaphysical conceptions of our capacities. It seems to me not to encourage the attempt to make computability "in principle" the measure of what is true or of what makes sense in mathematics.

Harvard University
Cambridge, MA

JUDEA PEARL

CAUSATION, ACTION, AND COUNTERFACTUALS

1. INTRODUCTION

The central aim of many empirical studies in the physical, behavioral, social, and biological sciences is the elucidation of cause-effect relationships among variables. It is through cause-effect relationships that we obtain a sense of a "deep understanding" of a given phenomenon, and it is through such relationships that we obtain a sense of being "in control," namely, that we are able to shape the course of events by deliberate actions or policies. It is for these two reasons, understanding and control, that causal thinking is so pervasive, popping up in everything from everyday activities to high-level decision-making: For example, every car owner wonders why an engine won't start; a cigarette smoker would like to know, given his/her specific characteristics, to what degree his/her health would be affected by refraining from further smoking; a policy maker would like to know to what degree anti-smoking advertising would reduce costs of health care; and so on. Although a plethora of data has been collected on cars and on smoking and health, the appropriate methodology for extracting answers to such questions from the data has been fiercely debated, partly because some fundamental questions of causality have not been given fully satisfactory answers.

The two fundamental questions of causality are:

1. What empirical evidence is required for legitimate inference of cause-effect relationships?
2. Given that we are willing to accept causal information about a certain phenomenon, what inferences can we draw from such information, and how?

The primary difficulty is that we do not have a clear empirical semantics for causality; statistics teaches us that causation cannot be defined in terms of statistical associations, while any philosophical analysis of causation in terms of deliberate control quickly reaches metaphysical deadends over the meaning of free will. Indeed, Bertrand Russell noted that causation plays no role in physics proper and offered to purge the word from the language of science.

Philosophical difficulties notwithstanding, scientific disciplines that must depend on causal thinking have developed paradigms and methodologies that successfully bypass the unsettled questions of causation and that provide acceptable answers to pressing problems of experimentation and inference. Social scientists, for example, have adopted path analysis and structural equations, and programs such as LISREL have become common tools in social science

M.L. Dalla Chiara et al. (eds.), Logic and Scientific Methods, 355–375.

research. Econometricians, likewise, have settled for stochastic simultaneous equations models as carriers of causal information and have focused most of their efforts on developing statistical techniques for estimating the parameters of these models. Statisticians, in contrast, have adopted Fisher's randomized experiment as the ruling paradigm for causal inference, with occasional excursions into its precursor – the Neyman-Rubin model of potential response.

None of these paradigms and methodologies can serve as an adequate substitute for a comprehensive theory of causation, suitable for AI purposes. The structural equations model is based largely on informal modeling assumptions and has hardly been applied beyond the boundaries of linear equations with Gaussian noise. The statisticians' paradigm, on the other hand, is too restrictive, as it does not allow for the integration of large body of substantive knowledge with statistical data. And philosophers have essentially abandoned the quest for the empirical basis of causation, focusing instead on semantical analysis of counterfactual and subjunctive conditionals, with little attention to issues of representation.

A new perspective on the problem of causation has recently emerged from a rather unexpected direction – computer science. When encoding and processing causal relationships on digital machines became necessary, the problems and assumptions that other disciplines could keep dormant and implicit had to be explicated in great detail, so as to meet the levels of precision necessary in programming.

The need to explicate cause-effect relationships arose in several areas of AI: automated diagnosis, robot planning, qualitative physics, and database updates. In the area of robotics, for example, the two fundamental problems of causation got translated into concrete and practical questions:

1. How should a robot acquire causal information through interaction with its environment?
2. How should a robot process the causal information it receives from its creator-programmer?

Attempts to gloss over previous difficulties with causation quickly resulted in a programmer's nightmare. For example, when given the information: "If the grass is wet, then the sprinkler must have been on" and "If I break this bottle, the grass will get wet", the computer will conclude: "If I break this bottle, the sprinkler must have been on". The swiftness and concreteness with which such bugs surface has enabled computer scientists to pinpoint loosely stated assumptions and then to assemble new and more coherent theories of actions, causation, and change.

The purpose of this paper is to summarize recent advances in causal reasoning, especially those that use causal graphs. Section 2 will survey some of the difficulties connected with the formalization of actions in AI. Building on the paradigm that actions are a form of surgeries performed on theories of mechanisms, we will describe the emergence of the causal relation as an abbreviation for the surgery process. Finally, we connect actions to Lewis' theory of counterfactuals using imaging to replace conditioning.

Section 3 will provide the formal underpinning for the discussion of Section 2 and will demonstrate how causal graphs can resolve many of the confusions and contradictions that have prevented a workable theory of actions from evolving. Specifically, in Section 3.1 it will be shown that if qualitative causal information is encoded in the form of a graph, then it is very simple to assess, from non-experimental data, both the strength with which causal influences operate among variables and how probabilities will change as a result of external interventions. Using graph encoding, it is also possible to specify conditions under which manipulative experiments are not necessary and where passive observations suffice. The graphs can also be queried to produce mathematical expressions for causal effects, or to suggest additional observations or auxiliary experiments from which the desired inferences can be obtained.

Finally, Section 3.2 will present a symbolic machinery that admits both probabilistic and causal information about a given domain, and produces probabilistic statements about the effect of actions and the impact of observations. The calculus admits two types of conditioning operators: ordinary Bayes conditioning, $P(y|X = x)$, which represents the observation $X = x$, and causal conditioning, $P(y|do(X = x))$, read: the probability of $Y = y$ conditioned on holding X constant (at x) by deliberate action. Given a mixture of such observational and causal sentences, together with the topology of the causal graph, the calculus derives new conditional probabilities of both types, thus enabling one to quantify the effects of actions and observations.

2. ACTION AS A LOCAL SURGERY

What gives us the audacity to expect that actions should have neat and compact representations? Why did the authors of STRIPS (Fikes & Nilsson, 1971) and BURIDAN (Kushmerick et al., 1993) believe they could get away with such short specification for actions?

Whether we take the probabilistic paradigm that actions are transformations from probability distributions to probability distributions, or the deterministic paradigm that actions are transformations from states to states, such transformations could in principle be infinitely complex. Yet, in practice, people teach each other rather quickly what actions normally do to the world, people predict the consequences of any given action without much hustle, and AI researchers are writing languages for actions as if it is a God given truth that action representation should be compact, elegant and meaningful. Why?

The paradigm I wish to explore in this paper is that these expectations are not only justified but, mainly, that once we understand the justification, we will be in better shape to craft effective representations for actions.

2.1. *Mechanisms and surgeries*

Why are the expectations justified? Because the actions we normally invoke in common reasoning tasks are *local surgeries*. The world consists of a huge number of autonomous and invariant linkages or mechanisms (to use Simon's word), each corresponding to a physical process that constrains the behavior of a relatively small groups of variables. In principle, then, the formalization of actions should not be difficult. If we understand how the linkages interact with each other, usually they simply share variables, we should also be able to understand what the effect of an action would be: Simply re-specify those few mechanisms that are perturbed by the action, then let the modified assembly of mechanisms interact with one another, and see what state will evolve at equilibrium. If the new specification is complete, a single state will evolve. If the specification is probabilistic, a new probability distribution will emerge and, if the specification is logical (possibly incomplete) a new, mutilated logical theory will then be created, capable of answering queries about post-action states of affair.

If this sounds so easy, why did AI ever get into trouble in the arena of action representation? Part of the answer is that what is local in one space may not be local in another. A speck of dust, for example, appears extremely diffused in the frequency (or Fourier) representation and, vice versa, a pure musical tone requires a long stretch of time to be appreciated. It is important therefore to emphasize that actions are local in the space of mechanisms and not in the space of variables or sentences or time slots. For example, tipping the left-most object in an array of domino tiles does not appear "local" in the spatial representation, because, in the tradition of domino theories, every tile might be affected by such action. Yet the action is quite local in the mechanism domain: Only one mechanism gets perturbed, the gravitational restoring force which normally keeps the left-most tile in a stable erect position. It takes no more than a second to describe this action on the phone, without enumerating all its ramifications. The listener, assuming she shares our understanding of domino physics, can figure out for herself the ramifications of this action, or any action of the type: "tip the *ith* domino tile to the right". By representing the domain in the form of an assembly of stable mechanisms, we have in fact created an oracle capable of answering queries about the effects of a huge set of actions and action combinations, without us having to explicate those effects.

2.2. *Laws vs. facts*

This surgical procedure still sounds easy and does not explain why AI got into trouble with action representation. The trouble begins with the realization that in order to implement surgical procedures in mechanism space, we need a language in which some sentences are given different status than others; sentences describing mechanisms should be treated differently than those describing other facts of life, such as observations, assumption and conclusions, because the former are presumed stable, while the latter are transitory. Indeed the mechanism which couples the state of the $(i+1)th$ domino tile to that of the *ith* domino tile remains unaltered (unless we set them apart by some action) whereas the states of the tiles themselves are free to vary with circumstances.

Admitting the need for this distinction has been a difficult cultural transition in the logical approach to actions, perhaps because much of the power of classical logic stems from its representational uniformity and syntactic invariance, where no sentence commands special status. Probabilists were much less reluctant to embrace the distinction between laws and facts, because this distinction has already been programmed into probability language by Reverend Bayes in 1763: Facts are expressed as ordinary propositions, hence they can obtain probability values and they can be conditioned on; laws, on the other hand, are expressed as conditional-probability sentences (e.g., $P(accident|careless\text{-}driving) = high$), hence they should not be assigned probabilities and cannot be conditioned on. It is due to this tradition that probabilists have always attributed nonpropositional character to conditional sentences (e.g., birds fly); refusing to allow nested conditionals (Levi, 1988), and insisting on interpreting one's confidence in a conditional sentence as a conditional probability judgment (Adams, 1975) (see also (Lewis, 1976). Remarkably, these constraints, which some philosophers view as limitations, are precisely the safeguards that have kept probabilists from confusing laws and facts, and have protected them from some of the traps that have lured logical approaches.[1]

2.3. *Mechanisms and causal relationships*

The next issue worth discussing is how causality enters into this surgical representation of actions. To understand the role of causality, we should note that most mechanisms do not have names in common everyday language. In the domino example above I had to struggle hard to name the mechanism which would be perturbed by the action "tip the left-most tile to the right". And there is really no need for the struggle; instead of telling you the name of the mechanism to be perturbed by the action, I might as well gloss over the details of the perturbation process and summarize its net result in the form of an *event*, e.g., "the left-most tile is tipped to right", which yields equivalent consequences as the perturbation summarized. After all, if you and I share the same understanding of physics, you should be able to figure out for yourself

[1] The distinction between laws and facts has been proposed by Poole (1985) and Geffner (1992) as a fundamental principle for nonmonotonic reasoning. It seems to be gaining broader support recently as a necessary requirement for formulating actions.

which mechanism it is that must be perturbed in order to realize the specified new event, and this should enable you to predict the rest of the scenario.

This linguistic abbreviation defines a new relation among events, a relation we normally call "causation": Event A causes B, if the perturbation needed for realizing A entails the realization of B.[2] Causal abbreviations of this sort are used very effectively for specifying domain knowledge. Complex descriptions of what relationships are stable and how mechanisms interact with one another are rarely communicated explicitly in terms of mechanisms. Rather, they are communicated in terms of cause-effect relationships between events or variables. We say, for example: "If tile i is tipped to the right, it causes tile $i+1$ to tip to the right as well"; we do not communicate such knowledge in terms of the tendencies of each domino tile to maintain its physical shape, to respond to gravitational pull and to obey Newtonian mechanics.

A formulation of action as a local surgery on causal theories has been developed in a number of recent papers [Goldszmidt & Pearl, 1992; Pearl, 1993a; Pearl, 1993b; Darwiche & Pearl, 1994; Pearl, 1994a; Goldszmidt & Darwiche, 1994]. Section 3.1 provides a brief summary of this formulation, together with a simple example that illustrates how the surgery semantics generalizes to nonprobabilistic formalisms.

2.4. *Causal ordering*

Our ability to talk directly in terms of one event causing another, (rather than an action altering a mechanism and the alteration, in turn, having an effect) is computationally very useful, but, at the same time it requires that the assembly of mechanisms in our domain satisfy certain conditions. Some of these conditions are structural, nicely formulated in Simon's "causal ordering" (Simon, 1953), and others are substantive – – invoking relative magnitudes of forces and powers.

The structural requirement is that there be a one-to-one correspondence between mechanisms and variables – a unique variable in each mechanism is designated as the output (or effect), and the other variables, as inputs (or causes). Indeed, the formal definition of causal theories given in Section 3.1 assumes that each equation is associated with a unique variable, situated on its left hand side. In general, a mechanism may be specified as a function

$$G_i(X_1, ..., X_n; U_1, \ldots, U_m) = 0$$

without identifying any so called "dependent" variable X_i. Simon's causal ordering provides a procedure for deciding whether a collection of such G functions has a unique preferred way of associating variables with mechanisms, based on the requirement that we should be able to solve for the *ith* variable without solving for its successors in the ordering.

[2]The word "needed" connotes minimality and can be translated to: "...if every minimal perturbation realizing A, entails B".

In certain structures, called *webs* (Dalkey, 1994; Dechter & Pearl, 1991), Simon's causal ordering determines a unique one-to-one correspondence, but in others, such as those involving feedback, the correspondence is not unique. Yet in examining feedback circuits, for example, people can assert categorically that the flow of causation goes clockwise, rather than counterclockwise. They make such assertions on the basis of relative magnitudes of forces; for example, it takes very little energy to make an input of a gate change its output, but no force applied to the output can influence the input. When such considerations are available, causal directionality can be determined by appealing again to the notion of hypothetical intervention and asking whether an external control over one variable in the mechanism necessarily affects the others. The variable which does not affect any of the others is the dependent variable. This then constitutes the operational semantics for identifying the dependent variables X_i in nonrecursive causal theories (see Section 3.1).

2.5. *Imaging vs. conditioning*

If action is a transformation from one probability function to another, one may ask whether every transformation corresponds to an action, or are there some constraints that are peculiar to exactly those transformations that originate from actions. Lewis (1976) formulation of counterfactuals indeed identifies such constraints: the transformation must be an *imaging* operator (Imaging is the probabilistic version of Winslett-Katsuno-Mendelzon possible worlds representation of "update").

Whereas Bayes conditioning $P(s|e)$ transfers the entire probability mass from states excluded by e to the remaining states (in proportion to their current $P(s)$), imaging works differently; each excluded state s transfers its mass individually to a select set of states $S^*(s)$, which are considered "closest" to s. The reason why imaging is a more adequate representation of transformations associated with actions can be seen more clearly through a representation theorem due to Gärdenfors [1988, Theorem 5.2 pp.113] (strangely, the connection to actions never appears in Gärdenfors' analysis). Gärdenfors' theorem states that a probability update operator $P(s) \rightarrow P_A(s)$ is an imaging operator iff it preserves mixtures, i.e.,

$$[\alpha P(s) + (1-\alpha)P'(s)]_A = \alpha P_A(s) + (1-\alpha)P'_A(s) \tag{1.1}$$

for all constants $1 > \alpha > 0$, all propositions A , and all probability functions P and P'. In other words, the update of any mixture is the mixture of the updates[3].

This property, called homomorphism, is what permits us to specify actions in terms of *transition probabilities*, as it is usually done in stochastic control

[3] Assumption (1) is reflected in the (U8) postulate of (Katsuno & Mendelzon, 1991): $(K_1 \vee K_2)o\mu = (K_1 o\mu) \vee (K_2 o\mu)$, where o is an update operator.

and Markov decision process. Denoting by $P_A(s|s')$ the probability resulting from acting A on a known state s', homomorphism (1.1) dictates:

$$P_A(s) = \sum_{s'} P_A(s|s')P(s') \tag{1.2}$$

saying that, whenever s' is not known with certainty, $P_A(s)$ is given by a weighted sum of $P_A(s|s')$ over s', with the weight being the current probability function $P(s')$.

This characterization, however, is too permissive; while it requires any action-based transformation to be describable in terms of transition probabilities, it also accepts any transition probability specification, however whimsical as a descriptor of some action. The valuable information that actions are defined as *local* surgeries, is totally ignored in this characterization. For example, the transition probability associated with the atomic action $A_i = do(X_i = x_i)$ originates from the deletion of just one mechanism in the assembly. Hence, one would expect that the transition probabilities associated with the set of atomic actions would not be totally arbitrary but would constrain one another.

We are currently exploring axiomatic characterizations of such constraints which we hope to use as a logic for sentences of the type: "X affects Y when we hold Z fixed". With the help of such logic we hope to be able to derive, refute or confirm sentences such as "If X has no effect on Y and Z affects Y, then Z will continue to affect Y when we fix X." The reader might find some challenge proving or refuting the sentence above, that is, testing whether it holds in every causal theory, when "affecting" and "fixing" are interpreted by the local-surgery semantics described in this paper.

3. FORMAL UNDERPINNING

3.1. *Causal theories, actions, causal effect, and identifiability*

Definition 1 *A causal theory is a 4-tuple*

$$T =< V, U, P(\mathbf{u}), \{f_i\} >$$

where

(i) $V = \{X_1, \ldots, X_n\}$ *is a set of observed variables*
(ii) $U = \{U_1, \ldots, U_m\}$ *is a set of unobserved variables which represent disturbances, abnormalities or assumptions,*
(iii) $P(\mathbf{u})$ *is a distribution function over* $U_1, \ldots, U_m$, *and*
(iv) $\{f_i\}$ *is a set of* n *deterministic functions, each of the form*

$$X_i = f_i(X_1, \ldots, X_n, U_1, \ldots, U_m) \quad i = 1, \ldots, n \tag{1.3}$$

We will assume that the set of equations in (iv) has a unique solution for $X_i, \ldots, X_n$, *given any value of the disturbances* $U_1, \ldots, U_m$. *Therefore the distribution* $P(\mathbf{u})$ *induces a unique distribution on the observables, which we denote by* $P_T(\mathbf{v})$.

We will consider concurrent actions of the form $do(X = x)$, where $X \subseteq V$ is a set of variables and x is a set of values from the domain of X. In other words, $do(X = x)$ represents a combination of actions that forces the variables in X to attain the values x.

Definition 2 *(Effect of actions) The effect of the action $do(X = x)$ on a causal theory T is given by a subtheory T_x of T, where T_x obtains by deleting from T all equations corresponding to variables in X and substituting the equations $X = x$ instead.*

Definition 3 *(causal effect) Given two disjoint subsets of variables, $X \subseteq V$ and $Y \subseteq V$, the causal effect of X on Y, denoted $P_T(y|\hat{x})$, is a function from the domain of X to the space of probability distributions on Y, such that*

$$P_T(y|\hat{x}) = P_{T_x}(y) \tag{1.4}$$

for each realization x of X. In other words for each $x \in dom(X)$, the causal effect $P_T(y|\hat{x})$ gives the distribution of Y induced by the action $do(X = x)$.

Note that causal effects are defined relative to a given causal theory T, though the subscript T is often suppressed for brevity.

Definition 4 (identifiability) *The causal effect of X on Y is said to be* identifiable *if the quantity $P(y|\hat{x})$ can be computed uniquely from any positive distribution of the observed variables, that is, if for every pair of theories T_1 an T_2 such that $P_{T_1}(\mathbf{v}) = P_{T_2}(\mathbf{v}) > 0$, we have $P_{T_1}(y|\hat{x}) = P_{T_2}(y|\hat{x})$*

Identifiability means that $P(y|\hat{x})$ can be estimated consistently from an arbitrarily large sample randomly drawn from the distribution of the observed variables. The notion of identifiability is central to much works in econometrics, where it has become synnanomous to the identification of the functions $\{f_i\}$ or some of their parameters (Koopman & Reiersol, 1950), mostly under conditions of additive noise. Definition 4, which does not assume any parametric represenatation of the functions $\{f_i\}$, extends the notion of identifiabiity to any quantity that can be computed from a given theory. In particular, it replaces the identification of structurral parameters with the nonparametric identification of causal effects, $P(y|\hat{x})$, which constitute the ultimate goal of computing structural parameters in policy analysis applications.

Figure 1 illustrates a simple causal theory in the form of a diagram. It describes the causal relationships among the season of the year (X_1), whether rain falls (X_2), whether the sprinkler is on (X_3), whether the pavement would get wet (X_4), and whether the pavement would be slippery (X_5). All variables in this figure are binary, taking a value of either true or false, except the root variable X_1 which can take one of four values: Spring, Summer, Fall, or Winter. Here, the absence of a direct link between X_1 and X_5, for example, captures our understanding that the influence of seasonal variations on the slipperiness of the pavement is mediated by other conditions (e.g., the wetness of the pavement).

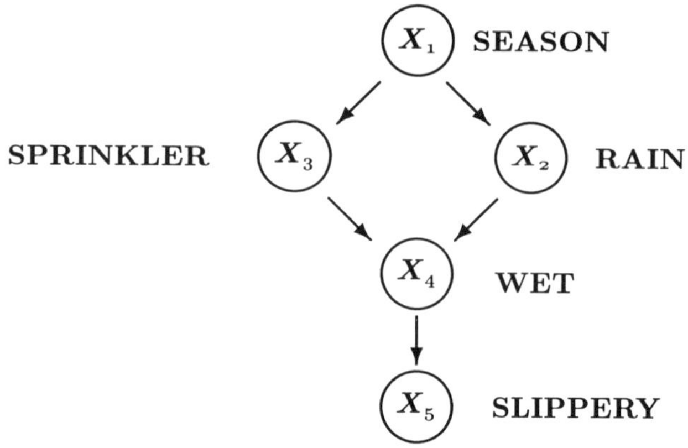

Fig. 1. A diagram representing a causal theory on five variables.

The theory corresponding to Figure 1 consists of five functions, each representing an autonomous mechanism:

$$\begin{aligned} X_1 &= U_1 \\ X_2 &= f_2(X_1, U_2) \\ X_3 &= f_3(X_1, U_3) \\ X_4 &= f_4(X_3, X_2, U_4) \\ X_5 &= f_5(X_4, U_5) \end{aligned} \tag{1.5}$$

To represent the action "turning the sprinkler ON", $do(X_3 = \mathrm{ON})$, we delete the equation $X_3 = f_3(x_1, u_3)$ from the theory of Eq. (1.5), and replace it with $X_3 = \mathrm{ON}$. The resulting subtheory, $T_{X_3=\mathrm{ON}}$, contains all the information needed for computing the effect of the actions on other variables. For example, it is easy to see from this subtheory that the only variables affected by the action are X_4 and X_5, that is, the descendant of the manipulated variable X_3.

The probabilistic analysis of causal theories becomes particularly simple when two conditions are satisfied:

1. The theory is recursive, i.e., there exists an ordering of the variables $V = \{X_1, \ldots, X_n\}$ such that each X_i is a function of a subset $\mathbf{pa}_i$ of its predecessors
$$X_i = f_i(\mathbf{pa}_i, U_i), \qquad \mathbf{pa}_i \subseteq \{X_1, \ldots, X_{i-1}\} \tag{1.6}$$
2. The disturbances $U_1, \ldots, U_n$ are mutually independent, which also implies (from the exogeneity of the U_i's)
$$U_i \perp\!\!\!\perp \{X_1, \ldots, X_{i-1}\} \tag{1.7}$$

These two conditions, also called Markovian, are the basis of Bayesian networks (Pearl, 1988) and they enable us to compute causal effects directly from the conditional probabilities $P(x_i|\mathbf{pa}_i)$, without specifying the functional form of the functions f_i, or the distributions $P(u_i)$ of the disturbances. This is seen immediately from the following observations:

The distribution induced by any Markovian theory T is given by the product

$$P_T(x_1, \ldots, x_n) = \prod_i P(x_i|\mathbf{pa}_i) \tag{1.8}$$

where $\mathbf{pa}_i$ are the direct predecessors (called *parents*) of X_i in the diagram. On the other hand the subtheory $T_{x'_j}$, representing the action $do(X_j = x'_j)$ is also Markovian, hence it also induces a product-decomposable distribution

$$P_{T_{x'_j}}(x_1, \ldots, x_n) = \begin{cases} \prod_{i \neq j} P(x_i|\mathbf{pa}_i) = \frac{P(x_1,\ldots,x_n)}{P(x_j|\mathbf{pa}_j)} & \text{if } x_j = x'_j \\ 0 & \text{if } x_j \neq x'_j \end{cases} \tag{1.9}$$

where the partial product reflects the surgical removal of the equation

$$X_j = f_j(\mathbf{pa}_j, U_j)$$

from the theory of equation (1.6). Given the post-action distribution, the effect of the action $do(X_j = x'_j)$ on any variable (or groups of variables) in the system can readily be computed.

In the example of Figure 1, the pre-action distribution is given by the product

$$P_T(x_1, x_2, x_3, x_4, x_5) = \\ P(x_1)P(x_2|x_1)P(x_3|x_1)P(x_4|x_2, x_3)P(x_5|x_4) \tag{1.10}$$

while the surgery corresponding to the action $do(X_3 = \text{ON})$ amounts to deleting the link $X_1 \rightarrow X_3$ from the graph and fixing the value of X_3 to ON, yielding the post-action distribution:

$$P_T(x_1, x_2, x_4, x_5|do(X_3 = \text{ON})) = \\ P(x_1)\ P(x_2|x_1)\ P(x_4|x_2, X_3 = \text{ON})\ P(x_5|x_4) \tag{1.11}$$

Note the difference between the action $do(X_3 = \text{ON})$ and the observation $X_3 = \text{ON}$. The latter is encoded by ordinary Bayesian conditioning, while the former by conditioning a mutilated graph, with the link $X_1 \rightarrow X_3$ removed. This mirrors indeed the difference between seeing and doing: after observing that the sprinkler is ON, we wish to infer that the season is dry, that it probably did not rain, and so on; no such inferences should be drawn in evaluating the effects of the deliberate action "turning the sprinkler ON". The amputation of $X_3 = f_3(X_1, U_3)$ from (1.5) ensures the suppression of any abductive inferences from any of the action's consequences.

Note also that Equations (1.8) through (1.11) are independent of T, in other words, the pre-actions and post-action distributions depend only on observed conditional probabilities but are independent of the particular functional form of $\{f_i\}$ or the distribution $P(\mathbf{u})$ which generate those probabilities. This is the essence of identifiability as given in Definition 4, which stems from the Markovian assumptions (1.6) and (1.7). Section 3.2 will demonstrate that certain causal effects, though not all, are identifiable even when the Markovian property is destroyed by introducing dependencies among the disturbance terms.

Generalization to multiple actions and conditional actions are straightforward. Multiple actions $do(X = x)$, where X is a compound variable result in a distribution similar to (1.9), except that all factors corresponding to the variables in X are removed from the product in (1.8). Stochastic conditional strategies of the form

$$do(X_j = x_j) \text{ with probability } P^*(x_j|\mathbf{pa}_j^*)$$

where $\mathbf{pa}_j^*$ is the support of the decision strategy, also result in a product decomposition similar to (1.9), except that each factor $P(x_j|\mathbf{pa}_j)$ is replaced with $P^*(x_j|\mathbf{pa}_j^*)$. The identification of conditional and multiple actions is treated in (Pearl, 1994b; Pearl & Robins, 1995).

The surgical procedure described above is not limited to probabilistic analysis. The causal knowledge represented in Figure 1 can be captured by logical theories as well, for example,

$$\begin{aligned}
x_2 &\iff [(X_1 = \text{Winter}) \vee (X_1 = \text{Fall}) \vee ab_2] \wedge \neg ab_2' \\
x_3 &\iff [(X_1 = \text{Summer}) \vee (X_1 = \text{Spring}) \vee ab_3] \wedge \neg ab_3' \\
x_4 &\iff (x_2 \vee x_3 \vee ab_4) \wedge \neg ab_4' \\
x_5 &\iff (x_4 \vee ab_5) \wedge \neg ab_5' \qquad (1.12)
\end{aligned}$$

where x_i stands for $X_i = true$, and ab_i and ab_i' stand, respectively, for triggering and inhibiting abnormalities. The double arrows represent the assumption that the events on the r.h.s. of each equation are the *only* causes for the l.h.s.

It should be emphasized though that the models of a causal theory are not made up merely of truth value assignments which satisfy the equations in the theory. Since each equation represents an autonomous process, the scope of each individual equation must be specified in any model of the theory, and this can be encoded using either the graph (as in Figure 1) or the generic description of the theory, as in (1.5). Alternatively, we can view a model of a causal theory to consist of a mutually consistent set of submodels, with each submodel being a standard model of a single equation in the theory.

3.2. *Action Calculus*

The identifiability of causal effects demonstrated in Section 3.1 relies critically on the Markovian assumptions (1.6) and (1.7). If a variable that has two descendants in the graph is unobserved, the disturbances in the two equations are no longer independent, the Markovian property (1.6) is violated and identifiability may be destroyed. This can be seen easily from Eq. (1.9); if any parent of the manipulated variable X_j is unobserved, one cannot estimate the conditional probability $P(x_j|\mathbf{pa}_j)$, and the effect of the action $do(X_j = x_j)$ may not be predictable from the observed distribution $P(x_1, \ldots, x_n)$. Fortunately, certain causal effects are identifiable even in situations where members of $\mathbf{pa}_j$ are unobservable and, moreover, polynomial tests are now available for deciding when $P(x_i|\hat{x}_j)$ is identifiable, and for deriving closed-form expressions for $P(x_i|\hat{x}_j)$ in terms of observed quantities (Pearl, 1995; Galles & Pearl, 1995).

These tests and derivations are based on a symbolic calculus to be described in the sequel, in which interventions, side by side with observations, are given explicit notation, and are permitted to transform probability expressions. The transformation rules of this calculus reflect the understanding that interventions perform "local surgeries" as described in Definition 2, i.e., they overrule equations that tie the manipulated variables to their pre-intervention causes.

Let X, Y, and Z be arbitrary disjoint sets of nodes in a DAG G. We write $(X \perp\!\!\!\perp Y|Z)_G$, if the set Z d-separates X from Y in G (see (Pearl, 1988) for a definition of d-separation). We denote by $G_{\overline{X}}$ the graph obtained by deleting from G all arrows pointing to nodes in X. Likewise, we denote by $G_{\underline{X}}$ the graph obtained by deleting from G all arrows emerging from nodes in X. To represent the deletion of both incoming and outgoing arrows, we use the notation $G_{\overline{X}\underline{Z}}$. Finally, the expression $P(y|\hat{x}, z) \triangleq P(y, z|\hat{x})/P(z|\hat{x})$ stands for the probability of $Y = y$ given that $Z = z$ is observed and X is held constant at x.

Theorem 1 Let G be the directed acyclic graph associated with a Markovian causal theory, and let $P(\cdot)$ stand for the probability distribution induced by that theory. For any disjoint subsets of variables X, Y, Z, and W we have:

Rule 1 Insertion/deletion of observations

$$P(y|\hat{x}, z, w) = P(y|\hat{x}, w) \text{ if } (Y \perp\!\!\!\perp Z|X, W)_{G_{\overline{X}}} \tag{1.13}$$

Rule 2 Action/observation exchange

$$P(y|\hat{x}, \hat{z}, w) = P(y|\hat{x}, z, w) \text{ if } (Y \perp\!\!\!\perp Z|X, W)_{G_{\overline{X}\underline{Z}}} \tag{1.14}$$

Rule 3 Insertion/deletion of actions

$$P(y|\hat{x}, \hat{z}, w) = P(y|\hat{x}, w) \text{ if } (Y \perp\!\!\!\perp Z|X, W)_{G_{\overline{X},\, \overline{Z(W)}}} \tag{1.15}$$

where $Z(W)$ is the set of Z-nodes that are not ancestors of any W-node in $G_{\overline{X}}$.

Each of the inference rules above follows from the basic interpretation of the "$\hat{x}$" operator as a replacement of the causal mechanism that connects X to its pre-action parents by a new mechanism $X = x$ introduced by the intervening force. The result is a submodel characterized by the subgraph $G_{\overline{X}}$ (named "manipulated graph" in Spirtes et al. (1993)) which supports all three rules.

Rule 1 reaffirms d-separation as a valid test for conditional independence in the distribution resulting from the intervention $do(X = x)$, hence the graph $G_{\overline{X}}$. This rule follows from the fact that deleting equations from the system does not introduce any dependencies among the remaining disturbance terms.

Rule 2 provides a condition for an external intervention $do(Z = z)$ to have the same effect on Y as the passive observation $Z = z$. The condition amounts to $\{X \cup W\}$ blocking all back-door paths from Z to Y (in $G_{\overline{X}}$), since $G_{\overline{X}\underline{Z}}$ retains all (and only) such paths.[4]

Rule 3 provides conditions for introducing (or deleting) an external intervention $do(Z = z)$ without affecting the probability of $Y = y$. The validity of this rule stems, again, from simulating the intervention $do(Z = z)$ by the deletion of all equations corresponding to the variables in Z (hence the graph $G_{\overline{XZ}}$).

Corollary 1 A causal effect q: $P(y_1, ..., y_k|\hat{x}_1, ..., \hat{x}_m)$ is identifiable in a model characterized by a graph G if there exists a finite sequence of transformations, each conforming to one of the inference rules in Theorem 1, which reduces q into a standard (i.e., hat-free) probability expression involving observed quantities. □

Although Theorem 1 and Corollary 1 require the Markovian property, they do not require all variables to be observable and, hence, they can be applied to non Markovian, recursive theories as well. To demonstrate, assume that variable X_1 in Figure 1 is unobserved, rendering the disturbances U_3 and U_2 dependent since these terms now include the common influence of X_1. Theorem 1 tells us that the causal effect $P(x_4|x_{\hat{3}})$ is identifiable, because:

$$P(x_4|\hat{x}_3) = \sum_{x_2} P(x_4|\hat{x}_3, x_2)P(x_2|\hat{x}_3)$$

Rule 3 permits the deletion

$$P(x_2|\hat{x}_3) = P(x_2), \text{ because } (X_2 \perp\!\!\!\perp X_3)_{G_{\overline{X_3}}},$$

while Rule 2 permits the exchange

$$P(x_4|\hat{x}_3, x_2) = P(x_4|x_3, x_2), \text{ because } (X_4 \perp\!\!\!\perp X_3|X_2)_{G_{\underline{X_3}}}.$$

This gives

$$P(x_4|\hat{x}_3) = \sum_{x_2} P(x_4|x_3, x_2)P(x_2)$$

which is a "hat-free" expression, involving only observed quantities.

In general, it can be shown (Pearl, 1995) that:

1. The effect of interventions can often be identified (from nonexperimental data) without resorting to parametric models,
2. The conditions under which such nonparametric identification is possible can be determined by simple graphical criteria, and,

[4]A path from X to Y is said to be a back-door path if it ends with arrows into both X and Y (Pearl, 1993b).

3. When the effect of interventions is not identifiable, the causal graph may suggest non-trivial experiments which, if performed, would render the effect identifiable.

The ability to assess the effect of interventions from nonexperimental data has immediate applications in the medical and social sciences, since subjects who undergo certain treatments often are not representative of the population as a whole. Such assessments are also important in AI applications where an agent needs to predict the effect of the next action on the basis of past performance records, and where that action has never been enacted out of free will, but in response to environmental needs or to other agent's requests.

3.3. *Historical background*

An explicit translation of interventions to "wiping out" equations from linear econometric models was first proposed by Strotz & Wold (1960) and later used in Fisher (1970) and Sobel (1990) . Extensions to action representation in nonmonotonic systems were reported in (Goldszmidt & Pearl, 1992; Pearl, 1993a). Graphical ramifications of this translation were explicated first in Spirtes et al. (1993) and later in Pearl (1993b). A related formulation of causal effects, based on event trees and counterfactual analysis was developed by Robins (1986, pp. 1422-25). Calculi for actions and counterfactuals based on this interpretation are developed in (Pearl, 1994b) and (Balke & Pearl, 1994a), respectively.

4. COUNTERFACTUALS

A counterfactual sentence has the form

If A were true, then C would have been true?

where A, the counterfactual antecedent, specifies an event that is contrary to one's real-world observations, and C, the counterfactual consequent, specifies a result that is expected to hold in the alternative world where the antecedent is true. A typical example is "If Oswald were not to have shot Kennedy, then Kennedy would still be alive" which presumes the factual knowledge of Oswald's shooting Kennedy, contrary to the antecedent of the sentence.

The majority of the philosophers who have examined the semantics of counterfactual sentences have resorted to some version of Lewis' "closest world" approach; "C if it were A" is true, if C is true in worlds that are "closest" to the real world yet consistent with the counterfactuals antecedent A (Lewis, 1973). Ginsberg (1986), followed a similar strategy. While the "closest world" approach leaves the precise specification of the closeness measure almost unconstrained, causal knowledge imposes very specific preferences as to which worlds should be considered closest to any given world. For example, considering an array of domino tiles standing close to each other. The manifestly closest world consistent with the antecedent "tile i is tipped to the right" would be a world in which just tile i is tipped, while all the others remain erect. Yet, we all accept the counterfactual sentence "Had tile i been tipped over to the right, tile $i + 1$ would be tipped as well" as plausible and valid. Thus, distances among worlds are not determined merely by surface similarities but require a distinction between disturbed mechanisms and naturally occurring transitions. The local surgery paradigm expounded in Section 3.1 offers a concrete explication of the closest world approach which respects causal considerations. A world w_1 is "closer" to w than a world w_2 is, if the the set of atomic surgeries needed for transforming w into w_1 is a subset of those needed for transforming w into w_2. In the domino example, finding tile i tipped and $i + 1$ erect requires the breakdown of two mechanism (e.g., by two external actions) compared with one mechanism for the world in which all j-tiles, $j > i$ are tipped. This paradigm conforms to our perception of causal influences and lends itself to economical machine representation.

4.1. *Formal underpinning*

The structural equation framework offers an ideal setting for counterfactual analysis.

Definition 5 *(context-based potential response): Given a causal theory T and two disjoint sets of variables, X and Y, the potential response of Y to X in a context u, denoted $Y(x,u)$ or $Y_x(u)$, is the solution for Y under $U = u$ in the subtheory T_x. $Y(x,u)$ can be taken as the formal definition of the counterfactual English phrase: "the value that Y would take in context u, had X been x,"* [5]

Note that this definition allows for the context $U = u$ and the proposition $X = x$ to be incompatible in T. For example, if T describes a logic circuit with input U it may well be reasonable to assert the counterfactual: "Given $U = u$, Y would be high if X were low", even though the input $U = u$ may preclude X from being low. It is for this reason that one must invoke some notion of intervention (alternatively, a theory change or a "miracle" (Lewis, 1973)) in the definition of counterfactuals.

If U is treated as a random variable, then the value of the counterfactual $Y(x,u)$ becomes a random variable as well, denoted as $Y(x)$ of Y_x. Moreover,

---

[5] The term *unit* instead of *context* is often used in the statistical literature (Rubin, 1974), where it normally stands for the identity of a specific individual in a population, namely, the set of attributes u that characterize that individual. In general, u may include the time of day, the experimental conditions under study, and so on. Practitioners of the counterfactual notation do not explicitly mention the notions of "solution" or "intervention" in the definition of $Y(x,u)$. Instead, the phrase "the value that Y would take in unit u, had X been x," viewed as basic, is posited as the definition of $Y(x,u)$.

the distribution of this random variable is easily seen to coincide with the causal effect $P(y|\hat{x})$, i.e.,

$$P((Y(x) = y) = P(y|\hat{x})$$

The probability of a counterfactual conditional $x \rightarrow y \mid o$ may then be evaluated by the following procedure:

- Use the observations o to update $P(u)$ thus forming a causal theory $T^o =< V, U, \{f_i\}, P(u|o) >$
- Form the mutilated theory T^o_x (by deleting the equations corresponding to variables in X) and compute the probability $P_{T^o}(y|\hat{x})$ which T^o_x induces on Y.

Unlike causal effect queries, counterfactual queries are generally not identifiable even in Markovian theories, but require that the functional-form of $\{f_i\}$ be specified. In (Balke & Pearl, 1994b) a method is devised for computing sharp bounds on counterfactual probabilities which, under certain circumstances may collapse to point estimates. This method has been applied to the evaluation of causal effects in studies involving noncompliance, and to the determination of legal liability.

4.2. *Applications to Policy Analysis*

Counterfactual reasoning is at the heart of every planning activity, especially real-time planning. When a planner discovers that the current state of affairs deviates from the one expected, a "plan repair" activity need be invoked to determine what went wrong and how it could be rectified. This activity amounts to an exercise of counterfactual thinking, as it calls for rolling back the natural course of events and determining, based on the factual observations at hand, whether the culprit lies in previous decisions or in some unexpected, external eventualities. Moreover, in reasoning forward to determine if things would have been different a new model of the world must be consulted, one that embodies hypothetical changes in decisions or eventualities, hence, a breakdown of the old model or theory.

The logic-based planning tools used in AI, such as STRIPS and its variants or those based on the situation calculus, do not readily lend themselves to counterfactual analysis; as they are not geared for coherent integration of abduction with prediction, and they do not readily handle theory changes. Remarkably, the formal system developed in economics and social sciences under the rubric "structural equations models" does offer such capabilities but, as will be discussed below, these capabilities are not well recognized by current practitioners of structural models. The analysis presented in this chapter could serve both to illustrate to philosophers and AI researchers the basic formal features needed for counterfactual and policy analysis, and to call the attention of economists and social scientists to capabilities that are dormant within structural equations models.

Counterfactual thinking dominates reasoning in political science and economics. We say, for example, "If Germany were not punished so severely at the end of World War I, Hitler would not have come to power," or "If Reagan did not lower taxes, our deficit would be lower today." Such thought experiments emphasize an understanding of generic laws in the domain and are aimed toward shaping future policy making, for example, "defeated countries should not be humiliated," or "lowering taxes (contrary to Reaganomics) tends to increase national debt."

Strangely, there is very little formal work on counterfactual reasoning or policy analysis in the behavioral science literature. An examination of a number of econometric journals and textbooks, for example, reveals a glaring imbalance: while an enormous mathematical machinery is brought to bear on problems of estimation and prediction, policy analysis (which is the ultimate goal of economic theories) receives almost no formal treatment. Currently, the most popular methods driving economic policy making are based on so-called *reduced-form* analysis: to find the impact of a policy involving decision variables X on outcome variables Y, one examines past data and estimates the conditional expectation $E(Y|X{=}x)$, where x is the particular instantiation of X under the policy studied.

The assumption underlying this method is that the data were generated under circumstances in which the decision variables X act as exogenous variables, that is, variables whose values are determined outside the system under analysis. However, while new decisions should indeed be considered exogenous for the purpose of evaluation, past decisions are rarely enacted in an exogenous manner. Almost every realistic policy (e.g., taxation) imposes control over some endogenous variables, that is, variables whose values are determined by other variables in the analysis. Let us take taxation policies as an example. Economic data are generated in a world in which the government is reacting to various indicators and various pressures; hence, taxation is endogenous in the data-analysis phase of the study. Taxation becomes exogenous when we wish to predict the impact of a specific decision to raise or lower taxes. The reduced-form method is valid only when past decisions are nonresponsive to other variables in the system, and this, unfortunately, eliminates most of the interesting control variables (e.g., tax rates, interest rates, quotas) from the analysis.

This difficulty is not unique to economic or social policy making; it appears whenever one wishes to evaluate the merit of a plan on the basis of the past performance of other agents. Even when the signals triggering the past actions of those agents are known with certainty, a systematic method must be devised for selectively ignoring the influence of those signals from the evaluation process. In fact, the very essence of *evaluation* is having the freedom to imagine and compare trajectories in various counterfactual worlds, where each world or trajectory is created by a hypothetical implementation of a policy that is free of the very pressures that compelled the implementation of such policies in the past.

Balke and Pearl (1995) demonstrate how linear, nonrecursive structural models with Gaussian noise can be used to compute counterfactual queries of the type: "Given an observation set O, find the probability that Y would have attained a value greater than y, had X been set to x". The task of inferring "causes of effects", that is, of finding the probability that $X = x$ is the cause for effect E, amounts to answering the counterfactual query: "Given effect E and observations O, find the probability that E would not have been realized, had X not been x". The technique developed in Balke and Pearl (1995) is based on probability propagation in dual networks, one representing the actual world, the other the counterfactual world. The method is not limited to linear functions but applies whenever we are willing to assume the functional form of the structural equations. The noisy OR-gate model (Pearl, 1988) is a canonical example where such functional form is normally specified. Likewise, causal theories based on Boolean functions (with exceptions), such as the one described in Eq. (1.12) lend themselves to counterfactual analysis in the framework of Definition 5.

Cognitive Systems Laboratory
Computer Science Department
University of California
Los Angeles

REFERENCES

Adams, E., *The Logic of Conditionals,* Chapter 2, D. Reidel, Dordrecht, Netherlands, 1975.

Balke, A. and Pearl, J., "Probabilistic evaluation of counterfactual queries," in *Proceedings of the Twelfth National Conference on Artificial Intelligence (AAAI-94)*, Seattle, WA, Volume I, 230-237, July 31 - August 4, 1994.

Balke, A. and Pearl, J., "Counterfactual probabilities: Computational methods, bounds, and applications," in R. Lopez de Mantaras and D. Poole (Eds.), *Proceedings of the Conference on Uncertainty in Artificial Intelligence (UAI-94)*, Morgan Kaufmann, San Mateo, CA, 46-54, July 29-31, 1994.

Balke, A. and Pearl, J., "Counterfactuals and Policy Analysis in Structural Models," in P. Besnard and S. Hanks (Eds.), *Uncertainty in Artificial Intelligence 11*, Morgan Kaufmann, San Francisco, CA, 11-18, 1995.

Dalkey, N., "Webs," UCLA Cognitive Systems Laboratory, *Technical Report (R-166),* Computer Science Department, University of California, Los Angeles, March 1994.

Darwiche, A., and Pearl, J., "Symbolic causal networks for planning under uncertainty," In *Symposium Notes of the 1994 AAAI Spring Symposium on Decision-Theoretic Planning*, Stanford, CA, 41-47, March 21-23, 1994.

Dechter, R. and Pearl, J., "Directed constraint networks: A relational framework for Causal Modeling," in *Proceedings, 12th International Joint Conference of Artificial Intelligence (IJCAI-91)*, Sydney, Australia, 1164-1170, August 24-30, 1991,

Fikes, R.E. and Nilsson, N.J., "STIRPS: A new approach to the application of theorem proving to problem solving," *Artificial Intelligence* 2(3/4), 189–208, 1971.

Fisher, F.M., "A correspondence principle for simultaneous equation models," *Econometrica*, 38, 73–92, 1970.

Galles, D. and Pearl, J., "Testing Identifiability of Causal Effects," in P. Besnard and S. Hanks (Eds.), *Uncertainty in Artificial Intelligence 11*, Morgan Kaufmann, San Francisco, CA, 185–195, 1995.

Gärdenfors, P., *Knowledge in Flux: Modeling the Dynamics of Epistemic States*, MIT Press, Cambridge, MA, 1988.

Geffner, H.A., *Default Reasoning: Causal and Conditional Theories*, MIT Press, Cambridge, MA, 1992.

Ginsberg, M.L., "Counterfactuals", *Artificial Intelligence*, 30, 35–79, 1986.

Goldszmidt, M. and Darwiche, A., "Action networks: A framework for reasoning about actions and change under uncertainty," in R. Lopez de Mantaras and D. Poole (Eds.), *Proceedings of the Tenth Conference on Uncertainty in Artificial Intelligence (UAI-94)*, Morgan Kaufmann, San Mateo, CA, 136–144, 1994.

Goldszmidt, M. and Pearl, J., "Rank-based systems: A simple approach to belief revision, belief update, and reasoning about evidence and actions," in B. Nebel, C. Rich, and W. Swartout (Eds.), *Proceedings of the Third International Conference on Knowledge Representation and Reasoning,* Morgan Kaufmann, San Mateo, CA, 661-672, October 1992.

Katsuno, H. and Mendelzon, A., "On the difference between updating a knowledge base and revising it," in *Principles of Knowledge Representation and Reasoning: Proceedings of the Second International Conference*, Boston, MA, 387–394, 1991.

Koopman, T.C. and Reiersol, O., "The identification of structural characteristics," *Annals of Mathematical Statistics*, 21, 165–181, 1950.

Kushmerick, N., Hanks, S., and Weld, D., "An algorithm for probabilistic planning," *Technical Report 93-06-03*, Department of Computer Science and Engineering, University of Washington, 1993.

Lewis D., *Counterfactuals*, Basil Blackwell, Oxford, UK, 1973.

Lewis, D., "Probabilities of conditionals and conditional probabilities," *Philosophical Review*, 85, 297–315, 1976.

Levi, I., "Iteration of conditionals and the Ramsey test," *Synthese*, 76, 49–81, 1988.

Pearl, J., *Probabilistic Reasoning in Intelligence Systems*, Morgan Kaufmann, San Mateo, CA, 1988.

Pearl, J., "From Conditional Oughts to Qualitative Decision Theory" in D. Heckerman and A. Mamdani (Eds.), *Proceedings of the Ninth Conference on Uncertainty in Artificial Intelligence*, Washington, D.C., Morgan Kaufmann, San Mateo, CA, 12–20, July 1993.

Pearl, J., "Graphical models, causality, and intervention," *Statistical Science*, 8 (3), 266–273, 1993.

Pearl, J., "From Adams' conditionals to default expressions, causal conditionals, and counterfactuals," in E. Eells and B. Skyrms (Eds.), *Probability and Conditionals*, Cambridge University Press, New York, NY, 47-74, 1994.

Pearl, J., "A probabilistic calculus of actions," in R. Lopez de Mantaras and D. Poole (Eds.), *Proceedings of the Tenth Conference on Uncertainty in Artificial Intelligence (UAI-94)*, Morgan Kaufmann, San Mateo, CA, 454-462, 1994.

Pearl, J., "Causal diagrams for experimental research," *Biometrika* 82 (4), 669–710, December, 1995.

Pearl, J. and Robins, J., "Probabilistic evaluation of sequential plans from causal models with hidden variables," in P. Besnard and S. Hanks (Eds.), *Uncertainty in Artificial Intelligence 11*, Morgan Kaufmann, San Francisco, CA, 444–453, 1995.

Poole, D., "On the comparison of theories: Preferring the most specific explanations," in *Proceedings of International Conference on Artificial Intelligence (IJCAI-85)*, Los Angeles, CA, 144–147, 1985.

Rubin, D.B., "Estimating causal effects of treatments in randomized and nonrandomized studies," *Journal of Educational Psychology*, 66, 688–701, 1974.

Simon, H., "Causal ordering and identifiability," in W.C. Hood and T.C. Koopmans (Eds.), *Studies in Econometric Method*, New York, NY, Chapter 3, 1953.

Sobel, M.E., "Effect analysis and causation in linear structural equation models," *Psychometrika*, 55(3), 495–515, 1990.

Spirtes, P., Glymour, C., and Schienes, R., *Causation, Prediction, and Search*, Springer-Verlag, New York, 1993.

Strotz, R.H. and Wold, H.O.A., "Recursive versus nonrecursive systems: An attempt at synthesis," *Econometrica* 28, 417–427, 1960.

Symposium 1
Logic and Philosophy of Science: Current Interfaces

LOGIC AND PHILOSOPHY OF SCIENCE: CURRENT INTERFACES

Logic and philosophy of science have had many alliances since the rise of modern logic. The Wiener Kreis was even based on the idea of permanent cooperation. However, the ties have not yet culminated in a strong interdiscipline, as for instance logic and linguistics have in logical semantics. To be sure, logicians and philosophers of science feel they belong to the same family, as is for instance clear from the very happening of the 10th LMPS international congress. But, curiously enough, besides the section General Methodology, there is no section called Logical, or Formal, Methodology.

One reason for the non-existence of an interdiscipline may be Suppes' well-known complaint that the logical study of science started to imitate meta-mathematics rather than mathematics. Another reason certainly is a general one, always threatening interdisciplinary cooperation. As Zandvoort[1] (1995) has demonstrated, successful cooperation usually has an asymmetric means-end nature: one programme plays the role of guide programme, it poses and determines the leading problems, the other plays the role of supply or service programme by helping to reformulate and solve these problems. However, when the potential supply discipline is well developed, as in the case of logic, the temptation is strong to create a problem area vaguely suggested by the guide discipline and to proceed by developing appropriate tools for that area. This may well be a useful strategy leading to interesting results, but not necessarily for the intended guide discipline, i.c. the philosophy of science. In sum, whereas philosophy of science has the lasting problem of keeping in touch with science, the logical study of science has the problem of keeping in touch with science or at least with the philosophy of science.

The foregoing somewhat pessimistic remarks do not preclude that there are interesting problems that might lead to an interdiscipline. One may discern four areas in the philosophy of science that naturally call for logical analysis:

- general methodology,
- theory structure and theory relations,
- computational discovery, evaluation and revision,
- epistemological-cum-ontological claims.

[1] H. Zandvoort (1995), Concepts of interdisciplinarity and environmental science, in *Cognitive Patterns in Science and Common Sense*, T.A.F. Kuipers and A.R. Mackor eds., Poznan Studies in the Philosophy of the Sciences and the Humanities, Vol. 45, Rodopi, Amsterdam/Atlanta, pp. 45–68.

M.L. Dalla Chiara et al. (eds.), Logic and Scientific Methods, 379–381.

The papers of the symposium illustrate that interesting work is done in these four areas.

General methodology concerns in particular the separate and comparative evaluation of theories and research programmes. As yet there is no generally accepted theory of confirmation, neither qualitative nor quantitative. Not even the standard riddles of confirmation have received an unequivocal solution, viz. Hempel's paradoxes and Goodman's queer predicates, both of which are of an elementary logical nature. As to the pro's and con's of the falsificationist and the instrumentalist methodology, and hence the purpose and role of falsification, it is clear that logical analysis is required. Similarly, for the comparative evaluation of theories and programmes. And, of course, in what sense can something like 'the truth' be the target of theory selection? Kevin Kelly and his coauthors offer an outlook on a general methodology of a logical nature.

The formal study of the structure of theories and their relations started with the syntactic approach, but due to the work of Beth, Suppes and Sneed, various versions of the semantic approach became popular. Though many problems can be dealt with without explicitly taking the language into account, many other problems ask for a dual, hence model-theoretical approach. For instance, what can scientists and philosophers of science mean when they postulate empirical identities of a general nature, and do these have non-circular and still manageable truth conditions? What are minimal conditions for a proper translation and which comparative concepts are then sensitive to translation? What is the precise logical relation between certain theories and is it possible to distinguish eliminative from non-eliminative kinds of reduction? Veikko Rantala presents a general approach to such questions and applies it to the relation between symbolic and subsymbolic theories in cognitive science.

Simon c.s. returned the topic of discovery to the agenda of the philosophy of science. The subsequent development of integrated computer programmes for discovery, evaluation and revision, such as BACON and PI, showed that there is room for something like computational philosophy of science. It has raised new challenges to the role of logic. If scientific research is governed by cognitive structures and arguments, the question is how much logic is needed to describe and programme them. Non-monotone reasoning became the paradigm of non-deductive reasoning in a logical garb. Many classical studies in the philosophy of science, e.g. about nomological and statistical explanation, gained a fruitful new perspective in this way. However, as Paul Thagard and his co-author argue, there are also argumentative patterns, such as abductive reasoning, which amount to coherence considerations that may not easily get a logical explication and implementation.

Attacking and defending epistemological-cum-ontological claims, be it of a realist, idealist, instrumentalist or empiricist nature, has become a popular theme in the philosophy of science. From time to time fundamental logical theorems have been used, such as that of Löwenheim-Skolem, not only in general debates but also, for instance, in the specifics of the mind-body problem. A related phenomenon is that positions and views are sometimes governed by logical slogans, such as physics is or should be axiomatizable or complete. One way to approach such claims is to first raise the question in what intuitive respects physics may or may not be axiomatizable or complete. In this way, Paul Weingartner shows the limits of completeness claims in physics.

All four areas of logical studies in the philosophy of science may lead to one or more interdisciplines. Besides guidance by a topic in philosophy of science, there are essentially two different heuristics for further exploration of such areas. One way is to bet on similarities between reasoning structures and processes in different fields, e.g. science, common sense, language and information processing, with the risk of loosing contact with typical features of science. The other way is to bet on typical features of science, with the risk of problematic isolation. Both ways are worth trying. Although the papers by Kelly and Thagard give priority to the first heuristic and Rantala and Weingartner to the second, none of them fully loses contact with science and the philosophy of science.

Theo A.F. Kuipers
Department of Philosophy
University of Groningen

KEVIN KELLY, OLIVER SCHULTE, VINCENT HENDRICKS

RELIABLE BELIEF REVISION

ABSTRACT.
Philosophical logicians proposing theories of rational belief revision have had little to say about whether their proposals assist or impede the agent's ability to reliably arrive at the truth as his beliefs change through time. On the other hand, reliability is the central concern of formal learning theory. In this paper we investigate the belief revision theory of Alchourrón, Gärdenfors and Makinson from a learning theoretic point of view.

1. CONSERVATISM AND RELIABILITY

There are two fundamentally different perspectives on the study of belief revision. A *conservative* methodologist seeks to minimize the damage done to his current beliefs by new information. A *reliabilist*, on the other hand, seeks to find the truth whatever the truth might be. Both aims support hypothetical imperatives about how inquiry ought to proceed, imperatives that might be considered principles of inductive rationality. It would seem that there is some tension between the two perspectives. The conservative sentiment is to smooth over the effects of new information, whereas reliable inquiry may require more radical changes. The inductive leap from a hundred black ravens to the universal generalization that all ravens are black is not conservative. Nor was Copernicus' revolutionary rejection of conservative tinkering within the Ptolemaic system. Conservatism and reliabilism are reflected in two equally distinctive logical perspectives on the problem of belief revision. Conservatism is the motivation behind the theory of belief revision proposed by Alchourrón, Makinson, and Gärdenfors (AGM). Reliabilism is the principal concern of formal learning theory.[1]

Although the AGM approach is not motivated by reliability considerations, its authors seem to hope for the best.

> Even if the analysis of belief systems presented in this book does not depend on any account of the relation between such systems and reality, I still believe that rational belief systems to a large extent mirror an actual world. (Gärdenfors, 1988, p.19)

But if this happy situation obtains, it must have been brought about by the process of belief revision. Therefore, an important question for any proposed theory of rational belief revision is whether its norms assist or interfere with an agent's ability to arrive at the truth as evidence accumulates. Indeed, when putative principles of inductive rationality can be shown to prevent one from

[1] For general introductions, cf. (Osherson *et al.*, 1986) and (Kelly, 1995).

M.L. Dalla Chiara et al. (eds.), Logic and Scientific Methods, 383–398.

being as reliable as one otherwise might have been, we are strongly inclined to side with reliability rather than with the principles.[2]

In this paper, we provide a learning theoretic analysis of the effects of the AGM axioms on reliabile inductive inquiry. We consider a variety of ways in which the AGM theory can be interpreted as constraining the course of inductive inquiry. A representation of the reliable AGM methods is established.

2. THEORY DISCOVERY

We will adopt a fairly simple and abstract construal of reliable theory discovery (Kelly, 1995). We assume that the scientist is studying some system with discrete observable states that may be encoded by natural numbers, so that in the limit the scientist receives an infinite stream ε of code numbers. Let $\mathcal{N}$ denote the set of all such data streams.

An *empirical proposition* is is a proposition whose truth or falsity depends only on the data stream. We therefore identify theories and hypotheses with sets of data streams. Let Λ be a set of empirical propositions. We may think of Λ as the set of all propositions we want to find the truth about. We assume only that Λ is closed under complementation. A Λ-theory is defined to be any result of intersecting some collection of propositions drawn from Λ. The complete Λ-theory of a data stream ε is just the intersection of all propositions in Λ that contain ε. Set-theoretic relations among propositions represent logical connectives and relations in the usual way: entailment is set inclusion, inconsistency is disjointness, conjunction is finite intersection, and so forth.

Let e be a finite data sequence. The empirical proposition expressed by e is just the set of all infinite data streams extending e, which will be denoted by $[e]$. We refer to such propositions as *fans.* $e^\frown q$ denotes the result of concatenating datum q onto the end of finite sequence e.

A *theory discovery method* δ takes finite data sequences as inputs and outputs propositions. In particular, δ produces an initial conjecture $\delta(\emptyset)$ on the empty sequence $\emptyset$. We don't require that the output proposition be in Λ. We will think of the method as reading ever longer initial segments of some infinite data stream ε. It is therefore convenient to let $\varepsilon|n$ denote the initial segment of ε of length n: i.e., $(\varepsilon(0), ..., \varepsilon(n-1))$. Now we define two concepts of reliable theory discovery.[3] Uniform discovery requires that the method be guaranteed to eventually produce only consistent conjectures entailing the complete Λ truth.[4] Non-uniform or piecemeal discovery requires that each proposition in

[2] Hilary Putnam (Putnam, 1963) responded to Rudolf Carnap's theory of confirmation in just this way.

[3] (Kelly and Glymour, 1990) and (Kelly, 1995) define these concepts in a first-order logical setting.

[4] This does not imply that the method's conjectures are ever true, since a false conjecture can entail each truth in Λ without entailing any falsehood in Λ.

Λ has its truth value eventually settled by the method's conjectures, but there may be no time by which all propositions in Λ have their truth values settled.

Definition 1 *Reliable Theory Discovery*

> δ uniformly discovers *the complete* Λ *truth* $\iff$ *for each data stream* ε *there is a time such that for each later time* m, $\delta(\varepsilon|m)$ *is consistent and entails the complete* Λ*-theory of* ε.
>
> δ piecemeal discovers *the complete* Λ *truth* $\iff$ *for each data stream* ε *and for each proposition* $\mathcal{P}$ *in* Λ *there is a time such that for each later time* m, $\delta(\varepsilon|m)$ *entails* $\mathcal{P}$ *just in case* $\mathcal{P}$ *is true of* ε.

Clearly, uniform success entails piecemeal success. Also, there is no guarantee that the successive theories produced by a piece-meal discovery method get ever more "verisimilar". A piece-meal method is permitted to add as much new falsehood as it pleases at each stage, so long as more and more propositions have their truth values correctly settled.

A few examples may clarify the difference between the two reliability concepts. Let Λ_0 be the set of all fans and their complements. It is trivial to find the complete Λ_0 truth in a piecemeal fashion simply by returning $[e]$ on input data sequence e. But no method can do so uniformly, since for each data stream ε, $\{\varepsilon\}$ is the complete Λ_0-truth, and whereas there are uncountably many such singleton theories, the range of each discovery method is countable, so most such theories cannot even be conjectured.

Even when there are only countably many distinct Λ-complete theories, piecemeal success may be possible when uniform success is not. Let the proposition $\mathcal{H}_n$ say that 0 occurs at stage n and forever after. Let Λ_1 be the set of all such propositions and their negations. Λ_1 requires piecemeal solutions to perform nontrivial inductive inferences, since each hypothesis $\mathcal{H}_n$ is a claim about the unbounded future. This time, there are only countably many distinct Λ_1-complete theories[5]. To succeed piecemeal, let method β conjecture $\mathcal{H}_{n+1} \cap \overline{\mathcal{H}_n}$ when the last non-0 occurring in the data occurs at position n. Nonetheless, uniform success is still impossible. For suppose for reductio that some method δ succeeds uniformly. Then a wily demon can present 0,0,0,... until δ's conjecture entails $\mathcal{H}_0$, which it must eventually do on that data stream. Then a 1 is fed (e.g., at stage m) followed by all 0s until δ's conjecture entails $\mathcal{H}_{m+1}$, which it must eventually do on that data stream, etc. The data stream ε so presented never stabilizes to 0, so δ produces infinitely many conjectures inconsistent with the complete Λ_1-theory of ε.

The same argument shows that no piecemeal solution could possibly succeed by eventually producing only true hypotheses, since the construction forces an arbitrary piecemeal solution to produce infinitely many false conjectures. Hence, piecemeal success is sometimes possible only if infinitely many

[5]Such a theory either says exactly when the data stream stabilizes to 0 or says that the data stream never stabilizes to 0.

false conjectures are produced. On the other hand, whenever uniform success is possible, it can be achieved by a method whose conjectures are eventually true. Given successful method δ, produce whatever δ says if it is not a consistent proposition entailing some Λ-complete theory and produce the (unique) Λ-complete theory entailed by the current conjecture of δ otherwise.

3. BELIEF REVISION

A belief revision operator $+$ takes a theory $\mathcal{T}$ and a proposition $\mathcal{P}$ and returns a revised theory $\mathcal{T}'$. Then we write

$$\mathcal{T} + \mathcal{P} = \mathcal{T}'$$

In our setting of empirical propositions, the AGM axioms amount to the following:

(AGM 1) $\mathcal{T} + \mathcal{P} \subseteq \mathcal{P}$.
(AGM 2) If $\mathcal{P} \neq \emptyset$, then $\mathcal{T} + \mathcal{P} \neq \emptyset$.
(AGM 3) If $\mathcal{P} \cap \mathcal{T} \neq \emptyset$, then $\mathcal{T} + \mathcal{P} = \mathcal{P} \cap \mathcal{T}$.
(AGM 4) If $\mathcal{P} \cap (\mathcal{T} + \mathcal{Q}) \neq \emptyset$, then $\mathcal{T} + (\mathcal{P} \cap \mathcal{Q}) = \mathcal{P} \cap (\mathcal{T} + \mathcal{Q})$.

We employ an elegant representation of these requirements due to A. Grove (1988). A *Grove system* for proposition $\mathcal{K}$ is a collection Γ of subsets of $\mathcal{N}$ such that

1. Γ is nested (i.e., totally ordered by $\subseteq$),
2. For each consistent proposition $\mathcal{P}$ there is a unique, $\subseteq$-minimal element of Γ whose intersection with $\mathcal{P}$ is nonempty.
3. $\mathcal{N} \in \Gamma$.
4. $\mathcal{K}$ is the least member of Γ (with respect to $\subseteq$).

Let $\Gamma_{\mathcal{P}}$ denote the least element of Γ whose intersection with $\mathcal{P}$ is nonempty. This is well-defined by the second condition on Grove systems. Let $\Gamma(\mathcal{P}) = \Gamma_{\mathcal{P}} \cap \mathcal{P}$. Then we have:

Proposition 1 (Grove 1988) *For each* AGM *belief revision operator* $+$ *and for each proposition* $\mathcal{K}$ *there is a Grove system* $\Gamma^{\mathcal{K}}$ *such that for each proposition* $\mathcal{P}$

$$\Gamma^{\mathcal{K}}(\mathcal{P}) = \mathcal{K} + \mathcal{P}.$$

For each proposition $\mathcal{K}$, *let* $\Gamma^{\mathcal{K}}$ *be a Grove system for* $\mathcal{K}$. *Then there exists a unique* AGM *belief revision operator* $+$ *such that for each* $\mathcal{K}, \mathcal{P}$

$$\Gamma^{\mathcal{K}}(\mathcal{P}) = \mathcal{K} + \mathcal{P}.$$

4. INDUCTION AS BELIEF REVISION

There are several ways in which the AGM theory of belief revision might interact with inductive methodology. An ambitious approach views induction as nothing but the revision of one's previous beliefs in accordance with the total evidence available. On this proposal, one's revision operator uniquely determines one's inductive method once an initial theory is specified. There are two ways in which this might happen. According to the first, inquiry begins with a fixed theory $\mathcal{K}$ and this theory is repeatedly updated on increasing data. We will refer to this as *repeated* revision. Repeated revision of $\mathcal{K}$ by a belief revision operator $+$ uniquely determines a discovery method δ according to the following relation:

1. $\delta(\emptyset) = \mathcal{K}$.
2. $\delta(e^\frown q) = \delta(\emptyset) + [e^\frown q]$.

Then we say that δ is *represented repetitively* by $\mathcal{K}$ and $+$. If δ is repetitively represented by some AGM operator starting with $\mathcal{K}$, then we say that δ is *repetitively* AGM. A second proposal is that it is always one's current theory that is revised as the data comes in. We will refer to this as *sequential* revision. Sequential revision starting with $\mathcal{K}$ uniquely determines a discovery method as follows:

1. $\delta(\emptyset) = \mathcal{K}$.
2. $\delta(e^\frown q) = \delta(e) + [e^\frown q]$.

Then we say that δ is *represented sequentially* by $\mathcal{K}$ and $+$.[6] If δ is sequentially represented by some AGM operator starting with $\mathcal{K}$, then we say that δ is *sequentially* AGM. The two concepts are equivalent for a given operator $+$ if the operator satisifes the following principle:

$$\mathcal{K} + [e^\frown q] = (\mathcal{K} + [e]) + [e^\frown q].$$

We refer to this as the *irrelevance of earlier subdata* (IES) principle. The IES principle does not follow from the AGM axioms, however, so repeated revision by a given operator is not the same as sequential revision by that operator.

It follows immediately from the AGM axioms that a sequentially AGM discovery method has various properties. First, it is *data retentive*, in the sense that its current conjecture always entails the data. Second, it is *consistent*, in the sense that its current conjecture is always consistent with its current data. Third, it is *stubborn* in the sense that its current conjecture always entails its preceding conjecture unless the preceding conjecture is refuted. Finally, it is *timid* in the sense that it only adds the data to its previous conjecture unless its preceding conjecture is refuted.

In fact, these properties characterize both the sequentially and the repetitively AGM discovery methods, so although it matters for a particular revision

[6] Both proposals are developed in (Martin and Osherson, 1995). Sequential revision is there referred to as 'iterated' revision.

operator whether it is interpreted sequentially or repetitively, the same methods are representable either way. We may now refer to discovery methods that are data retentive, consistent, stubborn and timid simply as AGM methods.

Proposition 2 *The following statements are equivalent:*

1. *δ is data retentive, consistent, stubborn, and timid.*
2. *δ is sequentially* AGM.
3. *δ is repetitively* AGM.

The proof is given in the appendix.

Data retentiveness is a fairly tame requirement for an ideal[7] logic of discovery: one may as well remember the data. Consistency is as well: while we produce contradictions, we couldn't possibly be on the path to the truth.[8] Stubbornness and timidity are another matter. Consider a stubborn scientist whose only belief is that a given coin will come up heads at least once. Suppose it never comes up heads. He can never retract his belief unless he adds other "auxilliary hypotheses" so that his whole belief set is eventually refuted, affording him an opportunity to retract the original false belief. But if this scientist is also timid, then he never gets an opportunity to add any such auxilliary hypotheses, so he is hopelessly frozen for eternity in a false belief. Since there is no problem finding the truth about this hypothesis (conjecture its negation until one head is observed and then conjecture it forever after), it is clear that there are some initial belief sets from which no AGM operator can recover, either in the repetitive or in the sequential sense.

So for some initial belief states, sequential and repetitive AGM updating does stand in the way of finding the truth. The question remaining is whether some AGM method with the right kind of initial belief set can succeed whenever success is possible. The preceding discussion already provides a clue: if one always produces refutable conjectures, one never ends up in the awkward situation of not getting an opportunity to add the right kind of auxilliary hypotheses. A trivial way to ensure this is to always produce an empirically complete hypothesis (i.e., a hypothesis that singles out a unique data stream).

Proposition 3 *Suppose that the complete Λ truth is uniformly [piecemeal] discoverable. Then the complete Λ truth is uniformly [piecemeal] discoverable by an* AGM *method.*

Proof: Let δ uniformly discover the complete Λ truth. Let $\phi(\mathcal{K}, e)$ denote the choice of some data stream in $\mathcal{K} \cap [e]$ if $\mathcal{K} \cap [e] \neq \emptyset$ and let $\phi(\mathcal{K}, e)$ denote an arbitrary data stream in $[e]$ otherwise. Then define

$$\gamma(\emptyset) = \{\phi(\delta(\emptyset), \emptyset)\};$$

[7]By "ideal" we mean "abstracted from all computability considerations".

[8]On the other hand, computable inquiry can be very severely restricted by imposing the consistency requirement (Kelly and Schulte, 1995a; Kelly and Schulte, 1995b).

$\gamma(e^\frown q) = \gamma(e)$ if $\gamma(e) \cap [e^\frown q] \neq \emptyset$;

$\gamma(e^\frown q) = \{\phi(\delta(e^\frown q), e^\frown q)\}$ otherwise.

By the definition of ϕ, γ is consistent and data-retentive. By the second clause of γ's definition, γ is stubborn and timid. So by proposition 2, γ is an AGM method. Since δ uniformly discovers the complete Λ truth, we have that on each data stream ε, there is a stage n after which δ always produces consistent conjectures entailing the complete Λ truth. Suppose that γ stabilizes to some conjecture $\{\tau\}$ along ε. Then $\tau = \varepsilon$, so γ succeeds uniformly. Suppose then that γ never stabilizes to any conjecture along ε. Then γ selects a new conjecture after stage n. Thereafter, γ's conjecture always entails the complete Λ truth because δ's conjecture does.

Now suppose δ discovers the complete Λ truth in the piecemeal sense. Construct γ just as before. Again, if γ happens to stabilize to some conjecture on ε, then γ succeeds uniformly. So suppose γ never stabilizes to a conjecture. For each $\mathcal{H} \in \Lambda$, there is a stage n by which δ's conjecture entails $\mathcal{H}$ if $\mathcal{H}$ is true of ε and entails $\overline{\mathcal{H}}$ otherwise. So there is a stage m after n such that each successive conjecture of γ also has this property. □

An AGM operator $+$ is said to be *maxichoice* just in case $\mathcal{K} + \mathcal{A}$ is the result of adding $\mathcal{A}$ to some minimal superset $\mathcal{S}$ of $\mathcal{K}$ consistent with $\mathcal{A}$, in the sense that any proper subset of $\mathcal{S}$ that includes $\mathcal{K}$ would not be consistent with $\mathcal{A}$. The method γ constructed in the preceding proof can be represented by a maxichoice operator, since $\gamma(e^\frown q) \cup \gamma(e)$ is always a minimal superset of $\gamma(e)$ consistent with $[e^\frown q]$. This is the strongest sense of "minimal change" considered by Gärdenfors (and rejected for being too strong), but even it does not interfere with the possibility of reliable inquiry (although we have seen that the AGM axioms themselves interfere with reliability for agents starting out with the wrong kinds of beliefs).[9]

On the other hand, conservative intuitions about minimal belief change are hardly comforted by the piecemeal discovery method γ just constructed. γ leaps from one complete theory to another in order to ensure that it can change its mind when necessary without violating stubbornness and timidity. This raises the question whether we can say something in general about how strong the conjectures of a reliable AGM method must be. A few topological concepts are helpful. A *limit point* of an empirical proposition $\mathcal{K}$ is a data stream along which $\mathcal{K}$ is never refuted. The *empirical closure* of $\mathcal{K}$ (denoted cl($\mathcal{K}$)) is the set of all limit points of $\mathcal{K}$ (i.e. the set of all data streams along which $\mathcal{K}$ is never refuted). $\mathcal{H}$ is *empirically decidable with certainty* or *empirically clopen* given $\mathcal{K}$ just in case the truth value of $\mathcal{H}$ is eventually decided by the data, assuming that $\mathcal{K}$ is true. That is, on each data stream in $\mathcal{K}$, there is a time after which the data together with $\mathcal{K}$ either entail $\mathcal{H}$ or entail the negation of

[9](Martin and Osherson, 1995, Prop. 23) show that if the notion of logical consequence is weakened, an initial belief set can be found that succeeds with any AGM operator in the repetitive sense. We do not consider weakened consequence relations here.

$\mathcal{H}$. $\mathcal{H}$ is *empirically decidable with certainty by stage n* or *empirically n-clopen* given $\mathcal{K}$ just in case the truth value of $\mathcal{H}$ is determined by stage n, given that $\mathcal{K}$ is true. Now we have:

Proposition 4 *Let δ be a discovery method.*

1. *If δ is data-retentive, stubborn and timid and δ piecemeal discovers the complete Λ truth, then for each e, each proposition in Λ is empirically clopen given cl($\delta(e)$).*
2. *If δ is data retentive, stubborn and timid and δ uniformly discovers the complete Λ truth, then for each e, there is an n such that each proposition in Λ is empirically n-clopen given cl($\delta(e)$).*

Proof: (1) Suppose otherwise. Then for some e, and for some $\mathcal{H} \in \Lambda, \mathcal{H}$ is not clopen in cl($\delta(e)$). Hence, ($*$) there is a data stream ε along which $\delta(e)$ is never refuted such that at no stage does the data read along ε entail either $\mathcal{H}$ or its complement. Since δ is data retentive, $\delta(e) \subseteq [e]$, so for some $n, e = \varepsilon|n$. Hence $\delta(\varepsilon|n) = \delta(e)$. But since δ is stubborn, timid and data retentive, for each $m \geq n, \delta(\varepsilon|m) = \delta(e) \cap [\varepsilon|m]$. So by ($*$), δ does not piecemeal discover the complete Λ truth.

The proof of (2) is similar. □

The first condition is equivalent to saying that the topological boundary of each proposition in Λ must be excluded from each conjecture produced by δ. In a setting with first-order sentences instead of propositions, this would mean that each of δ's conjectures entails that each sentence in Λ is equivalent to a quantifier-free sentence (Kelly, 1995, Thm. 12.7). This is a striking requirement, but in the propositional setting it can always be accomplished with conjectures of arbitrarily high probability if probability is countably additive.[10]

5. RELIABILITY AS A REVISION THEORETIC AXIOM

AGM belief revision theory makes no reference to reliability. What would happen if we were to add as an AGM axiom that each operator be a reliable solution to a specified, solvable inductive inference problem Λ? It would be nice to have an exact representation, analogous to Grove's, of the class of all AGM operators that are reliable solutions to Λ. We have not yet found such a representation for the notions of reliability defined above. But it is much easier to obtain such a result if we modify our notions of reliability so as to require in addition, that on each data stream there is a time after which all the theories produced by δ are *true* (recall that uniform and piecemeal discovery are both possible even when no conjecture is true). Then we speak, respectively, of *truly* discovering the complete Λ truth, uniformly or in a piecemeal fashion.

Recall from section 2 that the complete Λ truth is truly uniformly discoverable just in case it is uniformly discoverable. On the other hand, true

[10]Cf. (Kelly, 1995), proposition 13.15.

piecemeal discovery is properly easier than uniform discovery and piecemeal discovery is properly easier than true, piecemeal discovery.[11]

Two more concepts are necessary in order to represent the "truly" reliable AGM methods. $\mathcal{K}$ is *empirically closed* just in case $\mathcal{K}$ contains all its limit points (i.e. just in case $\mathcal{K}$ is identical to its own empirical closure). Then $\mathcal{K}$ is guaranteed to be refuted eventually by the data if it is false, for otherwise there would be a data stream making $\mathcal{K}$ false along which $\mathcal{K}$ is never refuted (i.e., $\mathcal{K}$ is missing one of its limit points). Let Γ be a Grove system and let $\mathcal{S} \in \Gamma$. Let core($\mathcal{S}$) denote the union of all $\mathcal{R} \in \Gamma$ such that $\mathcal{R}$ is a proper subset of $\mathcal{S}$. Then we have:

Proposition 5 *Truly Reliable* AGM *Learners*

1. *δ is a repetitively* AGM *method that truly, piecemeal discovers the complete Λ truth $\Longleftrightarrow$ there is a Grove system Γ such that*
   (a) $\forall e, \Gamma([e]) = \delta(e)$
   (b) $\forall \mathcal{S} \in \Gamma$, *core*($\mathcal{S}$) *is empirically closed, and*
   (c) $\forall e, \forall \mathcal{H} \in \Lambda, \mathcal{H}$ *is empirically clopen in* $\Gamma([e])$.
2. *δ is a repetitively* AGM *method that truly, uniformly discovers the complete Λ truth $\Longleftrightarrow$ there is a Grove system Γ such that*
   (a) $\forall e, \Gamma([e]) = \delta(e)$
   (b) $\forall \mathcal{S} \in \Gamma$, *core*($\mathcal{S}$) *is empirically closed, and*
   (c) $\forall e, \exists n$ *such that* $\forall \mathcal{H} \in \Lambda$, $\mathcal{H}$ *is empirically n-clopen in* $\Gamma([e])$.

Proof of (1): ($\Longrightarrow$) suppose δ is a repetitively AGM method. Then by proposition 1, there is a Grove system Γ such that $\forall e, \Gamma([e]) = \delta(e)$, so we have 1a. 1c follows from 1a and proposition 4. Finally, suppose for reductio that 1b is false. Then let $\mathcal{S} \in \Gamma$ be such that some $\varepsilon \notin$ core($\mathcal{S}$) is a limit point of core($\mathcal{S}$). But then for each n, $\Gamma_{[\varepsilon|n]} \subseteq$ core($\mathcal{S}$), so $\varepsilon \notin \Gamma([\varepsilon|n]) = \delta([\varepsilon|n])$. Hence, δ produces infinitely many false conjectures along ε and so δ fails to truly piecemeal identify the complete Λ truth.

($\Longleftarrow$) Suppose there is a Grove system Γ satisfying 1a–1c. Let ε be given. Since Γ is a Grove system, there is a least element $\mathcal{S} \in \Gamma$ such that $\varepsilon \in \mathcal{S}$. So $\varepsilon \notin$ core($\mathcal{S}$). There is an n such that $[\varepsilon|n] \cap$ core($\mathcal{S}$) $= \emptyset$, else ε is a missing limit point of core($\mathcal{S}$), contrary to 1b. But since $\varepsilon \in \mathcal{S}$, for each $m \geq n, \Gamma([\varepsilon|m]) = \Gamma([\varepsilon|n]) \cap [\varepsilon|m]$. By 1c, we have that for each $\mathcal{H} \in \Lambda, \exists k \geq$

[11]The first noninclusion is witnessed by the example Λ_0 provided at the beginning of the paper. We have already seen that this inductive problem is not uniformly solvable. But the trivial method that always repeats the current data succeeds truly in the nonuniform sense. The second noninclusion is witnessed by the example Λ_1 also presented at the beginning of the paper. It was shown above that this problem has a piecemeal solution. But the demonic argument against uniform solutions forces an arbitrary method to produce infinitely many false conjectures, so no true piecemeal solution is possible.

$n.\forall j \geq k, \Gamma([\varepsilon|n]) \cap [\varepsilon|k]$ either entails $\mathcal{H}$ if $\varepsilon \in \mathcal{H}$ or entails $\overline{\mathcal{H}}$ otherwise. By 1a, δ truly piecemeal discovers the complete Λ truth.

The proof of (2) is similar. □

The requirement that the method stabilize to the truth undercuts the piecemeal strategy of succeeding by always producing a complete, false theory. We leave open the interesting question whether AGM methods can solve every problem solvable in the truly piecemeal sense.

6. ARBITRARY REVISIONS

So far, we have examined an ambitious interpretation of belief revision theory, in which one's belief revision operator uniquely determines one's inductive inferences. It has been shown that even on this interpretation, the AGM axioms do not restrict the reliability of inquiry, at least for ideal methods that needn't worry about computational limitations. On the other hand, AGM methods have some awkward properties such as timidity and stubbornness, and these requirements must be "steered around" to arrive at the truth.

But there is a more tempered interpretation of belief revision theory, according to which a revision operator is just a way of consistently forcing into one's beliefs whatever one decides to force into them for whatever reason. Say that a method is repetitively AGM *with arbitrary revisions* just in case it always updates its initial beliefs on the conjunction of the total data and some arbitrarily selected proposition. The definition of sequentially AGM methods with arbitrary revisions is similar, except that it is the initial belief set that is revised.

Definition 2 AGM *Learners With Arbitrary Revisions*

1. *δ is repetitively* AGM *with arbitrary revisions* $\iff$ *there is an* AGM *revision operator* $+$ *such that for all* $e^\frown q$ *there is an* $\mathcal{A}$ *such that* $\delta(e^\frown q) = \delta(\emptyset) + ([e^\frown q] \cap \mathcal{A})$.

2. *δ is sequentially* AGM *with arbitrary revisions* $\iff$ *there is an* AGM *revision operator* $+$ *such that for all* $e^\frown q$ *there is an* $\mathcal{A}$ *such that* $\delta(e^\frown q) = \delta(e) + ([e^\frown q] \cap \mathcal{A})$.

Neither concept implies stubbornness or timidity, since $\mathcal{A}$ can be chosen to be inconsistent with $\delta(\emptyset)$ (or respectively, with $\delta(e)$). Data retentiveness still follows. So do weakened versions of stubbornness and consistency. Stubbornness obliges δ to move forward until contradicted by the data. An 'internal' version of stubbornness requires δ to move forward until contradicted either by the data or by δ's new beliefs. Accordingly, δ is *internally stubborn in the repetitive [sequential] sense* just in case δ's current beliefs either entail its initial [current] beliefs or are inconsistent with them. Finally, say that δ is *quasi-consistent* just in case δ's beliefs are never both inconsistent with the data and consistent. Then we have:

Proposition 6 *Let δ be a discovery method. Then*

1. *δ is repetitively [sequentially]* AGM *with arbitrary revisions $\Longleftrightarrow$ δ is data retentive and internally stubborn in the repetitive [sequential] sense.*

2. *This remains true if we add the condition that δ is quasi-consistent.*

Proof of 1, sequential case: ($\Longrightarrow$) Let δ be sequentially AGM with arbitrary revisions. Let $+$ be the AGM operator that witnesses this fact. Since δ is defined in terms of an AGM operator revising on at least the data, δ is data retentive. Now suppose that $\delta(e^\frown q)$ is consistent with $\delta(e)$. Since δ is sequentially AGM with arbitrary revisions, there is an $\mathcal{A}$ such that $\delta(e^\frown q) = \delta(e)+([e^\frown q]\cap\mathcal{A})$. By axiom (AGM 1), $\delta(e)+([e^\frown q]\cap\mathcal{A}) \subseteq ([e^\frown q]\cap\mathcal{A})$. Hence, $([e^\frown q]\cap\mathcal{A})$ is consistent with $\delta(e)$. So by axiom (AGM 3), $\delta(e)+([e^\frown q]\cap\mathcal{A}) = \delta(e)\cap([e^\frown q]\cap\mathcal{A}) \subseteq \delta(e)$. Hence, $\delta(e^\frown q) \subseteq \delta(e)$.

($\Longleftarrow$) Let δ be data retentive and internally stubborn. For each proposition $\mathcal{A}$ let Grove system $\Gamma^{\mathcal{A}} = \{\mathcal{N}, \mathcal{A}\}$. Let $+$ be the operator represented by the collection of all such $\Gamma^{\mathcal{A}}$ according to proposition 1. Then we have

$$(*) \text{ if } \mathcal{S}\cap\mathcal{R} = \emptyset \text{ then } \mathcal{S}+\mathcal{R} = \mathcal{R}.$$

By data retentiveness, $\delta(e^\frown q) \subseteq [e^\frown q]$, so

$$(**)\ \delta(e^\frown q)\cap[e^\frown q] = \delta(e^\frown q).$$

By internal stubbornness in the sequential sense, there are two cases. Case 1: $\delta(e)\cap\delta(e^\frown q) = \emptyset$. Then by $(**)$ and $(*)$, $\delta(e)+([e^\frown q]\cap\delta(e^\frown q)) = \delta(e)+\delta(e^\frown q) = \delta(e^\frown q)$.

Case 2: $\delta(e^\frown q) \subseteq \delta(e)$. Then by $(**)$, (AGM 3), and the case hypothesis, we have that $\delta(e)+([e^\frown q]\cap\delta(e^\frown q)) = \delta(e)+\delta(e^\frown q) = \delta(e)\cap\delta(e^\frown q) = \delta(e^\frown q)$.

Proof of 1, repetitive case: The ($\Longrightarrow$) side is just as before, with $\delta(\emptyset)$ in place of $\delta(e)$. The ($\Longleftarrow$) side is also just as before, except that we employ Grove system $\Gamma = \{\delta(\emptyset), N\}$, and choose $+$ to agree with Γ when $\delta(\emptyset)$ is being updated. The rest of the argument is as before, with $\delta(\emptyset)$ replacing $\delta(e)$ everywhere.

Proof of 2: ($\Longrightarrow$) Suppose that δ is sequentially AGM with arbitrary revisions. Suppose $\delta(e^\frown q) = \delta(e)+([e^\frown q]\cap\mathcal{A}) \neq \emptyset$. By (AGM 1), $\delta(e)+([e^\frown q]\cap\mathcal{A}) \subseteq ([e^\frown q]\cap\mathcal{A})$, so $\delta(e^\frown q)\cap([e^\frown q]\cap\mathcal{A}) \neq \emptyset$ and hence $\delta(e^\frown q)\cap[e^\frown q] \neq \emptyset$. The repetitive case is the same with $\delta(\emptyset)$ in place of $\delta(e)$.

($\Longleftarrow$) This side was already shown without appeal to quasi-consistency. □

In the proof, we showed that the extra set $\mathcal{A}$ chosen to revise on can always be $\delta(e^\frown q)$ itself. We also employed a trivial revision operator $+$ (i.e. the operator generated by full meet according to the Levi identity). Strengthening $+$ would endanger condition $(*)$ in the proof, possibly leading to a method whose conjectures are sometimes stronger than the corresponding conjectures of δ.

It is not intuitively clear to us why violations of internal stubbornness should be denounced as irrational, but it is at least straightforward (in the absence of computability considerations) to bring an arbitrary method into compliance with this principle without affecting its reliability or delaying its time of convergence. For suppose that method δ uniformly identifies the complete Λ truth, possibly violating internal stubbornness. Let γ agree with δ on data $\emptyset$. On data sequence $e^\frown q$, γ agrees with δ if this does not lead to a violation of internal stubbornness, and $\gamma(e^\frown q) = \delta(e^\frown q) - \gamma(e)$ otherwise. Then γ is internally stubborn in the sequential sense and γ uniformly identifies the complete Λ truth. The same modification works for a theory learner δ that piecemeal identifies the complete Λ truth.

7. MINIMAL REVISIONISM

There is an even weaker way for inductive methodology to interact with belief revision theory. According to this approach, revision theory is nothing but an explication of what it means to add a proposition to a theory that might be inconsistent with it, and an inductive method is allowed to retract on or to add whatever it pleases for whatever reason. Accordingly, δ is sequentially AGM in the *minimal sense* just in case there is some belief revision operator $+$ such that for each $e^\frown q$ there are $\mathcal{A}$ and $\mathcal{B}$ such that $\delta(e^\frown q)$ is the result of first contracting $\delta(e)$ by $\mathcal{A}$ and then revising the result by $\mathcal{B}$. The repetitive version of this concept is defined by replacing $\delta(e)$ with $\delta(\emptyset)$. On either interpretation, belief revision theory is entirely vacuous as a constraint on discovery methods.

Proposition 7 *Every method δ is a sequentially [repetitively]* AGM *method in the minimal sense.*

Proof: We begin with the sequential case. Let δ be given. Let $+$ be the operator represented by the trivial Grove systems defined in the proof of proposition 6. Let $-$ be defined according to the Harper identity as[12]

$$\mathcal{K} - \mathcal{A} = \mathcal{K} \cup (\mathcal{K} + \overline{\mathcal{A}}).$$

Then $\mathcal{K} - \mathcal{K} = \mathcal{K} \cup (\mathcal{K} + \overline{\mathcal{K}}) = \mathcal{K} \cup \overline{\mathcal{K}} = \mathcal{N}$ (the second to last identity is by $(*)$ in the proof of proposition 6.) Moreover, for each $\mathcal{A}, \mathcal{N} + \mathcal{A} = \mathcal{A}$. Hence, $\delta(e^\frown q) = (\delta(e) - \delta(e)) + \delta(e^\frown q)$. The repetitive case is similar. □

8. CONSERVATISM, RELIABILITY, AND LOGIC

We began with a fundamental distinction between conservative and reliabilist approaches to the problem of inductive inference. According to the former, our beliefs should be repaired in the most elegant possible manner when they are

[12]In our propositional setting, the intersection of the deductive closures of two propositions $\mathcal{X}, \mathcal{Y}$ is just the union of all propositions including both $\mathcal{X}$ and $\mathcal{Y}$.

revised. According to the latter, the aim is to find the right answer whatever that answer might be. Belief revision theory belongs to the former perspective and formal learning theory belongs to the latter. In this paper, we have examined various different ways in which belief revision theory could be viewed as as constraint on empirical inquiry, ranging from the view that inquiry is nothing but revision on the data to the proposal that revision is nothing but an explication of what it means to force a proposition into a system of beliefs, leaving inquiry to decide what to add or subtract. The result of our investigation is that the weaker interpretations impose few short-run restrictions on inquiry, whereas the stronger ones imply obstacles that must be carefully circumvented if inquiry is to be reliable.

This study illustrates two ways in which logic can be brought to bear on the problem of understanding inductive inquiry. One style of philosophical logic proposes intuitive normative principles and tests and refines these principles in a quasi-empirical way, by deducing particular consequences to compare against one's intuitions, by providing alternative representations of the principles, and by proving soundness and completeness theorems for the principles according to some illuminating semantic interpretation. The extensive work on belief revision theory fits into this approach. In this paper we have attempted to illustrate how the pertinent logical questions change when one shifts one's focus from coherence to reliability. Reliabilist applications of logic are more akin to the theory of computability than to usual work in philosophical logic. Whereas philosophical logicians tend to analyze systems reflecting norms derived from direct intuition or from particular case examples, the logic of reliability concerns the intrinsic solvability of inductive problems. The logical tools employed are programming, definability, and diagonalization rather than semantic interpretations, completeness and representations.

We hope to have illustrated how these two approaches can usefully interact. The logical analysis of intuitive axioms of rationality can motivate local side-constraints on inductive methods (e.g., internal stubbornness and timidity) that can enrich reliability analysis.[13] On the other side, we have shown in a preliminary way how reliability can be imposed as an extra axiom in a normative theory of belief revision, leading to enriched representations of acceptable inductive practice, as in the representation of truly reliable AGM methods by Grove systems of a special kind.

We hope that the range of questions raised by this preliminary study underscores the fruitfulness of reliability considerations in the study of belief revision. Is there an elegant representation of the reliable AGM methods that may produce infinitely many false conjectures? Are all problems solvable with finitely many false conjectures solvable by AGM methods? What happens to our results when more stringent constraints are imposed on revision than just

[13] It is interesting how naturally the Grove systems could be "programmed" as inductive methods.

the AGM axioms? What happens when we allow facts about the data stream to arrive in an arbitrary order?[14] And perhaps most importantly, how restrictive are the AGM principles for computable inquiry? Our non-restrictiveness results in this paper make use of highly idealized constructions in which uncomputable logical relations must be decided. In light of the strong negative results concerning consistency in similar settings (Kelly and Schulte, 1995a; Kelly and Schulte, 1995b), one would expect the AGM principles to be restrictive for computable inquiry for that reason alone.

APPENDIX: PROOF OF PROPOSITION 2

Proposition 2 *The following statements are equivalent:*

1. *δ is data retentive, consistent, stubborn, and timid.*
2. *δ is sequentially* AGM.
3. *δ is repetitively* AGM.

Proof: To see that (3) implies (1), let δ be repetitively AGM. Then for some AGM operator $+$, we have that for each $e, \delta(e) = \delta(\emptyset) + [e]$. δ is data retentive by (AGM 1) and is consistent by (AGM 2). Now suppose that $\delta(e) \cap [e^\frown q] \neq \emptyset$. Then by axiom (AGM 4), $\delta(e^\frown q) = \delta(\emptyset) + [e^\frown q] = \delta(\emptyset) + ([e] \cap [e^\frown q]) = (\delta(\emptyset) + [e]) \cap [e^\frown q] = \delta(e) \cap [e^\frown q]$. So δ is stubborn and timid. An even simpler argument establishes that (2) implies (1).

To see that (1) implies (3), let δ be data retentive, consistent, etc. To show that δ is also repetitively AGM, one must show not only that δ behaves like an AGM operator repeatedly updating $\mathcal{K}$, but that δ's domain can be extended to all empirical propositions in a manner that satisfies the AGM axioms. It is useful to employ the Grove representation for this purpose.

We will construct a single Grove system Γ for $\delta(\emptyset)$ that behaves just like δ, from which it follows by proposition 1 that δ is repetitively AGM. Define

$$\mathcal{S}_0 = \delta(\emptyset);$$

$$\mathcal{S}_{n+1} = \mathcal{S}_n \cup \bigcup_{lh(e)=n+1} \delta(e);$$

$$\mathcal{S}_\omega = \mathcal{N};$$

$$\Gamma = \{\mathcal{S}_\alpha : \alpha \leq \omega\}.$$

It is easy to verify that Γ is a Grove system for $\delta(\emptyset)$.

Next, we show by induction on the length of e that $\Gamma([e]) = \delta(e)$. It then follows immediately by proposition 1 that δ is repetitively AGM. In the base case, we have: $\Gamma([\emptyset]) = \Gamma(\mathcal{N}) = \mathcal{S}_0 = \delta(\emptyset)$.

For the inductive case, suppose $\Gamma([e]) = \delta(e)$. Let q be given.

Case 1: suppose that $[e^\frown q] \cap \Gamma([e]) \neq \emptyset$. Then $\Gamma_{[e^\frown q]} = \Gamma_{[e]}$, so $\Gamma([e^\frown q]) = \Gamma([e]) \cap [e^\frown q] = \delta(e) \cap [e^\frown q]$ by the inductive hypothesis. But since δ is both

[14] Cf. (Martin and Osherson, 1995).

timid and stubborn, the case hypothesis yields that $\delta(e^\frown q) = \delta(e) \cap [e^\frown q]$, and hence $\delta(e^\frown q) = \Gamma([e^\frown q])$.

Case 2: $[e^\frown q] \cap \Gamma([e]) = \emptyset$. Let $\mathrm{lh}(e) = n$. First we establish that

$$(*)\ \Gamma_{[e^\frown q]} = \mathcal{S}_{n+1}.$$

Since δ is consistent, $\delta(e^\frown q) \neq \emptyset$. Since δ is data retentive, $\delta(e^\frown q) \subseteq [e^\frown q]$. But $\delta(e^\frown q) \subseteq \mathcal{S}_{n+1}$, by definition of $\mathcal{S}_{n+1}$. Hence, $[e^\frown q] \cap \mathcal{S}_{n+1} \neq \emptyset$. Suppose for reductio that for some $i \leq n$ there is some $\varepsilon \in \mathcal{S}_i \cap [e^\frown q]$. Let k be the least such i. By choice of k and the definition of $\mathcal{S}_k$, there is an e' of length k such that $\varepsilon \in \delta(e')$. Since δ is data retentive, $\varepsilon \in [e']$. So $[\varepsilon|k], [\varepsilon|k+1], \ldots [\varepsilon|n], [\varepsilon|n+1]$ are all consistent with $\delta(\varepsilon|k)$. Since δ is both stubborn and timid, $\delta(e) = \delta(\varepsilon|n) = \delta(\varepsilon|k) \cap [\varepsilon|n]$. Hence, $[\varepsilon|n+1] \cap \delta(\varepsilon|n) \neq \emptyset$. But $\varepsilon|n = e, \varepsilon|n+1 = e^\frown q$ and by induction hypothesis, $\delta(e) = \Gamma([e])$, so $[e^\frown q] \cap \Gamma([e]) \neq \emptyset$, contrary to the case hypothesis.

Thus we have $(*)$. Hence, $\Gamma([e^\frown q]) = (\mathcal{S}_{n+1} - \mathcal{S}_n) \cap [e^\frown q]$. But by definition, $(\mathcal{S}_{n+1} - \mathcal{S}_n) = \bigcup_{lh(e)=n+1} \delta(e)$. Since δ is data retentive, $\delta(e^\frown q) \subseteq [e^\frown q]$. Let $e'' \neq e^\frown q$ be of length $n+1$. Then $[e''] \cap [e^\frown q] = \emptyset$.[15] Since δ is data retentive, $\delta(e'') \cap [e^\frown q] = \emptyset$. Hence, $\mathcal{S}_{n+1} \cap [e^\frown q] = \delta(e^\frown q)$.

Now we argue from (1) to (2). Let δ be data retentive, consistent, etc. Let proposition $\mathcal{K}$ be given. If $\mathcal{K}$ is not in the range of δ, then let $\Gamma_\mathcal{K}$ be an arbitrary Grove system. If $\mathcal{K}$ is in the range of δ then we define $\Gamma_\mathcal{K}$ as follows. First, define:

$D^0_\mathcal{K} = \{\emptyset\}$;

$D^{n+1}_\mathcal{K} = \{e^\frown q : \mathrm{lh}(e) = n \text{ and } \delta(e) = \mathcal{K}\}$.

Now define

$\mathcal{S}^0_\mathcal{K} = \mathcal{K}$;

$\mathcal{S}^{n+1}_\mathcal{K} = \mathcal{S}^n_\mathcal{K} \cup \bigcup_{e \in D^{n+1}_\mathcal{K}} \delta(e)$;

$\mathcal{S}^\omega_\mathcal{K} = \mathcal{N}$;

$\Gamma^\mathcal{K} = \{\mathcal{S}^\alpha_\mathcal{K} : \alpha \leq \omega\}$.

Since the $\delta(e)$'s added at each stage are mutually disjoint, one can show just as in the preceding argument that

$$\delta(\emptyset) = \Gamma^{\delta(\emptyset)}([\emptyset])$$

and

$$\delta(e^\frown q) = \Gamma^{\delta(e)}([e^\frown q]). \qquad \square$$

[15] Our argument turns heavily on the fact that two distinct data sequences e, e' of the same length generate disjoint propositions $[e], [e']$. (Martin and Osherson, 1995, Prop. 21) contains a construction of a reliable repetitive AGM method that does not require this assumption, employing a well-ordering of theories.

Kevin Kelly and Oliver Schulte
Department of Philosophy
Carnegie Mellon University, Pittsburgh

Vincent Hendricks
Department of Philosophy
University of Copenhagen

REFERENCES

Gärdenfors, P. (1988). *Knowledge In Flux: modeling the dynamics of epistemic states.* Cambridge: MIT Press.

Grove, A. (1988) "Two Modellings For Theory Change," *Journal of Philosophical Logic* 17: 157–170.

Kelly, K. (1995). *The Logic of Reliable Inquiry.* Oxford: Oxford University Press.

Kelly, K. and Glymour, C. (1990). "Theory Discovery from Data with Mixed Quantifiers," *Journal of Philosophical Logic* 19: 1–33.

Kelly, K. and Schulte, O. (1995a) "The Computable Testability of Theories Making Uncomputable Predictions," *Erkenntnis.* 43:29–66.

Kelly, K. and Schulte, O. (1995b) "Church's Thesis and Hume's Problem". *These proceedings.*

Martin, E. and Osherson, D. (1995). "Scientific discovery based on belief revision". *These proceedings.*

Osherson, D., Stob, M. and Weinstein, S (1986). *Systems That Learn.* Cambridge, Mass: MIT Press.

Popper, K. (1968). *The Logic Of Scientific Discovery.* New York: Harper.

Putnam, H. (1963). " 'Degree of Confirmation' and Inductive Logic," In *The Philosophy of Rudolph Carnap*, ed. A. Schilpp. La Salle, Ill.: Open Court.

Putnam, H. (1965). "Trial and Error Predicates and a Solution to a Problem of Mostowski," *Journal of Symbolic Logic* 30: 49–57.

VEIKKO RANTALA

EXPLANATORY TRANSLATION

Beyond Functionalism and Reductionism

1. INTRODUCTION

Many of the conflicts in the philosophy of cognitive science, such as those between the classical and connectionist paradigms and between functionalism and reductionism, have been largely discussed in the literature in terms of their logical features. However, the arguments presented have often been speculative, in the sense that they have not been based on, and they have not given rise to, any case studies. This seems to be true, in particular, about the arguments for and against reducibility of one cognitive model to another. One reason might be the fact that the case studies to be needed would often be logically extremely complicated if perfomed in detail; another the fact that they could not be worked out by means of classical notions of logic and reduction, but would rather require new and nonstandard insights.

The most controversial cases of intertheory relation in science seem to be such that their reconstruction and understanding requires an employment of logical principles in a very general sense, and, in addition, pragmatic exploration, involving hermeneutics and principles of interpretation. I try to show here how certain principles belonging to these different philosophical traditions — logic, pragmatics, and hermeneutics — which are often considered as being in conflict with each other, can merge into an approach to intertheory relations which is applicable to actual cases of theory change, and, in particular, to problems of cognitive science. Such a fusion seems to be necessary for a realistic approach; but then, somewhat unexpectedly perhaps, many philosophical conflicts about intertheory relations evaporate.

2. FUNCTIONALISM AND REDUCTIONISM

In the most general sense of the word, a view or discipline is functionalist if it says that something can be identified or characterized by means of its functional or causal role, rather than by means of its substance or stuff, or hidden essence, or the like. Therefore, the latter are not theoretically important and can be ignored. Currently, the term is most commonly used in cognitive science, but certain views within other disciplines can be counted functionalist in a broad sense. Thus, for example, in the philosophy of art, Goodman (1968) maintains the idea that something can be identified as a work of art only when it plays an appropriate symbolic role. This opposes

M.L. Dalla Chiara et al. (eds.), Logic and Scientific Methods, 399–412.

the view that a work is identified with, or reducible to, its material embodiment. In psychology, behaviorism has functionalist undertones; and in the philosophy of science, perhaps, instrumentalism.

Within cognitive science, functionalism has a variety of definitions and versions, and a variety of different applications. For example, according to Bechtel (1988, 112), "[f]unctionalism maintains that mental events are classified in terms of their causal roles", and, therefore, "...mental events can be recognized and classified independently of their physical constitution". According to Bechtel and Abrahamsen (1991, 282), the functionalist theory of mind says that "...mental states are characterized and to be understood in terms of their interactions with other mental states, not in terms of their physical embodiment...". Churchland (1992, 59) defines it as a discipline whose core idea "...is the thesis that mental states are defined in terms of their abstract causal roles within the wider information-processing system. A given mental state is characterized in terms of its abstract causal relations to environmental input, to other internal states, and to output"; and furthermore, "...functional kinds are specified by reference to their roles or relational profiles, not by reference to the material structure in which they are instantiated". Thus it seems that functionalism is a methodological rather than ontological discipline.

Since all these versions hold that mental states are not identifiable in terms of their embodiment, it follows that they can be embodied in many ways and in many stuffs. Churchland (1992, 60) and others see this feature as being the reason why functionalist have typically rejected reductionism. According to her construal, functionalists maintain that a reductive strategy could only succeed if a type of mental states were identical to a type of physical states, or rather, if there were a one-one relation between these two kinds of type. But this is not the case because there can be many embodiments. It is observed by several authors, however, that this is no reason for not having a reduction between descriptions of the two kinds available; on the contrary, multiple realizability would in fact be very much in agreement with certain views of the general notion of reduction.

The reasons for maintaining reductionist ideas are, in any case, often opposite to those motivating functionalism. In the philosophy of mind and in cognitive science, a reductionist may assume that the physical reality is fundamental, whence descriptions of the mental should or could be reduced to physics, and cognitive and psychological theories to neuroscience theories. Thus the former theories and the talk about the mental are less important; and can be ignored, even eliminated. As we saw, on the other hand, functionalists maintain that mental events can be characterized in functional terms, whence neuroscience is not that important for this purpose. Therefore, reductionism and functionalism usually represent conflicting views in cognitive science. This conflict can be roughly condensed as follows, if we speak about it in terms of two theories, say, T and T'. Let us say in what follows that reductionism assumes that the theory T' is more basic in cognitive science, whence T is to be reducible to T'. Then in some ontological or methodological sense the reducible theory T would be dispensable. Let functionalism maintain, on the other hand, that the alleged more basic theory T' is dispensable in the sense that cognition can be functionally characterized and understood by means of T alone.

This distinction between theories like T and T' can be made a little more fine-grained if we say that the theory T gives a functional and T' a structural characterization of the mind, or, rather, that T is *functional relative to* T' and T' *structural relative to* T.[1] Thus, for instance, we may assume that classical cognitive theories (symbolic theories in various forms) are functional relative to connectionist (subsymbolic) theories and to theories of neuroscience; connectionist ones in turn are functional relative to theories of neuroscience; and conversely for 'structural'. According to some views, the latter theories give more structural descriptions of mental events than do the former. The actual controversy between functionalism and reductionism usually refers to the relationship between these groups of theories. Roughly speaking, of these pairs of theories, the reductionist prefers the more structural one to the more functional one, and would like to see the latter as reduced to the former; the functionalist prefers the more functional one, and would like to dispense with the more structural one as far as cognition is to be characterized.

3. PROBLEMS WITH REDUCTION

Reduction has been extensively discussed in the philosophy of science and in science for various ontological and methodological reasons. For the ontological reductionist, a given phenomenon may be supported by a more fundamental or ontologically simpler reality, wherefore the descriptions of the former are reducible to descriptions of the latter. The methodological reductionist maintains that scientific research (in some particular area) is more feasible if it can be reduced to research which is more appropriate for the study of that fundamental reality. The notion has also been important in debates concerning scientific change, since it used to be a common view in the philosophy of science that one indication of scientific progress is that theories are reducible to their successors. It was even common to think that typically what was sometimes, unhappily, called "mature science" is progressive in this sense. Reducibility means that the successor theories are more comprehensive than their predecessors. The notion of reduction, in all these applications, was generally thought of as being closely related to explanation, in particular, to deductive-nomological explanation.

Nagel's (1961) (and others') classical notion of reduction has been referred to in cognitive science almost exclusively. It is well known, however, that the

[1]See Churchland, 1992; Bechtel and Abrahamsen, 1991.

relevance of this and some other, closely related notions has been heavily attacked in the philosophy of science. The criticism started by Kuhn and others in the sixties or so has purported to show that the roles which reduction and deductive explanation, in their standard senses, play in actual science is less important than philosophers used to think. It has been argued, for example, that there in fact exist no proper reductions in many important cases of scientific change where reductions were claimed to exist; or the notion of explanation which reduction involves is not appropriate for the human sciences. Furthermore, in many actual changes in physics auxiliary hypotheses that would be needed to establish reduction relations are counterfactual. In such a case the derivation needed to establish a reduction relation would be of no explanatory import in the purported sense, or it would amount to a reduction only in an approximate or limiting sense.

Because of this criticism, several new models of intertheory relation have been subsequently suggested in the literature. They have not been very much applied in cognitive science, however. Therefore, I shall next make a lengthy digression from cognitive science in order to discuss a general, nonstandard way to investigate conceptual change, and later on apply it to intertheory relations and, in particular, to cognitive science.

4. EXPLANATORY TRANSLATION AND CORRESPONDENCE

We start with the notion of translation. The standard view holds that a translation should preserve meaning. It is well known, however, that this ideal is in many cases unattainable and useless in practice, and must therefore be given up. In such cases, it may be replaced by the condition that there be syntactic and semantic transformations which cooperate in a systematic way. Versions of this generalized, nonstandard notion of translation occur in mathematical logic and philosophy of science, where they are known as 'interpretation' and 'reduction', or 'correspondence', respectively. I have argued elsewhere that such a nonstandard translation is implicitly employed in everyday discourse, and, furthermore, that Kuhn (1970) may also have had a similar notion implicitly available. Later on, Kuhn favors the standard notion, however.

If we study how nonstandard translation functions in everyday discourse, we may draw some conclusions for intertheory relations also; it enables us to see how their logical aspects are associated with contextual, paradigmatic, and even hermeneutic features. Some of these nonlogical elements can be made explicit when such relations are examined, whereas others just come from background principles. This association will become obvious from some actual and imaginary cases of conceptual change where communication breakdowns occur, and where the understanding of the other's position requires nontraditional kinds of explanation. I can only briefly touch on this issue here.[2]

How are syntactic and semantic transformations to cooperate? Let us first consider ordinary discourse, and single utterances with propositional contents. It is evident that in such a case a speaker intends to represent by his utterance a situation (or a possible world in a more comprehensive sense, or possibly a class or kind of situations or worlds) which satisfies the utterance. I shall call such a situation the *speaker's situation.*[3] A *hearer's situation*, on the other hand, is a situation which satisfies the hearer's translation and to which he intentionally refers by the translation, and it arises from his interpretation of the speaker's utterance. It is clear that when trying to understand the speaker's utterance the hearer must consider the speaker's position from his own presumptions. Therefore, it seems natural to assume that in order to understand the speaker's position, the hearer must be able to transform his situations (or some of them) into the speaker's, so that there will obtain a more or less specified *correlation* between them. The nature of the correlation is context-dependent, and it must be pragmatically warranted in order to help, together with the syntactic *translation*, the hearer to understand and explain the speaker's utterance. If the two transformations can be conjoined so as to be of *explanatory* import, I shall use the term explanatory translation. If the hearer considers the utterance false or otherwise inappropriate, his translation may correct the utterance. Then I speak of a *corrective translation.*

This is a generalization of the standard notion of translation, for we have a standard translation if the respective correlation is an identity mapping, i.e., the speaker's and hearer's situations coincide.[4] What I have said above about local translation can be generalized for more global cases, e.g., for texts and theories. Then I shall often speak of worlds and models, rather than of situations. The idea of semantic correlation has explicit predecessors in mathematical logic and in the philosophy of science, in connection with such notions as interpretation and reduction of theories,[5] and it is commonly, but only implicitly, present when expressions of natural languages or literary texts are translated with no chance to save meanings. When literary texts are translated, no semantic correlations are explicitly available, but the same effect

[2]I have elsewhere (Rantala, 1995, 199+) provided a detailed investigatation of everyday discourse and of some of Kuhn's well-known ideas, and argued that the notion of nonstandard translation I have in mind occurs implicitly in them.

[3]This notion is somewhat problematic, but I shall not discuss the problems here (see the references in Note 1). Because of intentionality, an utterance may be satisfied in a situation by the speaker's lights, even though it would not be satisfied in it by the hearer's, or according to general linguistic rules.

[4]I shall not discuss here the question of whether they can ever really coincide and what this would mean when no logical or mathematical framework is available. Let us just assume that it is more or less understandable what it means to say so.

[5]See, e.g., Enderton (1972), Adams (1955), Balzer, Moulines and Sneed (1987). For applications to actual cases of theory change, see Pearce and Rantala (1984ab), Rantala and Vadén (1995). For an extensive discussion, see Pearce (1987).

is often obtained by translators' explicit *commentaries*, which explain how meanings are changed when expressions of the target texts are ineffable. This is often the case, for example, when classical scholarly texts are translated.

In a global case, the respective correlation is a mapping from the hearer's worlds to those of the speaker, and the translation maps into the hearer's language at least the expressions of which it is meaningful to say that they are or are not satisfied at the speaker's worlds.[6] Now let F be a correlation and let I be a translation. We assumed that an utterance is satisfied at the speaker's worlds and its translation at the hearer's worlds. Therefore, we have to require that for all expressions φ in the domain of I, and for all worlds w in the domain of F,

(C) φ is satisfied at $F(w)$ if and only if $I(\varphi)$ is satisfied at w.

In addition, the two mappings have to satisfy logical and pragmatic conditions of adequacy, the nature of which depends on the case in question and on linguistic, aesthetic, and contextual details. They must yield sufficiently informative conceptual and ontological relationships between the two languages. The only explicitly stated condition, (C), is logical.[7]

Let us next consider two theories, T and T', which will be called a *secondary theory* and *primary theory*, respectively.[8] If T is not formal, we may not be able to speak of all sentences of the language of T in any exact way, but we may assume that there is a more exactly delineated subcollection S in which the principles and laws of T are included, and that only the sentences in S need to be translated into sentences of T'.[9] Let K' be some collection of possible worlds, determined by appropriate, logical or pragmatic, considerations, at which the principles and laws of T' are true. Then, let I map the sentences in S to sentences of the language of T', and let F map K' onto a collection K of possible worlds (likewise logically or pragmatically defined) at which the principles and laws of T are true. If the condition (C) is applied so as to hold for all sentences in S and all worlds in K', then the pair $\langle F, I\rangle$ is called a *correspondence* of T to T'.[10]

---

[6] How the domain of the translation is characterized depends on the case. It cannot be made explicit in nonlogical contexts.

[7] The definition and its applications also involve, explicitly or implicitly, what I call 'principles of interpretation'. I cannot discuss them here, but see my (1995), (199+). When narratives and other literary texts are explored, their holistic nature may require that one has to consider the whole text, in some way or other, before being able to adequately establish the mappings and the other details involved (see, e.g., Rantala (1987), Oesch and Rantala (199+)).

[8] These are terms used, e.g., by Nagel (1961) and Glymour (1970).

[9] We shall only consider scientific laws and other principles which are sentences, true at various worlds or situations.

[10] See, e.g., the works by Pearce and Rantala referred to above. A general notion of correspondence is logically weaker than reduction but stronger than supervenience. For the latter notion, see, e.g., Kim (1990). Sometimes it is useful to require that K and K' be characterizable or definable in the languages of T and T', respectively. This requirement may, of course, be vague if the theories are nonformal.

This is a global version of the nonstandard notion of translation. Notice that the elements of K and K' play the role of what we have called a speaker's or hearer's situations (or worlds), respectively.[11] To discuss correspondence in explicit terms, I shall next assume that the two theories are formal, represented in appropriate logics. Then correspondence can be defined model-theoretically, and instead of talking about situations or possible worlds we shall speak of models. It is easy to see that the following definition is in agreement with the above notions of global translation and correspondence.[12] So, let us assume that T and T' are formalized in some logics, L and L', respectively. That is, assume that their axioms are represented in these logics as sentences θ and θ', say. We assume, furthermore, that the classes K and K' are definable in L and L', respectively. Then the pair $\langle F, I\rangle$ is a *correspondence* of T to T' if the following condition holds for all models $\mathcal{M}$ in K' and all sentences φ of the language of T:[13]

(C*) $F(\mathcal{M}) \models_L \varphi$ if and only if $\mathcal{M} \models_{L'} I(\varphi)$.

5. COUNTERFACTUAL CORRESPONDENCE

There are cases of local, explanatory translation where a hearer considers a speaker's utterance false, and which involve contrary-to-fact assumptions leading to counterfactual conditionals in attempts to explain the speaker's position.[14] Such assumptions are often important for global translation as well; in particular, for the correspondence relation in some cases of actual theories. Consider a correspondence $\langle F, I\rangle$ of T to T', as above. If K' consists of possible worlds which are regarded counterfactual, i.e., in an appropriate sense nonstandard or deviating, by the scientific community holding the theory T', then the correspondence relation is called a *counterfactual correspondence*. If K' can be defined in L', as we assumed in the model-theoretic version, then the community considers some defining sentences false. They are false at those worlds described by T' which are regarded actual or intended. The actual or intended models of T' are not always definable. However, as can be seen from case studies, often it is enough to have pragmatic criteria of why some conditions are to be considered counterfactual and then express them, without any need to define the community-relative notion of actual model.

The most important, and the most controversial, special case of counterfactual correspondence is *limiting case correspondence*. Then the possible worlds

[11]If the collections of sentences and worlds indicated in the above definition cannot be exactly defined, so that a correspondence relation cannot be defined in any precise sense, we may still talk about a correspondence even if we only have a correspondence sketch.

[12]Since a more detailed formal definition is presented several times in earlier work, I shall not extensively comment it here.

[13]The translation could sometimes be defined as a partial mapping; then its domain S must comprise all the axioms of T.

[14]There are some examples in my (1995) and (199+). We cannot take them up here.

or models in K' are *almost identical* with their images in K and the translations of the laws of T are almost identical with the laws of T'. The idea of limiting case correspondence can be made exact for formal theories.[15] Then the exact meaning of 'almost identical' is 'infinitesimally close', in the sense of nonstandard analysis.

6. COUNTERFACTUAL EXPLANATION

A correlation mapping may be of explanatory import if it helps one to understand the relation of a hearer's situations to those of a speaker. This idea can be generalized for our global notions of translation and correspondence. For instance, it may help one to explain the position of the scientific community holding a secondary theory from a point of view of the community holding a primary theory. The notion of counterfactual correspondence is related to many actual, revolutionary scientific changes, and yet its explanatory function is not very clearly understood. I shall next make an overview of what seems to me to be its explanatory role.[16]

If $\langle F, I\rangle$ is (a model-theoretic version of) a correspondence of a theory T to a theory T', as indicated above, and if the logic L' is complete and the sentences $\theta', \sigma_0, \ldots, \sigma_n$ together define K' in L', then the following holds:[17]

$$\theta', \sigma_0, \ldots, \sigma_n \vdash_{L'} I(\theta).$$

If the correspondence is only a correspondence sketch, it may in some cases yield an analogous informal inference.

Despite the formal similarity, this inference does not yield a deductive-nomological explanation if one of the special assumptions is contrary-to-fact. Of what explanatory import is then the correspondence relation $\langle F, I\rangle$? Let us consider first Glymour's (1970) suggestion that in a counterfactual case one has to ask (instead of a why-question): 'Under what conditions would the secondary theory hold?'.[18] Then the theory is explained by showing under what conditions it would hold and contrasting these conditions with the conditions which actually obtain. Assume that the primary theory together with some special assumptions, some of which, say σ, is counterfactual, and together with definitions connecting the languages of the two theories, and possibly with limiting procedures, entails the secondary theory:

(D) $\theta', \sigma, \ldots \vdash \theta.$

By means of (D), the above explanation-seeking question can be answered, since it yields the truth of the following conditional:

(S) If it were the case that σ, then it would be the case that θ.

[15]It cannot be done here, but see Pearce and Rantala (1984a).

[16]For a more detailed, and less simplified, discussion, see my (1989) and (199+).

[17]If L' is not complete, either one has to be satisfied with logical consequence instead of deduction or to try to find another appropriate logic which is complete.

[18]For similar suggestions, see Rott (1987).

Furthermore, one should show that without the counterfactual assumption, the secondary theory is not in general entailed by the primary one.

However, Glymour's suggestion is not quite justified, for two reasons. First, (D) must be replaced by the following inference, as we saw:

(D*) $\theta', \sigma, \sigma_1, \ldots, \sigma_n \vdash I(\theta)$,

and, hence, (S) by the conditional

(S*) If it were the case that σ, then it would be the case that $I(\theta)$,

which provides an answer to the question 'Under what conditions would $I(\theta)$ hold?' Second, the truth (at a given world) of a conditional, such as (S*), is not always entailed by a respective argument, such as (D*), as is now well known.[19] Whether (D*) establishes the truth of (S*) depends on additional pragmatic and paradigmatic constraints.

I have elsewhere called *counterfactual explanation* a kind of explanation which may emerge when we correct and improve Glymour's model in the fashion I indicated above and use an appropriate intensional semantics of conditionals. It is a complex process, comprising, at least, a construction of a counterfactual correspondence, derivation of an argument of the form (D*), and showing that (D*) makes a conditional of the form (S*) true (at a world chosen).[20] It is clear in any case that an explanation of this kind can hardly be objective, since much depends on pragmatic constraints. There may even be individual disagreement, concerning, e.g., the question of what counterfactual conditions can be considered possible. For example, when the limiting case correspondence of classical particle mechanics to relativistic particle mechanics is studied, one has to consider the question whether it is in some sense possible that the velocity of light is infinite. It is well known that there are philosophers and physicists who hold that this assumption is so much contrary-to-fact that it cannot even be considered possible and others who consider it possible. On the former account, the correspondence relation is of no explanatory value.

Many of the most important intertheory relations involve counterfactuals. In these cases, familiar notions of explanation do not help us understand the conceptual changes involved, as we saw above; and, therefore, the model of counterfactual correspondence was introduced. In the model, logical investigations are, so to say, embedded in a pragmatic environment in an explicit way, and in such an environment problems of *interpretation* arise. Furthermore, explanation-seeking questions to which the counterfactual model may provide answers are different from the ones with which we are familiar. Therefore, whether we can speak of explanation and distinguish it from *understanding* is even more problematic here than with traditional cases of explanation. In the counterfactual model, syntactic, semantic, and pragmatic elements are closely

[19] See, e.g., Lewis (1973), Nute (1984).

[20] See Rantala (1989), (199+). If nonformal theories are at issue, these steps are not explicitly constructible, but must be replaced by a sketchier representation.

interconnected. Especially from its applications to actual theories its holistic and hermeneutic nature can be seen. Applications are processes which gradually increase one's understanding of intertheory relations in question, and involve logical and pragmatic steps, as already indicated. Because of their holistic character, there is some complexity in their logical and pragmatic features, and, as I have indicated elsewhere, the model is guided by certain principles of interpretation, which function in explanatory translations in general.[21]

7. SYMBOLIC VS. SUBSYMBOLIC THEORIES

After this digression, let us turn back to cognitive science. Consider first the question of how the symbolic (classical) and subsymbolic (connectionist) theories are related. It is suggested by Smolensky (1988) that this relation, which is an example of a relation between macrolevel and microlevel theories of cognition, is like the relation of classical mechanics and quantum mechanics in that the latter provides a refinement of the former, and the former is a limiting case of the latter. According to Smolensky, the microlevel is not an implementation of the macrolevel, nor does the microtheory eliminate the macrotheory. So it is obvious that Smolensky argues against reductionism and functionalism here, and considers both levels as playing an important cognitive role. Furthermore, the microtheory delineates the situations where symbolic approximations are valid and explains why they are valid.

Smolensky does not explicitly refer to similar views in the philosophy of science, and does not provide any construction of the limiting case correspondence he claims there is. And it seems obvious that there cannot exist such a comprehensive correspondence relation between the two paradigms as he claims. What can be done is to construct theories (or 'models') related to symbolic and subsymbolic representations, respectively, and point out that the former is in a limiting case correspondence to the latter. This is actually done in Rantala and Vadén (1994) and Vadén (1995), where it is also shown that the relationship between such theories (i.e., between descriptions of certain Turing and network representations) is much more involved, both logically and pragmatically, than Smolensky — and many others arguing for the existence of such relationships in different contexts — seem to expect. Its most important consequences for our purposes here are that, first, the correspondence relation takes care of the radical meaning variance involved, and, second, due to counterfactual assumptions involved, there cannot exist any reduction relation (that is not approximate) between the theories in the Nagelian sense — and hence no D-N explanation is forthcoming either. Rather, there can be what was in Section 6 called a counterfactual explana-

[21] For their logical complexity, see, e.g., Pearce and Rantala (1984a), and for pragmatic and interpretive elements, my (1989), (1995a), (199+).

tion of a symbolic representation by means of a symbolic representation. This is the case *if* appropriate pragmatic conditions are accepted, as we saw.

8. SUBSYMBOLIC VS. NEURAL THEORIES

The conceptual relationship between connectionist and neural theories is not very well understood; and most of what has been said of it in the literature is rather uncertain.[22] Smolensky (1988) suggests that this relation is in many respects similar to the relation of symbolic and connectionist theories. Let us study his suggestion more closely by first considering three features of the latter relation. Smolensky argues that:

(i) If a symbolic system (von Neumann computer) and a connectionist system are described at the conceptual level (and not at any lower levels), the descriptions are good approximations to each other.

As we saw, Smolensky also suggested that:

(ii) The relation between symbolic and subsymbolic theories (models) is that of limiting case correspondence. Hence, when certain idealizing assumptions are made, the latter approaches the former.

However, this proposal is very vague. On the other hand, it is pointed out in Rantala and Vadén (1995) that:

(iii) A certain conceptual level description of a Turing machine is in a limiting case correspondence to a similar description of a connectionist network.

Here the limiting case correspondence in question is a model-theoretic reconstruction in the sense of Section 5, above. It seems that the case study mentioned in (iii) to some extent confirms what is suggested in (i) and (ii).

Now consider the relation of connectionist and neural theories. According to Smolensky, the general principles of computation at the subconceptual level of description must apply to computation in the brain:

(iv) If a connectionist system and the brain are described at the *subconceptual level*,[23] the descriptions are good approximations to each other.

As far as I know, he does not propose, however, that there might be a limiting case correspondence between connectionist and neural theories of cognition.[24] The reason is that too little is known of the latter. However, it is known, and also mentioned by Smolensky, that connectionist models lack important features (e.g., spatial location) present in neural models, and include idealizations if compared with neural models. This and (iv) might suggest — and they even make it sound plausible — that the following will turn out to be true in the near future:

[22]This is, in fact, very much true about the whole discussion of functionalism and reductionism. What I shall say below is tentative as well.

[23]This is a "higher" level than the neural level.

[24]However, Smolenski says that subsymbolic models should be favored in the coming research that are more likely to be reducible to the neural level.

(v) A counterfactual (possibly a limiting case) correspondence of some subsymbolic theory to some neural theory can be constructed.

This would be a relation between subconceptual level descriptions; and it would again involve radical meaning variance. If (v) held, there would possibly be a counterfactual explanation of a subsymbolic theory by means of a neural one.

9. CONCLUSION

Given the above relationships, let us draw some conclusions concerning the debate between functionalists and reductionists. For instance, what are the implications of there being a counterfactual correspondence of a symbolic representation (or a description thereof) to a subsymbolic one? Analogous conclusions can be drawn if a similar relation between a subsymbolic and neural theory can be found — at least in so far as the former is not considered itself as a neural theory.

As we saw, the relation between the symbolic and subsymbolic representations is not that of reduction in the classical sense, since, first, one of the special assumptions needed is counterfactual,[25] and, second, not the symbolic representation itself can be inferred from the subsymbolic one plus special assumptions, but rather its translation. It follows that *if* this kind of relationship can be generalized, and *if* psychological laws are of a computational kind in the classical sense, these laws are not reducible in the standard sense to subsymbolic theories nor explainable in the classical D-N sense. Then psychological laws enjoy some degree of autonomy, and the hard-core reductionist is shown to be wrong, and the functionalist, at least in part, right. It is pointed out by Richardson (1979) that there is no inconsistency between computational models and classical reductionism.[26] But if what we have done can be generalized, it follows that no important (or nontrivial), classical reductions between symbolic and connectionist models in fact exist, and this may also make it implausible that there would be any important reductions between computational and neurological models either. Even though the counterfactual correspondence explored in Rantala and Vadén (1995) is only local, it is a relation connecting symbolic and subsymbolic representations which are of typical kinds. This typicality seems to make the idea that there exist no such reductions even more plausible.

On the other hand, if there only is a counterfactual correspondence, and hence only a counterfactual explanation, available, whether or not computational psychology is considered completely autonomous, seems to depend on pragmatic and paradigmatic elements, not merely on any objective or logical

[25] It says that the number of units is infinite.

[26] He shows, e.g., that the plurality of physical realizations is not inconsistent with reductionism in the classical sense. The nature of the correlation F makes this theoretically obvious in any case.

features of the relationship between the two kinds of theory. The classicist, maintaining that the symbolic representation in question (when generalized) yields correct psychological characterizations, need not worry about there being a limiting case correspondence, since it is not a reduction in the classical sense and does not yield an elimination. At least it does not yield any compelling, logical reasons to give up the belief in autonomy. But the connectionist, if believing in the subsymbolic nature of mental events, can as well argue that the classicist is wrong since the symbolic representation is not reducible. He could also claim (by using the notion of counterfactual explanation) that subsymbolic theories are good in the sense that they show under what conditions the classicist would be right (or 'almost' right) and why the classicist is in fact wrong.

On the other hand, our results makes the cognitive import of functionalism questionable. If a functional theory can be understood, even in a roundabout way, and counterfactually explained by means of a structural theory, it is kind of conceptually dependent on the latter. Thus it seems that Smolensky is correct as he argues that the different theories (he is talking here about classical and connectionist models) should intimately collaborate to provide an understanding of cognition. In view of our results — more or less hypothetical, though — it does not seem to make much sense to one-sidedly emphasize functionalist or reductionist attitudes.

Department of Mathematical Sciences Philosophy
University of Tampere

REFERENCES

Adams, E.W., *Axiomatic Foundations of Rigid Body Mechanics*, Unpublished diss., Stanford University, 1955.

Balzer, W., Moulines, C.U., and Sneed, J.D., *An Architectonic for Science*, D. Reidel Publishing Company, Dordrecht, 1987.

Bechtel, W., *Philosophy of Mind*, Lawrence Erlbaum Associates, Hillsdale, NJ, Hove, London, 1988.

Bechtel, W. and Abrahamsen, A., *Connectionism and the Mind*, Blackwell, Oxford, Cambridge, Mass., 1991.

Churchland, P.S., 'Reductionism and Antireductionism in Functionalist Theories of Mind', in B. Beakley and P. Ludlow (eds.), *The Philosophy of Mind*, MIT Press, Cambridge, Mass., London, 1992, pp. 59-67.

Glymour, C., 'On Some Patterns of Reduction', *Philosophy of Science* 7 (1970), 340-353.

Goodman, N., *Languages of Art*, Bobbs-Merrill, Indianapolis, 1968.

Glymour, C., 'On Some Patterns of Reduction', *Philosophy of Science* 37 (1970), 340-353.

Enderton, H., *A Mathematical Introduction to Logic*, Academic Press, New York, London, 1972.

Kim, J., 'Supervenience As a Philosophical Concept', *Metaphilosophy* 21 (1990), 1-27.

Kuhn, T.S., *The Structure of Scientific Revolutions* (2nd ed.), University of Chicago Press, Chicago, 1970.

Lewis, D., *Counterfactuals.*, Blackwell, Oxford, 1973.

Nagel, E., *The Structure of Science*, Routledge & Kegan Paul, London, 1961.

Nute, D., 'Conditional Logic', in D. Gabbay and F. Guenthner (eds.), *Handbook of Philosophical Logic*, Vol. II, D. Reidel Publishing Co., Dordrecht, 1984, pp. 387-439.
Oesch, E. and Rantala, V., 'Interpretation and Aesthetic Experience', forthcoming, 199+.
Pearce, D., *Roads to Commensurability*, D. Reidel Publishing Co., Dordrecht, 1987.
Pearce, D. and Rantala, V., 'A Logical Study of the Correspondence Relation', *Journal of Philosophical Logic* 13 (1984a), 47-84.
Pearce, D. and Rantala, V., 'Limiting-Case Correspondence between Physical Theories', in W. Balzer, D. Pearce, & H-J. Schmidt (eds.), *Reduction in Science*, D. Reidel Publishing Co., Dordrecht, 1984b, pp.153-185.
Rantala, V.,'Interpreting Narratives: Some Logical Questions', in I.Ruzsa and A.Szabolcsi (eds.), *Logic and Language*, Akadémiai Kiado, Budapest, 1987, pp. 121-132.
Rantala, V., 'Counterfactual Reduction', in K. Gavroglu, Y. Goudaroulis, and P. Nicolacopoulos (eds), *Imre Lakatos and Theories of Scientific Change*, Kluwer Academic Publishers, Dordrecht, 1989, pp. 347-360.
Rantala, V., 'Translation and Scientific Change', in W.E. Herfel, W. Krajewski, I. Niiniluoto and R. Wójcicki (eds.), *Theories and Models in Scientific Processes* (Poznan Studies in the Philosophy of Sciences and the Humanities, vol. 44), Rodopi, Amsterdam, Atlanta, GA, 1995, pp. 249-268.
Rantala, V., *Explanatory Translation. Beyond the Kuhnian Model of Conceptual Change*, forthcoming, 199+.
Rantala, V. and Vadén, T., 'Idealization in Cognitive Science. A Study in Counterfactual Correspondence', in M. Kuokkanen (ed.), *Idealization VII: Structuralism, Idealization and Approximation* (Poznan Studies in the Philosophy of the Sciences and the Humanities, Vol. 42), Rodopi, Amsterdam, Atlanta, GA, 1994, pp. 179-198.
Richardson, R.C., 'Functionalism and Reductionism', *Philosophy of Science* 46 (1979), 533-558.
Rott, H., 'On Relations between Successive Theories', in *Abstracts of the LMPS '87*, Vol. 4, Part 2, Moscow, 1987, pp. 123-127.
Smolensky, P., 'On the Proper Treatment of Connectionism' (with open peer commentary and author's response), *Behavioral and Brain Sciences* 11 (1988), 1-74.
Vadén, T., 'From Limitivism to Correspondence: The Symbolic-Subsymbolic Relation', in L. Niklasson and M. Bodén (eds.) *Current Trends in Connectionism*, Lawrence Erlbaum, New Jersey, 1995.

PAUL THAGARD AND CAMERON SHELLEY*

ABDUCTIVE REASONING: LOGIC, VISUAL THINKING, AND COHERENCE

ABSTRACT.

This paper discusses abductive reasoning—that is, reasoning in which explanatory hypotheses are formed and evaluated. First, it criticizes two recent formal logical models of abduction. An adequate formalization would have to take into account the following aspects of abduction: explanation is not deduction; hypotheses are layered; abduction is sometimes creative; hypotheses may be revolutionary; completeness is elusive; simplicity is complex; and abductive reasoning may be visual and non-sentential. Second, in order to illustrate visual aspects of hypothesis formation, the paper describes recent work on visual inference in archaeology. Third, in connection with the evaluation of explanatory hypotheses, the paper describes recent results on the computation of coherence.

There has been an explosion of recent work in artificial intelligence that recognizes the importance of abductive reasoning—that is, reasoning in which explanatory hypotheses are formed and evaluated. Many important kinds of intellectual tasks, including medical diagnosis, fault diagnosis, scientific discovery, legal reasoning, and natural language understanding have been characterized as abduction. Appropriately, attempts have been made to achieve a more exact understanding of abductive reasoning by developing formal models that can be used to analyze the computational properties of abductive reasoning and its relation to other kinds of inference. Bylander *et al.* [4] have used their formal model to analyze the computational complexity of abduction and show that in general it is NP-hard. Konolige [16] has used a similar formal model to derive results concerning the relation between abductive reasoning and Reiter's model of diagnosis [22]. While these formal results are interesting and useful, it would be unfortunate if researchers were to conclude that the analyses of Konolige, Bylander *et al.* have provided a precise understanding of abductive reasoning. We shall discuss numerous important aspects of inference to explanatory hypotheses that are not captured by one or both of the formalisms that these authors have proposed. In particular, we show how these models do not adequately capture abductive discovery using representations that are pictorial, and we argue that abductive evaluation should be conceived in terms of coherence rather than deduction.

1. THE MODELS

Konolige's model of abduction is contained in two definitions:

*We are grateful to John Josephson and Don Roberts for comments on an earlier draft. Thanks to John Ching and Andrew Wong for ideas about graphical representations. This research was supported by the Natural Sciences and Engineering Research Council of Canada.

M.L. Dalla Chiara et al. (eds.), Logic and Scientific Methods, 413–427.

Definition 1.1 *Let $\mathcal{L}$ be a first order language. A simple causal theory is a tuple $\langle C, E, \Sigma \rangle$ where:*

1. *C, a set of sentences of $\mathcal{L}$, are the causes;*
2. *E, a set of sentences of $\mathcal{L}$, are the effects;*
3. *Σ, a set of sentences of $\mathcal{L}$, is the domain theory.*

Definition 1.2 *Let $\langle C, E, \Sigma \rangle$ be a simple causal theory. An explanation of a set of observations $O \subseteq E$ is a finite set $A \subset C$ such that:*

1. *A is consistent with Σ;*
2. *$\Sigma \cup A \vdash O$;*
3. *A is subset-minimal over sets satisfying the first two conditions, i.e., there is no proper subset of A consistent with Σ that implies O.*

Bylander and his colleagues offer a similar definition:

Definition 1.3 *An abduction problem is a tuple $\langle D_{all}, H_{all}, e, pl \rangle$ where:*

1. *D_{all} is a finite set of all the data to be explained;*
2. *H_{all} is a finite set of all the individual hypotheses;*
3. *e is a map from all subsets of H_{all} to subsets of D_{all};*
4. *pl is a map from subsets of H_{all} to a partially ordered set representing the plausibility of various hypotheses.*

A set of hypotheses H is an *explanation* if it is complete and parsimonious, *i.e.*, if $e(H) = D_{all}$, and no proper subset of H explains all the data that H does.

The similarities between these two sets of definitions are clear. Konolige's [16] causes correspond to Bylander *et al.* [4] hypotheses; Konolige's effects correspond to Bylander *et al.*'s data. Konolige's requirement that a set of causes be subset-minimal is the same as Bylander *et al.*'s requirement that a set of hypotheses be parsimonious. (Peng and Reggia [21] call the same property irredundancy.) Bylander *et al.* [4] go beyond Konolige in requiring an explanation to explain all the data and in adding a plausibility ordering on hypotheses. On the other hand, their definition is more general than Konolige's in not restricting data and hypotheses to sentences.

These definitions usefully capture much of what is central in abductive reasoning, particularly the goal of assembling a set of hypotheses (causes) that provide good explanations of the data (effects). But each of them oversimplifies abductive reasoning in several important respects.

2. THE LIMITATIONS

2.1. *Explanation is not deduction*

First, let us consider the nature of explanation. The analysis of Bylander *et al.* contains the unanalyzed function e that specifies which hypotheses explain which data. Konolige's account has the apparent advantage that it spells out this relation by using the notion of logical consequence as represented by the turnstile "⊢". Hypotheses (causes) explain data (effects) if the latter can be deduced from the former and the domain theory. Konolige is thus assuming a deductive account of explanation that has been the subject of several decades of critical discussion in the philosophy of science.

In 1948, Hempel and Oppenheim proposed what has come to be known as the *deductive-nomological* model of explanation [12]. On this model, an explanation is a deductive argument in which a sentence representing a phenomenon to be explained is derived from a set of sentences that describe particular facts and general laws (*nomos* is Greek for law). This model provides a good approximation for many explanations, particularly in mathematical physics. But it is evident that the deductive model fails to provide either necessary or sufficient conditions for explanation. See [15] for a comprehensive review of several decades of philosophical discussions of the nature of explanation, and [18] for application to artificial intelligence.

First, there are many explanations in science and everyday life that do not conform to the deductive model. Hempel himself discussed at length statistical explanations in which what is explained follows only probabilistically, not deductively, from the laws and other sentences that do the explaining. Many critics have argued that explanations in such fields as history and evolutionary biology rarely have a deductive form. In both philosophy and AI, researchers have proposed that many explanations can be understood in terms of applications of schemas that fit a phenomenon into a pattern without producing a deductive argument. (For a review of different approaches to explanation in these fields, see [27, 28].) For example, a Darwinian explanation of how a species evolved by natural selection applies a general pattern that cites biological mechanisms and historical facts to suggest how an adaptation might have come about. But the historical record is too sparse and biological principles are too qualitative and imprecise for deductive derivation. Thus Konolige's use of deduction in his characterization of abduction arbitrarily excludes many domains in which hypotheses are formed and evaluated but do not provide deductive explanations.

Second, the deductive model of explanation does not even provide sufficient conditions for explanation, since there are examples that conform to the model but do not appear to constitute explanations. For example, we can deduce the height of a flagpole from information about its shadow along with trigonometry and laws of optics. But it seems odd to say that the length of a flagpole's shadow explains the flagpole's height. Konolige's subset-minimality requirement serves to rule out some of the cases of irrelevance that philosophers have discussed, for example the explanation that a man is not pregnant because he takes birth control pills. But other examples such as the flagpole show that some additional notion of causal relevance is crucial to many kinds of explanation, and there is little hope of capturing this notion using logic alone. Contrast Pearl's [20] work on Bayesian networks and Peng and Reggia's [21] model of abduction which employ ineliminably intuitive notions of causality. A general model of abduction requires an account of explanation that is richer than deduction.

2.2. *Hypotheses are layered*

Bylander *et al.* assume that abduction can be characterized by distinguishing sets of data from sets of hypotheses, with explanation consisting of a mapping from the latter to the former. This characterization is adequate for straightforward kinds of medical diagnosis involving a given set of symptoms (data) that can be explained by a given set of diseases. But it neglects the complexity of other kinds of abductive tasks in which the organization of hypotheses and data is more complex. In particular, it rules out a large set of cases where hypotheses explain other hypotheses and selection of the best overall explanation depends on taking these relations into account. For example, in legal inference the availability of a possible motive goes a long way to increasing the chance for a murder conviction. The data are the various pieces of physical evidence, *e.g.*, that the suspects' fingerprints were found on the murder weapon. The hypothesis is that the suspect is the murderer, and a higher level hypothesis might be that the suspect hated the victim because of some previous altercation. The plausibility of the lower-level hypothesis comes not only from what it explains, but also from it itself being explained. This kind of hierarchical explanation in which hypotheses explain other hypotheses that explain data is also found in science and medicine; it has been discussed both in the context of Bayesian models of probabilistic belief revision [20] and in the context of explanatory coherence accounts of scientific inference [26, 28].

Just as Bylander *et al.* assume that only data are explained, so Konolige assumes that only observed effects are explained by derivation from causes. But as both Bayesian and explanatory coherence analyses allow, causes are often themselves effects and assessment of overall acceptability of explanatory hypotheses must take this into account.

2.3. *Abduction is sometimes creative*

The definitions of Konolige and Bylander *et al.* have the unfortunate implication that abduction is always a matter of selection from a known set of hypotheses or causes. Creative abduction often involves the construction of novel hypotheses involving newly formed concepts such as natural selection or AIDS. For Charles Peirce, who coined the term "abduction" a century ago, the introduction of unifying conceptions was an important part of abduction [11, 25], and it would be unfortunate if our understanding of abduction were limited to more mundane cases where hypotheses are simply assembled. Abduction does not occur in the context of a fixed language, since the formation of new hypotheses often goes hand in hand with the development of new theoretical terms such as "atom," "electron," "quark," "gene," "neuron" and "AIDS."

2.4. *Hypotheses may be revolutionary*

The first clause of Konolige's second definition requires that a hypothesized cause A be consistent with a domain theory. While this requirement may be acceptable for mundane applications, it will not do for interesting cases of belief revision where the introduction of new hypotheses leads to rejection of previously held theories. For example, when Darwin proposed the hypothesis that species evolve as the result of natural selection, he was contradicting the dominant view that species were fixed by divine creation. Konolige's consistency requirement would prevent Darwin's hypothesis from ever being considered as part of an abductive explanation that would eventually supplant the accepted biological/theological domain theory. We can not simply delete a belief and then replace it with one inconsistent with it, because until the new belief comes in competition with the old one, there is no reason to delete the old one. Belief revision requires a complex balancing of a large number of beliefs and kinds of evidence. Algorithms for performing such balancing are available, such as Bayesian belief revision and connectionist judgments of explanatory coherence. Bylander *et al.* do not have a consistency condition like Konolige's.

2.5. *Completeness is elusive*

Bylander *et al.* do not count a set of hypotheses as an explanation unless it is complete, *i.e.*, explains all the data. They admit that there might be "partial" explanations, but want completeness as a requirement because of the goal that all the data should be explained. This is a laudable goal, but does not justify building completeness into the definition of an explanation, since the goal is so rarely accomplished in realistic situations. From medicine to science, it is not typically the case that everything can be explained even by the best of theories. Even the greatest scientific theories such as Newtonian mechanics have faced anomalies through their histories; for example, the perihelion of Mercury was not explained until Einstein developed the theory of relativity. Doctors need to make diagnoses even when some symptoms remain puzzling. The requirement of completeness makes sense only in limited closed domains such as simple circuits where one can be assured that everything *can* be explained given the known causes.

2.6. *Simplicity is complex*

Many researchers on abduction have seen the need to use simplicity as one of the constraints on inference to the best explanation, *i.e.*, we want not simply a set of hypotheses that explain as much as possible, but also the most economical set of assumptions possible. Hobbs *et al.* [13] sum this up by saying that we want the most bang (facts explained) for the buck (quantity of assumptions). But simplicity is an elusive notion that is not adequately captured by the relation of subset minimality which deals only with cases where we can prefer an explanation by a set of hypotheses to an explanation by a superset of those hypotheses. We need a broader notion of simplicity to handle cases where the competing explanations are accomplished by sets of hypotheses that are of different sizes but are not subsets of each other. Such cases are common in the history of science and require richer notions of simplicity than subset minimality [28, 25].

2.7. *Abductive reasoning may use non-sentential representations*

Konolige's definition, like virtually all available accounts of abduction, presumes that explanatory hypotheses are represented sententially. But we will now show that some abductive inference is better understood as using pictorial or other iconic representations.

3. VISUAL ABDUCTION

Peirce claimed that all thinking is in signs, and that signs can be icons, indices, or symbols [10, 11]. Icons are signs that resemble what they represent; examples include pictures, photographs, and geometrical diagrams. Peirce placed great emphasis on diagrammatic thinking and even developed a powerful system of predicate logic based on diagrams or "existential graphs" [23]. Surprisingly, Peirce does not seem to have connected his two highly original insights: we know of no text in which he discusses abduction as diagrammatic or iconic. But there are instances of abductive thinking that are most plausibly interpreted as pictorial.

Suppose you return to your car at the shopping center and find a big scratch on one door. Naturally, you wonder how it happened and start to generate hypotheses to explain how the scratch came to be. Your abductions may be purely verbal, if you start to apply rules such as "If a car door is opened and collides with another car door, the latter door is scratched." You could then verbally abduce that some other car door was opened and collided with yours. But a mode of thinking that is natural for many people is to perform the same kind of thinking pictorially. You can form a mental image of a car driving up beside yours and then its driver opening a door that scratches yours. Here the explanation is a kind of mental movie in which you imagine your door being scratched. The abductive inference that the accident happened this way involves a mental picture of the other car's door hitting yours. Such pictures provide an iconic representation of the event that you conjecture to have happened, since the picture you form resembles the hypothesized event in a much more direct way than a verbal/sentential representation would. Whenever our knowledge of how things work in the world involves dynamic pictorial representations, these representations can be used to generate iconic explanations of what occurs. Many scientists have reported that images played a crucial role in their most creative thinking: the most eminent include Bohr, Boltzmann, Einstein, Faraday, Feynman, Heisenberg, Helmholtz, Herschel, Kekulé, Maxwell, Poincaré, Tesla, Watson, and Watt.

Performing abduction visually may have strong cognitive advantages. With verbal representations such as rules, it may be necessary to search through many possible inferences before finding a plausible explanatory hypothesis. But a picture of a situation may immediately suggest a likely cause, if it vividly displays factors that are spatially contiguous and therefore more likely to be causally relevant. Artificial intelligence is still very limited in its ability to use such visuospatial information, although progress is being made in the direction of imagistic representations. Glasgow has argued for the importance of spatial reasoning in AI and proposed a representational scheme based on 3-dimensional arrays [9, 8]. Graphs provide visual representations that are more flexible than arrays [5, 30], so to show how abduction can be visual yet nevertheless amenable to formal treatment, we will now discuss graph grammars.

A simple graph can be thought of as a purely algebraic structure consisting of a set of vertices and edges, but uses of graphs often exploit their visual character. When we draw a graph representing, for example, the paths between various cities, we get a diagram or mental representation of the routes between cities that resembles the actual roads. The mathematical structure of a graph naturally translates into a graphical diagram that resembles what it represents much more directly than a set of sentences would. Graph grammars consist of sets of production rules that differ from standard verbal productions in that the left-hand sides (conditions) and right-hand sides (actions) are represented as graphs rather than as verbal structures [7, 14, 19]. A graphical production can be interpreted as saying that if you have one graphical structure and you apply a transformation to it, then you get another graphical structure. For abduction purposes, we can think of the right-hand side of a graph production as providing a visual representation of something to be explained, and the left-hand side and the transformation as providing a possible visual explanation.

To be more concrete, consider children's interlocking Lego blocks. We can represent each block as a graph whose vertices are connectors and sockets, where the connectors on one block fit into the sockets on another block. Figure 1 gives a graph for a simple 4-connector block, with 8 vertices and 12 edges. Transformations possible for Lego blocks include stacking one on top of the other, which produces a new structure in which the connectors on the bottom block go into the sockets on the top block. A graphical production representing this would have a left-hand side of two unconnected graphs and a transformation that produced new edges connecting the appropriate sockets

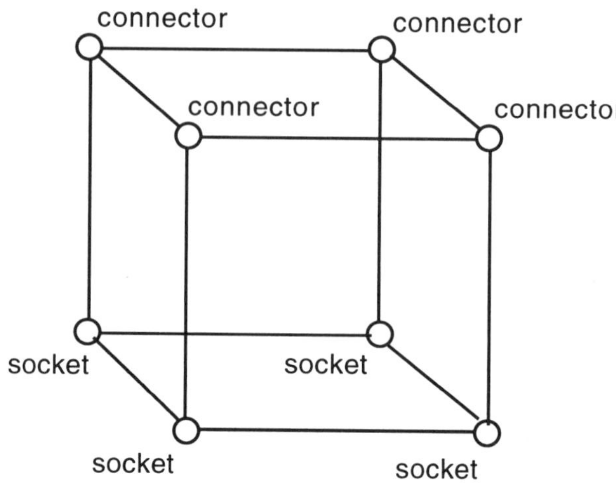

Fig. 1. Graph representation of a Lego block.

and connectors, producing a new connected graph with a total of 16 vertices and 28 edges, including 4 new ones.

Given such a production, we could explain the existence of a tower consisting of two blocks by hypothesizing that there were two independent blocks that had been transformed by joining. Abduction is then visual because the representations of both what gets explained and what does the explaining use structures that resemble what they represent.

Let us now state this more formally.

Definition 3.1 *A graph G is a tuple $\langle V, E\rangle$ where*

- *V is a set of vertices;*
- *$E \subseteq V \times V$ is a set of edges.*

Definition 3.2 *A graph grammar Γ is a finite set of productions P, where a production is a tuple $\langle G_l, G_r, T\rangle$ such that:*

- *G_l is the left-hand side;*
- *G_r is the right-hand side;*
- *T is the embedding transformation that specifies the relations between the vertices and edges of G_l and G_r.*

Definition 3.3 *A graph-grammatical abductive explanation of a target graph G^t is a hypothesis graph G^h such that there is a set of productions in P whose successive transformations transform G^h into G^t.*

Intuitively, the hypothesis graph G^h provides an explanation of the target graph G^t by virtue of a series of transformations that show how the target graph can be produced from the hypothesis graph.

Graph grammars are not the only possible basis for visual abduction. Leyton [17] presented a "process grammar" that could be used to infer the causal history of objects from their shapes. This grammar contains such productions as one that can be interpreted: "If a shape is squashed, it will indent." If the shapes and the processes of squashing and indenting are represented pictorially, then the inference that explains a shape's indentation by its having been squashed is an instance of visual abduction.

This paper does not attempt to present a general theory of visual abduction, but rather some instances that show the limitations of current formal models of abductive reasoning. A fully general characterization of abduction would have to allow representations of causes and effects that are pictorial as well as ones that are sentential. It would also have to admit forms of explanation that centrally employ visual transformations as well as ones that employ deduction and other verbal processes. In addition, we should not rule out the possibility of a multimodal theory of abduction that includes non-visual, nonverbal representations involving smell, touch, and emotion. For example, there are some diseases that physicians can diagnose based on a patient having a particular odor. The need for a multimodal theory is illustrated below with an archaeological example.

4. THE DEMISE OF SK54

Forming hypotheses to explain the unusual macroscopic properties of artifacts and skeletal remains is an important part of archaeology. This kind of abduction often requires the archaeologist to reconstruct the events in the history of an object which caused it to change in shape or structure from some initial form to the final, observed one. Since this kind of evidence for the history of an object is visual in nature, the archaeologist may find visual mental imagery useful in generating an explanation of it. (See [24] for further discussion.)

For example, in 1949 excavations of a cave at Swartkrans in South Africa yielded, among much other debris, the skullcap of an australopithecine (thereafter designated *SK54*). The distinctive features of this skullcap consisted of two notches, one on either side of the centerline, which had obviously been driven into the back of the skull by two pointed objects when the creature was still alive [2]. SK54 is pictured in figure 2, with the notches indicated by the two arrows. At first it was supposed that the notches had been inflicted by two *separate* blows from a weapon wielded by another hominid, because each notch had been made at *divergent* angles from the centerline [1]. This hypothesis accorded well with the prevailing theory that human evolution had been driven by murder and cannibalism—Dart's [6] "killer ape" hypothesis.

However, an alternative explanation has been offered by Brain [2]. Noting that the lower canine teeth of leopards diverge and are about the right distance apart, Brain hypothesized that the notches had been created by a leopard which had taken the australopithecine's head in its mouth, as dramatically

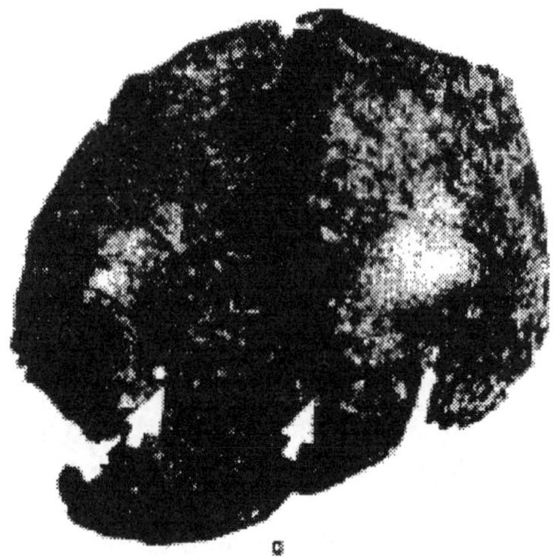

Fig. 2. The skullcap SK54 with notches indicated by white arrows [2, p. 1117].

Fig. 3. The abduced fate of SK54 in the jaws of a leopard [2, p. 1118].

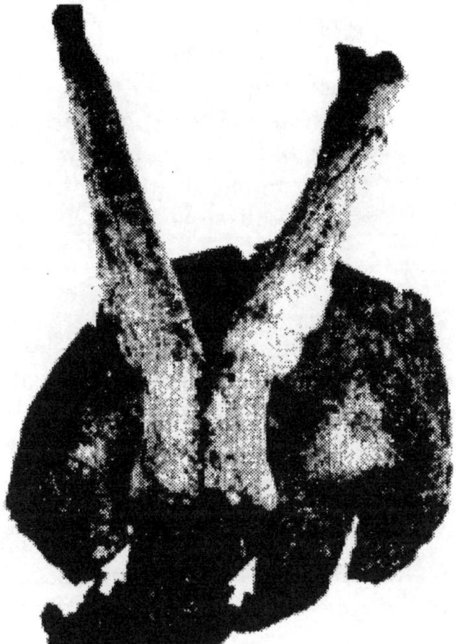

b

Fig. 4. Fossil leopard jaw SK349 fitted into the notches of SK54 [2, p. 1117].

illustrated in figure 3. Indeed, a fossil leopard jaw from the same site (SK349) fits the notches fairly well and shows that the hypothesis is a plausible one; see also [3]. This is shown in figure 4.

This explanation of the notches in SK54 also accords with several other facts about the debris in the Swartkrans cave. The entrance to the cave was a vertical shaft when the australopithecine remains were deposited and those remains are mostly comprised of skull fragments. Similar shaft caves in the area today are frequented by leopards which use the trees which grow around the entrances as places to safely consume prey out of the reach of hyenas. Since leopards tend to destroy the skeletal material of their primate prey with the exception of the skulls, the leopard-predator hypothesis would also explain why the australopithecine remains are mostly skull fragments—the skulls would simply have fallen from the trees and into the cave shafts [2]. This scenario, leopard predation of early hominids, is much different from the "killer ape" scenario favored by Dart.

Brain's leopard hypothesis exemplifies the use of visual abduction. The target of explanation—the unusual notches in the skullcap SK54—are highly visual in nature, consisting in their placement, depth, and direction. The hypothesis reconstructs the vital moment in the history of an unfortunate hominid, when its head was clenched in the jaws of a leopard to produce the notches. Because the relevant data are spatial, the hypothesis is most parsimoniously captured visually, as in figure 3, and may well have first occurred to Brain as just such a mental picture. The hypothesis thus abduced was then corroborated by further evidence of fossil leopard jaws and the feeding habits of modern leopards. Thus, this example illustrates that visual abduction fits the expanded criteria for abduction discussed in section 2.

5. COHERENCE

Abduction is often thought of as a kind of reverse deduction; see section 2.1. Whereas modus ponens infers from $p \rightarrow q$ and p to q, abduction infers from $p \rightarrow q$ and q to p. This formulation trivializes the point of abduction, which is to form and evaluate hypotheses that make sense of puzzling facts. Making sense is a holistic matter of fitting the puzzling facts into a coherent pattern of representations. The notion of coherence has usually remained vague, but Thagard and Verbeurgt [29] have recently defined coherence in terms of maximization of constraint satisfaction. After briefly stating their definition, we will show how abduction can be reconceptualized as a coherence problem.

Definition 5.1 *Let E be a finite set of elements e_i and C be a set of constraints on E understood as a set (e_i, e_j) of pairs of elements of E. C divides into $C+$, the positive constraints on E, and $C-$, the negative constraints on E. With each constraint is associated a number w, which is the weight (strength) of the constraint. The problem is to partition E into two sets, A (accepted) and R (rejected), in a way that maximizes compliance with the following two coherence conditions:*

1. *if (e_i, e_j) is in $C+$, then e_i is in A if and only if e_j is in A;*
2. *if (e_i, e_j) is in $C-$, then e_i is in A if and only if e_j is in R.*

Let W be the weight of the partition, that is, the sum of the weights of the satisfied constraints. The coherence problem is then to partition E into A and R in a way that maximizes W.

To understand abduction as a coherence problem, we need to specify the elements and the constraints. The elements are representations of causes and effects; in line with our discussion of visual abduction, we will allow the representations to include both sentential and nonsentential representations such as visual ones. The major positive constraint on abduction is that if one element explains another, then there is a symmetric positive constraint between them. In accord with the first of the two coherence constraints above, this will have the effect that if the explaining element and the explained element will tend to be accepted or rejected together. If, as often happens, more than one element is required to explain another element, the weight on the constraint between each explaining element and the explained element can be less than if there were only one explaining element, in keeping with the theory of simplicity of Thagard [28].

Two sorts of negative constraint are possible. If two elements contradict each other, then there is a symmetric negative constraint between them, but there is also a negative constraint if two elements offer competing explanations of the same fact [28]. Deciding which explanatory hypotheses to accept and which to reject is a matter of maximizing compliance with the two coherence conditions. Computationally, this is a very difficult problem and no tractable algorithm for solving it is available, although various approximation algorithms work quite well [29].

Our account of abduction as a coherence problem avoids all the limitations we discussed in section 2. We assume that the relation between elements is explanation, not deduction. Elements can explain other elements which explain other elements, so hypotheses can be layered and give rise to complex chains of constraints. Abduction is not defined in terms of a fixed set of elements but allows the possibility of creation of new elements that produce a new assessment of coherence. New hypotheses need not be consistent with existing ones, nor need the facts be completely explained. Maximizing coherence involves explaining as much as possible (positive constraints) and being as consistent as possible (negative constraints), but perfection is not to be sought in abductive reasoning. We are not urging inconsistency as a general epistemological strategy, only noting that it is sometimes necessary to form hypotheses inconsistent with what is currently accepted in order to provoke a general belief revision that can restore consistency.

Reducing the weights on constraints when explanation is accomplished by multiple hypotheses allows a complex assessment of simplicity. Finally, since elements can be visual or other nonverbal kinds of representation, we have transcended the limitation of abductive reasoning to sentences.

6. CONCLUSION

The definitions of abduction proposed by Konolige and by Bylander, Allemang, Tanner and Josephson are clearly inadequate as general characterizations of this important type of reasoning. These definitions made possible precise comparisons with other models and derivation of complexity theoretic results, but the results must be viewed as relative to the abstract model of abduction proposed. One might think that all complexity results would carry over to models of abduction more elaborate than the ones offered: if abduction is more complicated than Bylander *et al.* allow, then the computational complexity should be at least as great, so abduction would still be NP-hard. But the main results of Bylander *et al.* concern the difficulty of determining whether an explanation exists, where their notion of explanation is, as we saw, unrealistically restricted to cases where all the data are to be explained. In scientific reasoning, in contrast, the task is to explain as much of the data as you can and to come up with a set of explanatory hypotheses that is better than the current set. It is a matter of satisficing (choosing the most coherent set of hypotheses) rather than optimizing. Abduction characterized in terms of coherence is still intractable, but efficient approximation algorithms exist. There is also the possibility that heuristic search made possible by visual representations can greatly improve the computability of abductions, although this remains to be shown.

Thanks to important formal results such as the theory of NP-completeness, formal methods have achieved much well-deserved prestige in computer science and artificial intelligence. But an oversimplified formalization can distract from important aspects of the kinds of reasoning that underlie intelligence. Graph grammars and other techniques for visual representation offer the prospects of developing more general accounts of the nature of abductive reasoning, which should be construed not as a deviant form of deduction, but as a coherence problem.

Philosophy Department
University of Waterloo

REFERENCES

1. R. Ardrey. *African genesis: A personal investigation into the animal origins and nature of man.* Atheneum, New York, N.Y., 1961.
2. C. K. Brain. New finds at the Swartkrans australopithecine site. *Nature*, 225:1112–9, 1970.
3. C. K. Brain. *The hunters or the hunted? An introduction to African cave taphonomy.* The University of Chicago Press, Chicago, Ill., 1981.
4. T. Bylander, D. Allemang, M. C. Tanner, and J. R. Josephson. The computational complexity of abduction. *Artificial Intelligence*, 49:25–60, 1991.
5. J. Y. Ching. Computational imagery: An alternative framework. Ph.D. thesis proposal, Pattern Analysis and Machine Intelligence Lab, University of Waterloo, Waterloo, On., May 1994.
6. R. A. Dart. The predatory transition from ape to man. *International Anthropological Linguistic Review*, 1(4):201–18, 1953.
7. H. Ehrig, H. J. Kreowski, and G. Rozenberg, editors. *Graph-grammars and their application to computer science: 4th international workshop*, number 532 in Lecture notes in computer science, Berlin, Germany, March 1990. European Association for theoretical computer science, Springer Verlag.
8. J. I. Glasgow. The imagery debate revisited: A computational perspective. *Computational Intelligence*, 9:309–33, 1993.
9. J. I. Glasgow and D. Papadias. Computational imagery. *Cognitive Science*, 17(3):355–394, 1992.
10. T. A. Goudge. *The thought of C. S. Peirce.* University of Toronto Press, Toronto, On., Canada, 1950.
11. C. Hartshorne and P. Weiss, editors. *Collected papers of Charles Sanders Peirce.* Harvard University Press, Cambridge, Mass., U.S.A., 1958. Volumes 7–8 edited by A. Burks.
12. C. G. Hempel. *Aspects of scientific explanation.* The Free Press, New York, N.Y., 1965.

13. J. R. Hobbs, M. Stickel, D. Appelt, and P. Martin. Interpretation as abduction. Technical Note 499, AI Center, SRI International, 333 Ravenswood Ave., Menlo Park, CA 94025, December 1990.
14. C. V. Jones. An integrated modeling environment base on attributed graphs and graph-grammars. *Decision Support Systems: The international journal*, 9(10):255–75, Oct 1993.
15. P. Kitcher and W. Salmon. *Scientific explanation.* University of Minnesota Press, Minneapolis, M.N., 1989.
16. K. Konolige. Abduction versus closure in causal theories. *Artificial Intelligence*, 52:255–72, 1991.
17. M. Leyton. Inferring causal history from shape. *Cognitive Science*, 13:357–87, 1989.
18. D. V. McDermott. A critique of pure reason. *Computational Intelligence*, 3:151–60, 1987.
19. M. Nagl. Formal languages of labelled graphs. *Computing*, 16:113–37, 1976.
20. J. Pearl. *Probabilistic reasoning in intelligent systems.* Morgan Kaufman, San Mateo, CA, 1988.
21. Y. Peng and J. Reggia. *Abductive inference models for diagnostic problem solving.* Springer Verlag, New York, N.Y., 1990.
22. R. Reiter. A theory of diagnosis from first principles. *Artificial Intelligence*, 32:57–95, 1987.
23. D. D. Roberts. *The existential graphs of Charles S. Peirce.* Mouton, The Hague, 1973.
24. C. Shelley. Visual abductive reasoning in archaeology. *Philosophy of Science*, in press, 1996.
25. P. Thagard. *Computational philosophy of science.* MIT Press, Cambridge, Mass., 1988.
26. P. Thagard. Explanatory coherence. *Behavioral and Brain Sciences*, 12:435–67, 1989.
27. P. Thagard. Philosophical and computational models of explanation. *Philosophical Studies*, 64:87–104, 1991.
28. P. Thagard. *Conceptual revolutions.* Princeton University Press, Princeton, N.J., 1992.
29. P. Thagard and K. Verbeurgt. Coherence. Under review.
30. A. Wong, S. Lu, and M. Rioux. Recognition and shape synthesis of 3-D objects based on attributed hypergraphs. *IEEE Transactions on Pattern Analysis and Machine Intelligence*, 11(3):279–90, Jun 1989.

PAUL WEINGARTNER

CAN THE LAWS OF NATURE (PHYSICS) BE COMPLETE?

The purpose of the paper is to discuss the question of the completeness of physical laws (laws of nature) beginning with some related questions in the tradition and continuing with recent problems.

It will become clear throughout this essay that this question is related to several logical problems which have been solved only partially so far. However the physical questions might stimulate to find new ways for solutions. The concept of completeness used here is the rudimentary basic idea that a system of laws (axioms) is complete with respect to a given field (domain, research area) of reality iff all true sentences about that field are derivable from the laws (axioms). It should be emphasized however that results on completeness or incompleteness obtained in the field of logic and mathematics cannot automatically be transmitted to physical laws since it is very questionable whether important necessary conditions for these results (like axiomatic construction, closure conditions, definability of the concept of provability within the system etc.) are available for a system of physical laws. Therefore no more than the above basic idea of completeness will be used subsequently.

1. LEIBNIZ: THE KNOWN LAWS OF PHYSICS ARE NOT COMPLETE

According to Leibniz there are four great comprehensive areas of scientific research which can be built up according to the principle of rationalism and of more geometrico as axiomatic systems: logic and mathematics (LM), metaphysics (M), physics (PH), jurisprudence (including ethics) (JE). He thought that *in principle* these systems of truths are finitely axiomatizable, consistent and complete though for mankind the finite axiomatizability the consistency and the completeness are available only for logic, mathematics and metaphysics but not for the other two areas since they include contingent truths.[1] The principle for the consistency of axiomatic systems is according to Leibniz: All finitly analytic propositions are (necessarily) true (FAT). According to Leibniz a finitly analytic proposition is one that is either an 'identical' (i.e. one of the form A is A or AB is A like "a rational animal is an animal") or one that can be reduced to an identical proposition with the help of a finite number of steps using definitions.[2] Thus FAT is a soundness principle of the form:

[1]For a detailled discussion of this and the following see Weingartner (1983) part 2. where the respective passages in Leibniz's works are given.

[2]Leibniz (GP), Vol. 7, p. 195.

M.L. Dalla Chiara et al. (eds.), Logic and Scientific Methods, 429–446.

Every proposition which is derivable in the axiomatic system is (necessarily) true.[3]

Leibniz' principle of sufficient reason is a completeness principle. Leibniz gives different formulations of the principle. But he seems to understand them as different versions of the same underlying principle: "that nothing is without a reason, or that every truth has its a priori proof, drawn from the notion of the terms, although it is not always in our power to make this analysis."[4] "Generally, every true proposition (which is not identical or true *per se*) can be proved *a priori* by the help of axioms, or propositions true *per se*, and by the help of definitions or ideas. ... It is certain, therefore, that all truths, even the most contingent, have an *a priori* proof, or some reason why they are rather than are not. And this is itself what people commonly say, that nothing happens without a cause, or that nothing is without a reason."[5] The principle of sufficient reason can be formulated at least in three versions: (i) Nothing happens without a sufficient reason. (ii) Every true proposition has its a priori proof. (iii) Every true proposition is finitly or infinitly analytic. To understand the last version one has to know two things: (1) Leibniz' understanding of analysis: Analysis is a process of logical inference using definitional replacement and determination of predicational containment.[6] (2) Leibniz' view of two kinds of proof processes: In the area of truths of reason the proof-process is terminating after a finite number of steps, i.e. the proposition proved is finitly analytic. On the other hand in the realm of contingent truths the proof-process is not terminating after a finite number of steps, i.e. the proposition is infinitly analytic. In both cases however we have a genuine proof-process intended as an inferential procedure of the most rigorous kind, i.e. a proof of a proposition within a deductive system. My claim which I have substantiated in detail in my (1983) is that Leibniz' principle of sufficient reason is a completeness principle stating the completeness of a scientific field built up axiomatically (more geometrico). This can be seen already from some of the quotes: Every true proposition has an a priori proof (or is provable from the axioms in the respective deductive system which represents one of the four fields LM, M, PH and JE). But although all the four areas are in principle complete for God who knows all the right axioms, the completeness can be effectively shown by men only in those areas where every true proposition is finitly analytic, i.e. can be traced back in a finite number of steps to the axioms. According to Leibniz this is the case in LM and M which deal both only with "truth of reason".[7] However in PH and JE, since these areas contain contingent truths, not every true proposition is finitly analytic, some are, but

[3]For details see my (1983), Ch. 2.4.

[4]Leibniz (GP), Vol. 2, p. 62.

[5]Leibniz (GP), Vol. 7, p. 300, 301.

[6]Cf. Rescher (1967) p. 23.

[7]We know today that Leibniz was right with respect to Predicate Logic of First Order but not with respect to mathematics in general. He found an arithmetization and decision

some are only infinitly analytic. In these cases men cannot trace back the respective true proposition to the axioms (since the steps would be infinite) and he cannot find the right axioms (since they are not simple truths of reason but complicated truths of fact). Thus the axioms and laws of physics (and of JE) which are known by men are not complete.

An interesting related problem for logic is to investigate Leibniz's idea of ars inveniendi, i.e. a method to find laws or axioms or explanations if one has a lot of particular truths and lower level hypotheses within a field of research.

2. THOMAS AQUINAS: THE LAWS OF NATURE ABOUT THE WORLD (UNIVERSE) ARE NOT COMPLETE

The invariance with respect to displacement of time (and of place) are the oldest and perhaps most important invariance properties of physical laws and of laws of nature in general. It seems that the concept of law (of nature) is violated if these invariance conditions would not be satisfied. "The statement that absolute time and position are never essential initial conditions is the first and perhaps the most important theorem of invariance in physics. If it were not for it, it might have been impossible for us to discover laws of nature."[8] The first philosopher who seems to have realized this very clearly was Thomas Aquinas. In his quarrel with Bonaventura at the university of Paris he defended the view that the beginning in time of the world (universe) cannot be proved from universal principles (laws) of (about) this world. Because universal principles abstract from hic (place) et nunc (point of time): "That the world has not always existed cannot be demonstratively proved but is held by faith alone. ... The reason is this: the world considered in itself offers no ground for demonstrating that it was once all new. For the principle for demonstrating an object is its definition. Now the specific nature of each and every object abstracts from the here and now, which is why universals are described as being *everywhere and always*. Hence it cannot be demonstrated that man or the heavens or stone did not always exist."[9] The question whether it can be demonstrated that the world has always existed or that it has a beginning in (with) time — answered differently by competing theories of the universe — is a question about the completeness of the laws of nature. Or at least of that laws we know. Thomas Aquinas' standpoint was that the universal laws of nature (about this world) are not complete with respect to all questions (all truths) about this world. It is not just our insufficient knowledge of the laws of nature what he has in mind, but the true laws itself are incomplete according to him with respect to some special questions. Or in more modern terms: the laws of nature are incomplete with respect to some

procedure for syllogistics and seemed to be very optimistic to extend such a method further to mathematical proofs. Cf. my (1983), chapter 3.1.

[8] Wigner (1967), p. 4.

[9] Thomas Aquinas (STh) I, 46, 2.

important initial conditions. This problem plays an important role in the Big Bang Theory of the cosmological evolution in respect to (at least) the "first three minutes").[10] However important experiments since 1965 (Penzias and Wilson) revealed the cosmic background radiation which strongly supports a finite age (and beginning) of the universe. Even if Thomas Aquinas would perhaps say it is not a proof in the sense of a demonstration, i.e. a derivation from true or well confirmed laws, it seems still a strong experimental evidence in support to the Big Bang Theory and for a finite age of the universe. That the question is still open is shown by other theories who claim an universe infinite in time and try to be consistent or even to explain the same experimental data.[11]

A related logical problem is how many different theories are able to explain the same data and are at the same time consistent with the other laws and theories within a certain field. Another one is the logical status of physical equations. Equations are some sort of equivalences. Some of them may therefore be interpreted as (real) definitions which according to Aristotle and Thomas Aquinas are needed in a scientific explanation.

3. THE COMPLETENESS OF PHYSICAL LAWS ACCORDING TO LAPLACE'S IDEA.

The considerations in chapter 2 showed already that "completeness" as it is used in logic is not sufficient when applied to physical laws. Since from laws alone without initial conditions we cannot derive particular truths about F. Thus the question is rather whether a certain set of physical laws L + a certain set of initial conditions I could be complete with respect to a certain field F. But how many initial conditions do we need? Laplace thought that we need only one (the minimum) — more accurately: one state of the whole universe at a certain point of time — if the laws are dynamical (in this sense deterministic) laws. Dynamical laws like Newton's laws of motion characterize a physical system by three important conditions: (1) The state of the system at any given time t_i is a definite function of its state at an earlier time t_{i-1}. A unique earlier state (corresponding to a unique solution of the equation) leads under the time evolution to a unique final state (again corresponding to a unique solution of the equation). This property is usually taken as the defining condition for determinism. (2) The system is periodic, i.e. the state of the system repeats itself after a finite period of time and continues to do so in the absence of external disturbing forces. (3) The system has a certain type of stability. Assume we make very small changes in the initial states, say within a neighbourhood distance of ε. Then the distance of the state $h(\varepsilon)$ is proportionally small (no more than a linearly increasing function of

[10] Cf. Weinberg (1977).

[11] Cf. for instance Hawking (1988) who uses imaginary time and Linde (1990), whose theory avoids this rather unplausible assumption.

time). This kind of stability with respect to small perturbations is called "perturbative stability" which holds in all linear systems. Laplace's idea was that someone (a perfect intellect) knowing all the laws (of the universe) could calculate all its states just by knowing one (arbitrary) state.

"We ought then to regard the present state of the universe as the effect of its anterior state and as the cause of the one which is to follow. Given for one instant an intelligence which could comprehend all the forces by which nature is animated and the respective situation of the beings who compose it — an intelligence sufficiently vast to submit these data to analysis — it would embrace in the same formula the movements of the greatest bodies of the universe and those of the lightest atom; for it, nothing would be uncertain and the future, as the past, would be present to its eyes."[12]

This idea can be illustrated by the following example: Assume a film is made of the world, i.e. of the events happening in the whole universe. After the film is developed we cut it into pieces corresponding to single film-pictures. Now we put the single pictures successively in time (in the order of time) into a long card index box like the cards of a library catalogue. Then one special state of the universe at a certain time t corresponds to one such card (film picture) of the catalogue. One can follow one trajectory across the (perpendicular to the) catalogue-cards. Interpreted with the help of this illustration Laplace's determinism means that it suffices to know the law(s) of nature and one single catalogue card (film picture) in order to construct all other cards of the catalogue, i.e. to predict and to retrodict all the other states of the universe. That means that Laplace's idea is a completeness claim with respect to the laws of nature: the correct laws of nature (known to a perfect intellect) plus one initial state of the whole universe are together complete in the sense that all true descriptions of states of the universe follow from them.

There are some important points however which are related to logical problems: (1) If the laws in Laplace's sense, the differential equations of motion were not defined everywhere then also those laws would be incomplete. (2) Laplace's theory is still incomplete with respect to the question of Thomas Aquinas, i.e. it does not imply a finite age of the universe nor does it imply an infinite one in time. (3) If the universe is finite in time, then it is possible that some of the solutions of the equations could not be realized because the laws + one initial condition allow to determine more states than can happen in the finite time of the existence of the universe. This then is a sense of over-completeness of laws.

4. CAN STATISTICAL LAWS BE COMPLETE?

After the discovery of statistical laws in thermodynamics and later in other areas there was a general doubt as to an overall interpretation of the world

[12] Laplace (1951) p. 4.

by dynamical-deterministic laws.[13] It suffices to mention names like Maxwell, Boltzmann, Clausius, Poincaré, Gibbs and Carnot in order to remind us of very important new discoveries. Processes of heat, friction, diffusion, radiation, electric transport, absorption, dispersion, transmission, relaxation and phenomena like flow of water, flow of glacier, floods, avalanches, lightenings, growing, aging, propagation etc. cannot be described by dynamical laws in Newton's sense, although some essential feature of it, the application of the mechanical viewpoint, applied to a huge collection of very small parts of the system, was adopted and lead to a new theory: Statistical Mechanics. An important new property of these processes was their irreversibility and the discovery of the Second Law of Thermodynamics (the law of Entropy). For a number of such processes, especially those occurring in gases Boltzmann has given a theoretical explanation with his statistical mechanics. Entropy is a measure for the reversibility of a process. The law of entropy says that in a closed system the entropy can remain the same or must increase, it cannot decrease. Boltzmann connected the entropy with the atomic hypotheses, i.e., roughly with the assumption that matter is composed of a huge number of little balls dancing around wildly. At that time the atomic structure of matter was not generally accepted. Mach's scepticism about it is wellknown. Assume a litre of air consisting of about $2,7 \cdot 10^{22}$ molecules. Then it will be understandable that this system of $2,7 \cdot 10^{22}$ molecules can be in a huge number of different (micro-)states, the number is about $10^{5 \cdot 10^{22}}$ so as to realize the macrostate "litre of air". Thus the same (for our lungs the same) macrostate can be realized by a huge number of different microstates. Boltzmann's discovery was that the probability of such a macrostate can be defined as the number of microstates which can realize the macrostate and that this number (more accurately the logarithm of it) is the entropy. Thus the law of entropy has a new interpretation: The probability of a macrostate of a closed system can only increase or stay the same, it cannot decrease. That means that the time development of closed systems goes in the direction from less probable states to more probable states. This means of course that these processes are time irreversible (in this probabilistic sense). Recurrence of an earlier state is not impossible but highly unprobable.[14] The cosmological Big Bang Theory

[13]One of the first philosophers who noticed that a certain imperfection in all mechanical processes allows to enter chance was Charles Sanders Peirce. Cf. Peirce (1935, 1960), 6.47. Thomas Aquinas pointed out that the ethical and legal laws — except those very general and formal ones like "the good should be done the bad should be avoided" — are valid only for most cases and allow exceptions, i.e. are statistical in character. Cf. (STh) I–II, 96,6.

[14]Zermelo thought (1896) that he had proved that the statistical laws of Boltzmann's statistical mechanics (like the law of entropy) are not compatible with the basic dynamical laws of classical mechanics because it seemed to him that they are in conflict with Poincaré's Recurrence Theorem. Boltzmann showed in two articles (of 1896 amd 1897) against Zermelo's articles that Zermelo misunderstood important conditions for the applicability of Poincaré's Recurrence Theorem. For details see my (1996b).

is an example of such a description: the whole universe developes into more and more probable states.

(1) Statistical laws are incomplete because they don't describe individual processes. From what has been said so far one can grasp already one essential reason for the incompleteness of statistical laws. They describe ensembles and the behaviour of ensembles, they don't describe the individual microprocesses and microstates. The resulting realized macrostate (a litre of air or a litre of water at a certain temperature) is in fact an ambiguous thing it consists of a huge number of microstates of which the realized ones change quickly and permanently.

(2) The incompleteness of statistical laws increases with the complexity of the system. Consider the following example: Place a little beetle on the first square of a chess board with borders (so that it cannot leave the chess board). After some time of running around the beetle will again come back to the first square (recurrence). A measure for it will be the probability of recurrence. Place now 10 beetles on each square. The 640 beetles will run around randomly. The probability for recurrence for an individual state of the whole system (say that all beetles are coming back at the same time to their start positions) will be of course much lower. Thus the loss of information about a particular state of the system will increase. This can be interpreted as a kind of incompleteness of the statistical hypothesis (law) with respect to the individual (particular) state of the system.

Imagine now a system of 10^{22} molecules. The loss of information (and with it the corresponding incompleteness) about one particular microstate out of $10^{5\cdot 10^{22}}$ when the statistical hypothesis describes the behaviour of the macrostate is in fact enormous. Since non-recurrence and irreversibility are not equivalent notions one may explain the "arrow of time" more modestly by non-recurrence.

(3) What is the reason for the incompleteness of the statistical laws? An old question of the Logic of Probability has been whether probability and statistical laws have to be interpreted only subjectively (lack of knowledge) or can be interpreted also objectively (representing features of reality).

Is it a lack of our knowledge about physical reality (about nature) which permits us only to use statistical laws for the description of certain phenomena instead of dynamical laws? In this case there could be hidden parameters — unknown to us — which would allow a complete description and explanation (in principle) also for all individual states with the help of dynamical laws.

Or is it the indeterminateness of nature, of physical reality itself which forces the laws (and theories) to be incomplete. This question is one of the underlying main-questions of the EPR problem.

5. A NEW INCOMPLETENESS SHOWN BY CHAOTIC MOTION

The new discovery with respect to chaotic motion was that even within an area where dynamical laws are applicable the behaviour of the physical system can change radically such that the motion of a mechanical system may become chaotic. Thus even simple mechanical systems which have been paradigms for strictly obeying dynamical laws (like Newton's laws of motion) can turn into chaotic motion and become completely unpredictable just by changing slightly the initial conditions.

Experiments which prove that have been performed since the seventies. A very simple one has been done with the spherical pendulum. A spherical pendulum consists of a small weight attached to the lower end of a string (of length l) and has a period $T_0 = 2\pi\sqrt{l/g}$ of sinusodial oscillations (provided the oscillations are small). Unter normal conditions with small amplitudes it shows a regular behaviour with the three important characteristics given in chapter 3.

The important new discovery is now that this simple physical system becomes chaotic if the top end is forced to move back and forth (maximal displacement Δ) with a slightly different period T greater than T_0, provided that Δ is about $1/64$ of l and not more than about a tenth of the energy of motion is dissipated by damping (air resistance etc.). In 1984 Miles showed experimentally that the system is chaotic for values of $T = 1,00234T_0$. It has to be emphasized however that this does not just mean that the system becomes unstable in the sense of simple bifurcation which has been known for a long time. In this case the pendulum weigth makes a back and forth oszillation in the same plane and by forcing the upper end this movement begins to be unstable. Such a simple bifurcation where the plane is not changed occurs when $T = 0,989T_0$ and slightly above. But for $T = 1,00234T_0$ the pendulum is breaking out of the plane, the number of further bifurcations are arbitrarily increasing, the dependence on initial conditions is completely random such that there is no predictability (or only for very short times).[15]

Within the last twenty years important properties of chaotic motion (dynamical chaos)[16] have been discovered. One of the most important necessary conditions is the sensitive dependence on initial conditions. Traditionally this aspect is not new, it was known to Aristotle, attentive people have been aware of it and Maxwell discusses it explicitely with respect to the principle: Similar causes lead to similar effects: "The least initial deviation from the truth is multiplied later a thousandfold."[17] Experienced highlanders in mountainous countries know very well that extremely small initial events can lead to a bursting of an avalanche whereas another time the same small event leads to

[15] For a detailed description see Lighthill (1986).
[16] So called "Quantum Chaos" has different properties. Cf. Casati — Chirikov (1994).
[17] Aristotle (Heav) 271b8.

nothing (no comparable effect). A specific counterexample is due to Maxwell: "There is another maxime ..., which asserts 'That like causes produce like effects'. This is only true when small variations in the initial circumstances produce only small variations in the final state of the system. In a great many physical phenomena this condition is satisfied; but there are other cases in which a small initial variation may produce a very great change in the final state of the system, as when the displacement of the "points" causes a railway train to run into another instead of keeping its proper course."[18]

What these examples describe is a strong sensitivity with respect to initial conditions. And this is one of the most important necessary conditions of chaotic motion. Observe however that it is not sufficient. In Maxwell's example the running of the train in a completely different direction is not a chaotic phenomenon, though some aspects of the crash may have chaotic properties.

Hadamard and Poincaré were the first to mathematically attack the problem of the sensitivity with respect to initial conditions. Hadamard proved the sensitive dependence of a billiard game with negative (concave) curvature, i.e. with dents.[19] In connection with that Hadamard understood that the behaviour of a system becomes completely unpredictable if a small error is committed in the initial data. This result and also the socalled Hénon-attractor can be viewed as modern mathematical interpretations of Aristotle's passage above.[20]

The sensitive dependence with respect to initial conditions is measured by a positive Liapunov exponent. This exponent measures in fact two things: (1) First it measures the exponential separation of adjacent conjugate (with respect to the starting point) points. (2) Second it measures the average loss of information about the position of a point (in an interval) after one iteration. This loss of information is proportional to the socalled Kolmogorov entropy (K).[21]

Both the loss of information and the unpredictability when chaotic motion is present express another type of incompleteness; at first sight the dynamical

---

[18] Maxwell (MAM), p. 13.

[19] The respective proof for a billiard table with positive (convex) curvature (i.e. with hillocks or also with round corners) is more difficult and has been given only in the seventies by Sinai. Cf. Cornfeld-Fomin-Sinai (1982), Ruelle (1991).

[20] Calculations of Chirikov and Berry showed how drastic the sensitive dependence on initial conditions could be. Imagine the gravitational force of an electron (somewhere in the universe, say 10^{10} lightyears away) is taken away for a moment. Could there be a difference in the interaction (with respect to collision) of air-molecules on the earth? After how many collisions with other molecules the extremely small gravitational difference far away could have the effect of failing instead of meeting (collision) another molecule? The answer is very astonishing indeed, it is between 56 and 60 only. If we take a man being in 1m distance of a billiard table and his gravitational force on the billiard balls the respective answer is 9. Already Bernoulli in the 18th century made similar calculations with respect to games (known as "Bernoulli-shift"). Cf. Berry (1978).

[21] For details and references cf. Cornfeld et. al. (1982) or Schuster (1989).

laws seem to be not complete since it is not always possible to get correct predictions with them. The matter is however more complicated.

First it should be mentioned that there are different levels of chaotic behaviour which correspond to levels of increasing disorder. They begin with integrability and quasiperiodic motion. On a higher level there is KAM integrability and weak perturbation (the non-integrable part being a restricted chaotic layer). On a still higher level of disorder there is chaotic motion.[22] The type of incompleteness mentioned increases with these levels of disorder.

Secondly it should be remembered that the underlying dynamical laws of dynamical chaos are laws which obey the three conditions 3.1 – 3.3 given in chapter 3. With small amplitudes the pendulum (recall the experiment described above) obeys all three conditions 3.1 – 3.3 perfectly well. But even the forced pendulum obeys these conditions for most of the initial conditions. Should we now say that for very special asymmetrical initial conditions the (still) symmetric and complete laws lead to asymmetrical outputs?[23] Or should we say that the behaviour of the forced pendulum shows that we do not know the deeper underlying laws which might not have those symmetries which we assume for the dynamical laws. Is it just a question of complications of initial conditions or a question of the appropriate laws and their completeness?[24]

These questions are connected with the difficult problem of the distinction between laws and initial conditions — a distinction which mirrors our understanding of a law coming from the Greek ideal: To describe and explain the visible concrete particular changing material events of the world by non-visible abstract universal non-changing immaterial principles. The former are described by initial conditions the latter by laws. The problematic character of this distinction has been pointed out very clearly by Wigner.[25]

6. THE INCOMPLETENESS DISCUSSED IN EPR

6.1. *The basic assumptions of EPR*

The Incompleteness discussed in the paper of Einstein Podolsky and Rosen (EPR)[26] is concerned with the question whether a physical theory (especially Quantum Mechanics QM) can describe every part of physical reality.

[22]Cf. Chirikov (1991).

[23]Cf. chapter 4 of my (1996a).

[24]There are other important properties of chaotic motion besides the sensitive dependence on initial conditions and the non-predictability like non-linearity (no superposition), non-periodicity, boundedness, permanent change of the important variables, non-integrability, continuous spectrum which are also connected to incompleteness in different ways. For details see Weingartner (1996a).

[25]Wigner (1967).

[26]Einstein-Podolsky-Rosen (1935).

The argument of EPR with the conclusion that the quantum-mechanical description of physical reality given by the wave function is not complete, is based on four assumptions: (1) A necessary condition for "completeness" of a theory, (2) a criterion for "physical reality", (3) the principle of locality and (4) the correctness of the predictions of QM used in the argument.
Assumption 1: "... the following requirement of a complete theory seems to be a necessary one: *every element of the physical reality must have a counterpart in the physical theory.* We shall call this the condition of completeness."[27]

By "physical reality" the authors mean something objective which is independent of any theory: "Any serious consideration of a physical theory must take into account the distinction between the objective reality, which is independent of any theory, and the physical concepts with which the theory operates."[28]

This quote suggests an important point which seems to be implicit also in the locality-assumption: that it is meaningful to speak about properties of (physical) reality independently of an explicitly defined experiment.

Although "the elements of the physical reality cannot be determined by *a priori* philosophical considerations, but must be found by an appeal to results of experiments and measurements."[29] This passage is followed by a criterion of reality:
Assumption 2: "*If without in any way disturbing a system, we can predict with certainty (i.e., with probability equal to unity) the value of a physical quantity, then there exists an element of pyhsical reality corresponding to this physical quantity.*[30]

The assumption of locality is more suitably expressed in his autobiographical notes, where Einstein describes the standpoints of two physicists A (his opinion) and B (Bohr's opinion) and gives again a short description of essential points of the EPR argument:

Assume "there is to be a system which at the time t of our observation consists of two partial systems S_1 and S_2, which at this time are spatially separated ... If I make a complete measurement of S_1, I get from the results of the measurement and from ψ_{12} an entirely definite ψ-function ψ_2 of the system S_2. The character of ψ_2 then depends upon *what kind* of measurement I undertake on S_1".[31]
Assumption 3 (locality principle): "But on one supposition we should, in my opinion, absolutely hold fast: the real factual situation of the system S_2 is independent of what is done with the system S_1, which is spatially separated from the former. According to the type of measurement which I make of S_1, I get, however, a very different ψ_2 for the second partial system $(\Psi_2, \Psi_2' \ldots)$.

[27] Ibid. p. 777.
[28] Ibid.
[29] Ibid.
[30] Ibid.
[31] Einstein (1949) p. 83f.

Now, however, the real situation of S_2 must be independent of what happens to S_1. For the same real situation of S_2 it is possible therefore to find, according to one's choice, different types of ψ-function. (One can escape from this conclusion only be either assuming that the measurement of S_1 ((telepathically)) changes the real situation of S_2 or by denying independent real situations as such to things which are spatially separated from each other. Both alternatives appear to me entirely unacceptable.)[32]
Assumption 4: The respective predictions of QM are correct. The decisive point which leads back to assumption (3) is this: The two systems S_1 and S_2 can be arbitrarily far apart from each other (the first on earth, the second on a star, lightyears away). According to EPR the real properties of system S_2 should be independent of which measurement is done on system S_1. In particular the position and the momentum of system S_2 (on the star) should be already fixed independently of what we are going to do (measurement of position, of momentum, no measurement) with system S_1 on the earth. We could therefore, in principle predict the two magnitudes, position and momentum, of system S_2 far away. If not we must give up locality and assume that there is a ("spouky") influence from our measurement on system S_1 to system S_2 (arbitrarily far away). But since QM does not permit that a system has at the same time position and momentum QM must be incomplete; i.e. some element of physical reality (of system S_2 far away) does not have a counterpart in the physical theory.

6.2. *Bell's attack on locality*

Since the experiments, which violated Bell's inequalities and the GHZ-correlations, the assumption of locality in QM came in trouble.[33] Although that does not mean that anyone would have a good and more generally acceptable explanation (let alone solution) for it. "If I construct a usual quantum model where the effects could be non-local — in the sense of a sudden extension over arbitrarily large distances — then there are situations where the effect comes before the cause. Therefore we have to proceed somewhat more subtle — one has to find a way where something can extend very quickly without coming into conflict with Special Relativity — and this has not been done so far."[34]

The general and even harmless (and somewhat trivially-true) assumption underlying Bell's inequalities was given by Zeilinger[35] in the following principle:

---

[32]Ibid.

[33]The experiments are due to Clauser and later to Aspect et al. (1982). For a detailed overview of the experimental data cf. Redhead (1987). For the GHZ correlations see Greenberger-Horne-Zeilinger (1989).

[34]Bell (1992) p. 94 (my translation).

[35]Zeilinger (1992) p. 109.

For any arbitrary population the members of which can be classified by the presence or absence of at least three properties A, B, C it holds: The number of members (of the population) having both properties A and B is smaller or equal to the sum of the number of members having both properties A and C plus the number of members having property B but not C. The truth of this principle can be easily seen, since it corresponds to a set theoretical theorem.

In order to apply this to QM consider pairs of identical particles (i.e. particles which are permutational symmetric, like two or more electrons, protons, neutrons, neutrinos, photons etc.). Then it holds: If we find property A on particle 1 we know that particle 2 has also A. Assume now we investigate the different properties A, B and C on equally large sets (subsets) of (identical) particles. Then Bell's inequality holds in the following form:

The number of those pairs of identical particles (in a sufficiently large set) of which one has (property) A and the other has B is smaller or equal to the sum of the number of pairs (in an equally large set) of which one has A and the other has C plus the number of pairs (again in an equally large set) of which one has B and the other does not have C. In fact this principle holds for all observations in cases where QM is not used. But the strange thing is that there are experiments which violate it, i.e. Bell's inequalities (cf. note 33).[36]

Recently experiments which also strongly violate Bell's inequalities have been interpreted in a new way. Three kinds of different experiments have been carried out by Chiao, Kwiat and Steinberg.[37] (1) Measurement of the time of tunneling, (2) measurement of dispersion-balance, (3) measurement of coincidence occurrences according to Franson's experiment. However the three physicists underline that they interpret the results as violating locality in QM but not as violating Special Relativity or causality. For example in the first sort of experiment the result is that — unexpectedly — the tunnelling photons (of photon-pairs) are (on the statistical average) faster that the others (without tunnel). From the average arriving distance one can calculate a velocity inside the tunnel of 1,7 light velocity. This seems to contradict Special Relativity. But one has to remember that the position of a photon (particle) is quantum-mechanically a certain probability distribution (the maximum of the amplitude of the distribution represents the most probable position, the width represents the uncertainty of the position). Thus it could be that the wave packet is deformed by the tunnel (barrier) in such a way that the summit of the amplitude turns to the front of the distribution whereas this is not the case for the twin particle outside the tunnel. Thus the beginning of the distribution (representing the low probability for the earliest time to detect the photon) is the same for both particles of the pair and so does not violate

[36]For recent strong violations of the Bell inequalities see the experiments with pairs of photons with different colors and perpendicular polarizations: Kiess-Shih-Sergienko-Alley (1993).

[37]Kwiat-Steinberg-Chiao (1993), and Chiao (1994), Chiao-Kwiat-Steinberg (1993).

Special Relativity. Respective interpretations the authors gave also for the other two kinds of experiments.

6.3. *Consequences*

Is it possible that experiments can violate a logical theorem? This question came up already when classical logic was applied to the measurement problem in QM. In order to avoid such difficulties weaker logics, — so-called Quantum-Logics — have been proposed.[38] It is an open problem whether such logics can be applied also here to give some more adequate solution.

Concerning the incompleteness claim in the EPR-paper the main criticism so far was to attack some of the premisses: Bohr's attack was against assumptions 1 and 2, Bell's attack was against assumption 3 (locality). This does of course not prove the conclusion (claim of incompleteness) to be false, but could show only that it is not true in virtue of these questionable premisses. Thus the claim of the EPR-paper may be true on other grounds. 50 years after the EPR-paper Nathan Rosen (one of the authors) wrote: "Hence it is hard to believe that a theory will be found that will be complete, based on the criterion of an element of reality used in the EPR-paper."[39]

7. SYMMETRY OR COMPLETENESS

Invariance-principles or symmetry-principles describe properties of fundamental laws. Every invariance or symmetry means some kind of universality, abstraction and level of generality. At the same time it implies some unobservable and extinguishes some difference. Consequently with respect to the unobservable or extinguished difference produced by the symmetry there is a kind of incompleteness coming with it. Conversely if some difference or distinctive property is discovered some respective symmetry (invariance) will be broken (violated). There is a similiarity to the universality of the laws of logic. If we say that they are valid in all possible worlds then this means that they do not tell us something specific about this world (universe). Here are a few examples of symmetries:
(1) Permutational symmetry means that the laws remain the same if we interchange two particles of the same sort (say electrons, protons, neutrons ... or photons, π-mesons, gravitons ... according to the Fermi-Dirac or Bose-Einstein statistics). Thus laws are "incomplete" with respect to "individual" electrons.
(2) Translation-symmetry in space extinguishes absolute place leading to a level of generality of three conservation principles of momentum.

[38]Cf. Mittelstaedt (1978). An alternative to Quantum Logic are relevance-restrictions put on Classical Logic. They rule out redundant and not maximally compact (informative) parts in the consequence class. Cf. my (1993).
[39]Rosen (1985) p. 32.

(3) Translation-symmetry of time leads to the conservation of energy and makes unobservable an absolute point of time. If the gravitational constant is not really constant this symmetry will be violated.
(4) Galilean invariance: Inertial systems are indistinguishable.
(5) Charge symmetry implies that the absolute sign of electric charge is unobservable.
(6) Parity implies that right-left is indistinguishable.

As a biological example consider that some children are born with the heart on the right side which seems to show that the underlying biological law obeys parity and the dominating cases (heart on the left side) seems to be due to initial conditions (may be with an exponential building up effect) in the evolution.

The list could be continued with Lorentz invariance, invariance of General Relativity, with time-reversal, CPT-symmetry and Gauge symmetries. The consequence is clear and obvious: Symmetries (invariances) produce unobservables and extinguish differences thus destroying a certain type of completeness of the laws. That means that one cannot have both to a high degree: invariance (symmetry) and completeness. Observe however that this point is closely connected with the understanding of a law and with the above discussed problem of distinguishing the law from the initial conditions.

8. ARE THE LAWS OF NATURE VALID ONLY IN OUR UNIVERSE?

In 1959 Karl Popper proposed the following definition of natural (physical) necessity as an essential feature of laws of nature:

"A statement may be said to be naturally or physically necessary iff it is deducible from a statement function which is satisfied in all worlds that differ from our world (if at all) only with respect to initial conditions."[40]

Let us call the set of all the changes which do not change the laws (of nature) the symmetry group of nature.[41]

But how can we determine the set of all changes which leave the laws invariant? This would mean to know the line of demarcation between contingent initial conditions and necessary and invariant laws. It would mean to know which constants when changed do not affect the laws and which do; and which initial conditions and boundary conditions would affect the laws when changed and which would not. Are the laws of nature invariant with respect to a slight (drastical) change of the amount of energy (mass) of the whole universe (which is constant by the law of conservation of energy)? Or could we change the ratio of electron and proton mass slightly without changing laws? Very probably not. Could there be a constellation of the planets (or some

[40]Popper (1959) p. 433.
[41]Cf. Weinberg (1987) p. 73 who uses this concept in a similar way.

star-systems) which differs from the one realized now (at a certain time) or in other words: Are different constellations of systems of stars (of our planetary system) at a certain time after the Big Bang compatible with the laws?

What are the necessary conditions for the laws of nature to be valid just in our universe? I think there are at least four necessary conditions. If one of them is not satisfied the laws of nature are not entirely specific about our universe, they are then incomplete (with respect to this universe) in the sense of being too symmetric (recall chapter 7).

(1) All laws of nature are deterministic.

(2) Permutation change (interchange) of elementary particles of the same kind (see chapter 7(1)) does not change the world (universe).

(3) All sets of initial conditions compatible with the laws of nature occur as states (are played through) during the life time of our universe.

(4) All fundamental constants are ruled by laws of nature.

It is very questinable that condition (1) is satisfied. This is shown by processes of thermodynamics, radiation, dispersion, diffusion, of life, of cosmological evolution etc. Of course this holds under the assumption that not all randomness comes ultimately from inherent and hidden deterministic laws. This means that statistical laws have a realistic interpretation describing nature which has some degrees of freedom and not just describing our ignorance of nature.

In addition it is very unprobable that condition (3) is satisfied. Even if the choice of initial conditions at the beginning of the universe would have been rather small — a question which is hardly decidable — as soon as we take statistical laws seriously (cf. condition (1)) there will be a great number of states which have not been realized because of the degrees of freedom which allow different states by chance. But even if taking just deterministic laws, with asymmetric initial conditions asymmetric effects are produced: this special plane of the orbit of a planet (not that it is a plane which follows from the rotationally symmetric laws) is due to initial conditions and the plane could lie in a different angle to the one realized. Or take charge symmetry-violation. The ratio of the decay rate could be slightly different (due to some change in the initial conditions) from the one observed. Take parity: The ratio of the rates of snails having left screw houses to those having right screw houses (or the respective ratio of heart on the left side and on the right side) could be different without affecting biological laws. In radioactive decay phenomena parity violation could be more frequently than observed. Such examples (which could be continued) suggest that it is highly unprobable that all possible changes of initial conditions will be realized some time in this universe.

It is a very difficult question whether condition (4) is true. Can we change one of the fundamental constants without changing laws? This is connected with the question whether all of these constants are really constant or whether they change (slightly) with time (with the evolution of the universe). If Dirac would be right with his Large Numbers Hypothesis[42] or in general if the fundamental constants of nature change with time then one can imagine that by some change of initial conditions (say the amount of energy of the universe is higher) the change could be faster or slowlier. In this case again the laws would be satified by other universes which differ from ours just by the speed of change of those constants. From all what we know so far however it seems very unlikely that these fundamental constants are in some sense independent from the laws because they enter at least the fundamental laws. In this case they do not allow different universes from ours to satisfy the laws of nature.

Summarizing I think it is more likely that the laws of nature are valid also in other universes which differ from ours. The reason is that at least conditions (1) and (3) seem not to be satisfied.

Institute of Philosophy
Salzburg University

REFERENCES

Aristotle (Heav) "On the Heavens", in: *The Complete Works of Aristotle*, The revised Oxford Translation, Vol. 1, Barnes J. (ed.), Princeton University Press, Princeton 1985, p. 447–511.

Aspect, A. et al. (1982) "Experimental Tests of Bell's Inequalities using Time-varying Analysers", in: *Physical Review* Vol. 49, p. 1804–1807.

Bell, J. (1992) "Indeterminismus und Nichtlokalität", in: *Naturwissenschaft und Weltbild*, ed. H.C. Reichel und E.P. de la Riba, Vienna, p. 85–98.

Berry, M. (1978) "Regular and Irregular Motion" in: Jorna, S. (ed.) *Topics in Nonlinear Dynamics*, Amer. Inst. of Physics, New York 1978, p. 16–120.

Casati, G. – Chirikov, B.V. (1994) "The Legacy of Chaos in Quantum Mechanics", in: Casati, G. – Chirikov, B.V. (eds.) *Quantum Chaos Between Order and Disorder*, Cambridge University Press 1994.

Chiao, R.Y. – Kwiat, P.G. – Steinberg, A.M. (1993) "Faster Than Light?", in: *Scientific American* 10.

Chiao, R.Y. (1994) "Superluminal (but Causal) Propagation of Wave Packets in Transparent Media with Inverted Atomic Populations", in: *Phys. Review A*.

Chirikov, B.V. (1991a) "Time-Dependent Quantum Systems", in: Giannoni, M.J. – Voros, A. – Zinn-Justin, J. (eds.) *Chaos and Quantum Physics (Les Houches 1989)*, Elsevier, Amsterdam, p. 443–545.

Chirikov, B.V. (1991b) "Patterns in Chaos", in: *Chaos Solitons and Fractals* Vol. 1, No. 1, p. 79–103.

Cornfeld – Fomin – Sinai (1982) *Ergodic Theory*, Springer.

Dirac, P.A.M. (1973) "Fundamental Constants and Their Development in Time", in: Mehra, J. (ed.). *The Physicist's Conception of Nature*, Reidel, Dordrecht, p. 45–54.

Einstein, A. – Podolski, B. – Rosen, N. (1935) "Can Quantum-mechanical Description of Physical Reality be Considered Complete", in: *Phys. Review* 47, p. 777–780.

Einstein, A. (1949) "Autobiographical Notes", in: Schilpp, P.A. (ed.), *Albert Einstein. Philosopher – Scientist*, La Salle, Open Court, p. 1–94.

Greenberger, D.M. – Horne, M.A. – Zeilinger, A. (1989) "Going beyond Bell's Theorem", in: Kafatos, M. (ed.) *Bell's Theorem, Quantum Theory and Conceptions of the Universe*, Dordrecht, Kluwer, p. 73–76.

[42] Cf. Dirac (1973). For a discussion of these problems cf. Weingartner (1996a) ch. 6.

Greenberger, D.M. – Horne, M.A. – Shimony, A. – Zeilinger, A. (1990) "Bell's Theorem without Inequalities", in: *Amer. J. of Physics* 58, p. 1131f.

Hawking, St. (1988) *A Brief History of Time*, London, Bantam.

Kiess, T.E. – Shih, Y.H. – Sergienko, A.V. – Alley, C.O. (1993) "Einstein-Podolsky-Rosen-Bohm Experiment Using Pairs of Light Quanta Produced by Type-II Parametric Down-Conversion", in: *Rev. Phys. Lett* 71, p. 3893.

Kwiat, P.G. – Steinberg, A.M. – Chiao, R.Y. (1993) "High-Visibility Interference in a Bell-Inequality Experiment for Energy and Time", in: *Physical Review A* 47, 4, p. R2472–2475.

Laplace, P.S. (1951) *A Philosophical Essay on Probabilities*, New York, Dover.

Leibniz, G.W. (GP) *Die philosophischen Schriften von G.W. Leibniz*, Berlin 1875–90, Gerhardt, C.I. (ed.). Reprint: Hildesheim, Georg Olms.

Lighthill, J. (1986) "The Recently Recognized Failure of Predictability in Newtonian Dynamics", in: *Proceedings of the Royal Society London* A 407, p. 35–50.

Linde, A. (1990) *Inflation and Quantum Cosmology*, Academic Press.

Maxwell, J.C. (MAM) *Matter and Motion*, Dover, New York 1992.

Mittelstaedt (1978) *Quantum Logic*, Reidel, Dordrecht.

Peirce, Ch.S. (1935, 1960) *Collected Papers of Ch.S. Peirce* ed. by Ch. Hartshorne and P. Weiss. Harvard U.P., Cambridge Mass.

Popper, K.R. (1959) *The Logic of Scientific Discovery*, London, Hutchinson.

Rescher, N. (1967) *The Philosophy of Leibniz*, Englewood Cliffs.

Rosen, N. (1985) "Quantum Mechanics and Reality", in: Lahti, P. – Mittelstaedt, P. (eds.) *Symposium on the Foundations of Modern Physics*. Singapore, World Scientific, p. 32

Ruelle, D. (1991) *Chance and Chaos*, Princeton University Press.

Schuster, H.G. (1989) *Deterministic Chaos*, VCH, Weinheim.

Thomas Aquinas (STh) *Summa Theologiae*, Latin text and English translation, Balckfriars, London 1962 ff.

Weinberg, St. (1977) *The First Three Minutes*, London, Deutsch.

Weinberg, St. (1987) "Towards the Final Laws of Physics", in: *Elementary Particles and the Laws of Physics*, Cambridge 1987, p. 61–110.

Weingartner, P. (1983) "The Ideal of Mathematization of All Sciences and of 'More Geometrico' in Descartes and Leibniz", in: Shea, W.R. (ed.) *Nature Mathematized*, Dordrecht, Reidel, p. 151–195.

Weingartner, P. (1993) "A Logic for QM based on Classical Logic", in: de la Luiz Garcia Alonso, M. – Moutsopoulos, E. – Seel, G. (eds.), *L'Art, la Science et la Métaphysique*, Bern, P. Lang, p. 439–458.

Weingartner, P. (1996a) "Under What Transformations are Laws Invariant?", in: Weingartner, P. – Schurz, G. (eds.) *Law and Prediction in the Light of Chaos Research* (to appear).

Weingartner, P. (1996b) "Are Statistical Laws Genuine Laws? A concern of Poincaré and Boltzmann", in: *Grazer Philosophische Studien* (to appear).

Wigner (1967) *Symmetries and Reflections*, Bloomington, Indiana Univ. Press.

Zeilinger, A. (1992) "Physik und Wirklichkeit — Neuere Entwicklungen zum Einstein-Podolsky-Rosen Paradoxon", in: *Naturwissenschaft und Weltbild*, Reichel, H.C., Prat de la Riba, E. (eds.), Wien, p. 99–121.

Zermelo, E. (1896a) "Über einen Satz der Dynamik und die mechanische Wärmetheorie", in: *Wied. Annalen* 57.

Zermelo, E. (1896b) "Über die mechanische Erklärung irreversibler Vorgänge", in: *Wied. Annalen* 57.

Symposium 4
Logic in Central and Eastern Europe

LOGIC IN CENTRAL AND EASTERN EUROPE

The Special Symposium "Logic in Central and Eastern Europe" was held with Professors Vladimir A. Uspensky, Ryszard Wojcicki, Lev Bukovský, and Dimiter Vakarelov as invited speakers.

The reason for having a symposium on logic in central and eastern Europe was (understood to be) the great and substantial changes that happened in this area after the collapse of the communist system. The aim was not to describe or evaluate these changes but to identify their influence on logic, logical schools, groups and teams: what has happened with formerly famous groups, which are the new ones, where are the distinguished personalities, promising young logicians, what are the actual topics of research etc. The speakers were asked to pay attention to logic in full extent: not only to their own school but also to others, including both mathematical and philosophical logic. The talks could not go into too many details, but had to identify main streams of research as well as cooperation links. Needless to say, we had to discuss problems (e.g. extent of emigration or, in the case of Balkan, the unfortunate war). But the main gain expected was in showing (to us from Central and Eastern Europe as well as to interested people outside) that and how logic is alive in this area and inspire cooperation.

I am happy to say that these expectations were fulfilled. All four invited speakers accepted their invitation and the domain to be described and were able to attend the meeting and deliver their talks. The session was well attended, mainly but not only by logicians from the area concerned. There was a lively discussion, stressing, among other things, the thankfulness of logicians from the post-communist countries to the Western European logical community for their support during the past times, mainly (but not only) by sending invitations to Logic Colloquia (with financial support) to several logicians from these countries. Another fact was stressed, namely the importance of continuing scientific meetings concerning logic in these countries for keeping contacts.

One may well hope that in the new situation, without political barriers, the old cooperation and friendship wil survive and new will emerge, now fully in an all-European (and global) context.

Petr Hájek
Institute of Computer Science
Academy of Sciences, Prague

M.L. Dalla Chiara et al. (eds.), Logic and Scientific Methods, 449.

LEV BUKOVSKÝ

LOGIC IN CZECHOSLOVAKIA AND HUNGARY

1. INTRODUCTION

I have to speak about the development of research in logic in Czech Republic, Hungary and Slovak Republic. Since the development in Czech and Slovak Republics went for long time in former Czechoslovakia together —and still does— I will not distinguish between those two countries. Just speaking about the contemporary situation I will stress who is in which of the two republics.

I tried to obtain full information about logic in Czechoslovakia and Hungary, however, I was sure that it was not possible. So I tried to follow rather some general aspects of the development and not to give a complete survey of related activities. Moreover I will concentrate more on the development of logic in Czechoslovakia because essentially I have more personal information about it.

The word "logic" means both mathematical and philosophical logic. Mathematical logic is considered to the extent given by the Mathematical Reviews: general logic, model theory, recursion theory, set theory, proof theory and constructive mathematics, algebraic logic, nonstandard models. Set theory means axiomatic set theory and its problems, models of set theory and related independence results. Since in the considered countries the logicians tried to devote their research to computer science, I will emphasize also this area.

2. SPECIALITIES OF SOCIALISTIC COUNTRIES

I must recall some facts concerning life behind the iron curtain which had an influence on the development of research in general, and especially, research in logic.

Some six years ago I read a manuscript by A.R.D. Mathias of Cambridge [7] entitled "Logic and Terror" devoted to some sad stories of Stalin's times. I do not want to speak about this story but I only want to recall that, at the beginning of the fifties, some scientific domains were considered "false science", "pseudoscience" or "bourgeois science". The adjective "bourgeois" meant something that nobody explained but surely it meant something very bad and consequently dangerous for people. This concerned in particular genetics, cybernetics and partially mathematical logic. This happened in the Soviet Union and —unfortunately— sometimes the satellite countries accepted this attitude with still worse consequences.

M.L. Dalla Chiara et al. (eds.), Logic and Scientific Methods, 451–456.

The meaning of the word "emigration" had a special flavor. An emigrated person was considered an enemy of the country and it was rather obvious that he or she would have no possibility to visit the country he/she left, nor to come back.

Let me mention the problem of contacts with colleagues in the western countries. We were sure that our correspondence was read by the secret service, and traveling for some conference in the west was rather an exception. For a long time there were great difficulties with obtaining scientific information. In the fifties, Russian translations of scientific literature were the only sources of information. At least in Czechoslovakia, mathematicians used to buy plenty of Russian books (mostly translations) devoted to mathematics. E.g. the first contacts of many of them with mathematical logic were through A.S. Jesenin-Volpin's translation of S.C. Kleene's book [6] 1957. I suppose that we should express our thanks to our Russian colleagues for this good and very useful job.

I have to mention the following fact. Generally there was a great shortage of efficient computers. As a consequence, scientists interested in the computer sciences concentrated their research in theoretical areas, e.g. formal languages, computational complexity, artificial intelligence, etc. The result of this orientation is still visible.

3. THE FIRST LOGIC IN CZECHOSLOVAKIA

We could say that Czechoslovakia was very close to logic. Kurt Gödel was born in Brno and Gerhard Gentzen died in Prague. However their research had no direct influence on research in Czechoslovakia. We do not know about any research activities in logic in Czechoslovakia between the wars.

As I have already mentioned, the early fifties were not promoting any logical activity in Czechoslovakia and Hungary. Doing so was rather dangerous, as one would risk to be proclaimed a bourgeois scientist. Philosophical positivism was considered one of the main enemies of marxist philosophy, and mathematical logic was considered as something necessarily leading to positivism. Even later some caution was desirable: the book [9] contains chapters 1.4 and 1.9 entitled "The dialectic of the relation between mathematical and metamathematical aspects" and "Philosophy and mathematical logic", respectively.

Ladislav Svante Rieger was the first Czechoslovak mathematician to work systematically in mathematical logic, starting his work about 1957. Rieger began his investigations in algebraic logic giving a description of the algebra of intuitionistic logic with one free variable. Then he continued with the investigation of so-called cylindrical algebras. Finally he studied algebraic constructions of models for fragments of Gödel-Bernays set theory, using valued fields.

In the late fifties J. Hořejš (now a recognized specialist in computer science) in Brno and M. Kolibiar (the late prominent algebraist) in Bratislava tried to open logic seminaries. As far as I can remember (as a member of this seminar), Kolibiar's seminar was considered an ideologically non-desirable one. At the beginning of the sixties, L.S. Rieger started with a logic seminar (under name Axiomatic Set Theory) in Prague. P. Vopěnka and E. Kindler (later a specialist in computer science) were among the first members of this seminar. I obtained a research position in Rieger's section in the Mathematical Institute of the Czechoslovak Academy of Sciences in Prague and Petr Hájek became the first (and also the last) graduate student of Rieger. Unfortunately, Rieger died in 1963. P. Vopěnka —who already obtained the first important results in axiomatic set theory— joined the rising Rieger group in a big seminar in the academic year 1963/64. After one year of introductory studies devoted to basic set theory and its models, the seminar continued in studying Vopěnka's reworking of P.J. Cohen's proof of the independence of the continuum hypothesis. The history is well recorded by P. Hájek in [1] and [2].

In the early sixties, O. Demuth finished his studies of constructive mathematics with N.A. Shanin in Leningrad (= St. Petersburg) and returned to Prague. In philosophical logic, P. Tichý started research in transparent intensional logic.

4. THE FIRST LOGIC IN HUNGARY

In Hungary, research in logic started already between the wars. Both László Kalmár and Rózsa Péter wrote their first papers devoted to logic in 1934. The book of R. Péter [8] introduced many mathematicians of central Europe to recursion theory. L. Kálmar was known in logic in connection with his papers devoted to unsolvability of some mathematical problems — compare e.g. [6]. Later, with his papers [3, 4] he became a forerunner of artificial intelligence's aims.

A. Hajnal was the first important graduate student of L. Kalmár. In the second half of the fifties, he wrote his thesis devoted to axiomatic set theory.

5. LOGIC IN CZECHOSLOVAKIA NOWADAYS

Many Czechoslovak logicians (B. Balcar, M. Benda, L. Bukovský, K. Čuda, P. Hájek, K. Hrbáček, T. Jech, J. Mlček, partially K. Prikry, A. Sochor, P. Štěpánek) started their research in Vopěnka's seminar. I should like to stress the great interest in and help with the work of this seminar by Polish mathematicians, namely A. Mostowski and C. Ryll-Nardzewski. It was A. Mostowski who rather systematically informed P. Vopěnka and his fellows about all important results obtained in axiomatic set theory.

The year 1968 and everything that followed was a breaking point in the development of Vopěnka's seminar. The ideological pressure that replaced the freedom of Prague Spring caused emigration of many prominent scholars from Czechoslovakia. I cannot say that there were special attacks on logic.

However, because of their political activities in 1968 many logicians (mainly P. Vopěnka and some of his fellows) were considered enemies of the country. Vopěnka's group was practically destroyed. Consequently several members of his seminar emigrated. A. Kanamori in [5] considers the influence of this Czech emigration on the development of set theory in USA saying "their emigration was a Prague spring of sorts for set theory in the United States". In Prague it was B. Balcar who had the worst conditions for his research.

Of course ideology followed carefully the field of philosophical logic, and as a consequence the persona non grata P. Tichý emigrated to New Zealand.

Now I suppose I should tell another story. In the late sixties many logicians in western countries have understood the isolation of their colleagues in Czechoslovakia and in other east Europian countries and they began with their help. They found many financial sources to bring young mathematical logicians of east Europe to conferences in western countries, mainly to the logic colloquia, allowing them to be in contact with recent research. This maintenance, at least for mathematical logic in Czechoslovakia, was important in particular after 1968.

The collapse of the socialist system in 1989 has substantial consequences for the development of philosophical logic in Czechoslovakia. This domain practically did not exist. E.g. P. Cmoraj and his fellows in Bratislava could just try to follow some activities in western countries but did not participate. In the Academy of Sciences in Prague, a small group working in philosophical logic consisting mainly of young people was established in 1987 under a commonly used nickname "logic and artificial intelligence". After 1989 scientists of this group (T. Childers, M. Duží, P. Kolář, P. Materna, O. Majer, J. Peregrin, V. Svoboda) largely developed their research and other activities devoted mainly to non-classical logic. Their international cooperation is wonderful and the results obtained are a great hope for the future. In mathematical logic the development after 1989 went on rather continuously with increasing international cooperation and —as a consequence— grew both in quality as well as quantity.

Let me give a brief survey of topics of mathematical logic studied in Czech and Slovak Republics nowadays. The alternative set theory created by P. Vopěnka is investigated in Prague (K. Čuda, J. Mlček, J. Sgall, A. Sochor) and in Bratislava (P. Zlatoš and all.). There are at least two groups working in axiomatic set theory: B. Balcar and P. Simon in Prague close to topology and L. Bukovský, M. Repický and P. Vojtáš close to real analysis. The leading person in recursion theory is A. Kučera in Prague (O. Demuth died). I. Korec in Bratislava works between recursion theory and computer science. P. Štěpánek and O. Štěpánková in Prague work in logic programming. The strong Prague School of Arithmetics (and complexity) is represented at least by P. Hájek, J. Krajíček and P. Pudlák. There is a large group working in quantum logic: A. Dvurečenskij, R. Mäsiar, S. Pulmannová, Z. Riečanová and others in Bratislava and P. Pták and M. Navara in Prague. I am not able to give a list of people interested in fuzzy logic.

A great number of logicians came out from Vopěnka's seminar. P. Vopěnka spread the ideas and enthusiasm for research. However the leading person in logic is P. Hájek with results in many areas: models of set theory, theory of semisets, arithmetic, automatic generation of hypotheses, dynamic logic, modal calculus, approximative reasoning (i.e. reasoning under uncertainty) etc.

6. LOGIC IN HUNGARY NOWADAYS

As I have already said in the introduction, I cannot give a picture of logic in Hungary that is as detailed as the one sketched for Czechoslovakia. Anyway I will try not to forget anything important.

In Budapest, there is a strong group of mathematicians working in axiomatic set theory consisting mainly of recent students of A. Hajnal: J. Gerlits, I. Juhász, P. Komjáth, Zs. Nagy, L. Soukup, etc. According to them the main source of inspiration for their interest and research in logic, namely in axiomatic set theory, was a mathematician ignorant of logic: Paul Erdös.

Another group led by I. Németi started in algebraic logic. In papers written with H. Andréka and I. Sain he presents results devoted to the applications of non-standard models of arithmetic in verification theory.

The field of Artificial Intelligence is one of the most active fields of research leading to applications in Hungary today. The results obtained by F. Darvas, I. Futó, G. Rédey, I. Rudas, P. Szeredi and many others form both a theoretical contribution to knowledge and a practical efficient application. I recommend the survey paper by E. Såntåné-Tóth [10].

I finish with the assertion that there exists research in philosophical logic in Hungary.

7. SOME GENERAL CONCLUSIONS

Before 1989 the isolation of Czechoslovak scholars was much stronger than that of their Hungarian and Polish fellows. The quality of the research in logic that was done in Czechoslovakia in years 1964–1989 was strongly influenced by the following three factors: the Russian translations of scientific literature, the personal help of our Polish and Hungarian fellows, and the financial travelling support and invitations, which our western colleagues always were able to find.

We cannot say that the political changes of 1989 caused essential changes in the research in mathematical logic in the considered countries. The opening to the world, freedom in traveling and in other communication influenced research in logic. However, the field of philosophical logic started to really live just after 1989.

Katedra Matematickej Informatiky
Košice, Slovakia

REFERENCES

1. Hájek P., *Sets, semisets, models,* Axiomatic Set Theory (Scott D. S., ed.), Proceedings of Symposia in Pure Mathematics, Vol. XIII, Part I, American Mathematical Society, Providence, 1971
2. Hájek P., *Bibliography of the Prague seminar on foundations of set theory — part II,* Czech. Math. J. **23** (1973), 521–523
3. Kalmár L., *A new principle of construction of logical machines,* 2-ième Congrés International de Cybernétique, Namur, 1958, 458–463
4. Kalmár L., *Über einen Rechenautomaten, der eine mathematische Sprache versteht,* Zeitschr. Angew. Math. Mech. **40** (1960), 64–65
5. Kanamori A., *The Higher Infinite,* Springer Verlag, Berlin, 1994
6. Kleene S.C., *Introduction to Metamathematics,* van Nostrand com., New York, 1952
7. Mathias A.R.D., Logic and terror, *Physis, rivista internationale di storia della scienza,* XXVIII (1991), 557–578.
8. Péter R., *Rekursive Funktionen,* Akadémiai Kiadó, Budapest, 1951
9. Rieger L.S., *Algebraic Methods of Mathematical Logic,* Academia, Prague, 1967
10. Sântâné-Tóth E., *Survey of Hungarian KBS Tools and Applications in the Engineering Field,* Eng. Applic. Artif. Intell. **4** (1991), 409–416

VLADIMIR A. USPENSKY

MATHEMATICAL LOGIC IN THE FORMER SOVIET UNION: BRIEF HISTORY AND CURRENT TRENDS

1. NEW REALITIES IN SCIENTIFIC LIFE IN THE FORMER SOVIET UNION

Scientific life in the former Soviet Union (for short: *fSU*), as well as life in general, has changed drastically in recent years.[1] Those changes are not specific for Mathematical Logic but without taking them into consideration, our situation in ML cannot be understood properly.

Of course we have great advantages now: fall of the iron curtain, cease of ideological restrictions, free travel abroad (in 1993 the exit visas still existed!). Now our scientist can be a participant or even an organizer of an international conference, become a member of an editorial board or join an international association, and all those activities can be executed without seeking a permission from a Communist Party committee (and it was only in 1989 when the author got permission to cross the iron curtain for the first time). Mathematical logicians, as well as other scientists, are now allowed to get support from the West and even to work abroad. The current support from ISF (Soros Foundation), INTAS (EC Foundation) and some other international sources provides a vital though temporary aid. An important role is played also by the 3 years' old Russian Foundation for Basic Research.

Still, there are many buts and howevers, some of them very hard. The main ones are: (1) sharp decrease of state investments in science and education; (2) lack of literature; (3) transportation problems; (4) communication problems; (5) slump of living standard; (6) brain drain. Let us comment on those problems.

(1) One can say that scientists' complaints of government cuts in financing science are heard all over the world. But our situation is special. Indeed, in the Soviet Union there were no other sources of funding for research and education than the federal State budget, and the same mainly remains valid today in the fSU. So the cuts of federal support turn into a catastrophe for many traditional research centres.

(2) Because of the budget cuts, university libraries reduced or even ceased receiving foreign literature, including journals. In a tough competition, more traditional areas of science usually win. That is why subscriptions to journals in ML have been cancelled almost completely.

[1]All data mentioned below are up to the period of the congress, i.e., August 1995

M.L. Dalla Chiara et al. (eds.), Logic and Scientific Methods, 457–483.

(3) Growth of train and air tickets prices became a new barrier preventing scientists not only from visiting one another but even from participating in the fSU conferences; in many cases, it is less expensive to travel abroad than inside Russia. Not all borders inside the fSU are transparent; the border between the three Baltic states and the other 12 ones is rather stiff. Civil wars in the Caucasus area make communication with Transcaucasian republics extremely difficult.

(4) Preparing this talk required effective exchange of information between the author and mathematical logicians throughout the fSU. This exchange met serious technical obstacles. Here is a quotation from a letter of a leading Moldavian scientist: "We have no fax; we do have e-mail but it does not work". Because of money shortage, Moscow University has stopped mailing letters abroad (in particular, to any place of the fSU beyond the borders of Russia) and has practically stopped long distance telephone connection even inside Russia. All mailing and phoning needed for preparing this talk was granted by the courtesy of All-Russian Institute for Scientific and Technical Information (J. Arski, Director).

(5) In a Soviet movie a typical well-to-do cast character was a university professor. Currently in Russia, the salary of a full professor is less than $100 per month before taxes. Prices are close to the European ones, and just in August of 1995 the competent Russian newspaper "Izvestiya" declared that the prices for sugar, butter and gasoline had exceeded the Western ones. In Armenia and Georgia a professor's salary ranges from $2 to $4 per month which suffices only to pay the public transportation for one person.

(6) In the fSU, there are two kinds of brain drain: (a) the external one (leaving the country) and (b) the internal one (leaving science for more profitable activities). Why were Mathematics and Chess at such a high level in Soviet Union? Because they gave to ambitious young people a unique opportunity to exercise their intellect without any ideological control. Now those people move to business.

Nevertheless, as can be seen from the following pages, Mathematical Logic is still alive in the fSU. Moreover, despite extreme living and labor conditions some of the leading research groups and individuals celebrated the past 5 years with top class scientific results. Perhaps our researchers and scholars currently working in ML deserve the world community's encouragement and support.

How could the world logical community improve our uneasy situation? First, Western logicians have to answer the following straight question. What is more desirable for a scientist: to suppress a rival or to support a partner? Next, there is one practical observation: **one** dollar used for the support of researchers within the fSU is equivalent to **tens** of dollars used for the support **of the same researchers** abroad. The total annual expense for one PhD student in Western Europe is roughly equal to an annual budget of a medium size laboratory in Russia.

2. A GLIMPSE OF HISTORY OF MATHEMATICAL LOGIC IN THE FSU

Each history is preceded by a prehistory. Perhaps, the prehistory of ML in the fSU should include Mohammad al-Kwârizmi (IX c. A.D.) who gave his name to the word "algorithm". It does include two Professors of Kazan University: Platon S. Poretsky (15.10.1846–22.8.1907) who was the first in the fSU who lectured in ML and Nikolai A. Vasil'ev (11.7.1880–31.12.1940) who foresaw in a fashion some ideas of nonclassical logic. And at the border of prehistory and history Ivan I. Zhegalkin (3.8.1869–28.3.1947) is situated; he treated the propositional logic as a remainder arithmetic (modulo 2).

A special place in history is occupied by Sophia A. Yanovskaya (31.1.1896–24.10.1966). She made no research in ML by her own, but deserves credit for making it possible for others. She wrote surveys [133] and [134] (see also a survey [41] by B.A. Trakhtenbrot et al.). In 1943 Yanovskaya founded the Seminar in Mathematical Logic at Moscow University — the first such institution in SU. At the beginning, that Seminar was run by Yanovskaya, Zhegalkin, P.S. Novikov, later by A.A. Markov, and then by A.N. Kolmogorov, S.I. Adyan, V.A. Uspensky.

There are four main scientific schools in ML in the fSU. They are related to the names of Andrei N. Kolmogorov (25.4.1903–20.10.1967), Anatoly I. Mal't-sev (27.11.1909–7.7.1967), Pëtr S. Novikov (28.8.1901–9.1.1975), and Andrei A. Markov (22.9.1903–11.10.1979).

Kolmogorov published his first paper in ML in **1925**; that paper (in which, for the first time, intuitionistic logic became a subject of mathematical investigation) began the real history of ML in SU. In 1952 Kolmogorov suggested a very general notion of an algorithm ("Kolmogorov–Uspensky machine"). In 1954 Kolmogorov founded the theory of numberings; a *numbering* of a set M is an arbitrary mapping α from an arbitrary set of natural numbers onto M, so a number n, being a constructive object, can be treated as a *name* of $\alpha(n)$. In 1963–64 he came to the notion of Kolmogorov complexity. A survey of Kolmogorov's contribution to ML is given in [123]. The list of logicians coming from Kolmogorov's school includes: J. Barzdins, V.G. Kanovei, L. Levin, V.A. Lyubetsky, P. Martin-Löf, Yu.T. Medvedev, V.Ye. Plisko, A.Kh. Shen, V.A. Uspensky (who has now inherited Kolmogorov's Chair of Matematical Logic at Moscow University), N.K. Vereshchagin. The Kolmogorov school studies recursion theory and its applications (Kolmogorov complexity, algorithmic probability), computational complexity, nonclassical logic, mathematical linguistics.

In 1934–37 Mal'tsev was a postgraduate student of Kolmogorov. His first paper in ML is dated **1936**. He made significant contributions to model theory, recursion theory, study of decision problems for elementary theories (see a survey in [76]). He united model theory and Kolmogorov's theory of numberings, viz., he introduced and developed the notion of a numbered mathematical structure, i.e. of a structure together with a numbering (a *system of names*); since in a general case, elements of a structure are not constructive objects, only supplying them with names (numbers) makes it possible to speak on algorithms involving those elements. Among Mal'tsev's successors are: O.V. Belegradek, N.V. Belyakin, A.N. Degtev, Yu.L. Ershov (who has inherited Mal'tsev's position as the leader of Siberian scientific school in algebra and logic), S.S. Goncharov, B. Khoussainov, L.L. Maksimova, A.S. Morozov, A. Omarov, E.A. Palyutin, M.G. Peretyat'kin, V.V. Rybakov, M.A. Taitslin, B.I. Zilber. The areas of research of Mal'tsev's school are: model theory, recursion theory, first-order logic, nonclassical logic, theoretical CS.

Being a former student of Luzin, P.S. Novikov came to ML from descriptive set theory (and he, independently of others, noticed similarity between the projective and the arithmetical hierarchies). He published his first paper in ML in **1939**. His works contain outstanding results in proof theory, axiomatic set theory, algorithmic algebra: see a survey in [87]. His school includes: S.I. Adyan (who inherited Novikov's position as the Head of Department of Mathematical Logic at Steklov Mathematical Institute, Moscow), A.V. Gladkii, V.N. Grishin, A.A. Muchnik, An.A. Muchnik, A.A. Razborov. Among achievements of this school are: normalization of proofs in arithmetic, undecidability theorems for some problems of algorithmic algebra, solving the Post reducibility problem, studies in resource logic, superpolynomial lower bounds for monotone Boolean complexity (for the latter result Razborov was given the prestigious Nevanlinna Award of 1990 in Mathematics).

Markov's first publication in ML appeared in **1947**, and in that paper independently of Post, he demonstrated the undecidability of Thue word problem. Afterwards he obtained many other famous results in the theory of algorithms. A.A. Markov is the father of the Russian Constructivism. His personal achievements are reviewed in [80] and [81]. His successors in Foundations (i. e. in Constructivism) are B. Kushner, N.M. Nagorny, N.A. Shanin, I.D. Zaslavsky. Markov's school also includes: S.N. Artemov, A. Dragalin, D. Grigor'ev, M.I. Kanovich, G.S. Makanin, S.Yu. Maslov, Yu.V. Matiyasevich (now he is Director of the Euler Institute, St. Petersburg), G. Mints, V.P. Orevkov, A.L. Semenov, V.B. Shehtman, A. Slisenko, G.S. Tseytin. In the scope of interests of Markov's school there are proof theory, nonclassical logic, complexity of proofs and algorithms, automatic deduction, decidability or undecidability of algorithmic problems, logical problems of CS.

3. MAJOR SCIENTIFIC GROUPS IN MATHEMATICAL LOGIC IN THE FSU

Unlike many Western countries, in the Soviet Union scientific groups used to concentrate around few "centres of crystallization". The main groups in Logic were initiated by the four schools mentioned above. Now let us list the existing groups in ML and their specific areas of research in recent times.

RUSSIA

Moscow

1. *Department of Mathematical Logic, Steklov Mathematical Institute:* Sergei I. ADYAN (algorithmic algebra) — Head; Sergei N. ARTEMOV (proof theory, theoretical CS); Lev D. BEKLEMISHEV (proof theory, nonclassical logic); Vyacheslav N. GRISHIN (set theory, substructural logics); Gennady S. MAKANIN (algorithmic algebra); Alexander A. RAZBOROV (complexity theory, proof theory).

2. *Department of Mathematical Logic and Theory of Algorithms, Faculty of Mathematics and Mechanics, Moscow State University:* Vladimir A. USPENSKY(recursion theory, Kolmogorov complexity) — Head; Sergei I. ADYAN (algorithmic algebra); Sergei N. ARTEMOV (proof theory, theoretical CS); Vladimir N. KRUPSKI (complexity theory, theoretical CS); Elena Yu. NOGINA (recursive topology, provability logics); Valery Ye. PLISKO (constructive logic); Postdoc: Alexander D. YASHIN (intuitionistic logic); Ph.D. students: A. CHUPRINA, A. GOMOLKO, T. SIDON, M. PENTUS, R. YAVORSKY.

The Department contains Laboratory "Logic in Computer Science": Sergei N. ARTEMOV — Head; Lev D. BEKLEMISHEV (proof theory, nonclassical logic); Max I. KANOVICH (complexity theory, linear logic); Alexei P. KOPYLOV (linear logic); Mati R. PENTUS (mathematical linguistics, nonclassical logic); Valentin B. SHEHTMAN (nonclassical logic).

The Department also sponsors permanent seminars for the Moscow community in Logic and Theoretical Computer Science. The major of them are: "Moscow City Seminar on Mathematical Logic", Heads: S.I. ADYAN, V.A. USPENSKY; "Kolmogorov Seminar" on Kolmogorov complexity, computational complexity, and related topics, Heads: Alexei L. SEMENOV, Alexander Kh. SHEN, Nikolai K. VERESHCHAGIN; "Proof Theory", Heads: S.N. ARTEMOV, L.D. BEKLEMISHEV.

3. *All-Russian Institute for Scientific and Technical Information (VINITI):* Vladimir B. BORSHCHEV (theoretical CS); Viktor K. FINN (logic in AI); Elena V. PADUCHEVA (applications of ML to linguistics); Dmitry P. SKVORTSOV (nonclassical logic).

4. *"Moscow-Tver Group"* (nonclassical logic): Alexander V. CHAGROV — Head; Lilia A. CHAGROVA; Albert A. MUCHNIK; Valentin B. SHEHTMAN; Dmitry P. SKVORTSOV; Michael V. ZAKHARYASCHEV.

5. *Institute of Logic and Cognitology of Development:* Vladimir A. SMIRNOV (nonclassical logic, automated deduction) — Head; Peter I. BYSTROV (modal logic); Alexander S. KARPENKO (nonclassical logic); Vladimir MARKIN (automated deduction); Alexei NOVODVORSKY (automated deduction); Evgeny A. SIDORENKO (relevance logic); Alexei V. SMIRNOV (automated deduction).

6. *Department of Logic at Faculty of Philosophy, Moscow University:* Yuri V. IVLEV (modal and intensional logic); Elena D. SMIRNOVA (logical semantics); Evgeny K. VOISHVILLO (relevance logic).

7. *Other researchers:* Evgeny A. ASARIN (Kolmogorov complexity, algorithmic probability); Alexander Ya. DIKOVSKY (mathematical linguistics); Alexei V. GLADKII (mathematical linguistics); Konstantin Yu. GORBUNOV (complexity theory); Vladimir G. KANOVEI (axiomatic set theory); Valery Kh. KHAKHANYAN (intuitionistic set theory); Valentin I. KHOMICH (intermediate logics); Igor A. LAVROV (recursion theory); Vasily A. LUBETSKY (intuitionistic set theory, model-theoretic algebra); Sergei S. MARCHENKOV (recursion theory); Yuri T. MEDVEDEV (recursion theory, nonclassical logic); Andrei A. MUCHNIK (recursion theory, Kolmogorov complexity, algorithmic probability); Nikolai M. NAGORNY (constructive mathematics); Mars K. VALIEV (dynamic and temporal logics); Vladimir G. VOVK (Kolmogorov complexity, algorithmic probability); Vladimir V. V'YUGIN (Kolmogorov complexity, algorithmic probability).

St. Petersburg

1. *Laboratory at Steklov Institute, St. Petersburg:* This group has suffered significant losses in recent years: Gregory Mints, Anatol' Slisenko, Dmitry Grigor'ev have left the country, Sergei Yu. Maslov has tragically died.

The remaining logicians from this group are: Yuri V. MATIYASEVICH (recursion theory, proof theory, theoretical CS) — Head; Evgeny Ya. DANTSIN (proof theory); Eldar A. MUSAEV (recursion theory); Vladimir P. OREVKOV (proof theory); Nikolai A. SHANIN (constructive mathematics).

2. *Other logicians in St. Petersburg:* Nikolai K. KOSSOVSKY (complexity theory); Oleg F. SEREBRYANNIKOV (proof theory); Grigory S. TSEYTIN (recursion theory, theoretical CS).

Tver

1. *Department of Mathematics, Tver State University:* Alexander V. CHAGROV (nonclassical logic); Lilia A. CHAGROVA (nonclassical logic).

2. *Department of Computer Science, Tver State University:* Group in dynamic logics, finite model theory and linear logic: Michael A. TAITSLIN — Department Head; Dmitry A. ARCHANGELSKY; Michael I. DEKHTYAR; E. KRUGLOV; Ilmir Kh. MUSIKAEV. This group works in collaboration with Alexei STOLBOUSHKIN (who is now in USA).

Pereslavl-Zalessky Vladimir Yu. SAZONOV (finite model theory).

Ivanovo Sergei R. KOGALOVSKII (model theory, higher-order logic); Evgeny A. POLYAKOV (recursion theory); Mikhail G. ROZINAS (recursion theory); Tatiana Ya. SENKEVICH (nonstandard analysis); Alexei I. SHCHEGLOV (algebras of partial recursive functions); Boris Ya. SOLON (recursion theory).

Nizhny Novgorod Evgeny I. GORDON (nonstandard analysis).

Vladikavkaz Anatoly G. KUSRAEV (nonstandard analysis).

Ul'yanovsk Valentin A. BAZHANOV (history of logic).

Kazan

1. *Department of Algebra and Mathematical Logic, Kazan State University* (recursion theory): Marat M. ARSLANOV (degrees of unsolvability) — Head; Alexander A. EFREMOV; Vitaly V. OSIPOV; Victor V. VYACHESLAVOV.

2. *Department of Algebra and Mathematical Logic, N.G. Chebotarev Research Institute:* Marat M. ARSLANOV (recursion theory, logic in CS) — Head; Shamil T. ISHMUCHAMETOV (recursion theory, logic in CS); Ekaterina V. LIPACHEVA (recursive model theory); Valery D. SOLOV'EV (generalized recursion theory, logic in CS).

3. *Department of Applied Mathematics, Kazan State University* (proof theory, logic in CS): Nail K. ZAMOV — Head; Vadim Yu. BAZYLEV; Valery Yu. MIKHAILOV; Pavel V. PSHENICHNY.

Saratov Vladimir A. MOLCHANOV (nonstandard analysis); Dmitry A. BREDIKHIN (algebras of relations).

Izhevsk Nikolai N. NEPEIVODA (theoretical CS); Valentina BARKOVA (theoretical CS); Anatoly P. BELTYUKOV (theoretical CS); Alexander D. YASHIN (nonclassical logic).

Ekaterinburg Yuri M. VAZHENIN (algorithmic algebra).

Tumen Alexander N. DEGTEV (degrees of unsolvability).

Omsk Alexei G. MYASNIKOV (model-theoretic algebra); Vladimir N. REMESLENNIKOV (model-theoretic algebra).

Novosibirsk

1. Institute of Mathematics, Russian Academy of Sciences.

1.1. Department of Mathematical Logic. Head: Yuri L. ERSHOV.

1.1.1. *Laboratory of Mathematical Logic:* Yuri L. ERSHOV (model theory, resursion theory) — Head; Nikolai V. BELYAKIN (recursion theory); Larissa L. MAKSIMOVA (nonclassical logic); Sergei I. MARDAEV (modal logic); Victor L. SELIVANOV (recursion theory).

1.1.2. *Laboratory of Applied Logic* (constructive model theory): Sergei S. GONCHAROV — Head; Sergei T. FEDORYAEV; Andrei S. MOROZOV; Dmitry E. PAL'CHUNOV.

1.2. *Laboratory of Algebraic Systems at the Department of Algebraic Systems and Group Theory* (model theory): Evgeny A. PALYUTIN — Head; Alexander N. RYASKIN.

1.3. *Laboratory of Functional Analysis at the Department of Analysis and Geometry:* Semën S. KUTATELADZE (nonstandard analysis) — Head.

The Institute of Mathematics sponsors the seminar "Algebra and Logic" which was founded by A.I. Mal'tsev in 1960 and now is run by the board headed by Yu.L. ERSHOV.

2. *Institute for Information Systems, Russian Academy of Sciences:* Nikolai V. SHILOV (dynamic and temporal logics).

Kemerovo Boris I. ZILBER (model theory); Oleg V. BELEGRADEK (model-theoretic algebra).

Krasnoyarsk Vladimir V. RYBAKOV (nonclassical logic); Mikhail I. GOLOVANOV (modal logic).

Irkutsk Nikolai M. PERYAZEV (logic in CS); Andrei V. MANTSIVODA (logic in CS).

UKRAINE

Kiev Mikhail S. BURGIN (logic in AI); A.T. ISHMURATOV (nonclassical logic); Mikhail I. KRATKO (automata theory); Alexander A. LETICHEVSKY (algorithmic algebra); Leonid P. LISOVIK (recursion theory); V.I. OMEL'YANCHIK (modal logic, quantum logic); G.E. TSEITLIN (applied logic).

Lvov T.S. KUDRIK (nonstandard analysis); V.Eh. LYANTSEH (nonstandard analysis); Vladimir VASYUKOV (nonclassical logic).

Odessa Valery K. BULITKO (recursion theory).

Harkov V. NAVROTSKY (temporal logic).

Krivoi Rog Yaroslav SHRAMKO (nonclassical logic).

Uzhgorod Alexander CITKIN (nonclassical logic) — left in 1991.

KAZAKHSTAN

The Kazakhstan school of ML has been always greatly influenced and supported by Novosibirsk logicians: at first by A.D. Taimanov, then by Yu.L. Ershov, S.S. Goncharov, E.A. Palyutin. There are two informal groups in Kazakhstan: 1. *Algorithmic problems in logic:* Peretyat'kin (group leader), Badaev, Dobritsa, Drobotun, Karatabanova, N. Khisamiev, Z. Khisamiev, Omarova, Zakir'yanov, Zaurbekov. 2. *Model theory:* Omarov (group leader), Bekenov, Eshkeev, Kudaibergenov, Meirembekov, Nurmagambetov, Nauryzbaev, Shegirov, Zhetpisov.

Alma-Ata

1. *Laboratory of Mathematical Logic, Institute of Theoretical and Applied Mathematics, Academy of Science, Republic of Kazakhstan:* Michail G. PERETYAT'KIN (finitely axiomatizable theories, algorithmic problems)—Head; Serikzhan BADAEV (theory of algorithms, computable numberings); Kanat KUDAIBERGENOV (homogenous models).

2. *Department of Algebra and Logic, Kazakh State University:* Amangeldy OMAROV (model theory) — Head; Kuanysh MEIREMBEKOV (model theory); Kairtai SHEGIROV (model theory); Sapargali ZAURBEKOV (algorithmic problems).

3. *Alma-Ata State University:* Vyacheslav P. DOBRITSA (numbered structures); Soreiya KARATABANOVA (algorithmic problems); Kemesh NAURYZBAEV (algebraic systems); Gulya OMAROVA (constructive structures).

Karaganda

Department of Algebra and Logic, Karaganda State University: Tursynbek

NURMAGAMBETOV (stable theories) — Head; Maksut BEKENOV (model theory, stable theories); Aibat ESHKEEV (model theory, stable theories); Kabylda ZHETPISOV (model theory).

Ust'-Kamenogorsk

1. *East-Kazakhstan Division of Academy of Science, Republic of Kazakhstan:* Nazif KHISAMIEV (constructive groups).

2. *Ust'-Kamenogorsk State University:* Zarif KHISAMIEV (theory of algorithms); Kairat ZAKIR'YANOV (constructive groups).

Pavlodar

Pavlodar Pedagogical Institute: Boris N. DROBOTUN (constructive structures).

UZBEKISTAN **Tashkent**

N. Kh. KASYMOV (constructive structures); Bakhadur KHOUSSAINOV (constructive structures) — left in 1991.

MOLDOVA **Kishinev**

The founder of scientific school in Moldova was Alexander V. KUZNETSOV (28.10.1926–24.07.1984) who since 1946 till 1965 served as a scientific secretary of Moscow City Seminar on ML at Moscow University (see [67]). A group in nonclassical logic at the Institute of Mathematics, Kishinev, has lost much because of his death and also of the emigration (1992) of A. MURAVITSKY and G. GALANTER. Now the group makes research in expressive completeness problems.

Mephody RATSA — Head of the Laboratory; Ivan CUCU; O.I. KOVALZHIU.

ARMENIA **Erevan**

I.D. Zaslavsky's Laboratory, Computing Centre, Nat. Acad. Sci.

1. *Group in constructive logic and mathematics:* Igor' D. ZASLAVSKY; Misha H. KHACHATRIAN; S.G. KARAGEBAKIAN; Seda N. MANUKIAN; Sepukh M. MELIKIAN.

2. *Group in recursion theory:* Hrant B. MARANDJIAN; A.S. MASIIURIAN; Natasha P. TER-ZAKHARIAN.

GEORGIA **Tbilisi**

1. *Tbilisi Research Centre for Logical Analysis of Language* (modal and intermediate logics): Leo ESAKIA — Director; Merab ABASHIDZE; Guram BEZHANISHVILI; Revaz GRIGOLIA; Vyacheslav MESKHI.

2. *Institute of Philosophy:* Mikhail BEZHANISHVILI (nonclassical logic); Georgie DZHAPARIDZE (modal logic, proof theory) — left in 1991; Leri MCHEDLISHVILI (temporal logic, history of logic); Zurab MIKELADZE (nonclassical logic, history of logic) — died in 1993.

3. *Tbilisi University:* Roland OMANADZE (recursion theory).

LITHUANIA **Vilnius**

Institute of Mathematics, Vilnius: Regimantas PLIUSKEVICIUS (dynamic and temporal logics) — group leader; Valentinas KRIAUCIUKAS (logical programming); Stasis NORGELA (automated deduction); Aida PLIUSKEVICIENE (dynamic logic); Jurate SAKALAUSKAITE (dynamic and temporal logics); Rimantas VAICENAVICIUS (admissible sets).

LATVIA **Riga**

1. *Department of Informatics, Institute of Mathematics and Computer Science, University of Latvia* (inductive inference, complexity theory, probabilistic algorithms, theoretical CS): Janis BARZDINS — Head; Maris ALBERTS; Guntis BARZDINS; Alvis BRAZMA; Karlis CERANS; Ilona ETMANE; Rusins FREIVALDS; Efim KINBER — emigrated in 1992; Karlis PODNIEKS (foundations).

2. *B. I. Plotkin's informal group* (algebraic problems in logic, logic in CS, categorial logic, λ-calculus): Boris I. PLOTKIN — emigrated in 1993; Janis CIRULIS; Zinovy B. DISKIN; Nikolai D. VOLKOV.

ESTONIA

Tallinn

Group in logical programming and proof theory, Institute of Cybernetics: Sergei TUPAILO; Tanel TAMMET; Peeter LORENTS. In the 80-ies the research was essentially stimulated by Gregory Mints (now at Stanford University, USA). Mati Pentus began his way in mathematical logic there too.

Tartu

Tartu University: Rein PRANK (computer-aided teaching of logic); Ants TAUTS (intuitionistic logic).

4. JOURNALS, BOOKS, CONFERENCES

Journals publishing papers in mathematical logic:

Algebra i logika (Algebra and Logic); *Doklady RAN* (Russian Academy of Sciences. Doklady. Mathematics) (and Doklady of other Academies of independent states of the fSU); *Izvestiya RAN* (Russian Academy of Sciences. Izvestiya. Mathematics); *Matematicheskie zametki* (Mathematical Notes); *Vestnik MGU. Matematika, Mekhanika* (Moscow University Mathematics Bulletin); *Sibirskii matematicheskii zhurnal* (Siberian Mathematical Journal); *Matematicheskii sbornik* (Russian Academy of Sciences. Sbornik. Mathematics); *Uspekhi matematicheskikh nauk* (Russian Mathematical Surveys); *Izvestiya VUZov. Matematika* (Russian Mathematics (Izvestiya VUZ. Matematika)); *Ukrainskii matematicheskii zhurnal* (Ukrainian Mathematical Journal).

Other periodic issues:

Semiotika i informatika — VINITI (Semiotics and Informatics — All-Russian Institute for Scientific and Technical Information); *Logicheskie issledovaniya* — Institut logiki i kognitologii razvitiya (Logical Studies — Institute of Logic and Cognitology of Development); Steklov Mathematical Institute Preprint Series; and some other preprint series.

Recent books: [1, 40, 66, 69, 82, 83, 89, 90, 101, 124].

Conferences:

All-Union Conferences in Mathematical Logic ("Union" means Soviet Union). These conferences were organized rather regularly (the number of accepted abstracts is written in brackets):

1969 — Alma-ata, Kazakhstan (38)
1972 — Moscow, Russia (70)
1974 — Novosibirsk, Russia (108)
1976 — Kishinev, Moldova (179)
1979 — Novosibirsk, Russia (139)
1982 — Tbilisi, Georgia (215)
1984 — Novosibirsk, Russia (203)
1986 — Moscow, Russia (211)
1988 — Leningrad, Russia (191)
1990 — Alma-ata, Kazakhstan (173)
1992 — Kazan, Russia (176)

The last Conference was named Interrepublican, because Soviet Union ceased to exist. There is no idea, when and where the next Conference might be organized.

International Conferences on Logical Foundations of Computer Science (LFCS): 1989 — Pereslavl-Zalessky (Organizers: A. Meyer, M. Taitslin); 1992 — Tver (Organizers: A. Nerode, M. Taitslin); 1994 — St. Petersburg (Organizers: A. Nerode, Yu. Matiyasevich); 1997 — planned at Yaroslavl. The number of participants is about 150.

There were also several Conferences in Applied Logic organized by Novosibirsk logical community; in Logic, Methodology and Philosophy of Science (the fSU); two workshops on Non-Standard Logics and Logical Aspects of Computer Science.

5. NEW RESEARCHES

5.1. *PROOF THEORY AND RELATED TOPICS*

5.1.1. Provability Logics and Arithmetic. The idea of Provability Logic was suggested by Gödel in 1933. A provability logic is a classical logic enriched by a unary modality □ which stands for *provability* operator; in a regular case this □ is the usual Gödel's Provability operator *Bew* for Peano Arithmetic **PA**. In 1976 Solovay established that the set of all propositional modal formulas which are provable in **PA** under every arithmetical interpretation of propositional letters coincides with the modal system **GL**. This initiated intensive studies of Provability Logic in many research centres throughout the world. Some key results in this area were obtained in Russia: the uniformization of Solovay's arithmetical completeness theorem (Artemov, 1979; this was also independently established in 1979–82 by F. Montagna, A. Visser and G. Boolos), a negative solution to the axiomatization problem of the first order provability logic (Artemov and Vardanyan, 1985), provability logic for the

ω-rule in arithmetic (Dzhaparidze, 1985), the classification theorem for the propositional logics of provability (Beklemishev, 1989).

In the 90-ies the research is going on at the same high level. Rybakov proved decidability of admissible rules in **GL**[106]. Dzhaparidze found an axiomatization of provability logic with additional modalities for arithmetical complexities. Beklemishev gave a complete axiomatization of the Provability Logic for Turing progressions of iterating consistency.

Artemov and Montagna (University of Siena) suggested a new approach for building first order theories with provability operators. They demonstrated that unlike traditional first order provability logic, this approach leads in many interesting cases to decidable axiom systems.

Quite a new phenomenon was discovered by Shavrukov: the Lindenbaum algebra of any theory satisfying the conditions of the Second Gödel Incompleteness Theorem was known to be isomorphic to the countable atomless Boolean algebra; however, the Lindenbaum algebras with provability operator (Magari algebras) of **PA** and **ZF** turn out to be non-isomorphic.

In 1994 Beklemishev completed the efforts of his predecessors by discovering the precise theorem comparing two classical ways of proof-theoretical strengthening of a theory. He established that, modulo to mutual interpretability, α steps of extensions by the local reflection along natural paths of constructive ordinals are equivalent to ω^α steps of iterating consistency.

Being motivated by possible applications, Artemov in 1994 constructed a complete axiomatization of Logic of Proofs which is Provability Logic with labelled modalities for individual proofs. Introducing labelled modalities into the Provability Logic opens new possibilities for studying such notions as proofs, programs which compute theorems, complexity of proofs, abstract operations over proofs etc., by logical means. Nogina extended these results to logics with a so-called "strong provability" operator: $\boxdot A \equiv A \wedge \Box A$.

Also in 1994 Artemov and Chuprina gave a complete axiomatization of the logic of proofs with operators for the complexity of proofs.

Tatiana Sidon completely investigated Craig interpolation property in various provability logics with extra operators, such as Guaspari–Solovay logic, the logic of Rosser provability, the logic of proofs.

Labelled modal logic was used by Artemov and Krupski to describe Boolean data structures with storage and reading operations. Under this approach the language of labelled modal logic with reading operator may be considered as a sort of specification and programming language for special sorts of data structures; satisfiability algorithms for modal logics give methods of building and optimizing reference structures satisfying a given formula.

The importance and value of the work made by Russian provability logicians can be illustrated by the fact that George Boolos personally designed the cover of his recent book "Logic of Provability" (Cambridge University Press, 1993) in the colours of Russian national flag: white–blue–red.

5.1.2. Linear Logic. Linear logic was introduced by Girard in 1987 as a resource conscious logical calculus, which was also motivated by the idea of unity of proofs and programs. Moscow logicians joined studies in Girard's linear logic in 1990, and already in 1991 Kanovich solved the intriguing problem of NP-completeness of the multiplicative fragment of linear logic.

In 1994 Kopylov proved the decidability of Linear Affine Logic, which is Linear Logic with the Weakening rule. This logic has a very clear resource interpretation under which sequents describe logical inference with upper bounds of the resources without obligations to consume all of them. In 1995 Kopylov got an international Kleene award for this result. In 1992–94 Kanovich constructed concurrent semantics, as well as computational, and game semantics for Linear Logic. Some modifications of Linear Logic were studied by Archangelsky, Dekhtyar, Kruglov, Musikaev and Taitslin.

5.1.3. Complexity of Proofs. Dantsin proposed a randomized proof system for arithmetic theorems, and showed that its power is the same as of interactive proof systems. Orevkov compared lengths of direct and indirect proofs of existence theorems; for the predicate calculus the length of a direct proof has no upper bound expressed by a Kalmar elementary function of the length of an indirect proof of the same theorem (cf. his book of 1993); for a calculus with an induction schema, the length of a direct proof has no upper bound expressed by a primitive recursive function of the length of an indirect proof of the same theorem (talk at CSL'94, Poland).

5.1.4. Automated Deduction. In 1993 Gomolko discovered that intuitionistic logic provides a complete semantics for the negation as inconsistency in the system of logic programming PROLOG, thus solving a problem left open by Dov Gabbay et al. 7 years before.

Resolution method in quantifier-free fragments of set theory was introduced by Mikhailov and Zamov (cf. [40]). Formal models for knowledge acquisition are studied by Lorents.

5.2. *MODEL THEORY AND SET THEORY*

5.2.1. Model Theory and Algebra. There are few classes of fields with decidable elementary theories. Therefore any new example of such a class is an important achievement. In 1995 Ershov succeeded in finding a family of such examples. E.g. the field of all algebraic totally p-adic numbers (i.e. the largest Galois extension of $\mathbf{Q}$ in $\mathbf{Q}_p$) for any prime p has a decidable elementary theory. To obtain this result Ershov elaborated a sophisticated technique of studying valuated fields, which allowed to find new important axiomatizable classes of fields, as well as a criterion for elementary embeddings of such fields.

Starting from the end of the eighties Zilber made research in logical aspects of algebraic geometry. The most essential progress was made by Hrushovski (M.I.T., Cambridge Mass.) and Zilber in 1993. They proved that model-theoretic conditions together with some natural topological-type conditions (akin to properties of Zariski topologies) imply that a given structure is reducible to an algebraically closed field structure. In other words, topology completely defines algebra. This result is far from trivial, and it has already found some applications. Hrushovski has discovered a very elegant proof of the Mordell–Lang Conjecture for semi-Abelian varieties, and Zilber made progress in another direction: he showed how to use these ideas for construction of an analytic theory for an arbitrary algebraically closed field [138]. A very recent paper of Zilber (in preparation, 1995) shows how J. Tate's rigid analytic geometry over non-Archimedian fields can be treated within the framework of a generalized analytic theory.

In his recent works Belegradek studied correlations between the theory of a ring R and of the unitriangle group $\mathbf{UT}_n(\mathsf{R})$ of matrices over R.

Pal'chunov in 1992 obtained a complete classification of theories of Boolean algebras with finitely many distinguished ideals.

5.2.2. Models of Theories. A complete answer to the general question about expressive power of finitely axiomatizable first order theories was obtained by Peretyat'kin. He proved that in many aspects finitely axiomatizable theories are much as expressive as recursively axiomatizable ones, thus showing quite a new phenomenon in first order logic. An old problem of syntactic description of filtered formulas was solved by Omarov.

Kudaibergenov proved that if a countable totally transcendental theory has uncountably many homogeneous models in some uncountable cardinality, then it has a continuum of homogeneous models of every infinite cardinality.

5.2.3. Admissible Sets. Model-theoretic and algorithmic properties of systems defined in admissible sets of the form HF(M) were studied by Ershov [35]. An analogue of the Löwenheim–Skolem–Mal'tsev theorem was proved for uncountable structures definable in HF(M), for sufficiently saturated models M of a complete ω-categorical or ω-stable theory. It was established that for any linear order L, the field of reals is not Σ-definable in HF(L).

5.2.4. Numbered Structures. A numbering of a mathematical structure is called a *constructivization* (resp. *strong constructivization*) if the set of all quantifier-free sentences (resp. of all sentences) which are true in that structure is decidable (i.e. recursive); the numbered structure itself is called *constructive*, or *recursive* (resp. *strongly constructive*, or *decidable*). For every recursive structure M there exists a decidable structure N such that their groups of recursive automorphisms are isomorphic (Morozov, 1995). If a structure has strong and non-strong constructivizations then the class of its constructivizations is not computable and even effectively infinite (Goncharov, 1992–93).

Evidently, an algorithmic problem involving the structure's elements can be solvable for one numbering and unsolvable for another one. It was Mal'tsev who discovered this phenomenon for "good" numberings, e.g. constructivizations. That nice discovery started researches by Ershov, Goncharov and other logicians of Siberian school in comparing various numberings (esp. constructivizations) of the same structure w.r.t. preserving certain algorithmic properties. Given a structure M and a set A of its numberings, one can define 5 natural equivalence relations on A (see [124], no. 2.5.6 and References). If A is the set of all constructivizations, the 5-tuple of cardinalities of corresponding quotient sets is an important characteristic of M. Properties of these 5-tuples were investigated by Goncharov, Khoussainov, Fedoryaev.

Kasymov studied numbered algebras and topological spaces. Computable classes of numbered structures were investigated by Dobritsa.

5.2.5. Finite Model Theory. Stolboushkin and Taitslin in 1994 established that finite queries have no effective syntax. They also found a nice criterion of when all finite models of a given axiomatizable theory are rigid: there exists a finitely axiomatizable theory with a first order definable linear order, in which the given theory can be interpreted.

5.2.6. Heyting-valued (in particular, Boolean-valued) Models. Study of Boolean-valued models started with Gordon's paper of 1977.

Researches in nonstandard analysis using both infinitesimal (Robinson's) and Boolean-valued methods were made by Kusraev and Kutateladze. In particular, they considered the following topics: interpreting Lebesgue integral in a Boolean-valued model; a version of nonstandard convex programming based on the notion of an infinitesimal optimum.

In his recent papers Lyubetsky used Boolean- and Heyting-valued models for obtaining transfer theorems of the following two types: from classical to intuitionistic set theory and from a class of first-order structures to a broader one (e.g. from fields to regular rings).

5.2.7. Nonstandard Analysis and Its Applications. Gordon (1991, 1995) developed a nonstandard approach to harmonic analysis. He introduced *hyperfinite approximations* of locally compact Abelian (LCA) groups, defined Riemannian integration in these groups in a nonstandard way. This allowed him to prove some new facts about approximations of Fourier transformations in LCA-groups by Fourier transformations in finite groups.

Molchanov worked on a nonstandard approach to topology, treating a topological space as a relation "a point x is infinitely close to a standard point a".

5.2.8. Internal Set Theory. Internal Set Theory **IST** was introduced by Nelson in 1977; recall that it has two basic predicates: membership (ε) and standartness (**st**). This theory was found to be good enough to develop many branches of nonstandard mathematics. Nelson showed conservativity of **IST** w.r.t. **ZFC**; and recently Kanovei (1995) proved that **IST** does not extend **ZFC** too far: every set definable by a formula of the language of **IST** with only standard parameters involved, is standard. On the other hand Kanovei constructed a sentence Φ in **IST** which is non-equivalent to any ε-sentence.

Kanovei (1991) also introduced *bounded set theory* **BST** which is close to **IST** but in which paradoxical sentences like Φ in the above result, do not exist. Kanovei and Reeken (1995) proved that external sets can be legitimately "incorporated" into the universe of **BST**.

5.3. *RECURSION THEORY*

5.3.1. Theory of Domains. A topological approach in the theory of domains developed by Ershov, allowed him to introduce the notion of an α-space as a natural generalization of a domain, which retains all main properties of domains (cf. [32]).

5.3.2. Complexity Classes. After several attempts to attack the $\mathsf{P} \neq \mathsf{NP}$ problem had failed, people started to think on the reasons of that failure. Maybe this statement is independent from **PA** or even Set Theory? First independence results of this kind were obtained for rather weak theories by Sazonov (1980) who started the researches in the fSU in that direction.

The advance made by Boolean complexity theory in the 80-ies by Razborov and others researchers somewhat suspended this activity. Several strong lower bounds were proved for weak computational models (monotone circuits, restricted depth circuits etc.) The new wave of research on logical independence was mainly motivated by the intention to understand, why the methods used in lower bounds proofs for circuit complexity are inapplicable to $\mathsf{P} \neq \mathsf{NP}$ problem. Razborov examined, what appropriate "minimal" fragment of **PA** sufficed to carry out all known lower bound proofs in Boolean complexity. He suggested the second order bounded arithmetic V_1^1 as a conceivable answer. It would be nice to show that V_1^1 does not prove $\mathsf{P} \neq \mathsf{NP}$; however, it is not done yet. Such a result established (under a widely believed cryptographic assumption) by Razborov for a weaker theory called $\mathcal{S}S_2^2$ which nonetheless is strong enough to prove some non-trivial lower bounds.

5.3.3. Relativized Complexity Theory. It is well-known that in general recursion theory all natural theorems can be relativized, i.e. they remain true if all Turing machines are enriched by some (fixed) oracle A. This is not the case in polynomial recursion theory: Baker, Gill and Solovay (1975) showed that neither $\mathsf{P} = \mathsf{NP}$, nor $\mathsf{P} \neq \mathsf{NP}$ is relativizable. So it is natural to find out for various pairs of complexity classes K_1, K_2, if $\mathsf{K}_1^{\mathsf{A}} \subseteq \mathsf{K}_2^{\mathsf{A}}$ for all A, esp. in the cases when the truth of $\mathsf{K}_1 \subseteq \mathsf{K}_2$ is yet unknown. The work in this direction involved several authors; Vereshchagin proved that $\text{co-}\mathsf{R} \subseteq \mathsf{NP}$ and some other inclusions are non-relativizable. He also explained, why it was often easier to decide the question whether $\forall \mathsf{A}\, \mathsf{K}_1^{\mathsf{A}} \subseteq \mathsf{K}_2^{\mathsf{A}}$ than to decide the question whether $\mathsf{K}_1 \subseteq \mathsf{K}_2$. Namely, it was proved that $\forall \mathsf{A}\, \mathsf{K}_1^{\mathsf{A}} \subseteq \mathsf{K}_2^{\mathsf{A}}$ is equivalent to $\mathsf{K}_1\mathsf{LOGS} \subseteq \mathsf{K}_2\mathsf{LOGS}$; here KLOGS denotes the class obtained by replacing all polynomials in the definition of K by polynomials of logarithm, and decision problems by separation problems.

5.3.4. Degrees of Unsolvability. In recent years Arslanov (in collaboration with Efremov, Osipov, Lempp, Shore et. al.) investigated structures of n-r.e.

degrees of unsolvability, and their correlation. His most striking new results are: 1) for $n \geq 2$ an n-r.e. degree is REA only if it is 2-r.e.; 2) there is a non-recursive r.e. degree $a \neq 0, 0'$ such that any REA(a) (i.e. REA in a) degree $b \leq 0'$ is r.e. (see [55] for the definitions).

Answering Odifreddi's question of 1981, Degtev found a simple set with a special property related to m-reducibility. Various kinds of reducibility were studied by Omanadze and Bulitko; the latter also gave a solution to Lachlan's Problem: to find a completeness criterion for maximal sets (1991, 1995).

5.3.5. Properties of Recursive Functions were investigated by Marandjian who studied systems of functional equations for recursive functions, and by Marchenkov who studied superpositional bases for classes of recursive function.

5.3.6. Numberings. Goncharov and Badaev studied positive numberings. Badaev worked also on the problem by Ershov about the number of minimal computable numberings; he solved that problem for many families of r.e. sets, e.g. for families of recursive functions, finite families, etc.

5.3.7. Decision Problems. In Russia there is a strong tradition of studies of algorithmic problems in Algebra. In 1947 A.A. Markov and E. Post independently established unsolvability of the word problem for semigroups which was raised by A. Thue in 1914. In 1952 P.S. Novikov proved the unsolvability of the word problem for groups by constructing an example of a finitely presented group, for which the desired algorithm was impossible. In 1955 S.I. Adyan proved that for every finitely presented group the corresponding particular isomorphism problem is unsolvable. In 1958 Adyan and Rabin obtained a general theorem on the impossibility of recognition of invariant group properties (cf. [2]). In 1965 Ershov proved the decidability of the elementary theory of p-adic numbers. A complete solution to the problem of solvability of equations in a free semigroup was obtained by Makanin in 1977.

Developing his notion of *critical theory* (1988), Vazhenin recently described borders between decidability and undecidability for the ring **Z** and for free rings of finite ranks.

D. Gabbay (Imperial Colledge, London) and Shehtman constructed simple examples of undecidable intermediate and modal predicate logics with only two individual variables. This contrasts to the classical case where at least three variables are necessary to obtain undecidability.

Kratko studied algebraic automata theory. His recent result (1993) is a proof of undecidability of the completeness problem for the class of Redding automata.

Letichevsky studied decidable cases of equivalence problem for discrete transformers over semigroups and groups.

5.3.8. Kolmogorov Complexity. For any reference which is absent in *References* here, see [68]. The Kolmogorov complexity of a finite binary string was defined by Kolmogorov in 1965 as the length of the shortest program producing that string. Different versions of Kolmogorov complexity were introduced later. The basic results obtained during the sixties are covered by a survey by Zvonkin and Levin of 1970. That survey includes also the definition of an individual random infinite sequence given by Kolmogorov's student Martin-Löf (1966) and its relations to the sequence's complexity. A survey by V'yugin of 1981 (translated as [132]) covers next decade's main achievements. Kolmogorov's disciple Levin obtained two of them. The first one is a definition of new versions of complexity: *monotone* complexity (1973) and *prefix* complexity (1974). The second one (1973) is the characterization of a random (in Martin-Löf sense) infinite sequence as a sequence whose initial segments have maximal possible monotone complexities.

A general scheme for various complexities was suggested in 1979 first by Shen in terms of computable continuous mappings and then by Uspensky (cf. [124]) in terms of description modes (a *description mode* is a certain recursively enumerable set of pairs ⟨description, object⟩). During the next decade (1980–1990), Kolmogorov complexity was used to find new proofs for classical theorems of probability theory, e.g. for the law of iterated logarithm (Vovk). Asarin studied stochastic properties of finite objects with large complexity. He also gave a definition of randomness of a continuous function and proved the complexity criterion for it. An. Muchnik gave a characterization of Kolmogorov complexity relativized to $\mathbf{0}'$ in terms of frequencies. The definition of the so-called (α, β)-stochastic objects given by Kolmogorov (in terms of a finite object's randomness deficiency) was studied by Shen and V'yugin. Shen proved that the definition of randomness based on the von Mises frequency approach is not equivalent to the Martin-Löf definition. In their talk at a Stanford symposium, 1991, An. Muchnik, Semenov and Uspensky suggested a definition of randomness in terms of unpredictability.

Current researches establish a connection between Kolmogorov complexity and Kolmogorov's idea of 1932 to interpret the propositional connectives as operations on problems. For example, the complexity of the expression $A \Rightarrow B$ may be considered as the conditional complexity of B when A is known. Other topics are: (1) inequalities for complexities of finite sets of strings and their subsets; (2) the notion of "common information" in two objects; (3) applications to statistics and learning theory (Vovk and V'yugin). There are more questions than answers here.

5.3.9. Probabilistic Algorithms. What are advantages of using randomness in an algorithm if the statement of a problem does not involve randomness? Various situations when randomization helped significantly, and when it did not help at all, were studied by Freivalds et al. (cf. [42]). If randomization is allowed, the lower bounds for computation time and space become much more difficult to obtain. The first $\log n$ space lower bound for 2-way Monte Carlo Turing machines was obtained by Freivalds and M. Karpinski (1994).

5.3.10. Inductive Inference and Learning Theory. The research in this direction in the fSU was started by J. Barzdins in early seventies and is continued now mainly by the group of Freivalds. Freivalds and Smith used constructive ordinals to measure the complexity of a function. Freivalds and Jain considered the problem of minimal identifiability of certain classes of functions using the notion of the Kolmogorov complexity. Complexity of learning was studied in [45]. New models of inductive synthesis were studied by J. Barzdins, R. Freivalds, E. Kinber. A new approach to that synthesis based on data types and term rewriting systems was suggested by J. Barzdins, G. Barzdins and A. Brazma.

5.4. *NONCLASSICAL LOGIC*

5.4.1. Modal and Intermediate Propositional Logics. This area has been developed in our country since the sixties. Before 1990 some basic important results were obtained: construction of a continuum of intermediate logics (V.A. Yankov, 1968); construction of an intermediate propositional calculus without the finite model property (A.V. Kuznetsov, V.Ya. Gerchiu, 1970); non-arithmeticity of first-order realizability logic (V.E. Plisko, 1973); duality theory for modal algebras (L. Esakia, 1974); description of all intermediate logics with Craig interpolation property (L.L. Maksimova, 1977); construction of an undecidable intermediate calculus (V.B. Shehtman, 1978); proof of decidability of admissible rules in intuitionistic logic (V.V. Rybakov, 1984); proof of Dummet–Lemmon's conjecture on transfer of Kripke completeness from any intermediate logic to its minimal Gödel–Tarski modal translation (M.V. Zakharyaschev, 1989). The first research groups in the area were organized by A.V. Kuznetsov in Kishinev, L. Esakia in Tbilisi, L.L. Maksimova in Novosibirsk. Afterwards the field has grown much, under the influence of CS, Proof Theory and Mathematical Linguistics.

In recent time general positive results in propositional logic were obtained by Zakharyaschev; they are based on his notion of a *canonical formula* introduced at the beginning of the 80's. Every propositional formula (in intuitionistic logic or in modal logic **K4**) is deductively equivalent to a finite conjunction of canonical formulas (1992). This notion allowed him to obtain new sufficient conditions for the finite model property in modal logic (1995). Also recently Zakharyaschev and Alekseev proved that every finitely axiomatizable normal extension of **K4.3** is decidable.

Rybakov described in terms of Kripke models all logics with the finite model property inheriting all intuitionistically admissible rules (1993). In 1995 Rybakov solved positively the long-standing problem of recognizing hereditary structural completeness of modal logics above **K4**.

In 1995 Shehtman completed his result of 1990 solving McKinsey–Tarski's problem of 1944 on axiomatization and decidability of the modal logic of the derived set operation in $\mathbf{R}^n$.

In 1990 Chagrov suggested a general method which allowed to prove undecidability of many properties of modal and intermediate propositional logics (such as disjunctive property, Kripke completeness etc.); cf. Chagrov and Zakharyaschev, 1993; Chagrova, 1991. We can also mention two surprising counterexamples: (i) a decidable logic with an undecidable set of admissible rules (Chagrov, 1992); (ii) modal and intermediate logics without an independent set of axioms (Chagrov and Zakharyaschev, 1995).

Maksimova studied various kinds of definability in propositional modal logic. She proved that a weak form of Beth property holds for all logics above **K4** (1992), and that both Craig's interpolation property and weak Beth property fail for "tomorrow" logics (1991).

5.4.2. Nonclassical Predicate Logics. Studies in this area before 1990 led to many incompleteness results, and no plausible semantics was known to improve the situation. However in 1990–1991 Skvortsov introduced a new semantics of Kripke-type (called *Kripke metaframe semantics*). He showed that it is "maximal" i.e. non-weaker than any Kripke-type semantics satisfying some natural soundness conditions; Shehtman proved some general completeness results in metaframe semantics. (Cf. [115, 120].)

By using metaframes, Skvortsov [117] succeeded in constructing sophisticated counterexamples, viz. pairs of finitely axiomatizable intermediate predicate logics whose intersection is not finitely axiomatizable. In fact, one can take e.g. L_1 =(intuitionistic logic + $\neg p \vee \neg\neg p$) and L_2 =(intuitionistic logic + any propositional axiom), provided that L_1, L_2 are incomparable. Skvortsov also studied intermediate predicate logics of finite Kripke frames. He showed that sometimes they are finitely axiomatizable [118] and sometimes [119] are not.

Research of intermediate predicate logics is going also in Tbilisi group: L. Esakia introduced "amended intuitionistic logic", G. Bezhanishvili studied decidability for 1-variable fragments of quantified 3-valued intermediate logics.

Plisko proved several metatheorems, which in essence formalize the Tennenbaum Theorem about nonexistence of recursive nonstandard models of arithmetic, and applied these metatheorems to obtain bounds of the arithmetic complexity of constructive predicate logics based on modified realizability and Gödel interpretation semantics.

5.4.3. Temporal and Dynamic Logics. Shehtman (1993) axiomatized a temporal logic for progressive tenses and proved its decidability, thus answering a question of D. Scott of 1968. Pliuskevicius studied decidability and completeness for first order temporal calculi. Pliuskevicene studied proof theory for first order modal logic and Hoare-type logics of programs. Kriaucukas introduced denotational semantics of logical programs in Kripke models. Sakalauskaite obtained completeness theorems for some first order temporal logics. Valiev considered propositional dynamic logic with "shuffle" operator and proved Π^1_1-universality of its fragment without propositional letters.

5.4.4. Expressiveness in Nonclassical Logics. This area is developed in Kishinev starting from the 70-ies, when Ratsa found a criterion of expressive completeness of a system of formulas in intuitionistic propositional logic. On the other hand the corresponding problem for **S4** turned out to be undecidable. Cucu and Ratsa in 1992 proved "Parametric completeness" criterion for some particular intermediate logic (close to **LC**). In 1995 Yashin has completed a classification of all complete extensions of the intuitionistic logic by additional propositional constants. Expressive power of dynamic logics was studied by Musikaev and Taitslin.

5.5. *MATHEMATICAL LINGUISTICS*

In this area two break-through results have been obtained by Pentus (Moscow State University). In 1992 Pentus, at that time an undergraduate student of mathematics, succeeded in proving the famous 30-years old Chomsky conjecture about the equivalence of Lambek grammars and context-free grammars, thus solving one of the major problems in Mathematical Linguistics. This result is of great importance to the field by itself, as it connects the modern paradigm of logic-oriented categorial grammars with more traditional phrase structure based grammars in Mathematical Linguistics. In 1993 Pentus solved the main remaining problem here by proving the completeness of Lambek calculus with respect to the intended semantics in formal languages (so-called "language completeness problem"); this problem remained open during 35 years. Pentus was awarded the brand-new "Folli Prize" for the best paper of 1994 from the European Association for Logic, Language and Information.

Quite recently (1995) Kanovich established undecidability of the Lambek calculus with propositional quantifiers.

Department of Mathematical Logic and the Theory of Algorithms
Faculty of Mechanics and Mathematics
Moscow State University

REFERENCES

1. *Action and Norm (Logical and Onthological Aspects of the Theory of Action)* (collection of papers). Kiev, 1995 (in Russian: *Deistvie i norma (logicheskie i ontologicheskie aspekty teorii deistviya)*).
2. Adyan S.I., Makanin G.S. Investigations on algorithmic questions of algebra. *Proc. Steklov Inst. Math.,* **3** (1986), 207–226.
3. Archangelski D.A., Dekhtyar M.I., Kruglov E., Musikaev I.Kh., Taitslin M.A. Concurrency problem for Horn fragment of Girard's linear logic. *LN in CS,* **813** (1994), 18–22.
4. Archangelski D.A., Taitslin M.A. Linear logic with fixed resources. *Ann. Pure and Appl. Logic,* **67** (1994), 3–28.
5. Arslanov M.M. On the structure of degrees below $\mathbf{0}'$. *LN in Math.,* **1432** (1990), 23–32.

6. Arslanov M.M., Lempp L., Shore R.A. Interpolating d-r.e. and REA degrees between r.e. degrees. *Ann. Pure and Appl. Logic,* 1995, to appear.
7. Arslanov M.M., Lempp L., Shore R.A. On isolating r.e. and isolated d.-r.e. degrees. *Proc. of London Math. Society,* 1995, to appear.
8. Artemov S.N. Logic of proofs. *Ann. Pure and Appl. Logic,* **67** (1994), 29–59.
9. Artemov S.N, Chuprina A.V. Logic of proofs with complexity operators. *University of Amsterdam, ILLC Research report* ML–94–07, 1994. *Proc. Int. Conf. on Logic and Algebra in Memory of R.Magari, Siena, 1994,* to appear.
10. Artemov S.N., Krupski V.N. Data storage interpretation of labeled modal logic. *Techn. Rep. IAM 95–002, University of Berne* (1995), *Ann. Pure and Appl. Logic,* to appear.
11. Artemov S.N., Montagna F. On first order theories with provability operator. *J. Symb. Logic,* **59** (1994), no.4, 1139–1153.
12. Badaev S.A. Minimal numberings of positively computable families. *Algebra i Logika,* **33** (1994), no.4, 13–28 (in Russian).
13. Barzdins G. Inductive synthesis of term rewriting systems. *LN in CS,* **502** (1991), 253–285.
14. Barzdins J., Brazma A., Kinber E. Models of inductive syntactical synthesis. *Machine Intelligence,* **12** (1991), 139–148.
15. Beklemishev L.D. Provability logics for natural Turing progressions of arithmetical theories. *Studia Logica,* **50** (1991), 107–128.
16. Beklemishev L.D. Iterated local reflection vs iterated consistency. *Utrecht Research Inst. for Philosophy, Logic Group Preprint Series,* no.118, 1994.
17. Belegradek O.V. Quasiunitriangular groups. *J. Symb. Logic,* **58** (1993), no.1, 205–218.
18. Belegradek O.V. On the model theory of unitriangular groups. *Ann. Pure and Appl. Logic,* **68** (1994), 225–261.
19. Bezhanishvili G. The lattice of quantified extensions of the three-valued intermediate propositional logic. *Proc. LFMPS'95,* to appear.
20. Bulitko V.K. On characterizing complete sets. *Izvestiya Akademii Nauk SSSR, ser. Matematicheskaya,* **55** (1991), no.2, 227–253 (in Russian). English translation: *Mathematics of the USSR: Izvestija* **38** (1992), no.2, 225–249.
21. Bulitko V.K. On some complexity characteristics of immune sets. *Math. Logic Quarterly,* **41** (1995), no.3, to appear.
22. Chagrov A.V. Decidable modal logic with undecidable problem of admissibility for inference rules. *Algebra i Logika,* **31** (1992), no.1, 83–93 (in Russian). English translation: *Algebra and Logic,* **31** (1992), no.1, 53–61.
23. Chagrov A., Zakharyaschev M. The undecidability of the disjunction property of propositional logics and the other related problems. *J. Symb. Logic,* **58** (1993), 967–1002.
24. Chagrov A., Zakharyaschev M. On the independent axiomatizability of modal and intermediate logics. *J. Logic and Computation,* 1995, to appear.
25. Chagrova L.A. An undecidable problem in correspondence theory. *J. Symb. Logic,* **56** (1991), no.4, 1261–1272.
26. Cucu I.V., Ratsa M.F. Parametric completeness in the logic of simpliest pseudo-Boolean algebra with incomparable elements. *Buletinul Academie de Ştiinţe a Republicii Moldova. Matematica,* 1992, no.1, 46–51 (in Russian).
27. Dantsin E. Probabilistic logic programs and their semantics. *LN in CS,* **592** (1992), 152–164.
28. Dantsin E. Randomized proofs in arithmetic. *Zapiski Nauchnykh Seminarov POMI,* **220** (1995), 49–71 (in Russian).
29. Degtev A.N. A problem of P. Odifreddi. *Sibirskii Matematicheskii Zhurnal,* **30** (1989), no.1, 185–187 (in Russian). English translation: *Siberian Mathematical Journal,* **30** (1989), no.1, 144–146.

30. Dobritsa V.P. Computable semilattices of indexations for classes of constructive models. *Matematicheskie Zametki,* **55** (1994), no.5, 43–50 (in Russian).
31. Dzhaparidze G.K. Provability logic with modalities for arithmetical complexities. *Bull. Acad. Sci. Georgian SSR,* **138** (1990), no.3, 481–484. An extended version of that work is: Dzhaparidze G. The logic of the arithmetical hierarchy. *Ann. Pure and Appl. Logic,* **66** (1994), 89–112.
32. Ershov Yu.L. Theory of domains and nearby. *LN in CS,* **735** (1993), 1–7.
33. Ershov Yu.L. RC*-fields. *Algebra i Logika,* **33** (1994), no.4, 367–386 (in Russian).
34. Ershov Yu.L. Fields with continuous locally elementary properties, I, II. *Algebra i Logika,* **33** (1994), no.6, 628–653; **34**, 1995, no.3, 262-273 (in Russian).
35. Ershov Yu.L. Definability in hereditary finite superstructures. *Doklady Akademii Nauk,* **340** (1995), no.1, 12–14 (in Russian).
36. Ershov Yu.L. Nice local-global fields. *Doklady Akademii Nauk,* 1995, to appear (in Russian).
37. Ershov Yu.L. Model theory of nice global fields. *Vol. Abstracts, 10th Int. Congress of LMPS, Florence, 1995*, p.72.
38. Esakia L. Amended intuitionistic logic and formal provability. *Proc. LFMPS'95,* 1995, to appear.
39. Fedoryaev S.T. Countability of widths of algebraic reducibility structures for models in some classes. *Siberian Adv. Math.,* **3** (1993), no.2, 81–102.
40. Fermuller C., Leitsch A., Tammet T., Zamov N. Resolution methods for the decision problem. *LN in AI,* **679** (1993).
41. "Foundations and mathematical logic". In: *Istoriya Otechestvennoi Matematiki,* **4**, book 2, Naukova dumka, Kiev, 1970, 392–454 (in Russian).
42. Freivalds R. Complexity of probabilistic versus deterministic automata. *LN in CS,* **502** (1991), 565–613.
43. Freivalds R., Jain S. Kolmogorov numberings and minimal identification. *LN in CS,* **904** (1995), 182–195.
44. Freivalds R., Karpinski M. Lower space bounds for randomized computation. *LN in CS,* **820** (1994), 580–592.
45. Freivalds R., Kinber E., Smith C.H. On the intrinsic complexity of learning. *LN in CS,* **904** (1995), 154–168.
46. Freivalds R., Smith C.H. On the role of procrastination in machine learning. *Information and Computation*, **107** (1993), no.2, 237–271.
47. Gabbay D., Shehtman V.B. Undecidability of modal and intermediate first-order logics with two individual variables. *J. Symb. Logic,* **58** (1993), no.3, 800–823.
48. Gomolko A.L. Negation as inconsistency. *LN in CS,* **832** (1994), 128–138.
49. Goncharov S.S. Positive computable numberings. *Doklady Akademii Nauk,* **332** (1993), no.2, 142–143 (in Russian). English translation: *Russian Academy of Sciences: Doklady. Mathematics*, **48** (1994), no.2, 268–270.
50. Goncharov S.S. A unique positive numbering. *Siberian Adv. Math.,* **4** (1994), no.1, 52–64.
51. Gordon E.I. Reals in Boolean-valued models of set theory and K-spaces. *Doklady Akademii Nauk SSSR,* **237** (1977), no.4, 773–775 (in Russian). English translation: *Soviet Mathematics Doklady,* **18** (1977), no.6, 1481–1484.
52. Gordon E.I. Nonstandard analysis and compact abelian groups. *Sibirskii Matematicheskii Zhurnal,* **32** (1991), no.2, 26–40 (in Russian). English translation: *Siberian Mathematical Journal,* **32** (1991), no.1, 197–209.
53. Gordon E.I. Nonstandard analysis and locally compact abelian groups. *Acta Appl. Math.,* **25** (1991), 221–239.
54. Hrushovski E., Zilber B.I. Zariski geometries. *Bull. AMS,* **28** (1993), 315–323.
55. Jockusch C.G., Jr., Shore R.A. Pseudo-jump operators II. *J. Symb. Logic,* **49** (1984), 1205–1236.
56. Kanovei V.G. Undecidable hypotheses in Edward Nelson's internal set theory. *Russ. Math. Surveys,* **46** (1991), no.6, 1–54.

57. Kanovei V.G. Uniqueness, collection, and external collapse of cardinals in IST and models of Peano arithmetic. *J. Symb. Logic,* **60** (1995), no.1, 318–324.
58. Kanovei V.G., Reeken M. Internal approach to external sets and universies. *Studia Logica,* **54** (1995), no.5, no.6; **55** (1996), no.4, to appear.
59. Kanovich M.I. The multiplicative fragment of linear logic is NP-complete. *University of Amsterdam, ITLI Technical Report X–91–13,* 1991.
60. Kanovich M.I. Petri nets, Horn programs, linear logic, and vector games. *Ann. Pure and Appl. Logic,* **75** (1995), 105–135.
61. Kasymov N.Kh. On algebras over negative equivalences. *Algebra i Logika,* 1995, to appear (in Russian).
62. Kopylov A.P. Decidability of linear affine logic. *CNRS, Laboratoire de Math. Discrètes,* Pretirage 94–32, 1994.
63. Kratko M.I., Khruzin A.N. Unsolvability of certain problems in the theory of Redding automata. *Doklady Akademii Nauk,* **329** (1993), no.4, 416–418 (in Russian). English translation: *Russian Academy of Sciences: Doklady. Mathematics,* **47** (1993), no.2, 285–288.
64. Kriaciukas V. Nonclassical models for logic programs. *LN in AI,* **567** (1991), 179–190.
65. Kudaibergenov K.Zh. Homogeneous models of stable theories. *Siberian Adv. Math.,* **3** (1993), no.3, 1–33.
66. Kusraev A.G., Kutateladze S.S. *Nonstandard Methods in Analysis.* Kluwer Acad. Publ., 1994.
67. "Alexandr Vladimirovich Kuznetsov (obituary)" *Russ. Math. Surveys,* **41** (1986), no.2, 185–187.
68. Li M., Vitányi P. *An Introduction to Kolmogorov Complexity and Its Applications.* Springer-Verlag, N.Y. e.a., 1993.
69. *Logic and Problems of Rationality* (collection of papers). Kiev, 1993 (in Russian: *Logika i problemy ratsional'nosti*).
70. Lorents P. *Algebraic framework for knowledge acquisition.* Technical report. TRITA–ITR–93:10, ISSN 1103–534X, ISRN KTH/IT/R–93/10–SE. December 1993.
71. Lyubetsky V.A. Heyting valued analysis: P. Novikov's hypothesis. *Contemporary Math.,* **131** (1992), part 1, 565–583.
72. Lyubetsky V.A. Transfer theorems and modal operators. *Logicheskie Issledovaniya,* **3**, Nauka, Moscow, 1995, 210–235 (in Russian).
73. Maksimova L.L. Amalgamation and interpolation in normal modal logics. *Studia Logica,* **50** (1991), no.3/4.
74. Maksimova L.L. An analogue of Beth's theorem in normal extensions of the modal logic K4. *Sibirskii Matematicheskii Zhurnal,* **33** (1992), no.6, 118–130 (in Russian). English translation: *Siberian Mathematical Journal,* **33** (1993), no.6, 1052–1065.
75. Maksimova L.L. Definability and interpolation in classical modal logic. *Contemporary Mathematics,* **131** (1992), part 3, 583–599.
76. "Anatolii Ivanovich Mal'tsev (obituary)" *Uspekhi Matematich. Nauk,* **23** (1968), no.3, 159–170 (in Russian). English translation: *Russian Mathematical Surveys,* **23** (1968), no.3, 157–168.
77. Marandjian H.B. *Selected Topics in Recursive Function Theory in Computer Science.* DTH, Lyngby, 1990.
78. Marandjian H.B. General form recursive equations I. *LN in CS,* 1995, to appear.
79. Marchenkov S.S. Basis by superposition for classes of recursive functions. *Matematicheskie Voprosy Kibernetiki,* issue 3, Nauka, Moskow, 1991, 115–139 (in Russian).
80. "Andrei Andreevich Markov (on his 60-th birthday)" *Russ. Math. Surveys,* **19** (1964), no.3, 181–196.
81. "Andrei Andreevich Markov (on his seventieth birthday)". *Russ. Math. Surveys,* **29** (1974), no.6, 171–175.
82. Matiyasevich Yu.V. *Hilbert's 10th Problem.* MIT Press, 1993.
83. Meskhi V. *Heyting Algebras with a Regular Involution.* Metsniereba, Tbilisi, 1990 (in Russian: *Algebry Geitinga s regulyarnoi involyutsiyei*).

84. Molchanov V.A. Nonstandard extensions of uniform algebraic systems. *Sibirskii Matematicheskii Zhurnal,* **35** (1994), no.5, 1094–1105 (in Russian).
85. Musikaev I.Kh., Taitslin M.A. Limitations of the program memory and the expressive power of dynamic logics. *Information and Computation,* **103** (1993), no.2, 195–203.
86. Nogina E.Yu. Logic of proofs with the strong provability operator. *University of Amsterdam, ILLC Research report* ML–94–10, 1994.
87. "Petr Sergeevich Novikov (on his seventieth birthday)" *Uspekhi Matematicheskikh Nauk,* **26** (1971), no.5, 231–242 (in Russian). English translation: *Russian Mathematical Surveys,* **26** (1971), no.5, 165–176.
88. Omarov A.I. Syntactical description of filtered formulas. *Soviet Math. Dokl.,* **139** (1991), no.1, 58–61.
89. Omelyanchik V.I. *Logical Foundations of Modal Realism.* Kiev, 1993 (in Russian: *Logicheskie osnovaniya modal'nogo realizma*).
90. Orevkov V.P. *Complexity of Proofs, and their Transformations in Axiomatic Theories.* AMS translations, **128** (1993).
91. Pentus M. Lambek grammars are context free. *Proc. of the 8th Symposium LICS'93,* IEEE Computer Society Press, Montreal, 1993.
92. Pentus M. Lambek calculus is L-complete. *Proc. 9th Symposium LICS'94,* IEEE Computer Society Press, Paris, 1994.
93. Pentus M. Models for the Lambek calculus. *Ann. Pure and Appl. Logic,* **75** (1995), 177–212.
94. Peretyat'kin M.G. Expressive power of finitely axiomatizable theories. Parts I, II, III. *Siberian Adv. Math.,* **3** (1993), no.2, 153–197; **3** (1993), no.3, 123–145; **3** (1993), no.4, 131–201.
95. Plisko V. A formalization of the Tennenbaum Theorem and its applications to modified realizability and Gödel interpretation. *Vol. Abstracts, 10th Int. Congress of LMPS, Florence, 1995*, p.57.
96. Pliuskeviciene A. Gentzen-type calculi for modal logic S4 with Barcan formula. *LN in AI,* **592** (1991), 381–390.
97. Pliuskeviciene A. Specialization of derivations in Hoare-like logic with recursive procedures. *Proc. Int. Conf. "Software Engineering and Knowlege Engineering",* 1994, 124–130.
98. Pliuskevicius R. Investigation of finitary calculus for discrete linear time logic. *LN in CS,* **502** (1991), 504–528.
99. Pliuskevicius R. On saturated calculi for linear temporal logic. *LN in CS,* **711** (1993), 640–649.
100. Pliuskevicius R. On the saturation principle for linear temporal logic. *LN in CS,* **713** (1993), 289–300.
101. Ratsa M. *Expressiveness in Propositional Calculi.* Shtiintsa, Kishinev, 1991 (in Russian: *Vyrazimost' v ischislenii vyskazyvanii*).
102. Razborov A. An equivalence beetween second order bounded domain arithmetic and first order bounded arithmetic. In: *Arithmetic, Proof Theory and Computational Complexity,* (Clote P. and Krajicek J. eds.), Oxford University Press, 1992, 247–277.
103. Razborov A. Bounded Arithmetic and lower bounds in Boolean complexity. *Feasible Mathematics II* (1993), should have appeared.
104. Razborov A. On provability disjoint NP-pairs. *Basic Research in CS Center, Technical Report* RS–94–36, Aarhus, Denmark, 1994.
105. Razborov A. Unprovabilty of lower bounds on circuit size in certain fragments of Bounded Arithmetic. *Izvestiya Academii Nauk, ser. Matematika,* **59** (1995), no.1, 201–224.
106. Rybakov V.V. Logical equations and admissible rules with parameters in modal logic of provability. *Studia Logica,* **49** (1990), no.2, 215–239.
107. Rybakov V.V. Intermediate logics preserving admissible inference rules of Heyting calculus. *Math. Logic Quarterly,* **39** (1993), 403–415.

108. Rybakov V.V. Hereditarily structurally complete modal logics. *J. Symb. Logic,* **60** (1995), no.1, 266–288.
109. Sakalauskaite J. A sequent calculus for a first order linear temporal logic with explicit time. *LN in AI,* **592** (1991), 442–451.
110. Sakalauskaite J. A sequent calculus for a first order linear temporal logic with equality. *LN in AI,* **620** (1992), 430–440.
111. Sazonov V.Yu. A logical approach to the problem P=NP? *LN in CS,* **88** (1980), 562–575.
112. Shavrukov V.Yu. A note on the diagonalizable algebras of PA and ZF. *Ann. Pure and Appl. Logic,* **61** (1993), 161–173.
113. Shehtman V.B. Derived sets in Euclidean spaces and modal logic. *University of Amsterdam, ITLI Prepublication Series, X-90-05,* 1990.
114. Shehtman V.B. A logic with progressive tenses. In: *Diamonds and Defaults: Studies in Pure and Applied Intensional Logic.* (M. De Rijke, ed.), Kluwer Acad. Publ., 1993, 255–285.
115. Shehtman V.B. Sahlqvist completeness theorem in modal predicate logic. *Vol. Abstracts, 10th Int. Congress of LMPS, Florence, 1995*, p.173.
116. Sidon T. Craig interpolation property in modal logics with provability interpretations. *LN in CS,* **813** (1994), 329–330.
117. Skvortsov D.P. On finite intersections of intermediate predicate logics. *Proc. Int. Conf. on Logic and Algebra in Memory of R.Magari, Siena, 1994,* to appear.
118. Skvortsov D.P. On the predicate logic of finite Kripke frames. *Studia Logica,* **54** (1995), no.1, 79–88.
119. Skvortsov D.P. Not every intermediate predicate logic over a finite frame is finitely axiomatizable. *Vol. Abstracts, 10th Int. Congress of LMPS, Florence, 1995*, p.175.
120. Skvortsov D.P., Shehtman V.B. Maximal Kripke-type semantics for modal and superintuitionistic predicate logics. *Ann. Pure and Appl. Logic,* **63** (1993), no.1, 69–101.
121. Stolboushkin A.P., Taitslin M.A. Finite queries do not have effective syntax. *UCLA, preprint,* 1994.
122. Stolboushkin A.P., Taitslin M.A. Normalizable linear orders and generic computations in finite models. *UCLA, preprint,* 1994 (submitted to LICS'95).
123. Uspensky V.A. Kolmogorov and mathematical logic. *J. Symb. Logic,* **57** (1992), no.2, 385–412.
124. Uspensky V., Semenov A. *Algorithms: Main Ideas and Applications.* Kluwer Acad. Publ., 1993.
125. Valiev M. Π^1_1-universality of some propositional logics of concurrent programs. *Theor. CS,* **119** (1993), 223–232.
126. Vazhenin Yu.M. Critical theories. *Sibirskii Matematicheskii Zhurnal,* **29** (1988), no.1, 23–31 (in Russian). English translation: *Siberian Mathematical Journal,* **29** (1988), no.1, 17–23.
127. Vereshchagin N.K. Relationships between NP-sets, Co-NP-sets and P-sets relative to random oracles. *Proc. 8th Annual IEEE Conf. on Structure in Complexity Theory, 1993, San-Diego CA,* 132–138.
128. Vereshchagin N.K. Relativizable and non-relativizable theorems in the polynomial theory of algorithms. *Izvestiya Akademii Nauk, ser. Matematicheskaya,* **57** (1993), no.2, 51–90 (in Russian).
129. Vereshchagin N.K. NP-sets are Co-NP-immune relative to a random oracle. *Proc. Third Israel Symposium on Theory of Computing and Systems, 1995,* 40–45.
130. Vovk V.G., V'yugin V.V. On the empirical validity of the Bayesian method. *J. Royal Statist. Soc. B,* **55** (1993), no.1, 253–266.
131. Vovk V.G., V'yugin V.V. Frequential level of impossibility with some applications. *J. Royal Statist. Soc. B,* **56** (1994), no.1, 115–123.
132. V'yugin V.V. Algorithmic entropy (complexity) of finite objects and its application to defining randomness and amount of information. *Selecta Mathematica (formerly Sovietica),* **13** (1994), no.4, 357–389.

133. Yanovskaya S.A. Foundations and mathematical logic. In: *Matematika v SSSR za Tridtsat' Let,* Gostekhizdat, Moscow, 1948, 11–50 (in Russian).
134. Yanovskaya S.A. Mathematical logic and foundations. In: *Matematika v SSSR za Sorok Let,* **1**, Fizmatlit, Moscow, 1959, 13–120 (in Russian).
135. Zakharyaschev M. Canonical formulas for K4. Part I: basic results. *J. Symb. Logic,* **57** (1992), 1377–1402.
136. Zakharyaschev M. A sufficient condition fot the finite model property of modal logics above K4. *Bull. of the IGPL,* **1** (1993), 13–22.
137. Zakharyaschev M., Alekseev A. The finitely axiomatizable normal extension of K4.3 are decidable. *Math. Logic Quarterly,* 1995, to appear.
138. Zilber B. Quazi-Riemann manifolds. In: *Proc. Europ. Logic Coll., Kiel, 1993,* to appear.

DIMITER VAKARELOV

LOGIC IN CENTRAL AND EASTERN EUROPE: BALKAN REGION

The main aim of this lecture is to give some information about Logic in the Balkan region before and after the collapse of the communist system. In my talk I will not discuss the complicated political situation here. I would rather draw your interest to how this situation and radical social change reflects on the development of Logic, logical groups and the links with the international logical community.

The countries I will consider are Bulgaria, Romania and former Yugoslavia. Albania should have also been included here, but I did not find any information about Logic there. Excluding Bulgaria, where I come from, I realized that my information about the other countries was rather incomplete. Our University (and probably those of the other Balkan countries) still not having regular e-mail, the only way for me to collect some new information was to write as many letters as possible. Some of the people I wrote to were very punctual, but unfortunately, most of my letters have not been answered. Probably this was so because some of the addresses I used are not actual now and the addressees are somewhere abroad, or because of the difficult political situation on the Balkans. That is why I still consider my information about the Logic on the Balkans incomplete, partial and probably in some places incorrect, for which I very much apologize.

Here is the moment for me to mention the names of the people who helped me very much with this difficult task. In some sense I consider them as co-authors of these notes and express my sincere gratitude to them.

From Yugoslavia:

- Aleksandar Kron, Belgrade, Faculty of Philosophy,
- Andrea Prijatelj, Ljublljana /temporarily in the Netherlands, at the University of Amsterdam/,
- Bosko Zivaljević, Sarajevo /now in USA, at the University of Illinois at Urbana-Champaign/,
- Branislav Borizić, Belgrade /now in Greece, at the University of Crete/,
- Dragić Banković, Kragujevać, Faculty of Science,
- Gradimir Voivodić, Novi Sad, Institute of mathematics,
- Kosta Došen, Belgrade, /now in France, at the University of Paul Sabatier, Toulouse/,
- Vladeta Vučković, Belgrade /now in USA, at the University of Notre Dame, South Bend/.
- Zoran Marković, Belgrade, Institute of Mathematics.

M.L. Dalla Chiara et al. (eds.), Logic and Scientific Methods, 485–495.

From Romania:

- Cristian Calude, Bucharest, /now in the University of Auckland, New Zealand/,
- Gabriel Ciobanu, Iasi, A. I. Cuza University,
- Sergiu Rudeanu, Bucharest, Bucharest University,
- Şerban Basarab, Bucharest, Institute of Mathematics.

Also many thanks to all my Bulgarian colleagues from the Department of Logic and its Applications at Sofia University and from the Section of Logic at the Bulgarian Academy of Sciences, especially to Professors Dimiter Skordev and Ljubomir Ivanov.

The next part of my talk is divided into three sections, devoted to Logic in Yugoslavia, Romania and Bulgaria respectively.

1. ON LOGIC IN FORMER YUGOSLAVIA

After World War II, Yugoslavia was the biggest state in the Balkans. Now as a result of the collapse of the communist system and a long civil war Yugoslavia is divided into several new states: Federal Republic of Yugoslavia /including Serbia and Montenegro/, called sometimes "New Yugoslavia", Slovenia, Croatia, Macedonia and Bosnia and Hercegovina /here the war is not finished yet/. Some of these new independent countries, such as Slovenia and Croatia, do not consider themselves part of the Balkan region now. In my talk "Yugoslavia" will allways mean "former Yugoslavia".

I will start with some data on the early history of Logic in Yugoslavia.

According to Professor Vladeta Vučković, after World War II, in a period, say, from 1945 to 1955, the only official logic in Yugoslavia was the so called Dialectical Logic, considered as a part of Marxist-Leninist philosophy. A tragic joke from that time was that mathematical logic had been considered as a "gun of western imperialism". Nevertheless, a course on Mathematical Logic, probably the first in Yugoslavia, was given by Professor Kajica Milanov in the period 1945–47 at the Faculty of Philosophy of Belgrade University. Professor Kajica Milanov (1905–1986) studied philosophy in Vienna (1924–8) where he obtained his Ph.D. in 1932. Until 1947 he was Professor in the Philosophy department of the Faculty of Philosophy at the Belgrade University. In 1947, as a result of fierce communist terror he lost his job and was forced to emigrate.

In the period of 1959–60 a course on Mathematical Logic in the Philosophy Department at Belgrade University was given by Professor Mihailo Marković and his assistant Svetlana Knjazeva. Professor Marković (born 1922) obtained his Ph.D. in Logic, first in Belgrade and then at London University (his promotor was A.J. Ayer).

The most important person in Mathematical Logic at the beginning of the sixties was Professor Vladeta Vučković. He obtained his Ph.D. in the field of mathematical analysis and studied logic by himself. For a long period of time he could not find a job in mathematics for political reasons. In 1960 he became a Professor in the Faculty of Engineering at the University of Belgrade and attended the philosophical group "Praxis". He emigrated to the USA in 1963 and was Professor of Mathematics and Mathematical Logic at the University Notre Dame, Indiana. The field of interest of Vladeta Vučković in Mathematical Logic is Recursion Theory and its generalizations. In 1962 he gave the first course on Recursion Theory in the Philosophy Department of the Faculty of Philosophy in Belgrade University.

Since 1965 Mathematical logic has been taught regularly in the Philosophy Department of Belgrade University and some years later similar courses have been introduced in the department of Mathematics. The first to introduce such a course in the Philosophy Department was Professor Aleksandar Kron, studying Mathematical Logic in Amsterdam (1963–4) under the supervision of Professors E.W. Beth and A. Heyting. His field of interest is Relevant Logic.

According to Professor Dragić Banković, "the father" of Serbian logic is Slaviša Prešic, Professor in the Department of Mathematics, University of Belgrade. "The other logicians in Serbia" — says Banković — "are his students or students of his students".

One of the leading institutions having a central position in the development of Mathematical Logic in Yugoslavia is the Mathematical Institute of the Serbian Academy of Sciences and Arts. From 1964, a regular Seminar for Mathematical Logic works in this institute, called sometimes Belgrade Logic Seminar. It brings together logicians not only from the Mathematical Institute and the University of Belgrade but also from some other institutions and places: the University of Niš, the University of Novi Sad, the University of Kraguevac, the University of Montenegro, Podgorica. The seminar was founded by Professors Slaviša Prešić, Aleksandar Kron and Svetlana Knjazeva.

Besides the founders, some of the active leaders of the Seminar were Professor Djuro Kurepa (Department of Mathematics (DM), University of Belgrade), Professor Zarko Mijajlović (DM), Professor Kosta Došen (Mathematical Institute (MI)), Professor Zoran Marković (MI).

During the work of the Seminar all fields of logic have been covered: general and philosophical logic, proof theory, model theory, set theory and recursion theory. The seminar organizes also advanced courses in various areas of logic. The last decade saw courses in non-standard analysis, descriptive set theory, infinitary combinatorics, cylindric algebras, universal algebra, category theory, abstract model theory, models of arithmetic, stability theory, intuitionistic logic, modal logic and programming languages. In its expansive periods the seminar had up to twenty active members.

The seminar is now under the chairmanship of Professor Aleksandar Kron and Professor Djordje Vukomanović. Here is a list of senior logicians active in the Seminar during the last several years with their affiliation and area of current research:

- Branislav Borizić (University of Belgrade, now Visiting Professor at the University of Crete, Greece) — intermediate logics.

- Kosta Došen (Mathematical Institute, Belgrade) — nonclassical logics, substructural logics.
- Radosav Dordević (Faculty of Science, University of Kragujevac) — model theory.
- Milan Grulović (Department of Mathematics, University of Novi Sad) — forcing.
- Aleksandar Jovanović (Department of Mathematics, University of Niš) — set theory.
- Miodrag Kapetanović (Mathematical Institute, Belgrade) — proof theory.
- Aleksandar Kron (Faculty of Philosophy, University of Belgrade) — relevant logic.
- Zoran Marković (Mathematical Institute, Belgrade) — intuitionistic logic.
- Zarko Mijajlović (Faculty of Mathematics, University of Belgrade) — model theory.
- Zikica Perović (Department of Mathematics, University of Niš) — Boolean algebras.
- Miodrag Rašković (Faculty of Science, University of Kragujevac) — model-theoretic logics.
- Predrag Tanović (Mathjematical Institute, Belgrade) — stability theory.
- Djordje Vukomanović (Faculty of Civil Engineering, University of Belgrade) — algebraic logic.
- Slobodan Vujošević (Institute of Mathematics, University of Montenegro) — model theory, Boolean algebra.

The juniors in the seminar are the following: Mirjana Borisavlević, Dragan Cvetković, Ilias Farah, Silvia Gilezan, Zoran Ognjanović, Zoran Petrić, Nataša Rakić, Dragan Stankov, Dragan Urošević.

In former Yugoslavia, a very important role in Logic have been played a Group, informally called "Beograd-Zagreb group", with meetings organized almost every year. The members of this Group are mainly logicians from Belgrade Seminar and logicians from Zagreb — Professor K. Šeper, Professor Mihaljinec, Professor Dean Rozenzweig, Professor Zvonko Šikić. Later on this Group was transformed into a General Yugoslav Seminar having meetings every year, sometimes with algebraists and sometimes with computer-scientists. After 1980 this seminar grew into a yearly Yugoslav conference on "Logic and Computer Science" (LIRA), generally organized by the University of Novi Sad. Other meetings with a logical part are: the conference "Algebra and Logic", organized by the University of Niš, and PRIM — Applied Mathematics Conference, organized by the Institute of Mathematics in Novi Sad.

Another group of people working in the field of Logic and its applications is in the Institute of Mathematics, Novi Sad:

- Gradimir Vojvodić — universal algebra and logic,
- Ratko Tošić — combinatorial algorithms, discrete mathematics, computational geometry, many-valued logics.
- Siniša Crvenković — algebra and automata theory, algebraic logic.

• Rozália Sz. Madarasz — universal algebra, algorithmic problems, algebraic logic.

• Milan Grulović — model theory, forcing.

This is in short the information about Logic in "New Yugoslavia".

As for Logic in the other new countries coming from former Yugoslavia I can say only a few words.

I have no information if there are working logicians in Macedonia, except Professor Smile Markovski from the Faculty of Mathematics at the University of Skopje.

Probably in Croatia there is an active logic group. I have mentioned some names from Zagreb in the Belgrade-Zagreb Group, but my information is very incomplete.

About logic in Slovenia I can say the following. In the University of Ljubljana Mathematical Logic is taught by Professor Niko Prijatelj in the Department of Mathematics and by Professor Izidor Hafner in the Faculty of Computer Science. His Ph.D. thesis is on Lesnewski's Logic. In 1995, Andreja Prijatelj obtained her Ph.D. degree at Amsterdam University in the field of Linear Logic. Professor Niko Prijatelj is an author of three text-books on Mathematical logic for undergraduate students. It may be interesting to note that Competitions in Logic for primary school pupils have been widespread in Slovenia for some 5 years already, organized by Izidor Hafner.

Probably there are no logicians in Bosnja and Hercegovina now. The only logician from Sarajevo, Bosko Zivaljević emigrated in 1991, just before the war, and from that time he has been a permanent resident of the USA. He obtained his Ph.D. in 1989 in the University of Illinois at Urbana-Champaign, USA. The title of his dissertation is *Hyperfinite Transversal Theory.* Now he is doing very active research on Descriptive Set Theoretical questions of internal sets in the context of Nonstandard Analysis.

The new political and social situation in Yugoslavia did not reflect radically on the situation in Logic. According to some members of Belgrade Logical Group, the Logical Seminar is as lively and active as ever. The new thing is that now many logicians are working abroad, but some younger logicians have come instead. Almost all links with the former socialist countries fell down and many new links, mostly personal, with the international logical community have been established. The civil war in Yugoslavia and the subsequent economic embargo have some negative influence on the Logic. For instance the Belgrade Group, together with similar groups from various former socialist countries, was included in 1991 in a TEMPUS project on Logic and its applications, but then the Yugoslav part was excluded from the project.

2. ON LOGIC IN ROMANIA

After World War II and after establishing the communist system, Logic in Romania was understood mainly as Dialectical Theory of Thinking and a special Center of Logic in the Romanian Academy of Sciences was organized.

The father of Romanian Modern Logic was Grigore Moisil, working in the field of algebraic and many-valued logic. His first paper "Recherche sur la logique" appeared in 1935. In the bibliography of his main book *Essais sur les logiques non-chrysippiennes* (1972) more than 50 works of Moisil have been cited. It is interesting to observe a break between 1945 and 1958, when probably mathematical logic was prohibited.

One of the main fields of the logic developed in Romania is Algebraic Logic. The following authors have contributed to the field of Algebraic Logic: Vlad Boicescu, Alexandru Brezuleanu, George Georgescu, Grigore Moisil, Afrodita Iorgulescu. We have to mention that Moisil was one of the founders of the algebraic approach to many-valued logic. He introduced the notion of Łukasiewicz algebras as an algebraic counterpart for many-valued logic to Boolean algebras for classical logic. Now Łukasiewicz algebras are frequently called Łukasiewicz-Moisil algebras. They have been studied by a number of his followers such as V. B̃oicescu, A. F̃ilipoiu, G. G̃eorgescu and S. R̃udeanu who have collected the main results in this field in the book *Łukasiewicz-Moisil Algebras* (1991).

In the field of classical logic we can mention the name of Eugen Mihailesku; in model theory — Şerban Basarab, George Georgescu, Ioana Petrescu--Voiculescu; connections with universal algebra — A.F̃ilipoiu, Sergiu Rudeanu; Logic and Computing Science — Christian Calude, Virgil Cãzãnescu, Gabriel Ciobanu, Ion Filoty, Gheorghe Pãnu; logic in natural languages — Solomon Marcus; connections with philosophy — Anton Dimitriu, Octav Onicescu, Eugen Radu. Thanks to Professor Cristian Calude, the information about recursive function theory is more detailed.

The most important result in recursive function theory obtained by a Romanian scientist is Gabriel Sudan's example of a recursive function which is not primitive recursive, published in 1927. It has been proven by Calude, Marcus and Tevy that Sudan and Ackermann have to share the priority of such an example (see C. C̃alude, S. M̃arcus, I. T̃evy, The first example of a recursive function which is not primitive recursive. *Historia Math.*, 9(1979), 380–384.)

According to Professor Moisil, Romanian scientists have become interested in recursive function theory indirectly, via problems related to the programming language ALGOL, since the mid 60-s.

The first courses in recursive function theory taught at Bucharest University date from that period; the first textbook published in Romanian is Mariusciac — *Theory of Algorithms*, Editura Didacticã, si Pedagogicã, Bucuresti 1966 (in Romanian).

A textbook due to Popovici — *Theory of Algorithms, Recursive Functions and Turing Machines*, Bucharest University Press, Bucharest 1976 (in Romanian) — has been used for many years at Bucharest University.

Calude's monograph — *Computational Complexity. Qualitative Aspects*, Ed. stintific, si Enciclopedica, Bucharest, 1982, 300 pp. (in Romanian) — was the first published in Romanian. An almost complete bibliography for the period 1928–1987 can be found in C. Calude — *Romanian results in recursive function theory: 1927–1987*, proc. Computer Center Anniversary Symposium, Bucharest University, 1987, 16–22 (in Romanian).

We are now enumerating the main directions in which Romanian scientists have contributed:

- Algorithmic information theory — Calude, Chitescu, Staiger, Câmpeanu, Zimand.
- Generalizations of Kleene's theorem — Breazu, Stãnãsila, Buzeteanu.
- Size of recursive functions — Calude, Buzeteanu.
- Gödel incompleteness — Calude, Pãum, Jürgensen, Zimand, Streinu.
- Blum complexity theory — Calude, Zimand.
- Structural complexity — Zimand.
- Topological methods — Calude, Câmpeanu, Zimand [35]).
- Subrecursive hierarchies — Calude, Tãtãrãm, Sântean.
- Undecidability — Artemi, Calude, Vaida.

The general conclusion about Logic in Romania is the following. There has been a stable tradition in Logic in Romania and people who are founders of schools in some directions: in the algebrization of Many-valued Logic — the school of G. M̃oisil, in Recursion Theory — the school of C. Calude and probably some others.

3. ON LOGIC IN BULGARIA

The state of Logic in Bulgaria after the World War II was similar to that of the other socialist countries on the Balkans — for a period of time, approximately till 1955 the only official logic was Dialectical Logic, considered as "the Higher and the only True Logic", "Logic with a capital letter" and so on. Even so called "Traditional Logic" or "Elementary Logic" was reconstructed and decorated with dialectical terminology.

Probably at around 1958–59 it was recognized that contemporary Symbolic Logic had to be included in university programs and scientific plans.

The first course of Mathematical Logic at the Faculty of Mathematics in Sofia University was given by Professor Bojan Petkanchin in the academic 1959–60 year. It contained standard material for such courses from Propositional and Predicate Calculus, including Gödel's Completeness Theorem.

Bojan Petkanchin was a Professor in Geometry. His interests in Mathematical Logic were mainly connected with his course on Foundation of Mathematics, which for many years was one of the best courses in Bulgaria, introducing the standard axiomatic method.

The first course on Recursion theory was presented in 1962 at the Faculty of Mathematics by Professor Dimiter Skordev. The course was organized in a very interesting and unusual way. It starts with an axiomatic characterization of universal partial recursive functions based on some ideas of Combinatory Logic and then all of recursion theory was derived from the properties of the universal functions.

Dimiter Skordev obtained his Ph.D. in Functional Analysis, but his main scientific interests are in Recursion theory and its generalizations. He is the founder of a school in this field in Bulgaria. His main result is an axiomatic generalization of Recursion Theory in algebraic terms, called "Combinatory Space".

The first course on Symbolic Logic at the Philosophy Department in Sofia University was given by Professor Dobrin Spasov in 1960–61 academic year. It contained some elementary information from propositional and predicate calculus with some naive critical comments from the point of view of the official Marxist-Leninist Philosophy.

A very important thing which stimulated the development of Mathematical Logic in Bulgaria was the establishment in 1972 of the "Sector of Mathematical Logic", an integrated institution in the Faculty of Mathematics and Institute of Mathematics in the Bulgarian Academy of Sciences. In 1990 the Sector of Logic was split into a "Department of Mathematical Logic with Laboratory of Applied Logic" at the Faculty of Mathematics and a "Section of Mathematical Logic" at the Institute of Mathematics in the Bulgarian Academy of Sciences. In the next pages I present some information about these two main institutions of Mathematical Logic in Bulgaria.

DEPT. OF MATH. LOGIC WITH LAB. FOR APPLIED LOGIC, FAC. OF MATH. AND CS, SOFIA UNIVERSITY

Head of the Department: Dimiter Skordev, Professor, Dr Habil. Head of the Laboratory: Dimiter Vakarelov, Assoc. Professor, Ph.D.

List of Professors and their principal research interests

• Anatoly Buda — System programming: theory of compilation. Theory of program schemata: decidability problems. Tree devices and languages: program machines, global data flow analysis, tree rewriting systems, unification, tree structures. Logic programming: unification, semi-unification and generalized unification. Parallel devices: decidability problems and formal languages.

• Angel Ditchev — Recursion theory. Computability on abstract structures. Effectively enumerable structures and effectively enumerable families of r.e. sets. Strong degrees of unsolvability. Recursive models. Effective computation.

• Rousanka Loukanova — Natural language semantics. Categorial grammars. Situation theory. Mathematical linguistics.

• Stela Nikolova — Non-deterministic computabilities. Computable functionals. Computability via enumerations. Skordev's Combinatory spaces. Abstract Church Thesis. Computability in abstract structures of arbitrary power. Computability in multiple-valued structures.

• Petio Petkov — Constructive mathematics, Mathematical Logic (Proof theory), History of mathematics.

• Dimiter Skordev — Mathematical Logic: Classical and generalized recursive function theory; Algebraic methods in logic and in the theory of computability;Constructive mathematics. Theoretical Computer Science: Theory of programs and computation; Program verification; Detection of loops in computational processes.

• Ivan Soskov — Mathematical Logic: Computability on abstract structures; Maximal concepts of computability on abstract structures; Non-deterministic and sequential computabilities; External definitions of computabilities; Abstract Church Thesis; Effectively enumerable structures; Algorithmic complete structures; Skordev's Combinatory spaces; Prime and Search computability of Moschovakis. Theoretical Computer Science:: Semantics of Program Constructs; Maximal programming languages; Sequential and non-deterministic programming languages; Generalized semantics of Horn-clause programs; Schematology.

• Alexandra Soskova — Mathematical Logic: Computability on abstract structures; Computability on many-sorted structures; Effective data types; Logic definability; Effectively enumerable structures Skordev's Combinatory spaces; Abstract data types.

• Tinko Tinchev — Mathematical Logic: modal, temporal and other non-classical logics — algebraic and model theoretic approaches, semantics, definability, axiomatizations. Applied logic: logic programming; dynamic logics, non-monotonic logics.

• Dimiter Vakarelov — Non-classical Logics: Modal and polymodal logic, Intuitionistic logic, Many-valued logic, Modal logic with intuitionistic and many-valued base, Constructive logic with strong negation, Logics of negation, Algebraic logic. Applied Non-classical Logic: Dynamic logic, Modal logics for knowledge representation, Modal logics for approximate reasoning, Information logics, Logics for reasoning about knowledge, Arrow logics.

• Jordan Zashev — Axiomatic and Abstract Recursion Theory; Algebraic Recursion Theory and its connections with λ-calculus Models. Computable functionals — domains of Scott, Girard and others. Categorial recursion theory — least fixed points of endofunctors in enriched categories and their expressibility with each others.

Basic fields of teaching

(Constructive) Mathematical Logic, Foundations of Mathematics, Set Theory, Foundations of Mathematics and Non-standard Analysis, Constructive Mathematical Analysis, Foundations of Arithmetics, Gödel's Incompleteness

Theorems, Universal Algebra and Model Theory, Recursive Function Theory, Theory of Programs and Computability, Algebraic Theory of Computability; Abstract Recursion Theory; Finite Automata and Formal Languages, λ-Calculus, Foundations of Logic Programming, Program Verification, Algebraic Logic, Non-classical Logics, Applied Modal Logic.

Scientific colaboration with some related institutions abroad

1. Laboratory of Mathematical Logic, Institute of Mathematics in Siberian Branch of Soviet Academy of Sciences — till 1990,

2. Department of Mathematical Logic at the Faculty of Mathematics and Computer Sciences in Warsaw University — till 1990,

3. Department of Mathematical Logic, Faculty of Mathematics in Humboldt University — till 1990.

4. TEMPUS project 1991–1994, coordinated by the University of Amsterdam.

Seminars

There have been several regular seminars at the Department of Mathematical Logic:

Colloquium on Mathematical Logic,
Seminar on Theory of Computability,
Seminar on Nonclassical Logic.

Conferences on Mathematical Logic

1980, dedicated to Andrej Markov,

1983 - Summer School on Mathematical Logic and its Applications. Three editions of "Summer School and Conference on Mathematical Logic":

1986, dedicated to Kurt Gödel, 1988, dedicated to Arend Heyting, 1990, dedicated to Stephen Kleene.

SEC. OF LOGIC, INST. OF MATH., BULGARIAN AC. OF SCI.

- Ljubomir Ivanov, head of the section — Algebraic and Axiomatic Recursion Theory.
- Vladimir Sotirov — Non-classical Logic, Modal logics on intuitionistic base
- Marion Mircheva — Logic Programming, Non-Monotonic Logics

Other people who contributed to the field of mathematical logic

Krasimir Atanasov — fuzzy sets. Kiril Chimev — functions in k-valued logics. Jordan Denev — Boolean functions, functions in k-valued logics. Georgi Gargov — Intuitionistic Mathematics, Intuitionistic and superintuitionistic logics, Modal and Multimodal Logics, Provability logics, Dynamic Logic, Many valued Logics. Nadejda Georgieva — Classical recursion theory, Algebraic Recursion Theory. Valentin Goranko (now in South Africa) intuitionistic, modal, polymodal, temporal and other non-classical logics, refutation systems, dynamic logics, logics of programs, logics of knowledge, non-monotonic logics. Solomon Passy — Dynamic Logic, Modal and Polymodal Logic. Radoslav Pavlov — Algorithmic problems in Algebra. Slavian Radev — Dynamic Logics, Infinitary Modal Logic, Decision Logic. Chrasimir Tarkalanov — Algorithmic Problems in Algebra.

Philosophical Logic in Bulgaria is represented by the following people:

From the Faculty of Philosophy, Sofia University: Angel Bankov, Nedjalko Merdzanov, Dobrin Spasov, Vitan Stefanov.

From the Institute of Philosophy, Bulgarian Academy of Sciences: Boris Chendov, Bogdan Djankov, Nikolaj Obreshkov, Martin Tabakov, Chrastiu Vakarelov, Jana Yaneva.

The general conclusion about Logic in Bulgaria is the following. Till 1959 the Logic in Bulgaria was developed mainly as a part of the official Marxist-Leninist philosophy. Mathematical Logic started in 1960 and is now a well-established branch of mathematics represented by a Department with Laboratory at Sofia University and a Section at the Academy of Scienses. There was a well established system of scientific links mainly between some former socialist countries: Poland, USSR, DDR. The links with the European logical community were informal and realized mainly by logical conferences and private initiatives.

The political and economical change in 1990 reflects on Logic in Bulgaria in several aspects. First, due to economic reasons, the financial support of Science by the Government has approached the absolute minimum. As a consequence, now it is very difficult to appoint and find a job for young logicians, to organize conferences and some other meetings. Second, scientific links between the former socialist countries have been interrupted. Now new links have been established using mainly financial support from the European Community, e.g. TEMPUS, and individual invitations and grants from some institutions. So, in my view, there is a need for a more perfect and flexible system of collaboration between different logical groups and institutions allowing us to preserve and develop the former links and to establish new ones.

Department of Mathematical Logic with Laboratory for Applied Logic
Faculty of Mathematics and Computer Science
Sofia University

RYSZARD WÓJCICKI

THE POSTWAR PANORAMA OF LOGIC IN POLAND

In this essay[1] I briefly discuss some developments in logic in Poland after the Second World War at the period of the communist domination. By no means the essay is meant to be a technical report on problems and results. Rather, I shall try to present some of the main trends and ideas, as well as to give some flavor of those times which happily enough are over.

You should not expect me to go into any political considerations; I am certainly not going to advocate any political tenet. But the fact is that the collapse of the communist system brought radical changes which affected academic life in East-European countries, Poland included, in various ways. It is only natural to ask how the challenges and opportunities of the present time may stimulate or, quite to the contrary, hinder growth of science, growth of logic in particular in these countries. And it seems to be natural to discuss this question here, on the occasion of a large international congress gathering people with so many countries and, thus, with so different backgrounds and so different experience.

What I am going to say about logic will be restricted to logic construed in a rather narrow way. By logic I shall mean the theory of formalized reasoning, thus a theory whose chief concern is the conditions which an argument should satisfy in order to be both stated and carried out in a sound way. I believe that the above definition of logic is well anchored in Polish tradition, although it should not necessarily be accepted without any proviso by logicians from other countries. Logic is often considered to be the theory of logical truth; accordingly the main task of the logicians is considered to be to state the rules which, for a given formalized language, allow one to separate all the logically true sentences from the remaining ones.

Note that already in the early thirties, Tarski, at that time a young docent at Warsaw University, formed his theory of deductive systems. It was intended to be the most general theory of logical reasoning. The central notion of Tarski's theory is a consequence operation, thus an operation which applied to any set of premises yields the set of all the conclusions derivable from (or, under semantic interpretation of the consequence operation entailed by) those premises. Since, depending on one's logical preferences, logical validity (and hence both derivability and entailment) can be understood in different

[1] An abbreviated version of this essay was presented at the 10th Congress of Logic Methodology and Philosophy of Sciences (Florence, August 19–25, 1995) on a special symposium *Logic in Central and Eastern Europe*.

M.L. Dalla Chiara et al. (eds.), Logic and Scientific Methods, 497–508.

ways, Tarski focused his attention on those characteristics of the consequence operation which are by and large independent of our logical preferences.

Curiously enough the theory of deductive systems, known also as the theory of the consequence operation, has attracted little attention outside of Tarski's native country. In this way the theory of the consequence operation became a Polish speciality, not to say idiosyncrasy, often treated as some strange, not to say redundant, province of logical investigations. Only now, after non-monotonic logic has been invented, the significance of Tarski's idea of studying the consequence operation in an abstract way, independent of any rules of inference derived from a specific system of logic, has become obvious.

Another characteristic feature of logic in Poland has been a considerable interest in applied problems. This is not to say that purely theoretical ideas were ignored, still Polish logicians viewed logic as a discipline whose primary task is to provide tools for clear, precise and sound presentation of both views and arguments, especially those applied in scientific discourse.

Well, as is customary, I shall view logic as a part of a larger set of disciplines known as the foundations of mathematics (set theory, model theory, recursion theory, decidability theory, etc.). Whichever way logic is defined it certainly cannot be sharply separated from any of the above mentioned areas of inquiry. For instance, many results of model theory are part of logic in the most obvious and direct way. On the other hand, the chief concern of model theory is models, thus certain set theoretical structures, examined independently of any language which may serve to describe and to argue about them. Model theory separated from any syntactical problems is part of the theory of abstract algebra rather than logic. The point I am trying to make is that, although logic is both strongly related to and often overlaps with foundational disciplines, it clearly is an autonomous discipline with a well defined subject matter.

Right or wrong, my definition of logic will confine the area of my considerations concerning logic in Poland. I should also mention some other restrictions. I shall not be able, except in a very sketchy and incomplete way, to cover the work done by historians of logic or philosophers of the discipline. This omission is not as natural as it may seem to be at first glance. Both historical and philosophical studies have influenced Polish logic very strongly. An obvious example is Łukasiewicz's investigations into Greek logic which, among others, resulted in both his challenging the Inconsistency Principle and in his and Kotarbiński putting forward the idea of many-valued logic.

One may also point out that an interwar (1920–1939) formation widely known as the "Polish School of Logic" overlapped in a very substantial way with another formation known as the "Lwow–Warsaw School of Philosophy". The members of the latter not only considered logic to be the main tool of philosophical analyses but they often contributed to logical investigations by themselves.

The time of the "Polish School of Logic" was marked by such outstanding achievements as Ajdukiewicz's categorial grammar, Łukasiewicz's idea of many valued–logic, Lindenbaum's and Tarski's method of algebra of language, Tarski's conception of truth and Tarski's investigations of deductive systems, Leśniewski's attempt to form ontology meant to become a system alternative to set theory, Chwistek's simplified theory of types, Hosiasson–Lindenbaum's investigations into methodological and logical aspects of probability theory, Łukasiewicz's and Tarski's algebraic treatment of logical matrix, Jaśkowski's system of natural deduction and many others. For a survey of this period see Zygmunt (199?).

The second World war and then German and initially also Soviet occupation of Poland was a disaster for both the country and its academic life. For the very large Jewish community of the time it brought agony, humiliation and the holocaust. Such eminent representatives of Polish logic as Adolf Lindenbaum, his wife Janina Hossasion Lindenbaum, Mojżesz Presburger, Józef Pepis, Jan Salamucha, Z. Schmierer, Mordchaj Wajsberg shared the fate of millions of Jews murdered on Polish soil by Nazi's occupants. Some logicians perished in the Soviet Union. Others, not necessarily of Jewish origin, who were lucky enough to leave Poland, often decided to stay abroad rather than go back to the country which under the Jalta agreement was left under Soviet domination. Thus, for instance, Józef M. Bocheński settled in Switzerland, Czesław Lejewski in England, Jan Łukasiewicz in Ireland, Henryk Hiż, Bolesław Sobociński, and Tarski in the USA.

To a large extent the postwar history of Logic in Poland was formed by logicians of new generations: L. Borkowski, A. Ehrenfeucht, A. Grzegorczyk, L. Gumański, S. Jaśkowski, J. Kotas, J. Łoś, A. Mostowski, J. Mycielski, A. Pawlak, W.A. Pogorzelski, H. Rasiowa, C. Rauszer, Cz. Ryll-Nardzewski, R. Sikorski, J. Słupecki, S. Surma, R. Suszko, W. Szmielew, L. Szczerba., T. Traczyk and still younger Z. Adamowicz, K.R. Apt, J. Czelakowski, W. Dziobiak, W. Guzicki, H. Kotlarski, S. Krajewski, M. Krynicki, G. Malinowski, W. Marek, G. Mirkowska, M. Mostowski, R. Murawski, E. Orłowska, L. Pacholski, J. Perzanowski, T. Prucnal, A. Salwicki, A. Skowron, M. Tokarz, A. Wasilewska, P. Wojtylak, A. Wroński, P. Zbierski, J. Zielonka, J. Zygmunt and many others.

Needless to say that the revival of logic in postwar Poland was possible only because continuity was not entirely broken. Some outstanding scholars of the Lwow–Warsaw School like K. Ajdukiewicz, T. Kotarbiński, M. Kokoszyńska, gained fairly influential positions in academic life, they were able to raise students, promote logical investigations, and even to assure contacts with the world outside of the Iron curtain. Thus for instance, at a time, though at differet periods, Kazimierz Ajdukiewicz was the Rector of Poznań University, Tadeusz Kotarbiński was the President of the Polish Academy of Sciences, Maria Kokoszyńska was a Pro-Rector of Wrocław University, Klemens Szaniawski was the Dean of Department of Philosophy of Warsaw University and Jerzy Słupecki was the Rector of the Opole Educational University. Still, from the very beginning the Communist Party started to form the system of political control of all aspects of intellectual and cultural life, university education included. I hate to go further with political considerations. It would force me to discuss the complexity of the political landscape of those days. After the tremendous losses Poland suffered during the Second World War, any unreasonable boycott of communist rulers of the country, for instance by declining from accepting nominations to exposed academic positions, could result in affecting restoration of the country, its educational system in particular. Polish intellectuals, attached to their traditional values, had no other way than to start a long and complicated history of 'cohabitation' with the communist rulers of the country. None of the above mentioned persons has ever accepted communist ideology or even has ever paid lip service to it.

Efforts to spread Marxist ideology in Poland never resulted in eradication of alternative views. Whichever could be their motives, people responsible for ideological purity avoided to resort to drastic measures. For instance, the demand to suppress formal logic by so-called dialectical logic (the latter was to be rediscovered from the writings of Marx, Engels, Lenin and Stalin) was never executed with full vigor in Poland. In particular no Polish university has ever had a chair of dialectical logic and nobody was offered an eminent academic position because of her or his devotion to the idea of such "progressive" logic. The campaign for dialectical logic was undertaken, but its results were not necessarily of the kind the true believers of the communist doctrine could consider satisfactory.

Nowadays, hardly anybody is aware that Jaśkowski's discussive logic put forward by him in 1952, was an attempt to form a logical system which admits controversies and thus in some way resembles dialectical logic. Jaśkowski did not care to call those controversies "dialectical" or to use the word "dialectical" on whichever occasion in his paper. Not to mention that he never cared to quote any of the so called "classics" (i.e. Marks, Engels, Lenin and Stalin) in order to support his views. Jaśkowski's paper as well as a few more published at that time by logicians (M. Kokoszyńska, T. Kubiński) were an attempt to implicitly demonstrate that some ideas of dialectical logic, can perhaps be stated reasonably and if so, the logicians are open to do some job consistent with the ideas of "dialectical approach". On the other hand, both in its style and its essence those papers were miles away from the idiocy of the politically correct logic.

The public criticism of Lwow-Warsaw School (directed especially against K. Ajdukiewicz, T. Kotarbiński, M. Ossowska, S. Ossowski, and W. Tatarkiewicz) launched by the Party's people in 1953 could be the beginning of something much more serious than the above mentioned campaign for dialectical logic. It certainly could open the process of full eradication of any politically wrong philosophy from academic life. Luckily enough only a few years later the death of Stalin, followed by Khrushchev's revelations, gave way to liberalization of the communist system. Like other communist countries Poland took her chance to widen the margin of intellectual freedom.

To make a long story short: Logic in Poland was never affected in a drastic and direct way by political principles and based on bureaucratic measures. During the whole period of so called People's Democracy in Poland there were both chairs and sections of foundations of mathematics at mathematical departments, and chairs of formal logic at the humanity departments, philosophy departments in particular. What is more, Ajdukiewicz (whose efforts were tacitly backed by Adam Schaff, his chief opponent in the 1953 debate and at that time the chief Party's ideologist) succeeded to organize several conferences on formal logic, to held a regular seminar on philosophical logic, to form Section of Logic at Polish Academy of Sciences (the Section eventually became a part of the Institute of Philosophy and Sociology of the Academy) and, in 1953, to start *Studia Logica* which still remains the leading Polish logical journal.

This seems to be the right time to mention briefly some other logical journals published in Poland. In 1973 *Reports on Mathematical Logic, RML* was formed by Stanisław Surma (at that time a Pro-Rector of the Jagiellonian University) with the help of his younger colleagues (notably Jerzy Perzanowski and Andrzej Wroński). Somewhat later the journal became the common enterprise of two Universities: the Jagiellonian University in Cracow and the Silesian University in Katowice. The editorial representatives of the latter were W.A. Pogorzelski, T. Prucnal and P. Wojtylak.

The idea to start *RML* was part of the efforts to have an international logical journal in Poland, entirely published in congress languages, notably English. The fact that at that time, most of the papers published in *Studia Logica* appeared in Polish drastically narrowed the audience of the journal and consequently discouraged the top logicians, also Poles, from submitting their papers to it. Those of Polish logicians who wanted to present their results in a language intelligible for sufficiently large international audience preferred to choose either *Fundamenta Mathematicae* (an eminent Polish journal on foundations of mathematics established in 1920) or *Bulletin of the Polish Academy of Sciences* or just to send the paper for publication abroad.

As a matter of fact, the above mentioned state of affair reflected the political restrictions imposed on academic life in Poland. By no means international cooperation was encouraged; both Party's ideologist and bureaucrats considered it risky and sought to keep it under control. Under these circumstances, the initiative to start a new international journal was not merely a practical question; it had some political dimension. The decision to give it green light had to be approved by the right department of the Central Committee of the Party. Moreover, if the journal was expected to have some international Editorial Board (eventually this was the case of *Studia Logica*) the initiative had to be cleared by the State Security Police.

At about the same time when the idea to start *Reports on Mathematical Logic* was put forward, I initiated the process of transforming *Studia Logica* in an international quarterly. The project would never have taken off the ground, had it not been strongly backed by several eminent logicians (Klemens Szaniawski, Roman Suszko, Andrzej Mostowski, Helena Rasiowa and others) who were able to persuade the right people to support the idea.

Some other editorial initiatives should also be mentioned in this context. In 1972 I was able to start publishing a newsletter called *Bulletin of the Section of Logic of the Institute of Philosophy and Sociology of the Polish Academy of Sciences, BSL* for short. Although *BSL* is no more published by the Institute of Philosophy and Sociology of the Polish Academy of Sciences, it continues to appear under the editorship of Grzegorz Malinowski as a publication of Łódź University. In 1974 Helena Rasiowa and a group of her collaborators started *Fundamenta Informaticae*, the journal that covers applications of logic to computer science. A quite recent editorial event is *Logic and Logical Philosophy*, started by Jerzy Perzanowski. Its first issue appeared in 1993.

The leading centers of logic (let me state once more that I refrain from reporting on foundations of mathematics) in the postwar period of logic were active at the Departments of Mathematics of Warsaw, Wrocław, and Poznań Universities. Surma was able to form a strong group of logic at the Philosophy Department of the Jagiellonian University. A considerable role in logical life was always played by the Section of Logic of the Institute of Philosophy and Sociology of the Polish Academy of Sciences. The Section was initially headed by K. Ajdukiewicz, then R. Suszko and eventually by the author of this report. There were periods of considerable activity of Słupecki's group of logicians from the Opole Educational University, Pogorzelski's group at the Silesian University (one of its members was P. Wojtylak), J. Kotas's and L. Gumański groups at the Toruń University and others.

It is not an easy task to provide an adequate account of both activities and theoretical achievements of all the above mentioned entities. At different periods of time, different research programs were pursued, both formal and informal, different editorial initiatives were sought to be implemented, different forms of cooperation were promoted. I am sure that many participants of those events will find my survey incomplete and inaccurate in various respects. I apologize in advance for any such failures, even though I am afraid hardly anybody would be able to do justice to all those versatile and rich developments.

Let me start my account with mentioning briefly two academic units which, at least formally were expected to play some unifying role for logical investigations. One of them was the already mentioned Section of Logic of the Polish Academy of Sciences. Being for a long time the Head of the Section, I am not the right person to evaluate the outcome of work done by it. But besides editing two journals *Studia Logica* and *Bulletin of the Section of Logic*, the Section (R. Wójcicki, J. Czelakowski, G. Malinowski, M. Tokarz, J. Zygmunt and others) were in various ways active in promoting several research programs, both formal and informal. Notably, during the period of 1985–89, I was the chief coordinator of the research program *Formal aspects of Informal*

Reasoning. The coordinators of specific parts of the program were: H. Rasiowa, Z. Pawlak, W. Buszkowski, J. Pogonowski, M. Tokarz, J. Czelakowski and L. Szczerba. The program was designed on a rather large scale and sponsored by a governmental agency for scientific investigations formed at about that time.

Another institution of the Polish Academy of Sciences which fulfilled an eminent role in both initiating and promoting logical investigations was the Stefan Banach Mathematical Center. Although the activity of the Banach Center covers mathematics as a whole, quite many "semesters", workshops, conferences and other meetings (most of them international) held at the Center were devoted to foundations of mathematics, logic in particular. The organizers were A. Grzegorczyk, H. Rasiowa, C. Rauszer and others.

The above was meant to give an idea of the formal structure of the logical life in Poland: journals, academic units, research centers. But, as we all know, progress in any branch of science, logic in particular, depends much less on any formal arrangements than informal activity of creative personalities.

The most prominent figure of the postwar logical life in Poland was Andrzej Mostowski. With his outstanding theoretical achievements, and well established international position, he was an unquestionable leader of Polish logic. His unexpected death in 1975 during his visit to Canada was one of the greatest losses suffered by Polish logic in the postwar period.

We were lucky enough to have quite a number of logicians whose international reputation was well established, and who, like Andrzej Mostowski, were allowed by the communist rulers to go abroad. But even people of Mostowski's format had to apply for their passports every time they wanted to leave the country, and occasionally the application was turned down.

As I have already mentioned, international cooperations were controlled and often discouraged but never completely blocked. One might say, not being entirely wrong, that for a long period of time the chief Polish seminar on logic was held in Berkeley, California, in the residence of Alfred Tarski who kept very close cooperation with his friends and colleagues from Poland, Andrzej Mostowski in particular.

By no means it would be a mistake to say that Mostowski's bibliography includes positions which from a theoretical point of view are more important than his *Logika Matematyczna* ("Mathematical Logic") published in 1948 and *Thirty years of Foundational Studies* which appeared in 1965. But these two books played a distinguished role: they summarize, even though in a selective manner, the past period of investigations and suggest new ones. In particular *Logika Matematyczna* was a bridge linking the interwar and postwar periods of developments in Polish Logic.

In a sense all the leading logicians of the post war period were students of Mostowski. Of those who were very close to him one should mention Z. Adamowicz, A. Ehrenfeucht, W. Guzicki, W. Marek, L. Pacholski, H. Rasiowa, R. Sikorski, P. Zbierski, and others. The theoretical achievements of the group chiefly belong to the foundations of mathematics.

From the point of view of this survey, of special significance were all the results which contributed to the development of the algebraic method in logic. In fact, the algebraic chapter of Polish logic deserves our special attention. All the results in that area go back to Lindenbaum–Tarski's idea of the algebra of language and Lindenbaum's observation that for each set of propositional (i.e., 0th-order) formulas closed under substitution there is its adequate algebraic representation. The latter was generalized in many ways. In particular Rasiowa–Sikorski's method of Boolean algebras (see 1963) extends Lindenbaum ideas on both first order predicate logic (by putting forward the theory of Boolean valued models) and on first order intuitionistic logic (the corresponding theory set for that purpose was that of pseudo–Boolean valued models). Rasiowa (1974) provides a generalization of the above mentioned method to all of implicative logics (all logics that involve implication defined to be a logical operator satisfying some postulated in advance conditions).

Of course, algebraic methods have been both applied and pursued not only by Rasiowa's group. The relevance of those methods for the theory of the consequence operation was noted by Jerzy Łoś as early as in (1953) and later was fully acknowledged by almost all the chief contributors to that theory. Thus, for instance, in the early seventies, at the Jagiellonian University (Cracow) investigations into algebraic aspects of logical systems were initiated by A. Wroński. In particular, the people grouped around Wroński made some important contributions into the theory of BCK algebras and logical systems related to them. In Toruń, algebraic methods were applied by a group headed by J. Kotas.

More recent activity of Rasiowa and her group (Grażyna Mirkowska, Ewa Orłowska, Cecylia Rauszer, Andrzej Salwicki, Andrzej Skowron, Anita Wasilewska, to mention but a few) was oriented toward developing applied logic, such as algorithmic logic, logic of programming, logic of information and logic of knowledge. I am afraid I am not the right person to to provide an account of the results obtained by Rasiowa and her collaborators in the above mentioned areas. This makes my report incomplete in some essential way, for those results seem to belong to that part of logic which will largely determine the future growth of the discipline. The unexpected death of Rasiowa in 1994 was a blow comparable to the death of Mostowski.

Wrocław was another important center of postwar logic. The most prominent figure in foundations of mathematics has been Czesław Ryll-Nardzewski. He grouped around himself several very talented young peoples (L. Pacholski, B. Węglorz, A. Wojciechowska, and others) but that group was much more oriented towards foundational than purely logical problems. Also Jerzy Łoś, known for a number of outstanding theoretical results, notably his famous ultraproduct theorem, started his academic career in Wrocław. Another eminent Wrocław logician was Jerzy Słupecki. His most prominent collaborators were Ludwik Borkowski, Witold A. Pogorzelski and Bogusław Iwanuś.

Since my own academic biography starts in Wrocław too, please allow me at this juncture to say a few words about my theoretic interest, and about the work I have done with some colleagues of mine. To begin with I am certainly indebted a lot to J. Słupecki. In particular, the fact that I was able to attend his seminar allowed me to become acquainted with Tarski's theory of deductive systems as well as with the newest, at that time, contributions to that theory, notably Słupecki's, Borkowski's and Pogorzelski's investigations into the problem of defining logical constants by means of conditions imposed on consequence operation (somewhat later some results in that area was also achieved by A. Grzegorczyk), and Słupecki's idea of inverse consequence.

Another important source of my interest in consequence operation was Łoś and Suszko's (1958) paper on matrix representation of some consequence operations. In particular, by generalizing Łoś and Suszko's result I was able to discover the inferential counterpart of the Lindenbaum's representation theorem and hence I was able to define a procedure which, given any "structural" (i.e. substitution-closed) set of inferences, meant to be "figures" of the form $(X \vdash \alpha)$ —where X is a set of formulas and α is a single formula— determines an adequate algebraic representation for that set. The algebraic representation in question is available in a form of a class of logical matrices in the sense of Łukasiewicz and Tarski, thus not so much "truth tables" but rather algebraic structures of a kind.

As I have already mentioned at the beginning of my talk, on the received view, a system of logic is tantamount to the set of all its formulas which are logically valid. Appealing as it might be, this view is rather narrow. If logic is to be identified with any set, it should be the set of all the logically valid inferences. Regardless of any intuitive advantages or disadvantages of these two approaches, the latter is more general; logic meant to be a set of formulas can be viewed as the set of all the logically valid inferences of the form $\emptyset \vdash \alpha$. Consequently, which was fully appreciated by most of Polish logicians, the inferential approach allows for more subtle treatment of various problems. The following example illustrates the point.

As has been proved by Gödel (the result was later rediscovered by Łukasiewicz, and sometime is incorrectly attributed to him) classical connectives can be defined in terms of intuitionistic ones. The theorem is correct with some qualification however. The defined connectives are adequate counterparts of classical connectives, if we do not care about their inferential properties. In other words: the set of classical tautologies can be defined in terms of intuitionistically valid sentences (this is Gödel's result), but (cf. Wójcicki 1970) one cannot define the classical consequence operation in terms of the intuitionistic one.

I apologize for dwelling on the problems that were my special concern. I studied them along with a quite large group of students and colleagues

(J. Czelakowski, W. Dziobiak, M. Tokarz, J. Prucnal, R. Suszko, J. Zygmunt, A. Wroński, P. Wojtylak) in a quite systematic way for a fairly long period of time. The investigations were intended to examine the differences between the sentential and inferential approaches and more importantly to present the inferential approach in a form of a full fledged and well developed theory. The results obtained, combined with the relevant results of some other authors (W. Blok, H. Hiż, J. Łoś, D. Pigozzi, H. Rasiowa, W. Rautenberg, W.A. Pogorzelski, and others) are to be found in Wójcicki (1988). Czelakowski's monograph *Protalgebraic Logics* which is being prepared for print should cover some recent developments in the area.

Next, let us move to the Philosophy Department of the Jagiellonian University at Cracow. In the late sixties S. Surma, who collected a group of young and talented people around himself, originated an ambitious program of reconstruction of various results obtained in past in logic but presented in an incomplete or inaccurate way. Surma's team was able to offer a complete and rigorous setting for some of Lindenbaum's, Post's, Wajsberg's and others' results. A side effect of that activity was raising a talented group of historians of logic (J. Woleński, J. Zygmunt, R. Murawski). A. Wroński's activity, which has been already mentioned, belongs to a later period; Wroński has succeeded Surma when the latter settled abroad, first in Nigeria and later in New Zealand.

I am about to close my survey. But before I do this, let me try to give at least some idea of various investigations and results which were not or very unsatisfactorily covered above.

Note for instance that such an eminent Polish logician as A. Grzegorczyk was hardly mentioned. True, his main interest and his chief achievements were in recursion theory (Grzegorczyk is widely known for his hierarchy of recursive functions). On the other hand, however, various of Grzegorczyk's papers fall directly in the area of logical investigations. For instance, one of the most widely studied systems of modal logic has been Grzegorczyk's *Grz.*

R. Suszko's results belong to various areas. But he himself considered his system of "non-Fregean logic", as he called two–valued logic supplemented by the sentential identity connective, his chief theoretic achievement.

Incidentally, I have said almost nothing about those numerous investigations which concerned specific logical systems, such as for instance many-valued logics, where the work done by the Polish logicians (J. Słupecki, H. Rasiowa, T. Traczyk, G. Malinowski, T. Prucnal, M. Tokarz, Trzęsicki, R. Wójcicki, ..., not to mention Łukasiewicz, Tarski, Wajsberg and other prominent predecessors of the topic) has been really enormous. A good introduction into this domain is Malinowski (1993). Similar remarks apply to intuitionistic logic (H. Rasiowa, R. Sikorski, P. Wojtylak, A. Wroński), modal logic (A. Grzegorczyk, H. Rasiowa, J. Perzanowski, C. Rauszer, J. Zygmunt), relevance logic (W. Dziobiak, M. Tokarz).

One area of intense investigations (L. Szczerba, M. Krynicki, M. Mostowski, A. Krawczyk, M. Zawadowski and others) has been logical problems derived from Mostowski's results on generalized quantifiers (Mostowski was one of the chief founders of that idea. Some computational aspects of those problems were examined by A. Pawlak, H. Rasiowa, E. Orłowska. Also their semantic aspects have been studied. Notably, since logics of languages with generalized quantifiers can be non-elementary, in order to examine their semantic properties one may need to resort to non-standard semantic tools, such as the Ehrenfeucht–Fraïssé games.

Let me also give some idea of investigations in philosophical logic, another domain which I have excluded from the main part of my presentation. Investigations in deontic logic and its relevance to juridical problems have been made by quite a large number of logicians (J. Gregorowicz, L. Gumański, J. Kalinowski, T. Kubiński, W. Suchoń, K. Świrydowicz, J. Woleński, Z. Ziemba, Z. Ziembiński). Another important direction of philosophically oriented studies was causal logic (H. Greniewski, A. Pieczkowski, A. Trybulec, K. Trzęsicki and M. Urchs), mostly based on some ideas exposed by S. Jaśkowski in his (1951) paper.

Several Polish authors contributed in various way to Tarski's theory of truth. Various expository and historical papers concerning that theory were published by J. Woleński, some formal investigations into this theory concerning non-standard conceptions of truth were carried out by S. Krajewski and H. Kotlarski. On the other hand, M. Przełęcki and R. Wójcicki examined possibilities of expanding that theory so that it could also partially cover interpreted languages.

Finally, let me mention some model-theoretic results of obvious relevance to more narrowly construed logic. For a survey of this area see W. Marek (1976).

Besides J. Łoś' ultraproduct theorem and Rasiowa and Sikorski's results on generalized (e.g. Boolean-valued) models, perhaps the most important were the following: A. Mostowski and A. Ehrenfeucht's method of indiscernibles, Cz. Ryll-Nardzewski's characterization of $\aleph_0$-categorical theories, Łoś, Suszko and Słomczyński results on extending models. Some model theoretic dimension occurs also in Szczerba's investigations into the notion of interpretability of a theory in another one.

Large-scale investigations were made in the foundations of arithmetic, a domain in which my competence is almost nil. Let me restrict myself to mentioning some names, occasionally supplied by a short description of the corresponding results. They are the following: Cz. Ryll-Nardzewski (Peano arithmetic is not finitely axiomatizable); A. Mostowski (Peano arithmetic is reflexive); A. Grzegorczyk, A. Mostowski, H. Kotlarski, R. Murawski (some extensions of Gödel incompleteness theorem); Z. Adamowicz (arithmetic with open induction); A. Erenfeucht, A. Grzegorczyk, A. Mostowski, Cz. Ryll-Nardzewski, K.R. Apt, M. Dubiel, W. Guzicki, H. Kotlarski, S. Krajewski,

W. Marek, R. Murawski, Z. Ratajczak, M. Srebrny, P. Zbierski, Z. Vetulani (foundations of second and higher-order arithmetic and satisfaction classes).

I may only hope that, defective as it is, this survey can be of some use for those who want to learn a bit more about "Polish Logic" or rather Logic in Poland (for, happily enough, logic remains international) within the period marked by the 1945 collapse of Hitler's Berlin headquarter and ending with the 1989 collapse of Honecker's Berlin wall.

Institute of Philosophy and Sociology, PAN
Warszawa

ACKNOWLEDGMENTS

I wish to thank Janusz Czelakowski, Andrzej Grzegorczyk, Jacek Malinowski, Marcin Mostowski, Roman Murawski, Ewa Orłowska, Witold A. Pogorzelski, Kazimierz Świrydowicz, Maks Urchs, Jan Woleński, Andrzej Wójcik, Jan Zygmunt, who were kind enough to offer me various suggestions or materials relevant to this survey. I am sorry that I was able to exploit only some of them. It goes without saying that the full responsibility for the final content of this essay is mine.

REFERENCES

S. Jaśkowski, 'On the modal and causal functions in symbolic logic', *Studia Philosophica* **4** (1951), pp.71–92.

J. Łoś, 'The algebraic treatment of the methodology of elementary deductive systems', *Studia Logica*, **2**(1955), pp.151–212.

J. Łoś, R. Suszko, 'Remarks on sentential logic', *Indagation es Mathematicae*, **20**(1958), pp.177–183.

G. Malinowski, *Many-Valued Logics*, Clarendon Press, Oxford, 1993

W. Marek, "The foundations of mathematics in Poland after World War II" in R.O. Gandy and J.M.E. Hyland (eds), *Logical Colloquium '76*, pp. 129–38, North-Holland, Amsterdam.

A. Mostowski, *Logika matematyczna* ('Mathematical Logic, in Polish), PWN, Warszawa 1948

A. Mostowski, 'Thirty Years of Foundational Studies; Lectures on the Development of Mathematical Logic and the Study of the Foundations of Mathematics in 1930–1964', *Acta Philosophica Fennica* **17**(1965), pp.5–179

H. Rasiowa, *An Algebraic Approach to Non-Classical Logic*, PWN and North Holland, Warszawa and Amsterdam 1974

H. Rasiowa, R. Sikorski, *The Mathematics of Metamathematics*, PWN, Warszawa 1963

A. Tarski, *Logic, Semantics, Metamathematics; papers from 1923 to 1938*, Clarendon Press, Oxford 1956

R. Wójcicki, 'On reconstructability of classical propositional logic in intuitionistic logic', *Bulletin de l' Académie Polonaise des Sciences, Série des Sciences Mathématiques, Astronomiques and Physiques*, **8**(1970), pp.333–335.

R. Wójcicki, *Theory of Logical Calculi; Basic theory of Consequence Operations*, Kluwer, Dordrecht 1984

J. Zygmunt, "Polish Logic", *The Routlege Encylopedia of Philosophy*, vol. The Philosophy of Mathematics; Formal Logic, in print.

Closing Address

BAS C. VAN FRAASSEN

STRUCTURE AND PERSPECTIVE: PHILOSOPHICAL PERPLEXITY AND PARADOX

ABSTRACT.

My topic today is structuralism in the philosophy of science. The name "structuralism" is often used in a narrow sense to denote specific programs, such as Bourbaki's in mathematics and the approach to science due to Sneed and Stegmüller. I will use it in a broad sense, for any of a variety of views to the effect that science describes only structure, that scientific theories give us information only about the structure of processes in nature, or even that all we can know is structure. Since the very term "structuralism" comes from a by now fragile traditional distinction between form and structure, I will focus on the precise formulation of such views, with little regard to whether they honor that putative distinction. I shall describe some of the history of the general idea and its problems, and then ask what form of structuralism could be viable today.

1. SCIENCE AS REPRESENTATION

The prehistory of structuralist views is found, it seems to me, in the nineteenth century discussions of the extent and manner in which science represents natural phenomena: selectively, as representation must always be, yet accurately to some 'relevant' extent.

1.1. *The philosophical consensus ca. 1900*

At the end of the 19th century, the common view of science was that science is representation. Helmholtz, Hertz, Boltzmann, Mach, and Duhem are now all often seen as forerunners of anti-realism about science. It does not surprise us that they speak of 'mere' representation, 'thought-pictures', vindicated only by their role in the classification of observable phenomena. But their main opponent Planck, who espoused an explicitly realist view in his controversy with Mach, wrote in the same terms.

Planck began by insisting that unlike 'ordinary' representations such as drawings or paintings, scientific theories or models must "correspond ... literally in all respects with" their subjects (quoted Blackmore p. 127).

Why does Planck not say simply that science must be accurate and true? Silence is often the tacit admission of a problem, and so it is here. When science is admitted to be viewable as giving us pictures, likenesses, models (in the sense of similarities or analogies to familiar processes), the verbal concession lets in a problem. For if two things are independently known and independently described, then it makes sense to claim that they are like or unlike in some respect, and the claim can be evaluated. But if only one is

M.L. Dalla Chiara et al. (eds.), Logic and Scientific Methods, 511–530.

described, what is meant by a bare claim of similarity? How can that be non-trivial or informative — how could it even be true or false?

As example, take the anti-realist sounding dictum: "there is nothing to Maxwell's theory of electro-magnetism except Maxwell's equations." The realist wants to add something. The equations in question resemble those which describe certain familiar processes involving heat, sound, and motion in a fluid. The scientific realist wishes to add that the theory says that this resemblance is no accident, that electromagnetic phenomena are due to real waves in a real material medium. What is meant by this addition? With no independent description available of those waves and that medium, claims that go beyond the noted resemblance (that Maxwell's equations have a certain formal character) cannot be evaluated. But in that case, after all, there is nothing more to the theory than those equations — is there?

What Planck clearly had to do was to confront the question of what representation is, i.e. the conditions under which a theoretical model does in fact represent some part or aspect of nature. But he was so little able to give content or depth to his insistence on literal correspondence that he retreated to a neo-Kantian idealism. First he gives a list of famous scientists who were all, according to him, staunch realists. But then, to head off the philosophical problem of giving this realism some substance, he says:

> Is there any knowable [erkennbarer] difference between their world and our "world picture of the future"? Surely not. For there is not even a method to let us examine such a difference, [a thought] which through Immanuel Kant has become the common property of all thinkers. (tr. in Blackmore p. 132)

Can realists do better? I will leave this aside for now, to address the prior question: why was there this consensus, this agreement to talk about science as representation, and just what did this consensus signify?

We need to read between the lines of history here. I think this consensus resulted from reflections on geometry, which forced scientists to become conscious of the intellectual medium in which they do their creative work.

1.2. *The problem of multiple geometries*

The story of how geometry, and indeed mathematics over all, was revolutionized in the 19th century is a familiar one. Can we tell it as a prehistory of structuralism in the philosophy of science?

When non-Euclidean geometries were first created, the main reaction was that to understand them they had to be interpreted within Euclidean geometry — that is, as strangely worded descriptions of parts or aspects of Euclidean space. This was done initially by Beltrami in his "Saggio di Interpretatione della Geometria Non-Euclidia" (1868) for hyperbolic geometry.

Note that this was interpretation in the strictest, most minimal sense. Beltrami himself expressed his goal as maintaining Euclidean geometry as the one true theory of space (cf. Coffa p. 9). Felix Klein perceived the limits of this effort and first attempted to interpret both Euclidean and non-Euclidean geometries within project- ive geometry. But very soon through Klein's work, and then even more radically through Riemann's, there came into being such a cornucopia of geometries that any true theory of space underlying all of them could be very little more than pure logic.

The mathematicians' response was that none were privileged, all geometries were on a par and certain to remain parts of mathematics. What was the philosophers' reaction? Everyone today knows of the reaction of Poincaré, who was both scientist and philosopher: he embraced the mathematicians' liberalism and held that physics could in principle be written in the language provided by any geometry. Not so well remembered is the reactionary philosophical response of two of our other heroes, Frege and Russell. For reasons which I'll explain later, I will call their view 'real property' realism.

Both Frege and Russell saw what had happened as a sort of betrayal: geometry had forsaken the task of describing space, and turned to exercises in pure logic. In response to Klein's initial ranging of Euclidean and non-Euclidean geometries under projective geometry, Russell held that there must be a unique real relation of congruence. Geometry, at least the geometry used in physics, must be a non-vacuous theory of real spatial relations.

But how are those real relations identified? We are directly acquainted with them, through intuition. When pressed to elaborate on this by Poincaré, Russell dismissed the question. That is, he said, like asking me to spell the letter 'A' (cited Coffa, pp. 20–21).

Frege elaborated his similar response in a controversy with Hilbert. The theory of space must be non-vacuous and true. But the question of truth can't arise unless the primitive terms have an independent meaning which fixes their reference. They certainly cannot get that meaning from the axioms or theorems, as Hilbert asserted. For those axioms and theorems are incapable of having a truth-value unless their terms have referents. Pressed to explain how we identify those real spatial relations, Frege also retreated to intuition or direct acquaintance — just the sort of thing which neither Poincaré nor Hilbert professed to understand.

Now it was not very much later that Russell appeared to change his mind, and began to develop a structuralist view. Before explaining that, we should ask why 'real property' realism had so little appeal at the time. The answer is exactly what usually precipitates relativism — namely, that realism about geometry precipitated objectively unanswerable questions, questions that seemed to make a mockery of the subject altogether.

1.3. *Relativity of representation*

As mentioned, Beltrami had interpreted hyperbolic geometry within Euclidean geometry. In view of this he thought to have justified Lobachevskii's non-Euclidean geometry by identifying "a real substratum for that doctrine, instead of admitting the necessity of a new order of entities and concepts", and thus leaving Euclidean geometry as the true doctrine of space.

If this sort of interpretation could proceed only in a single direction, it would indeed have established an important asymmetry in the representation of nature. But that was a big "if", and it is not how things turned out.

The point was graphically illustrated by Helmholtz. His example concerns two geometries which result from a common basis by distinct choices of a metric, as in Klein's initial results. Helmholtz imagines that we are making measurements to determine whether our space is Euclidean. Do the interior angles of a triangle add up to 180 degrees? Suppose they do. At the same time he imagines that we are reflected in a huge concave mirror. In that mirror we see little people moving around with rulers and 'doing the same thing' as we do. Of course, they get the 'same' results, and announce that they live in a Euclidean space.

We want to disagree; they are moving around on a concave surface, and their measuring rods change length as they work. They, on the other hand, could say something similar to us! The fact is of course that there is a simple function which establishes a correspondence between points in their space and in ours. It carries our geometric planar structure over to their surface. What is the true geometric description is relative to the choice of which measuring rods shall be regarded as remaining of the same length. But with that choice brought to light, both descriptions are adequate for each of the two spaces.

The isomorphism defeats any attempt to correct the mirror people's judgment within geometric language. Now a realist like Frege or the early Russell can insist that one description is true and the other false. But we can retort that whichever alternative is chosen, the other will be true under proper interpretation, and in any case just as useful. Nor will there be any experiments or empirical predictions which will rule out one in favor of the other. Frege and Russell may insist that there are real spatial relations, which are the semantic values of terms in one geometric language and not in the other. But the question 'Which ones?' (i.e. the challenge to identify those relations and single them out from among all sets of pairs, triples, and so on) is beyond human ken, and will be one of those bits of ontological fact that makes no difference to science at all.

This is an uncomfortable ending for the realist story. It leaves one with that single consolation all philosophers have, of consistency and logical invulnerability; but that is cold comfort indeed. Thus it will not surprise us that Russell wished to move to a more comfortable position — and structuralism is the position to which he moved.

2. STRUCTURALISM, ITS INITIAL (MIS)FORTUNES

The history of the structuralist idea in 20th century philosophy of science is a complicated story of many characters and many adventures. I shall here focus on Russell's very moderate structuralism and its collapse, partly because it was moderate but mainly because it gave us a very precise formulation of such a view. As has been pointed out repeatedly in the literature, the argument precipitating its collapse is related to Putnam's more recent 'model-theoretic argument' against metaphysical realism. Some see a paradox there, inescapable except by realism — but I shall suggest that the paradox can be dissolved. However, this will still leave us with the question whether structuralism can survive in some contemporary form.

2.1. *Russell's structuralist turn*

It is really no wonder that Russell began to move in the direction of a less stringent representationalism for science. When physics describes the world, how much does it describe? Perhaps there are real spatial relations and gerrymandered, artificial metrics. Which of them are denoted by the geometric language used in physics seems not to matter, in principle anyway. But then, better to say, perhaps, that science does not describe nature down to that level of detail. Better to say, perhaps, that science describes only structure without content.

We see Russell taking this line, though not explicitly with reference to geometry, in his *Problems of Philosophy* (1912). He insists, as before, that to understand a proposition (so that it may be capable of being judged true or false) we must be acquainted with every item in it. This includes concrete individual entities, but also the properties and relations expressed by its predicates. However, we are acquainted only with those things which are part of our direct experience. Everything else, including material bodies, belongs to the External World. Unfortunately, it is exactly that External World which science purports to describe. How is this possible?

Russell could have said that science describes nature simply by saying that there exist entities, with which we are not acquainted, but which have the same properties and stand in the same relations, as enter our direct experience. But he does not say that, for science had moved too far and too fast for this to be plausible. Every analogy with familiar things, like waves in water, planets, and billiard balls was already heavy with disanalogies. So Russell says: we can only infer the properties of the properties, and the properties of the relations — the type of structure:

> We can know the properties of the relations required to preserve the correspondence with sense data, but we cannot know the nature of the terms between which the relations hold. "[...] although the relations of physical objects have all sorts of knowable properties,

> ... the physical objects themselves remain unknown in their intrinsic nature ..." (Russell 1912, pp. 32, 34)

When he makes this precise in *The Analysis of Matter* (1927), he says explicitly that this structure is exactly, no more and no less, what can be described entirely in terms of mathematical logic. The logic in question is strong, and today we would say higher order logic or set theory. But still note how little this is! Science is now interpreted as saying that the entities stand in relations which are transitive, reflexive, etc. but as giving no further clue as to what those relations are.

> ... whenever we infer from perceptions it is only structure that we can validly infer; and structure is what can be expressed by mathematical logic. The only legitimate attitude about the physical world seems to be one of complete agnosticism as regards all but its mathematical properties. (Russell 1927, pp. 254, 270)

Russell's younger contemporary Ramsey made explicit what this amounts to if a theory is formalized. On the face of it, the theory does describe all sorts of entities in terms of unfamiliar properties and relations. But really the physicist who says "atom", "electron", "ether", "charge" and "spin" is like the mathematician who says "there must be a fixed point, call it x". The only difference is that "atom" and "spin" are more mnemonic than "x" and "y". So the theory, formally stated, says only that there exist properties and relations instantiating a certain higher-order structure. That formal statement is generally called the theory's Ramsey sentence.

For a quick if simplistic example, consider the theory axiomatized by "Water is composed of hydrogen atoms and oxygen atoms". Its Ramsey sentence is "There exist three properties, of which the first and second are mutually exclusive, such that water is composed of things which have the first and third property and things which have the second and third property." Even adding in the initially tacit logical relationship between "hydrogen" and "oxygen" does not seem to make the Ramsey sentence very informative, and this impression obstinately remains in less simplistic examples. Is the impression correct or mistaken?

2.2. *The collapse of structuralism?*

The mathematician M.H.A. Newman made the crucial critical point in a review article concerning *The Analysis of Matter*:

> ... it is meaningless to speak of the structure of a mere collection of things ... Further, no important information about the aggregate A, except its cardinal number, is contained in the statement that there exists a system of relations, with A as field, whose structure is an assigned one. For given any aggregate A, a system of relations

> between its members can be found having any assigned structure compatible with the cardinal number of A. (Newman p. 140)

The importance of this article, its devastating import, and its relation to more recent discussions of realism, was pointed out by Demopoulos and Friedman.

We can put Newman's point in terms of models and equations quite easily. Suppose Maxwell's equations have a model in which there are N distinct entities. Choose a set of the same cardinality in the world. Because same cardinality implies the existence of a correspondence, we have an implicit transfer of the relations in the model to that chosen set. Therefore the world satisfies Maxwell's equations! The reasoning is more abstract, but the point is no different from that of Helmholtz' concave mirror. But for Russell's view of science it has a corollary: if that view is correct, it seems that we can assert Maxwell's theory of electromagnetism to be true practically without engaging in empirical investigation.

Russell capitulated. In a letter to Newman, he reverted to the 'real property' realism of his early days. The only difference is an update from real spatial relations to real spatio-temporal relations:

> Dear Newman, [...] It was quite clear to me, as I read your article, that I had not really intended to say what in fact I did say, that nothing is known about the physical world except its structure. I had always assumed spatio- temporal continuity with the world of percepts, that is to say, I had assumed that there might be co-punctuality between percepts and non-percepts ... And co-punctuality I regarded as a relation which might exist among percepts and is itself perceptible. (Russell 1968, p. 176)

Russell's response clearly re-introduces his earlier 'real property' realism, slightly updated, with the claim that we are directly acquainted with the crucial spatio-temporal — rather than spatial — relations. Science does not merely assert that there are relations which have the requisite formal character, but gives information about those very relations themselves.

Newman had suggested that Russell could repair his position in this way, by distinguishing between 'important' and 'unimportant' relations. The structuralism would of course have to be restricted to the 'unimportant' part! The effect of the modification would be to charge the transfer of the model's structure, which "reads it into" the world, with mistakenly treating arbitrary sets as real properties and relations. A realist can object to that. Russell accepted this suggestion, adding that the 'important' relations are exactly those of which we have direct acquaintance, and postulating that they include enough relations to make his construal of science a success. Newman had also derided his own suggestion as a counsel of despair, and in fact Russell appears to have had nothing to present by way of support for his postulate.

The form of answer Newman suggested could of course be left more abstract, as purely ontological distinction. If only certain relations are important and quantification in the Ramsey sentence is restricted to those, then that sentence is no longer vacuously true. This was the escape Russell utilized; but one could take it without adding some criterion, like direct acquaintance, to single out the membership of the domain of quantification. David Lewis took this course. The theory's Ramsey sentence says that the relational structure described by the theory is instantiated. Lewis suggested that we identify the content of the theory with something somewhat stronger, namely the assertion that the described structure is uniquely instantiated. This is clearly still a purely structural assertion.

If the quantifiers in that assertion range over all classes (i.e. relations extensionally construed as is usual in mathematics) then Newman's argument establishes that, while the Ramsey sentence will be almost always true, Lewis' strengthening will be almost always false. But if the quantifiers are restricted in the way Newman suggested, then those conclusions cannot be drawn. Lewis too accepts Newman's suggestion, in effect, with his distinction of 'natural' versus 'arbitrary' classes. The natural classes are the extensions of the real properties and relations; the structure described by physics is uniquely instantiated among them if and only if physics is successful. But which classes are natural? Well, if we assume that physics is successful then they are such as to yield a unique instantiation of the structures described in physics — and that, I suppose, is pretty well all we can know about them, even on that assumption. Is this still structuralism? No, it is not, since among the 'arbitrary' classes there will be many relational systems isomorphic to those which are constituted by the 'natural' classes. In other words, the 'important' relational systems in the world, those which successful science describes, are not singled out through structure alone.

2.3. *Putnam's model-theoretic argument — a paradox?*

It is now already twenty years ago that Putnam presented his model-theoretic argument against metaphysical realism, but there is no consensus on what it establishes.[1] Various writers have pointed out connections between this argument and Newman's objection to Russell. You can see for yourself. Here is the argument as presented in 1976:

> So let T_1 be an ideal theory, by our lights. Lifting restrictions to our actual all-too-finite powers, we can imagine T_1 to have every property except objective truth — which is left open — that we like. E.g. T_1 can be imagined complete, consistent, to predict correctly all observation sentences (as far as we can tell), to meet whatever 'op-

[1] For the following discussion I am especially indebted to the papers by Lewis and Elgin, whose conclusions are of course diametrically opposite to each other.

> erational constraints' there are ... to be 'beautiful', 'simple', 'plausible', etc. [...] I imagine that THE WORLD has (or can be broken into) infinitely many pieces. I also assume T_1 says there are infinitely many things (so in this respect T_1 is 'objectively right' about THE WORLD). Now T_1 is consistent ... and has (only) infinite models. So by the completeness theorem ... T_1 has a model of every infinite cardinality. Pick a model M of the same cardinality as THE WORLD. Map the individuals of M one-to-one into the pieces of THE WORLD, and use the mapping to define relations of M directly in THE WORLD. The result is a satisfaction relation SAT — a 'correspondence' between the terms of [the language] L and sets of pieces of THE WORLD — such that theory T_1 comes out true — true of THE WORLD — provided we just interpret 'true' as $TRUE$ (SAT). So whatever becomes of the claim that even the ideal theory T_1 might really be false? (Putnam 1978, pp. 125–126; italics omitted)

I display the original text, for two reasons. The first is that it is couched in terms of the so-called syntactic, axiomatic view of scientific theories. It relies on identification of a theory as a set of sentences in a vocabulary which is divided into 'observation terms' and 'theoretical terms'. That aligns Putnam's argument with the Ramsey sentence formulation of Russell's structuralism — and may in the end also be our main clue for how we can move philosophy of science out of this so very arid puzzle area.

The second and more important reason is to highlight the anthropomorphic way of talking about mathematical entities. Putnam commands us: "Pick a model M ... Map the individuals of M one-to-one into the pieces of THE WORLD". That way of speaking, as if we humans are actually carrying out specific tasks, may be harmless when we are doing pure classical mathematics, but is it really so harmless here?

What he commands us to do here may or may not be possible. He has said that model M and THE WORLD have the same cardinality. Therefore there certainly exists a one-to-one map between them — indeed, very many such maps exist. But can we identify or pick one of them?

Here is an analogous example. Consider a geometric object, a sphere in Euclidean space. Can we coordinatize the surface of this sphere? Our first inclination is to say Yes; for their certainly exist many functions that map the surface points into triples of real numbers in the right way. But I asked: can *we* do this? The answer is No, for this object has perfect symmetry. We would like to give one point the coordinates $(0, 1, 0)$ and call it the North Pole. But this point is not distinguished from any other on the sphere, so we cannot do it. Of course, if the sphere is not a mathematical object but, say, the Earth, then we can do it. The reason is that we can independently identify a spot to designate as North Pole. Similarly, we may be able to do it if the sphere is already related to some other mathematical object, has some functions defined it, and so on — for the same reason. But we cannot do it unless we have some independent way to describe points on that sphere.

Does this distinction — between the existence of a function and our being able to carry out the mapping — matter here? It does indeed if Putnam's argument is meant to apply to real science formulated in real language. We shall be able to grasp that theory, on Putnam's implicit view of understanding language, if we can grasp even one interpretation of the sort he describes. But whether we can do that depends on whether or not we can independently describe THE WORLD.

With this clarification in hand, we are ready to answer Putnam. To Putnam we must say:

> (A) if we cannot describe the elements of THE WORLD, neither can we describe/define/identify any function that assigns extensions to our predicates in THE WORLD;
> (B) if we can describe those elements then we can also distinguish between right and wrong assignments of extensions to our predicates in THE WORLD.

Indeed, if we can describe those elements we can say that an interpretation, if applied to our language, is wrong unless it assigns the set of cats to the words "cat", "gatto", "kat" etc. We learned more from Tarski than Putnam lets on here![2]

What I have just done is not to solve Putnam's paradox, but to dissolve it. But it went a little fast, so I suppose I had better explain just what I did and what I assumed.

What I did was to read Putnam's argument as applying to real scientific theories formulated in *our* language, the very language in which we state his argument. If we read his argument as applying to an arbitrary given language, it may be objective and universal in form, but it also restricts itself to semantics, and abstracts from all pragmatic aspects. In that case, if we try to draw consequences for our own case, we deal with our language as if it were an alien syntax. If we do so and then try to discuss truth, we can indeed get no further than truth-relative-to some interpretation. But the situation changes entirely if we understand the argument throughout as concerning our language, in which we state it, and on which we rely implicitly.

In that case, the discussion of truth is not elliptic, and 'true' does not mean 'true relative to some interpretation'. But while "true" is then not elliptic, it is indexical — tacitly indexical. The tacit indexical reference is to our language.

[2]Perhaps Putnam is supposing that we can describe THE WORLD when he writes "I imagine that THE WORLD has (or can be broken into) infinitely many pieces", for "broken into" may be a metaphor for "described". This would then be the assumption that we can describe THE WORLD, but only to the very limited extent of singling out individuals and counting them. Not surprisingly then, our theories about the world contain no real information which goes beyond cardinality!

The criteria for proper understanding of our own language express themselves in pragmatic tautologies. Consider the sentence:

"cat" denotes cats.

This is undeniable by me if I acknowledge "cat" to be a word in my language. The semantic content of this (to me undeniable) assertion is however not a necessary proposition (for if our language had developed differently in a certain way then "cat" would have denoted gnats, rats, or bats). The example belongs to the class highlighted by Moore's paradox ("It is snowing in Peru but I do not believe that it is snowing in Peru", which is often true, so its semantic content is a contingent proposition; yet I cannot coherently and truly assert it).

It is only if we lose sight of this pragmatic aspect of the argument that we are led into the fallacy of thinking that there is a gap to be filled by metaphysics. For then the adequacy of a theory seems to have to derive in part from the adequacy of the language in which it is stated, which must then seemingly derive from something that makes our language especially blessed among languages or privileged in some objective, use-independent way. The road into this fallacy is clear, and it is quite clear how the fallacy engenders metaphysics.

Even if we see through this, there are problems left for us of course — for example, to explain how such pragmatic tautologies can be combined with coherent admissions of our fallibility. It is not easy to state such admissions coherently. That is a common problem with pragmatic tautologies: once again, exactly the problem high-lighted by Moore's paradox. I will assume that we can solve that without creating more ontological debris. In any case, these problems are not peculiar to our subject matter here.

So I have asserted that there is a paradox only if we try to treat a problem in pragmatics on the level of semantics, i.e. by illegitimate abstraction from pragmatic aspects. The dissolution of Putnam's paradox is not a solution based on a specific ontology. Does it still belong to some sort of philosophical position? If so, it is not to be called absolute relativism (to coin a phrase), since it does not refuse to acknowledge any and all sorts of distinctions among languages. But still one might call it a sort of relativism, or of perpectivism perhaps, because the privilege it notes is the historically conditioned, contingent status of being our own language. This privilege obviously bestows no security and no guarantee of adequacy; it cannot function as foundation. So perhaps "anti-foundationalism" is the best label; though best of all, I think, would be to discard such labels altogether.

If a paradox is dissolved, it does not need a solution. The large question that remains for us now is: did we only dissolve a problem peculiar to the syntactic/axiomatic view of theories? Does Newman's problem persist, so as to defeat any other form of structuralism, mutatis mutandis? Putnam in effect put the problem raised by Helmholtz and Newman in explicitly linguistic form

— and my dissolution pertained to it in that form. Will the problem just re-emerge if we leave this (in my opinion, overly) linguistic approach behind?

3. STRUCTURALISM: THE SEMANTIC APPROACH

In the second half of this century an alternative approach has developed within philosophy of science: the so-called semantic approach or semantic view of theories. Its antecedents go back fifty odd years to Herman Weyl and Evert Beth, as well as to work in the fifties and sixties by Patrick Suppes, and in fact to a large diversity of related work by philosophers and scientists in Poland, Italy, Germany, and other countries. After that initial period we have seen two distinct developments, on the one hand by Fred Suppe, Ronald Giere, myself, and others, and on the other hand, by Sneed, Stegmüller, Moulines, and others. The latter group actually uses the name "Structuralism". Finally, and this is very heartening, during the last ten years a new group has drawn on all these developments and taken them further. I am referring to Da Costa and French and their collaborators.

By common, if often tacit, consent among all these rather disparate writers, the semantic approach is the current form of the general idea of structuralism. Let us first of all examine why.

3.1. *Structural description of nature*

According to the semantic approach, to present a scientific theory is, in the first instance, to present a family of models — that is, mathematical structures offered for the representation of the theory's subject matter. Within mathematics, isomorphic objects are not relevantly different; so it is especially appropriate to refer to mathematical objects as "structures". Given that the models used in science are mathematical objects, therefore, scientific theoretical descriptions are structural; they do not "cut through" isomorphism. So the semantic approach implies a structuralist position: science's description of its subject matter is solely of structure.

A fine point is in order here, and must be noted. If one structure can represent the phenomena, then so can any isomorphic structure, mutatis mutandis. This last clause is important. We can represent temporal sequences by means of the less-than-or-equal relation on the real numbers — as we do in fact in the calendar — to take one example. Now the greater-than-or-equal relation, i.e. the converse of the former, is isomorphic. So it too can be used to represent temporal sequences of events (we could use a reverse order calendar, for instance by multiplying our dates by -1). But the use must be 'mutatis mutandis': this point about isomorphism certainly does not mean that if A precedes B then we can also say, without furthermore, that B precedes A!

Properly understood, it is entirely correct to say that models represent nature only up to isomorphism — they only represent structure. Therefore, it really is a consequence of the semantic view that science describes only structure. But what is the import of that fine point about representation by distinct but isomorphic structures being "mutatis mutandis"? We distinguish the two relations of less and greater — or more generally two isomorphic models — by noting their relations to each other in a larger structure in which they appear, while calling them isomorphic when considering each solely by itself. What this brings out very graphically is that representation is using (something) to represent (something). There is no such thing as 'representation in nature' or 'representation tout court'; the question whether one given object is a representation of another is an incomplete question. Specifically, in science models are used to represent nature, used by us, and of the many possible ways to use them, the actual way matters and fixes the relevant relation between model and nature — relevant, that is, to the evaluation as well as application of that theory.

This point, though somewhat banal at first sight, leads to a qualification of the structuralism here noted; a qualification, but not, I shall insist, a weakening. I will have to discuss this issue in two stages. Philosophers have in the past wanted to ask immediately, "But is the theory true?", "Under what conditions is a theory true?" and so on. The elaboration of the semantic view has, however, mainly concentrated not on truth in toto, but on some lesser relation of greater practical interest. We must discuss both.

3.2. *Adequacy to the phenomena*

Within the semantic approach, attention has focused not so much on truth per se, truth in toto, but on lesser relations of adequacy. Polish and Italian writers used terms like "empirical truth"; I have used the term "empirical adequacy"; Da Costa and French use "partial truth". These terms are not synonymous but closely related, though the guiding idea is that of adequacy in respects which can be made subject of practical investigation.

Patrick Suppes had long emphasized that theories do not confront the data bare and raw. The experimental report is already a selective and refined representation, a 'data model' as he calls it. This is especially true today, as Fred Suppe has emphasized, now that scientists routinely process gigabytes of data. It was already true in Newton's time when he claimed to deduce laws from the phenomena — for of course he used as basis very smooth functions distilled from thousands of astronomical observations. But it is true even of the idealized, simple observation report discussed by the logical positivists, as they themselves came to agree after some debate.

When we claim adequacy of a theory to the phenomena, that is adequacy to the phenomena as thus described. Note well that this claim is not restricted to actual data models or actual descriptions issuing from observation. It makes sense to claim that a certain graph depicts fluctuations in the dinosaur population, which was certainly never observed. That claim is true or false, whether we know it or not. The theory's adequacy requires that the correct graph — whichever it be — matches a theoretical model. More generally then, those structures which represent phenomena in terms of the relevant parameters must "fit" a theoretical model — that is what constitutes adequacy with respect to the phenomena.

Obviously this is a fallible process, and not only fallible but history and theory conditioned. The phenomena are in principle observable by anyone, but the form of description is chosen, taught, and learned by human beings who are thoroughly immersed in their inherited background theories, opinions, and assumptions. That there is danger of failure inherent in this situation goes without saying, and I won't say it again.

The important point for us here is that this claim of adequacy too is in the first instance a structural claim. To be matched are two models, a data model and a theoretical model. The matching in question may be as simple as an embedding or partial isomorphism, or it may need to be some measure-theoretic refinement thereof to allow for approximation. But is in any case a mathematical relationship, and therefore purely structural. The claim of adequacy is in the first instance a claim about how two structures are structurally related.

Do we have Newman's problem here? Let us take an example. The same exponential curve might be the shape of two data models, one from bacterial population growth and the other from radio-active decay. But data from radium samples are not relevant to a theory about bacteria. It is simply not relevant that a data model obtained in studies of radio-activity is thus structurally related to the bacterial population model, even if in this respect it is not different from what was found in bacteriological experiments.

Thus, something more about the phenomena, besides their structure, matters to the claim of adequacy for the theory. The data model is important not in itself, but in its role of representation of the phenomena. As with Putnam's paradox, we must insist that this role does not consist simply in having a certain structure. The claim of adequacy is with respect to the structure of real phenomena described in terms of the relevant parameters of the theory.

3.3. *Structuralism lost after all?*

This last point may sound like an admission of defeat. Is structuralism lost here, and must it give way to 'real property' realism? I do not think so. One way to state the challenge is to say:

> "Oh, so the theory does not confront the observable phenomena, (those things, events, and processes out there) in and by themselves, but only certain descriptions of them. Empirical adequacy is not adequacy to the phenomena pure and simple, but to the phenomena as described!"

There is something wrong with this objection, once again an error which cannot show up if we keep our discussion on the level of semantics.

On the one hand: the claim that the theory is adequate to the phenomena is not the same as the claim that it is adequate to the phenomena as described by someone, nor by everyone, or anyone. On the other hand: if we try to check a claim of adequacy, then we will compare one representation or description with another — namely, the theoretical model and the data model. The tension between these two obvious points can quickly induce the temptation to think that there must be a specially right, objectively right, or naturally privileged description. This privilege will then be a great mystery, begging for special ontological status.

But if we put ourselves into the picture — us, the users of the theories, the ones who use mathematical structures to represent phenomena — then the picture changes radically. For us the claim (A) that the theory is adequate to the phenomena and the claim (B) that it is adequate to the phenomena as described, i.e. as described *by us*, are indeed the same! That is a pragmatic tautology. It is a truism of the rank of such sayings as that any description we give, we give in our language in use, that we can't describe without using language or some equivalent medium of representation. From a tautology, even a pragmatic tautology, nothing follows for ontology.

So what is there to the data model, besides its structure, which makes it important to a scientific theory? We must add, using our own language, that for example this data model summarizes certain findings about bacteria, or about radio-active decay, as the case may be. Because representation is something we do, and not something that exists in nature independently of what we do, our claim of adequacy for the theory must involve reference to how we are using both models — data model and theoretical model — to represent our subject matter.

The best analogy I know is still between a person looking at a theory and one looking at a map. The latter has to be able to say "this is what things are like around *here* where we are — so *that* is where we are on the map". It would be no use putting extra informa tion into the map (such as an inscription "you are here") to try and obviate the need for such self-attributions by the user, for the same point would still apply to that "extended" map. Because the needed sort of judgement is indexical, it lies outside the domain of semantics. This point is of crucial importance when any sort of theoretical product is to be related to our practice.

3.4. *Truth beyond the phenomena*

Questions concerning the adequacy of a theory may be restricted or (putatively) unrestricted. The former, discussed in the preceding section, may be philosophically motivated — naturally, empiricists favor a focus on empirical adequacy, i.e. adequacy with respect to the observable phenomena. A restricted question of adequacy may also be philosophically neutral, e.g. a question concerning adequacy with respect to processes of a certain type, or in a limited domain. The form of question will be the same in the two cases. So will be the form of answer, according to the semantic approach: whether the claim is one of isomorphic embeddability of the phenomena or processes in the theoretical model, or some less strict relation, that claim will be construed as asserting a structural relationship.

Now we have seen that this view needs a qualification, though at the pragmatic level. To make the claim (that there is such a structural relation) relevant, we must know how the models in question are being used to represent something. The required extra knowledge is a self-attribution of the user, expressed by him or her in indexical language, and cannot be obviated by building more into the models or theories themselves.

But what if the question concerning the adequacy of the theory purports to be unrestricted? What if it is, so to speak, the question of truth in toto? The suspicion has been raised in several quarters that within the semantic approach to science, at least when allied with empiricism, we cannot make sense of truth (simpliciter) at all.

If that were so it would indeed be a disaster for empiricism. For the basic idea of empiricist philosophies of science, that theory is in general underdetermined by empirical fact, requires a meaningful distinction between truth and empirical adequacy. I will try to show that the suspicion is not well founded. Indeed, I shall argue that it may derive from the fact that today, paradoxically enough, it is exactly the scientific realists who still yearn for an 'internal' logical connection between truth and verification.

Let us examine this suspicion, that we cannot, on our approach, make sense of a theory being empirically adequate but not true. At first sight the suspicion may seem easy to alleviate. Suppose a theory is empirically adequate exactly if it has some model in which all the observable phenomena are isomorphically embeddable. Can we then not equally and meaningfully assert that the theory is true exactly if it has a model in which not only all observable phenomena, but all systems and processes to be found in nature, can be so embedded? We would of course have to add the same discussion of pragmatics: an actual claim of truth, by us, for a theory, relates the theory's models to nature as described in our language. But at this point — when we mention description of nature in our language — the suspicion pricks up its ears, so to speak, and rears its ugly head much higher; for now we have touched on its very sources in the recent history of philosophy.

We must admit that it has for the most part been empiricists who tried to keep a logical connection between truth and empirical verifiability or access. Among not only the logical positivists but also among their antecedents and heirs there has been a distinct tendency to tie the possibility of description to the possibility of knowledge or even acquaintance. Here is a very early example, in the writings of Boltzmann:

> Formally ... we can raise questions of this kind: does only matter exist and is force one of its properties or does the latter exist independently of matter or, conversely, is matter a product of force? None of these questions, however, has any meaning, since all of these concepts are merely thought-pictures that have the purpose of correctly representing appearances. (tr. in Danto, p. 245)

In this passage Boltzmann implicitly shoulders us with a dichotomy. If what he says is right there must be a dichotomy between concepts that are theoretical and those which are purely neutral with respect to all theory. For if those questions about matter and force have no meaning and do not arise exactly because the concepts involved are mere 'thought pictures', then what? Then we must also have concepts which are not like that; else no question at all would have any meaning! Today we cannot accept such a dichotomy, for we realize that all description of nature is non-trivially theory-laden. That is history: empiricists and scientific realists alike have learned that lesson. But representation makes sense; even representation of things never encountered in experience.

So what prompted Boltzmann's difficulty? We have such terms as "force", "position", and "momentum" in use, and have personally encountered some things to which these terms apply. Whether they apply or not depends on what those things are like. The question is now: do these terms apply to unobservable entities in the same way?

It would be better for us to ask what that question is supposed to mean! It sounds like the echo of the medieval debate about whether human predicates apply to God literally or analogically. The consequent theory of analogical predication was, in my opinion, an untenable answer to an incoherent question, for there cannot be a meaningful analogy if there is no available literal description. The theory becomes no more coherent if extended to electrons and quarks rather than God or gods. In any case, the very discussion seems fueled by a 'thought picture' of what it means for language to apply to things — something like labels being stuck on to things according to some rule followable by a user of that language. The point is that whatever Boltzmann's personal difficulties were (and whether or not they were the very fallacies from which Neurath had to rescue the early Carnap), neither empiricism nor the semantic approach to science is still tied up in these knots.

The entanglement of semantics and epistemology did not die easily, and it took constant re-immersion in Tarski's work to keep the two even reasonably separate. It is a credit to the realists of the sixties that they showed us the logical fallacies involved when the two are entangled. But the sources of metaphysical realism also exhibit a deep desire to keep the very idea of truth intimately allied with knowledge. Russell, after all, was a leader of the revolt against idealism. But Russell immediately opted for the doctrine that we can understand a proposition only if we are directly acquainted with each constituent. With such a doctrine, which remained influential in philosophy of language to this very day, semantics is grounded in epistemology and truth is logically linked to human access.

This story continues. Today's metaphysical realism derives in part from a theory of language which did indeed lead in the revolt against positivism. But the theory entails that we understand a proposition only if, after all definitions and definite descriptions are eliminated, reference of other singular and general terms is to entities with which we are 'causally connected' — a mysterious place holder for 'whatever it takes' to secure our epistemic access to the world.

In my opinion, these epistemic motives and their metaphysical train have no place in semantics. They enter as surrogates when it is attempted to simultaneously make semantics autonomous from pragmatics and also to infuse it with some independent cogency. Much better, I say, to study empirical adequacy overtly and explicitly for epistemic reasons, and to let truth have its sense derivative from our actual language in use, accepting without qualms our historical contingency and fragility.

So, what about truth? What about the sense of questions such as whether current atomic physics is true? Given the character of today's physics, and its highly mathematical form, we cannot sidestep issues of interpretation altogether. But we must approach them exactly as in our discussion of Putnam's paradox. Consider the statement

(*) Electrons always have a (precise) position.

Epistemologically this is an interesting example, for it is false according to certain interpretations and true according to others (such as Bohm's). Given that those interpretations yield empirically equivalent versions of quantum mechanics, it appears that if quantum mechanics is true (under any of those interpretations) then there is no way for us to know or confirm whether or not (*) is true!

But this does not affect in any way the question of whether it makes sense to ask whether (*) is true. If this is a statement in our language in use, then whether it is true or not simply depends on what electrons are like. That is, in that case,

"Electrons always have a (precise) position" is true if and only if electrons always have a (precise) position.

Of course, if the statement does not belong to our language, then the question of whether it is true (as opposed to "true under some interpretation") does not and cannot arise at all. This becomes clear when we replace "electrons" by "sharks", "boojums", or "jabberwocks" (with a bow to David Mermin as well as Lewis Carroll of course).

Let me sum up this conclusion about structure and truth. A scientific theory gives us a family of models to represent the phenomena. But it represents the phenomena as fragments of a larger and simpler structure — the world as it is according to the theory, if you like. These models are mathematical entities, so all they have is structure, the only thing they can represent is structure. There is nevertheless a right and a wrong in representation.

The question whether a theory as a whole is true or false makes sense. It just doesn't make sense if we separate this question from the language in which we ask it, our language in use.

If we cannot describe the world to be represented, then we cannot understand how a theory represents it at all. If we can describe the world, then we can understand that; but then we automatically have also a criterion of correctness or accuracy; that is, of truth. More moderately, questions of truth make sense exactly and only to the extent that we have words to describe what things are like. At this point we can rely on our own language or we can distrust it, but we cannot get behind it — at this point, as Carnap used to say, the question we face is no longer theoretical but practical — or as perhaps we should say, more precisely and less faintheartedly, existential.

Department of Philosophy
Princeton University

REFERENCES

1. Beltrami, E. (1868) "Saggio di Interpretatione della Geometria Non-Euclidia", in his *Opere Mathematiche*, vol. 1, pp. 406–429. Milan: V. Hoepli, 1902.
2. Blackmore, J. (ed.) *Ernst Mach — A Deeper Look.* Boston Studies in the Philosophy of Science, vol. 143. Dordrecht: Kluwer, 1992.
3. Boltzmann, L. (1905a) "Theories as representations", translation of Boltzmann (1905b), pp. 253–269 in Danto and Morgenbesser, pp. 245–252.
4. Boltzmann, L. (1905b) *Populaere Schriften.* Leipzig: J.A. Barth.
5. Coffa, A. "From geometry to tolerance: sources of conventionalism in nineteenth-century geometry", in Colodny, pp. 3–70.
6. Colodny, R.G. (1986) *From Quarks to Quasars. Philosophical Problems of Modern Physics.* Pittsburgh: University of Pittsburgh Press.
7. Da Costa, N.C.A. and French, S. "The model-theoretic approach in the philosophy of science", *Philosophy of Science* 57 (1990), pp. 248–265.
8. Danto, A. and Morgenbesser, S. (eds.) *Philosophy of Science.* New York: Meridian Books, 1960.
9. Demopoulos, W. and Friedman, M. "Critical Notice: Bertrand Russell's *The Analysis of Matter*: its Historical Context and Contemporary Interest", *Philosophy of Science* 52 (1985), 621–639.
10. Elgin, C.Z. "Unnatural Science", *Journal of Philosophy* 92 (1995), 289–302.
11. Hull, D., Forbes, M. and Ohkruhlik, K. (eds.) PSA 1992, Vol. 2. *Proceedings of the Philosophy of Science Association Conference 1992.* Evanston: Northwestern University Press, 1993.
12. David K. Lewis "Putnam's Paradox", *Australasian Journal of Philosophy* 62 (1984), 221–236.
13. Newman, M.H.A. "Mr. Russell's causal theory of perception", *Mind* n.s. 37 (1928), 137–148.
14. Planck, M. "Die Einheit des physikalischen Weltbildes", *Physikalisches Zeitschrift* 10 (1909), 62–75. A new English translation of section 4 is provided in Blackmore, who notes that the "Weltbild" terminology was apparently introduced by Hertz in the 1890's.

15. Putnam, H. (1976) "Realism and reason", in his (1978), pp. 123–140.
16. Putnam, H. (1978) *Meaning and the Moral Sciences.* N.Y.: Routledge.
17. Russell, B. (1912) *Problems of Philosophy.* Reissued Oxford: Oxford University Press, 1959.
18. Russell, B. (1927) *The Analysis of Matter.* London: Allen and Unwin.
19. Russell, B. (1968) *The Autobiography of Bertrand Russell*, vol. 2. London: Allen and Unwin.
20. Suppe, F. *The Semantic Conception of Theories and Scientific Realism.* Urbana: University of Illinois Press, 1989.
21. van Fraassen, B. (1989) *Laws and Symmetry.* Oxford University Press.
22. van Fraassen, B. (1993) "From vicious circle to infinite regress, and back again", in Hull, Forbes, and Ohkruhlik, pp. 6–29.

TABLE OF CONTENTS VOLUME II

190. W. Callebaut and R. Pinxten, *Evolutionary Epistemology.* A Multiparadigm Program, with a Complete Evolutionary Epistemology Bibliograph. 1987 ISBN 90-277-2582-9
191. J. Kmita, *Problems in Historical Epistemology.* 1988 ISBN 90-277-2199-8
192. J. H. Fetzer (ed.), *Probability and Causality.* Essays in Honor of Wesley C. Salmon, with an Annotated Bibliography. 1988 ISBN 90-277-2607-8; Pb 1-5560-8052-2
193. A. Donovan, L. Laudan and R. Laudan (eds.), *Scrutinizing Science.* Empirical Studies of Scientific Change. 1988 ISBN 90-277-2608-6
194. H.R. Otto and J.A. Tuedio (eds.), *Perspectives on Mind.* 1988 ISBN 90-277-2640-X
195. D. Batens and J.P. van Bendegem (eds.), *Theory and Experiment.* Recent Insights and New Perspectives on Their Relation. 1988 ISBN 90-277-2645-0
196. J. Österberg, *Self and Others.* A Study of Ethical Egoism. 1988 ISBN 90-277-2648-5
197. D.H. Helman (ed.), *Analogical Reasoning.* Perspectives of Artificial Intelligence, Cognitive Science, and Philosophy. 1988 ISBN 90-277-2711-2
198. J. Wolenski, *Logic and Philosophy in the Lvov-Warsaw School.* 1989 ISBN 90-277-2749-X
199. R. Wójcicki, *Theory of Logical Calculi.* Basic Theory of Consequence Operations. 1988 ISBN 90-277-2785-6
200. J. Hintikka and M.B. Hintikka, *The Logic of Epistemology and the Epistemology of Logic.* Selected Essays. 1989 ISBN 0-7923-0040-8; Pb 0-7923-0041-6
201. E. Agazzi (ed.), *Probability in the Sciences.* 1988 ISBN 90-277-2808-9
202. M. Meyer (ed.), *From Metaphysics to Rhetoric.* 1989 ISBN 90-277-2814-3
203. R.L. Tieszen, *Mathematical Intuition.* Phenomenology and Mathematical Knowledge. 1989 ISBN 0-7923-0131-5
204. A. Melnick, *Space, Time, and Thought in Kant.* 1989 ISBN 0-7923-0135-8
205. D.W. Smith, *The Circle of Acquaintance.* Perception, Consciousness, and Empathy. 1989 ISBN 0-7923-0252-4
206. M.H. Salmon (ed.), *The Philosophy of Logical Mechanism.* Essays in Honor of Arthur W. Burks. With his Responses, and with a Bibliography of Burk's Work. 1990 ISBN 0-7923-0325-3
207. M. Kusch, *Language as Calculus vs. Language as Universal Medium.* A Study in Husserl, Heidegger, and Gadamer. 1989 ISBN 0-7923-0333-4
208. T.C. Meyering, *Historical Roots of Cognitive Science.* The Rise of a Cognitive Theory of Perception from Antiquity to the Nineteenth Century. 1989 ISBN 0-7923-0349-0
209. P. Kosso, *Observability and Observation in Physical Science.* 1989 ISBN 0-7923-0389-X
210. J. Kmita, *Essays on the Theory of Scientific Cognition.* 1990 ISBN 0-7923-0441-1
211. W. Sieg (ed.), *Acting and Reflecting.* The Interdisciplinary Turn in Philosophy. 1990 ISBN 0-7923-0512-4
212. J. Karpiński, *Causality in Sociological Research.* 1990 ISBN 0-7923-0546-9
213. H.A. Lewis (ed.), *Peter Geach: Philosophical Encounters.* 1991 ISBN 0-7923-0823-9
214. M. Ter Hark, *Beyond the Inner and the Outer.* Wittgenstein's Philosophy of Psychology. 1990 ISBN 0-7923-0850-6
215. M. Gosselin, *Nominalism and Contemporary Nominalism.* Ontological and Epistemological Implications of the Work of W.V.O. Quine and of N. Goodman. 1990 ISBN 0-7923-0904-9
216. J.H. Fetzer, D. Shatz and G. Schlesinger (eds.), *Definitions and Definability.* Philosophical Perspectives. 1991 ISBN 0-7923-1046-2
217. E. Agazzi and A. Cordero (eds.), *Philosophy and the Origin and Evolution of the Universe.* 1991 ISBN 0-7923-1322-4
218. M. Kusch, *Foucault's Strata and Fields.* An Investigation into Archaeological and Genealogical Science Studies. 1991 ISBN 0-7923-1462-X

219. C.J. Posy, *Kant's Philosophy of Mathematics.* Modern Essays. 1992 ISBN 0-7923-1495-6
220. G. Van de Vijver, *New Perspectives on Cybernetics.* Self-Organization, Autonomy and Connectionism. 1992 ISBN 0-7923-1519-7
221. J.C. Nyíri, *Tradition and Individuality.* Essays. 1992 ISBN 0-7923-1566-9
222. R. Howell, *Kant's Transcendental Deduction.* An Analysis of Main Themes in His Critical Philosophy. 1992 ISBN 0-7923-1571-5
223. A. García de la Sienra, *The Logical Foundations of the Marxian Theory of Value.* 1992 ISBN 0-7923-1778-5
224. D.S. Shwayder, *Statement and Referent.* An Inquiry into the Foundations of Our Conceptual Order. 1992 ISBN 0-7923-1803-X
225. M. Rosen, *Problems of the Hegelian Dialectic.* Dialectic Reconstructed as a Logic of Human Reality. 1993 ISBN 0-7923-2047-6
226. P. Suppes, *Models and Methods in the Philosophy of Science: Selected Essays.* 1993 ISBN 0-7923-2211-8
227. R. M. Dancy (ed.), *Kant and Critique: New Essays in Honor of W. H. Werkmeister.* 1993 ISBN 0-7923-2244-4
228. J. Woleński (ed.), *Philosophical Logic in Poland.* 1993 ISBN 0-7923-2293-2
229. M. De Rijke (ed.), *Diamonds and Defaults.* Studies in Pure and Applied Intensional Logic. 1993 ISBN 0-7923-2342-4
230. B.K. Matilal and A. Chakrabarti (eds.), *Knowing from Words.* Western and Indian Philosophical Analysis of Understanding and Testimony. 1994 ISBN 0-7923-2345-9
231. S.A. Kleiner, *The Logic of Discovery.* A Theory of the Rationality of Scientific Research. 1993 ISBN 0-7923-2371-8
232. R. Festa, *Optimum Inductive Methods.* A Study in Inductive Probability, Bayesian Statistics, and Verisimilitude. 1993 ISBN 0-7923-2460-9
233. P. Humphreys (ed.), *Patrick Suppes: Scientific Philosopher.* Vol. 1: Probability and Probabilistic Causality. 1994 ISBN 0-7923-2552-4
234. P. Humphreys (ed.), *Patrick Suppes: Scientific Philosopher.* Vol. 2: Philosophy of Physics, Theory Structure, and Measurement Theory. 1994 ISBN 0-7923-2553-2
235. P. Humphreys (ed.), *Patrick Suppes: Scientific Philosopher.* Vol. 3: Language, Logic, and Psychology. 1994 ISBN 0-7923-2862-0
Set ISBN (Vols 233–235) 0-7923-2554-0
236. D. Prawitz and D. Westerståhl (eds.), *Logic and Philosophy of Science in Uppsala.* Papers from the 9th International Congress of Logic, Methodology, and Philosophy of Science. 1994 ISBN 0-7923-2702-0
237. L. Haaparanta (ed.), *Mind, Meaning and Mathematics.* Essays on the Philosophical Views of Husserl and Frege. 1994 ISBN 0-7923-2703-9
238. J. Hintikka (ed.), *Aspects of Metaphor.* 1994 ISBN 0-7923-2786-1
239. B. McGuinness and G. Oliveri (eds.), *The Philosophy of Michael Dummett.* With Replies from Michael Dummett. 1994 ISBN 0-7923-2804-3
240. D. Jamieson (ed.), *Language, Mind, and Art.* Essays in Appreciation and Analysis, In Honor of Paul Ziff. 1994 ISBN 0-7923-2810-8
241. G. Preyer, F. Siebelt and A. Ulfig (eds.), *Language, Mind and Epistemology.* On Donald Davidson's Philosophy. 1994 ISBN 0-7923-2811-6
242. P. Ehrlich (ed.), *Real Numbers, Generalizations of the Reals, and Theories of Continua.* 1994 ISBN 0-7923-2689-X
243. G. Debrock and M. Hulswit (eds.), *Living Doubt.* Essays concerning the epistemology of Charles Sanders Peirce. 1994 ISBN 0-7923-2898-1
244. J. Srzednicki, *To Know or Not to Know.* Beyond Realism and Anti-Realism. 1994 ISBN 0-7923-2909-0
245. R. Egidi (ed.), *Wittgenstein: Mind and Language.* 1995 ISBN 0-7923-3171-0